# Spacecraft Engineering: Systems and Design

# Spacecraft Engineering: Systems and Design

Edited by Corey Reid

Clanrye International
www.clanryeinternational.com

Clanrye International,
750 Third Avenue, 9th Floor,
New York, NY 10017, USA



ISBN: 978-1-64726-125-2

**Cataloging-in-Publication Data**

Spacecraft engineering : systems and design / edited by Corey Reid.
p. cm.
Includes bibliographical references and index.
ISBN 978-1-64726-125-2
1. Space vehicles. 2. Aerospace engineering. 3. Space vehicles--Design and construction.
4. Aeronautics. I. Reid, Corey.
TL795 .S63 2022
629.47--dc23

For information on all Clanrye International publications
visit our website at www.clanryeinternational.com

Clanrye International

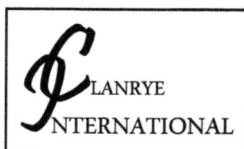

# Contents

# Preface

This book aims to highlight the current researches and provides a platform to further the scope of innovations in this area. This book is a product of the combined efforts of many researchers and scientists, after going through thorough studies and analysis from different parts of the world. The objective of this book is to provide the readers with the latest information of the field.

A spacecraft is a machine that is created to fly in the outer space. There are numerous purposes for which spacecrafts are used, such as space colonization, communications, meteorology, navigation, transportation, Earth observation, etc. There are two types of spacecrafts- crewed and unmanned. Spacecraft engineering is a branch of engineering that deals with the development and creation of spacecraft. The designing of spacecrafts involves various technological and engineering disciplines such as aerodynamics, propulsion, avionics, materials science, manufacturing and structural analysis. The interaction between these technologies is known as spacecraft engineering. Some of the elements of spacecraft engineering include astrodynamics, aircraft structures, aeroelasticity, fluid dynamics, aeroacoustics, etc. This book contains some path-breaking studies in the field of spacecraft engineering. It is a compilation of chapters that discuss the most vital concepts and emerging trends in the field of spacecraft engineering. This book is appropriate for students seeking detailed information in this area as well as for experts.

I would like to express my sincere thanks to the authors for their dedicated efforts in the completion of this book. I acknowledge the efforts of the publisher for providing constant support. Lastly, I would like to thank my family for their support in all academic endeavors.

**Editor**

# Numerical Simulation of Projectile Oblique Impact on Microspacecraft Structure

**Zhiyuan Zhang, Runqiang Chi, Baojun Pang, and Gongshun Guan**

*Hypervelocity Impact Research Center, Harbin Institute of Technology, Harbin 150080, China*

Correspondence should be addressed to Runqiang Chi; chirq@hit.edu.cn

Academic Editor: Enrico C. Lorenzini

In the present study, the microspacecraft bulkhead was reduced to the double honeycomb panel, and the projectile oblique hypervelocity impact on the double honeycomb panel was simulated. The distribution of the debris cloud and the damage of a honeycomb sandwich panel were investigated when the incident angles were set to be 60°, 45°, and 30°. The results showed that as incident angle decreased, the distribution of debris cloud was increased gradually, while the maximum perforation size of the rear face sheet was firstly increased with the decrease of the incident angle and then decreased. On the other hand, the damage area and the damage degree of the front face sheet of the second honeycomb panel layer were increased with the decrease of the incident angle. Finally, the critical angle of front and rear face sheets of the honeycomb sandwich panel was obtained under oblique hypervelocity impact.

## 1. Introduction

A large amount of the space debris has been accumulated as the increase of the human space activities within the scope of the near-earth space, which posed a serious threat to the safe operation of the spacecraft in orbit [1, 2]. Honeycomb sandwich panel is commonly used as a kind of structure material of the spacecraft bulkhead which is made up of panels and honeycomb core sticking together [3, 4]. The space debris can easily penetrate honeycomb sandwich panel with an average speed of 10 km/s in the earth orbit [1, 3]. As for the spacecraft bulkhead, the honeycomb sandwich panel will be firstly suffered the impact of the space debris [5, 6].

A lot of researches have analyzed the damage characteristics induced by the impact of the space debris on the monolayer honeycomb sandwich panel [7] by experiments and numerical simulations by considering the effects of size [8], the impact velocity [8, 9], the materials [10], and the collision limit [11]. But the hypervelocity impact on double honeycomb sandwich panels which is different from two layer honeycomb sandwich panels being bonded together [5] has not yet been reported. One side of the spacecraft bulkhead will be penetrated; then the inside the equipment and the other side of the bulkhead can also be damaged due to the high speed and the large kinetic energy of the space debris. The study of the effects of the high speed impact on the front and rear bulkheads is necessary.

In view of the space debris impact on both sides of the spacecraft bulkhead, the spacecraft was simplified to the structure of double honeycomb sandwich panel, which used to study the breaking condition of the projectile and the damage of the honeycomb sandwich panel under the oblique impact. Finally, the critical incident angle of the single honeycomb sandwich panel is calculated in this paper.

## 2. Model of Simulation

Figure 1 shows a double honeycomb sandwich panel structure of the simplified model of the spacecraft. The distance between two layers of the honeycomb sandwich plates is 500 mm and the space debris is simplified as spherical aluminum alloy projectiles. Smoothed Particle Hydrodynamics (SPH) is used as aluminum alloy projectile model which can solve the problem that the mesh deformation of projectile is too large. Finite element method (FE) is used in the honeycomb sandwich panel model and the shell element is used for the honeycomb core. Particle size of SPH is set to be 0.2 mm. Grid size near impact point and outside

TABLE 1: Description of the simulation model.

| | Material | Dimension/mm | EOS | Strength | Failure |
|---|---|---|---|---|---|
| Projectile | Al 2017 | Sphere/$D = 5$ | Shock | Johnson-Cook | Principal stress |
| Face sheet | Al 5A06 | $h = 0.8$ | Shock | Johnson-Cook | Plastic strain |
| Honeycomb core | Al 5A06 | $L = 4,\ H = 20,\ t = 0.025$ | Linear | Johnson-Cook | Plastic strain |

TABLE 2: Parameters of materials [12].

| | $A$/MPa | $B$/MPa | $n$ | $C$ | $m$ | $T_m$/K | Density/g·cm$^{-3}$ |
|---|---|---|---|---|---|---|---|
| Al 5A06 | 235.4 | 622.3 | 0.58 | 0.0174 | 1.05 | 853 | 2.64 |
| Al 2017 | 249.9 | 426.0 | 0.34 | 0.015 | 1.0 | 775 | 2.79 |

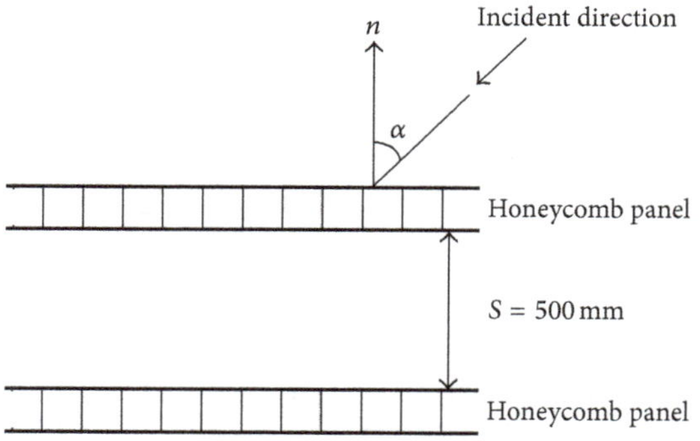


FIGURE 1: Double honeycomb sandwich panel structure diagram.

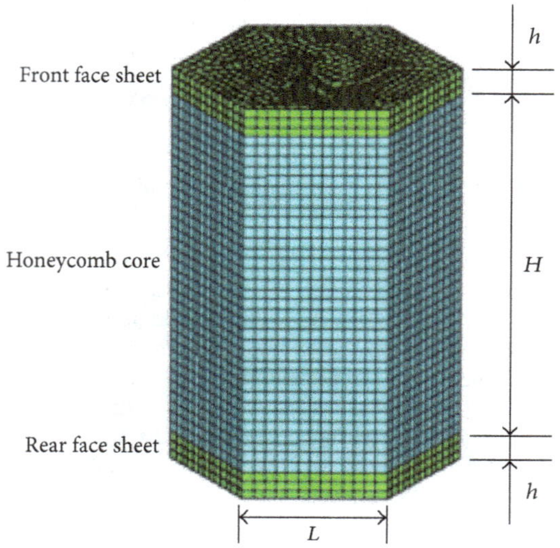


FIGURE 2: Honeycomb panel model.

are set to be 0.3 mm and 0.8 mm. Each honeycomb core is equal to the grid on the panel correspondingly. By using the model in Figure 2, the nodes of honeycomb core and panel grids are processed. The process of the projectile impact on honeycomb sandwich panel at a high speed was simulated using AUTODYN 15.0 and the simulation parameters were shown in Table 1. Geometric strain was used as the erosion model, and the value was set 2. In this model, $D$ is the projectile diameter, $h$ is thickness of honeycomb sandwich panel, $L$ is a side length of hexagonal honeycomb core, $H$ is height of the honeycomb core, and $t$ is thickness of cellular wall. The model parameters of the honeycomb sandwich panel in this paper were based on data from microspacecraft project, which can apply to communication, ground remote sensing, the interplanetary exploration, scientific research, and so forth.

Johnson-Cook model can describe the nonlinear process of the high speed impact, which can be written as

$$\sigma = \left(A + B\varepsilon^{n}\right)\left(1 + C\ln\dot{\varepsilon}^{*}\right)\left(1 - T^{*m}\right) \quad (1)$$

in which $A$, $B$, $n$, $C$, $m$ are constants related to materials, $\dot{\varepsilon}^{*}$ is the ratio of reference strain rate and strain rate, $T^{*} = (T - T_r)/(T_m - T_r)$, $T_r$ is room temperature (300 K), and $T_m$ is the melting point of materials. Material parameters of numerical simulation are shown in Table 2.

We have validated the accuracy of the simulation model by experiment as shown in Figure 3 [13] and Table 2. The parameters of the projectile and the honeycomb sandwich panel were shown in Table 1. The impact velocity of projectile was 1.915 km/s. The perforation diameter of rear face sheet is about 14 mm in both experimental and simulation results. The others experiments were shown in Table 3. The experimental results coincide with the simulation results, which verified the correctness of the simulation model.

## 3. Projectile Oblique Impact on Structure of Double Honeycomb Sandwich Panel

*3.1. The Configuration of Debris Cloud.* Half symmetry model was adopted, including 145164 solid elements, 239056 shell elements, and 4072 SPH particles. The projectile velocity is set to be 3.07 km/s. 3 km/s (or lower) was studied because it is in the extent of speed of orbital debris, and less than or equal to 3 km/s is an important part of the ballistic limit. Every SPH particle contains various kinds of physical quantities, such as mass and speed, and so forth. The distribution of debris cloud can be described by the analysis of the projectile mass fraction of debris cloud. The distribution of debris cloud after the projectile impacted the first layer of honeycomb sandwich panel was shown in Figure 4. In Figure 4(a), the debris cloud distribution area was divided into three regions. Region I was the part above the front face sheet of honeycomb sandwich panel, which was the backwash debris cloud. Region II was the interior of the honeycomb sandwich panel, which was

TABLE 3: Experiment and simulation results of rear face sheet of honeycomb sandwich panel.

| | Projectile diameter/mm | Velocity/km·s$^{-1}$ | Experiment result/mm | Simulation result/mm | Error |
|---|---|---|---|---|---|
| Test 1 | 5 | 1.915 | 14.1 | 15.3 | 7.8% |
| Test 2 | 8 | 1.7 | 22.5 | 24.6 | 8.5% |
| Test 3 | 5 | 3.065 | 25.4 | 23.9 | 5.9% |

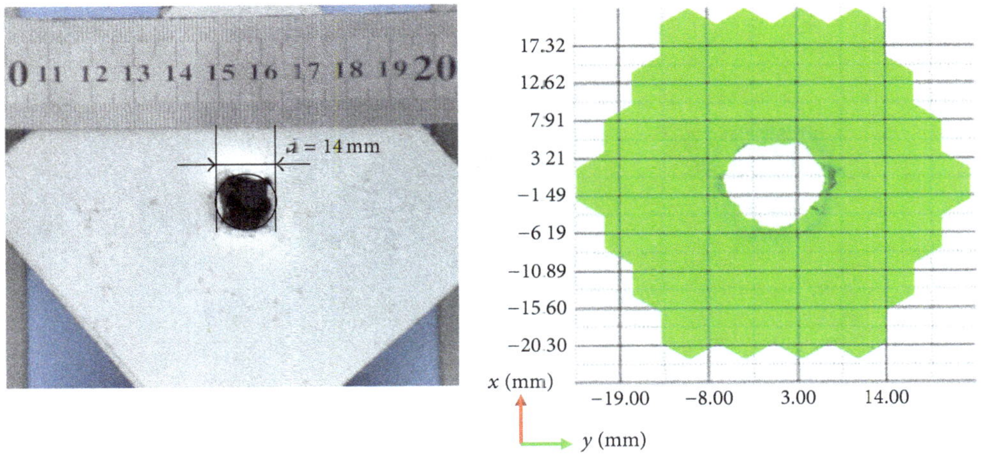


(a) Experimental result (b) Simulation result

FIGURE 3: Experiment and simulation results [12].

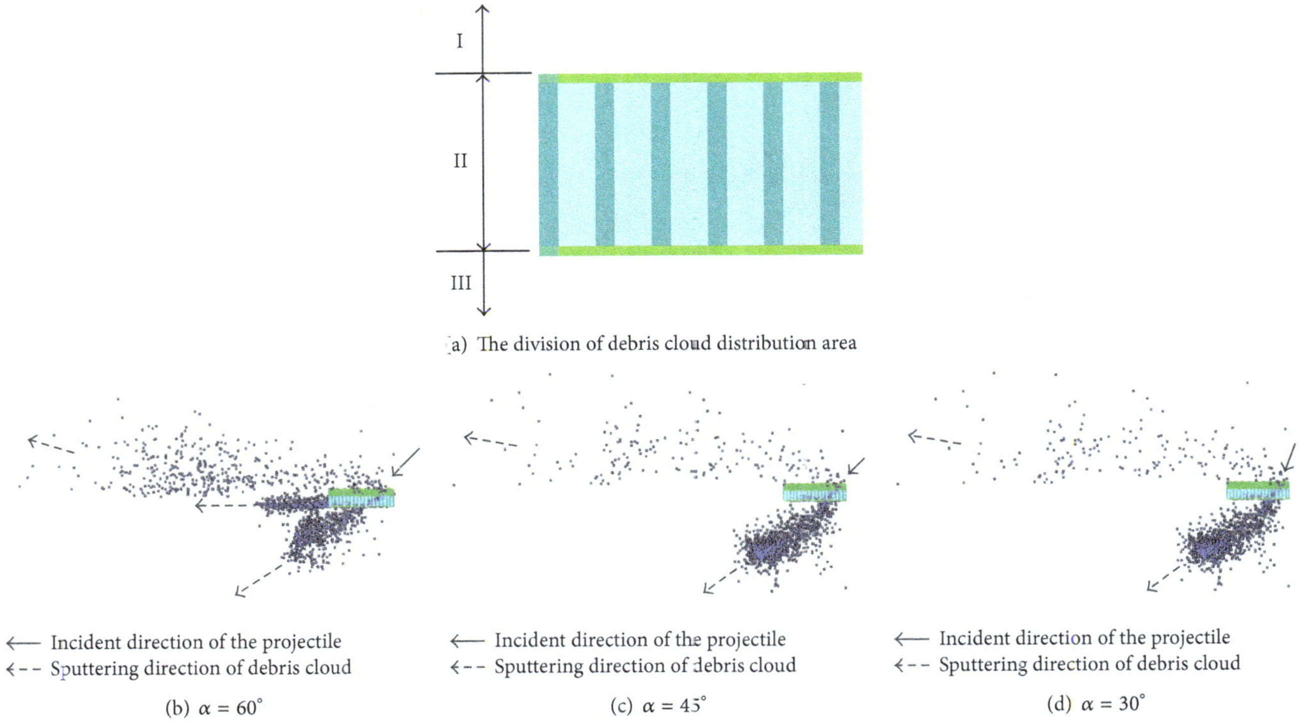


FIGURE 4: The distribution of the debris cloud in 9.5$E$ − 002 ms.

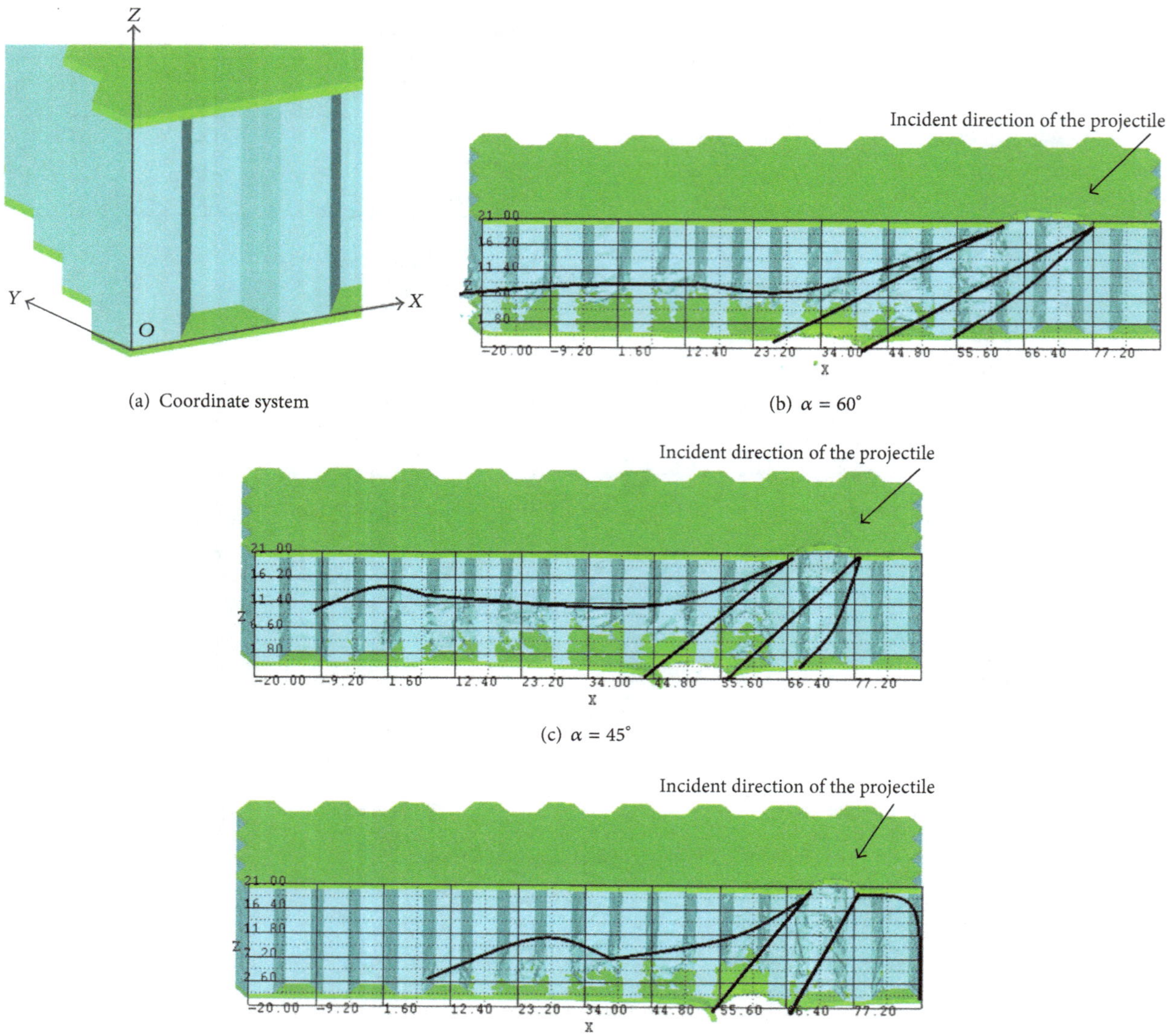


(a) Coordinate system

(b) $\alpha = 60^{\circ}$

(c) $\alpha = 45^{\circ}$

(d) $\alpha = 30^{\circ}$

FIGURE 5: Damage of the first layer of honeycomb sandwich panel.

made of the radial motion parts of projectile debris cloud and the backwash debris cloud which formed due to influence of the rear face sheet. Region III was the part below rear face sheet, which was the debris cloud part which go through the rear face sheet.

Figures 4(b), 4(c), and 4(d) show the distribution of debris cloud of projectile when the impact angle was 60°, 45°, and 30°, respectively. The debris cloud distribution was obviously influenced by the impact angles. In order to analyze the distribution of debris cloud accurately, the projectile mass fraction of the debris cloud in different debris cloud distribution areas was obtained as shown in Table 4. From Figure 4 and Table 4, it can be seen that Region III increased gradually with the decrease of the impact angle. The projectile debris cloud was mainly distributed in Region III when the impact angle was 30°, and the main movement direction of the debris cloud was in the vertical direction.

TABLE 4: Projectile mass fraction of debris cloud.

| Incident angle | Region I | Region II | Region III |
|---|---|---|---|
| 60° | 18.33% | 33.33% | 48.33% |
| 45° | 4.76% | 23.67% | 71.67% |
| 30° | 1.67% | 18.33% | 80.00% |

*3.2. Damage of the First Layer of Double Honeycomb Sandwich Panel.* Damage of the panel and honeycomb core after the projectile impact the first layer of double honeycomb sandwich panel was formed as shown in Figure 5. The debris cloud of projectile enters the honeycomb sandwich panel inside can be divided into two parts, one was the "main part" which was between the two straight lines as shown in Figure 5, and the other part was the "secondary part" which was between the two curves except "main part"; this part

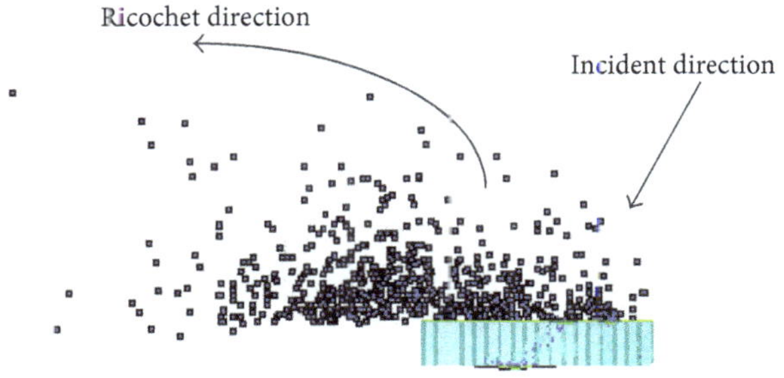


(a) The configuration of debris cloud

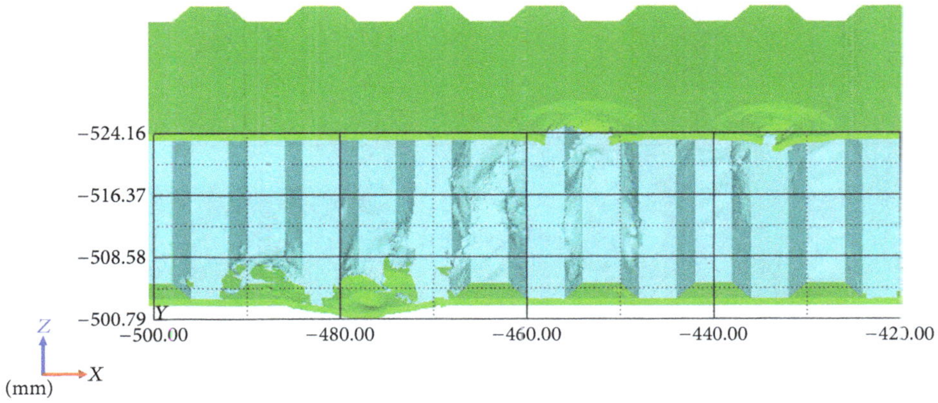


(b) Damage of the second layer of double honeycomb sandwich panel

FIGURE 6: Second layer of double honeycomb sandwich panels (incidence angle is 45°).

of debris cloud did not have enough kinetic energy to get through the panel but only expanded inside the honeycomb core after the backwash and then penetrated into the cellular wall. Since penetrating the cellular wall will consume the kinetic energy of debris cloud, the energy of the debris cloud decreased so that it cannot penetrate into the honeycomb core again after penetrating a certain number of the cellular walls.

A coordinate system was given to describe the damage degree of honeycomb panel (Figure 5(a)), and the coordinate origin can move along $x$-axis. With the decrease of the projectile impact angle, the projectile velocity along $x$-axis was reduced accordingly. The projectile mass fraction of inflation debris cloud along $x$-axis was also reduced. As a result, damage area of the honeycomb core caused by debris cloud was decreased. It can be seen from Figure 6 that the expansion scope of debris cloud was reduced with the decrease of the impact angle. When the impact angle was 30° (Figure 5(d)), the damage range of the honeycomb core was reduced significantly compared with 60° (Figure 5(b)) and 45° (Figure 5(c)). Thus, there are obvious relationships between the damage of honeycomb sandwich panel and the impact angle.

Different impact angles lead to various perforation shape and size of rear face sheet as shown in Table 5. It can be found that the perforation shape can be described by an ellipse approximately. The perforation shape was more likely an oval shape when the impact angle was 60° compared with that of 45° and 30°; this was induced by the degeneration of elliptic equations.

As shown in Table 5, the largest perforation size of the front face sheet reduces with the decrease of the impact angle. Meanwhile, the largest perforation size of the rear face sheet was increased firstly and then decreased. The largest perforation size of the front face sheet was slightly greater than the rear face sheet when the impact angle was 60°, but the former is significantly smaller than the latter when the impact angle was 45° and 30°. This is because the velocity component of $z$-axis was larger when the impact angle was smaller and more debris cloud particles impact the rear face sheet, resulting in the greater damage in the rear face sheet.

*3.3. Damage of the Second Layer of Double Honeycomb Sandwich Panel.* For the double honeycomb sandwich panel structure, the projectile was broken after it penetrated the first layer of the honeycomb sandwich panel, and then the projectile continued to move in the form of debris cloud until it impacted the second layer of honeycomb sandwich panel. The shape of debris cloud was related to the incident angle of the projectile, including the projectile broken partly or completely. The damage degree of rear face sheet was associated with the shape of projectile debris cloud. Because of the size and velocity, the nonsignificant fragment of front face sheet of the first layer of honeycomb sandwich panel was observed, so we ignored the effect of fragment of face sheet on damage in this paper.

The debris cloud, which formed after the projectile penetrated into the first layer of the honeycomb sandwich panel, impacted the second layer of honeycomb sandwich panel,

Table 5: The perforation shape of panel.

| | Incident angle 60° | Incident angle 45° | Incident angle 30° |
|---|---|---|---|
| Front face sheet | 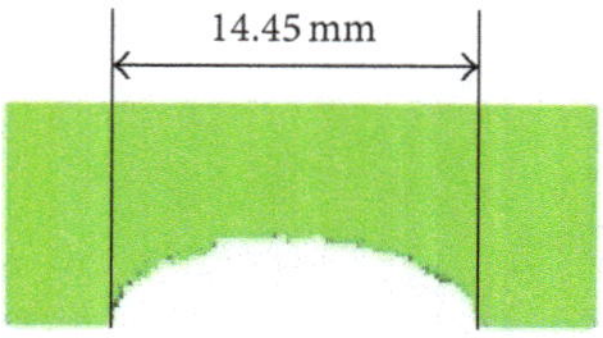 14.45 mm | 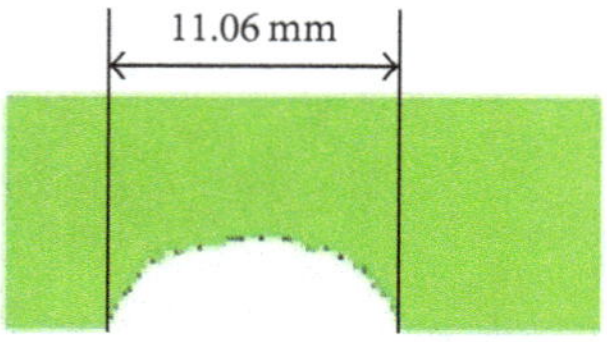 11.06 mm | 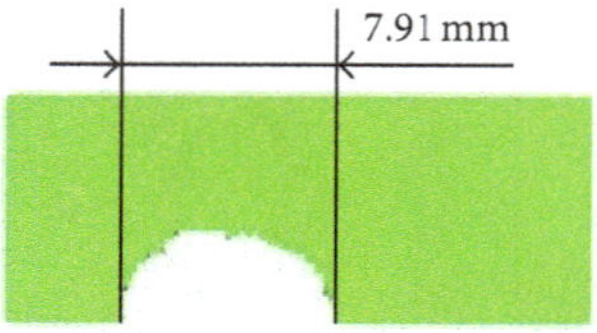 7.91 mm |
| Rear face sheet | 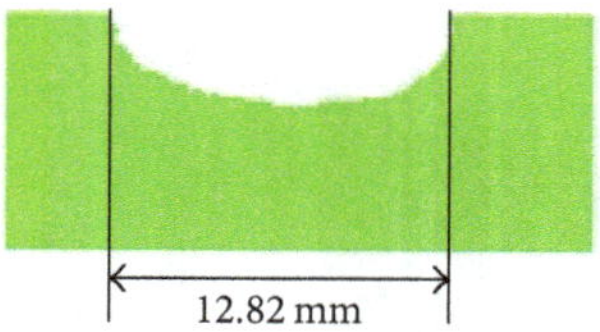 12.82 mm | 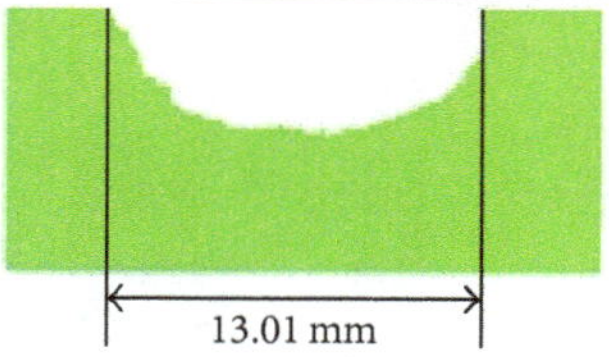 13.01 mm | 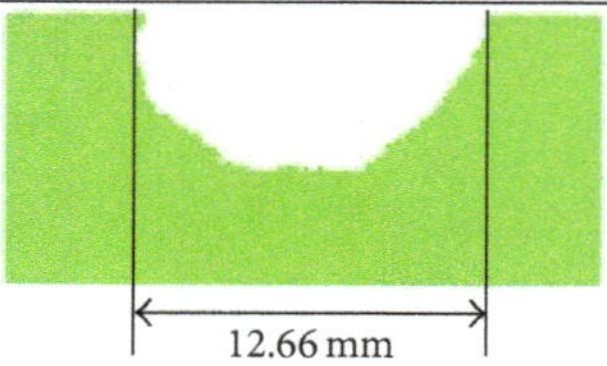 12.66 mm |

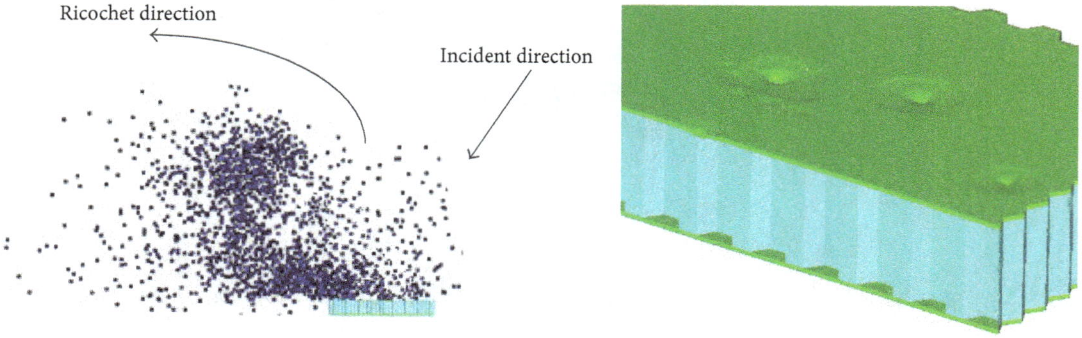


(a) The configuration of debris cloud

(b) Damage of the second layer of honeycomb sandwich panel

Figure 7: Second layer of honeycomb sandwich panels (incidence angle is 30°).

where distance to the first layer is 500 mm. The projectile mass fraction of debris cloud which penetrated the first layer of the honeycomb sandwich panel was not much and broken fully (Figure 4(b)); the kinetic energy of debris cloud was too small to induce obvious damage on the honeycomb sandwich panel plate when impact angle was 60°. Therefore, damage of the second layer of honeycomb sandwich panel was studied only under the case of the impact angles of 45° and 30° in this paper, as shown in Figures 6 and 7.

Figure 4(c) shows the projectile debris cloud form on the situation that the projectile has penetrated the first layer of honeycomb sandwich panel but did not impact the second layer when the impact angle was 45°. Big pieces did not break in the front of debris cloud, which are the main factors that lead to the perforation on front face sheet and then impacted the second layer of double honeycomb sandwich panel. Most of the scattered debris cloud cannot penetrate the front face sheet of second layer of double honeycomb sandwich panel, but splash after impacted the front face sheet as shown in Figure 6(a). Figure 6(b) shows the damage of the second layer of honeycomb sandwich panel. According to the two perforations in front face sheet, we can know that there were two pieces of debris that impacted the front face sheet and formed punches and smaller piece of debris fully broken after getting through the front face sheet so that it cannot cause damage to the rear face sheet again. Large pieces of debris impact the rear face sheet after penetrating the front face sheet. The kinetic energy of debris cloud was too small to penetrate the back panel at that moment and only craters in the rear face sheet were formed.

Distribution and damage of debris cloud after the projectile impact the second layer of honeycomb sandwich panel of double when impact angle was 30° as shown in Figure 7. There are three obvious perforations on the rear face sheet. The projectile was further broken when debris cloud penetrated the second layer of honeycomb sandwich panel so that the rear face sheet without obvious damage due to the velocity component of $z$-axis at the impact angle of 30° is larger than 45°.

According to the analysis of the damage of projectile impact double honeycomb sandwich panel, it can be found that the impact damage was closely related to the oblique impact angle. The projectile broken degree, the debris cloud distribution, and the damage on each layer of honeycomb sandwich panel were different under different impact angles. Ruining of the first layer of honeycomb sandwich panel was the most serious when projectile oblique impact double honeycomb sandwich panel structure. Debris cloud forms

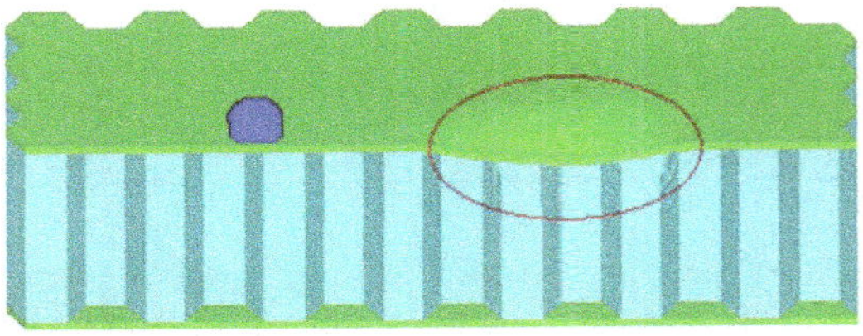

(a) Impact angle is 87°

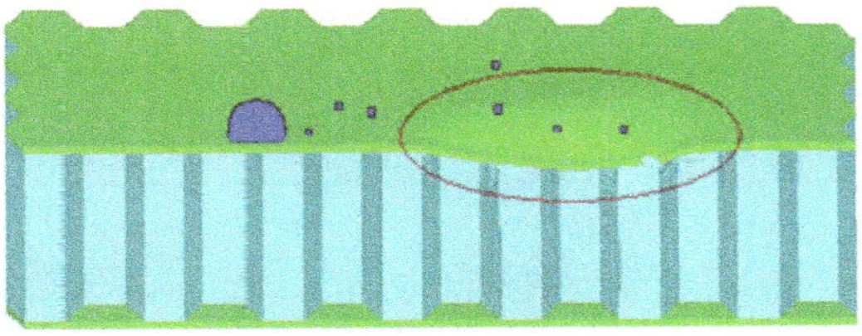

(b) Impact angle is 85°

FIGURE 8: Critical penetration of the front face sheet.

(a) Impact angle is 72°

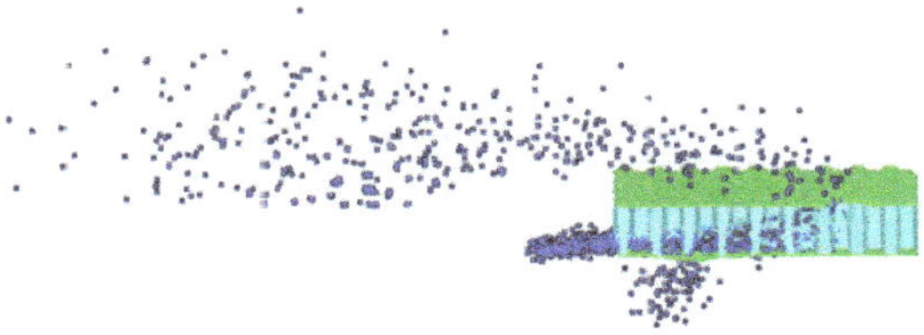

(b) Impact angle is 70°

FIGURE 9: Critical penetration of the rear face sheet.

were uncommon after the projectile penetrated the first layer of honeycomb sandwich panel at different impact angles, resulting the different damage shapes on the second layer of honeycomb sandwich panel. Incident angle of space debris impact the spacecraft is also different because the space debris and spacecraft orbit were different in the real space environment. It is terrible that the internal equipment of spacecraft and the other side of the bulkhead will be damaged if the space debris penetrated one side of the bulkhead. Critical penetration angle of the projectile oblique impact single honeycomb sandwich panel was studied in this paper based on using single honeycomb sandwich panel to simulate one side of the spacecraft bulkhead.

## 4. Critical Penetration Angle of Projectile Oblique Impact on Honeycomb Sandwich Panel

*4.1. Critical Penetration Angle of the Front Face Sheet.* The projectile penetrated first layer honeycomb sandwich panel would lead to more or less damage on the second layer honeycomb sandwich panel according to the study of the double honeycomb sandwich panel structure. On the other hand, critical perforation angle of single layer honeycomb sandwich panel is a part of a double honeycomb sandwich panel structure, so the critical penetration angle of projectile oblique impact on honeycomb sandwich panel was studied firstly.

The critical penetration is infinitely close to penetration and unpenetration. The critical penetration angle of the front face sheet of a honeycomb sandwich panel was found through a lot of simulation tests based on dichotomy. Impact speed is 3 km/s, for example, to analyze the critical penetration angle of the front face sheet. The critical angle that the front face sheet cannot be penetrated was 87° as shown in Figure 8(a). The projectile was not broken, and only plastic deformation occurred at that time. Impact craters were formed on the front face sheet of the honeycomb sandwich panel, but it was unable to be punched. Critical penetration angle of the front face sheet was 85° as shown in Figure 8(b). Local plastic deformation and broken of projectile occurred, and there were holes on the front face sheet at the same moment, so the critical penetration angle of the front face sheet was 86°.

It can be viewed according to Figure 9 that projectile was not broken when projectile angle was greater than the critical penetration angle of the front face sheet. Therefore, critical penetration angle of the front face sheet is also the critical value of projectile broken. Symbol $C$ in this formula represents the collection of projectiles, which were not broken (see (2)) and Symbol $\overline{C}$ represents the collection of projectiles, which were broken (see (3)), $\theta$ represents the projectile incident angle, and $\widehat{\theta}$ represents critical penetration angle of the front face sheet:

$$C = \left\{\theta > \widehat{\theta} \mid 0^\circ \leq \theta \leq 90^\circ\right\} \quad (2)$$

$$\overline{C} = \left\{\theta > \widehat{\theta} \mid 0^\circ \leq \theta \leq 90^\circ\right\}. \quad (3)$$

*4.2. Critical Penetration Angle of the Rear Face Sheet.* The critical angle of the rear face sheet of honeycomb sandwich panel was 72° at the impact speed of 3 km/s as shown in Figures 9(a) and 10(a). Projectile was broken after impacting the front face sheet and part of the debris cloud backwash then moved along the horizontal direction. Meanwhile, another part of the debris cloud moves into the honeycomb core layer. The motion of the debris cloud lateral leaded to the rupture and the collapse of the honeycomb core layer.

The critical angle of the rear face sheet of honeycomb sandwich panel was 70° as shown in Figures 9(b) and 10(b). Compared to the incidence angle of 72°, the number of SPH

TABLE 6: Particle diameter at ballistic limit (the particle diameter of simulation was 3 mm).

| Velocity/km·s$^{-1}$ | 1.8 | 2 | 2.2 | 2.4 | 2.6 | 2.8 | 3 |
|---|---|---|---|---|---|---|---|
| Calculation results of (4)/mm | 3.53 | 3.36 | 3.34 | 3.22 | 3.07 | 2.92 | 2.79 |

(a) Impact angle is 72°

(b) Impact angle is 70°

FIGURE 10: Damage of rear face sheet with critical perforated.

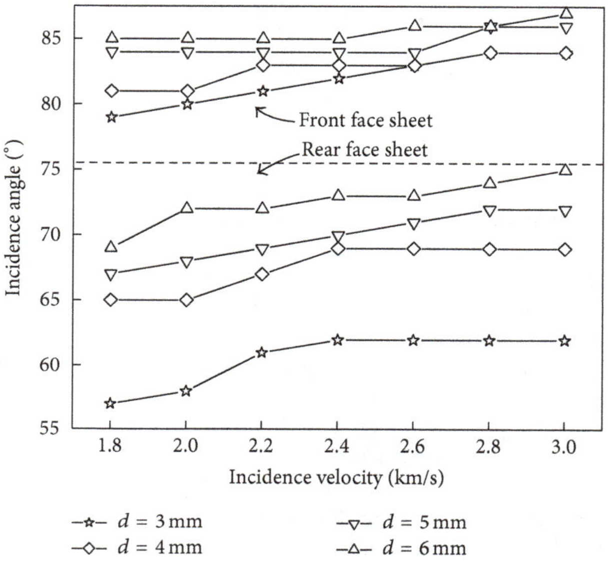


FIGURE 11: Critical perforation angles under various diameter and velocity of projectile.

particles that enter inside of honeycomb core and vertical component of the projectile velocity was increased. Damage on the rear face sheet was also increased so that the critical perforation at the incidence angle of 70° occurred in rear face sheet. Therefore, the critical perforation angle of the rear face sheet was 71°.

The ability of projectile penetrating honeycomb sandwich panel is different when the impact speed of projectile or diameter is different. In order to study the critical perforation angle of front and rear face sheets of honeycomb sandwich panel at different impact speeds or diameters when oblique impacting, the impact test was simulated within the impacting speed range of 1.8~3.4 km/s and the diameter is 3 mm, 4 mm, 5 mm, and 6 mm as shown in Figure 11.

Simulation results were compared with the ballistic limit of a sandwich panel in order to testify the accuracy of the simulation. Equation (4) is an example of the generic form of the ballistic limit of a sandwich panel [14]. Particle diameter at ballistic limit was calculated at a diameter of 3 mm as shown in Table 6. The simulation result showed good agreement with the calculation results.

$$d_c(v_n) = \left[ \frac{(t_w/K_{3S})(\tau/40000)^{0.5} + t_b}{0.6 \cos\theta v_n^{0.677} \rho_p^{0.5}} \right]^{0.947}. \quad (4)$$

In (4), $d_c$ is a particle diameter at ballistic limit, $t_w$ is thickness of rear wall (cm), $K_{3S}$ is a baseline (1.4 [15]), $\tau$ is rear wall yield stress, $t_b$ is thickness of bumper, $\theta$ is incidence angle, $v_n$ is normal component of velocity, and $\rho_p$ is projectile density.

It can be seen that the critical perforation angle of front and rear face sheet increased slowly or are unchanged with the increase of impact velocity. The critical perforation angle of rear face sheet increased dramatically with the increase of diameter, especially the diameter from 3 mm to 4 mm, but the front face sheet increased slowly.

## 5. Conclusions

The distribution of the debris cloud and the damage of double honeycomb sandwich panel were simulated under oblique hypervelocity impact. In addition, the critical penetration angle of front and rear face sheets of honeycomb sandwich panel was obtained. The conclusions are as follows.

Projectile mass fraction of debris cloud decreases with the decrease of the impact angle which back-splashed and stayed inside of the first layer of honeycomb sandwich panel and then across the honeycomb core. However, the damage of first layer of honeycomb sandwich panel increased gradually. That result from the biggest perforation size of the front face sheet decreased accordingly, and the biggest perforation size of the rear face sheet increased firstly but then decreased.

The projectile mass fraction of debris cloud which get through the first layer of honeycomb sandwich panel was too little to damage the second layer when the impact angle was larger, but it increased gradually with the decrease of impact angle so that the damage of the front face sheet of second layer honeycomb panel became more and more obvious, and the damage degree of the second layer was related to crushing degree of projectile after penetrated the first layer. The projectile will not break when projectile angle was greater than the critical perforation angle of the front face sheet, but it will break gradually when projectile angle is less than the critical perforation angle.

The critical perforation angle of front and rear face sheet increased slowly or is unchanged with the increase of impact velocity. The critical perforation angle of rear face sheet increased dramatically with the increase of diameter, especially the diameter from 3 mm to 4 mm, but the front face sheet increased slowly.

For the research on the projectile oblique hypervelocity impact on double honeycomb panel, we only embarked on the study the distribution of projectile debris cloud and the level of damage of honeycomb panel under a velocity (3.07 km/s) and three incidence angles (60°, 45°, and 30°) of projectile. Therefore, future studies are worthy to be carried out to explore this subject further, such that a wide range of the speed conditions will be investigated, especially at higher velocity (about 10 km/s). On the other hand, different geometric models that will affect the fragmentation degree of projectile, which in turn will affect the damage degree of honeycomb panel, will also be studied. At last, we expect to obtain the critical penetration angle of projectile oblique impact on double honeycomb sandwich panel by more experiments.

## Conflicts of Interest

The authors declare that they have no conflicts of interest.

## Acknowledgments

This work was supported by the National Natural Science Foundation of China (11172083), the Aeronautical Science Foundation of China (2016ZD53036), and the Fundamental Research Funds for the Center Universities (HIT.NSRIF.2015029).

## References

[1] J.-C. Liou and N. L. Johnson, "Risks in space from orbiting debris," *Science*, vol. 311, no. 5759, pp. 340–341, 2006.

[2] F. Zuiani and M. Vasile, "Preliminary design of debris removal missions by means of simplified models for low-thrust, many-revolution transfers," *International Journal of Aerospace Engineering*, vol. 2012, Article ID 836250, 22 pages, 2012.

[3] S. Ryan, F. Schaefer, R. Destefanis, and M. Lambert, "A ballistic limit equation for hypervelocity impacts on composite honeycomb sandwich panel satellite structures," *Advances in Space Research*, vol. 41, no. 7, pp. 1152–1166, 2008.

[4] W. Schonberg, F. Schäfer, and R. Putzar, "Hypervelocity impact response of honeycomb sandwich panels," *Acta Astronautica*, vol. 66, no. 3-4, pp. 455–466, 2010

[5] E. A. Taylor, J. P. Glanville, R. A. Clegg, and R. G. Turner, "Hypervelocity impact on spacecraft honeycomb: hydrocode simulation and damage laws," *International Journal of Impact Engineering*, vol. 29, no. 1, pp. 691–702, 2003.

[6] M. Higashide, N. Onose, and S. Hasegawa, "Sub-millimeter debris impact damage of unmanned spacecraft structure panel," *Procedia Engineering*, vol. 58, pp. 517–525, 2013.

[7] K. Nitta, M. Higashide, Y. Kitazawa, A. Tekaba, M. Katayama, and H. Matsumoto, "Response of a aluminum honeycomb subjected to hypervelocity impacts," *Procedia Engineering*, vol. 58, pp. 709–714, 2013.

[8] P. Kang, S.-K. Youn, and J. H. Lim, "Modification of the critical projectile diameter of honeycomb sandwich panel considering the channeling effect in hypervelocity impact," *Aerospace Science and Technology*, vol. 29, no. 1, pp. 413–425, 2013.

[9] S. Ryan, F. Schaefer, and W. Riedel, "Numerical simulation of hypervelocity impact on CFRP/Al HC SP spacecraft structures causing penetration and fragment ejection," *International Journal of Impact Engineering*, vol. 33, no. 1–12, pp. 703–712, 2006.

[10] M. Wicklein, S. Ryan, D. M. White, and R. A. Clegg, "Hypervelocity impact on CFRP: testing, material modelling, and numerical simulation," *International Journal of Impact Engineering*, vol. 35, no. 12, pp. 1861–1869, 2008.

[11] F. K. Schäfer, S. Ryan, M. Lambert, and R. Putzar, "Ballistic limit equation for equipment placed behind satellite structure walls," *International Journal of Impact Engineering*, vol. 35, no. 12, pp. 1784–1791, 2008.

[12] B. Jia, Z.-T. Ma, W. Zhang, and B.-J. Pang, "Numerical simulation investigation in ballistic limit of Whipple shield structure with Al-foam bumperb," *Material Science and Technology*, vol. 18, no. 3, pp. 368–372, 2010.

[13] Z. Zhang, R. Chi, B. Pang, and G. Guan, "A study on the split effect of projectile debris on honeycomb core," in *Proceedings of the 4th International Conference on Sustainable Energy and Environmental Engineering*, vol. 53, pp. 766–731, Shenzhen, China, December 2015.

[14] E. L. Christiansen, "Design and performance equations for advanced meteoroid and debris shields," *International Journal of Impact Engineering*, vol. 14, no. 1, pp. 145–156, 1993.

[15] W. P. Schonberg, F. Schäfer, and R. Putzar, "Predicting the perforation response of honeycomb sandwich panels using ballistic limit equations," *Journal of Spacecraft and Rockets*, vol. 46, no. 5, pp. 976–981, 2009.

2

# Fast Image Registration for Spacecraft Autonomous Navigation Using Natural Landmarks

**Yun-Hua Wu**,[1] **Lin-Lin Ge**,[1] **Feng Wang**,[2] **Bing Hua**,[1] **Zhi-Ming Chen**,[1] **and Feng Yu**[1]

[1]*Micro-Satellite Research Center, Nanjing University of Aeronautics and Astronautics, Nanjing 210016, China*
[2]*Research Center of Satellite Technology, Harbin Institute of Technology, Harbin 150001, China*

Correspondence should be addressed to Yun-Hua Wu; yunhuawu@nuaa.edu.cn

Academic Editor: Linda L. Vahala

In order to satisfy the real-time requirement of spacecraft autonomous navigation using natural landmarks, a novel algorithm called CSA-SURF (chessboard segmentation algorithm and speeded up robust features) is proposed to improve the speed without loss of repeatability performance of image registration progress. It is a combination of chessboard segmentation algorithm and SURF. Here, SURF is used to extract the features from satellite images because of its scale- and rotation-invariant properties and low computational cost. CSA is based on image segmentation technology, aiming to find representative blocks, which will be allocated to different tasks to speed up the image registration progress. To illustrate the advantages of the proposed algorithm, PCA-SURF, which is the combination of principle component analysis and SURF, is also analyzed in this paper for comparison. Furthermore, random sample consensus (RANSAC) algorithm is applied to eliminate the false matches for further accuracy improvement. The simulation results show that the proposed strategy obtains good results, especially in scaling and rotation variation. Besides, CSA-SURF decreased 50% of the time in extraction and 90% of the time in matching without losing the repeatability performance by comparing with SURF algorithm. The proposed method has been demonstrated as an alternative way for image registration of spacecraft autonomous navigation using natural landmarks.

## 1. Introduction

Spacecraft autonomous navigation only needs to regularly check the spacecraft working conditions, eliminating the complex navigation computing tasks, which greatly reduces the manpower and ground facility requirements and the cost of space projects [1]. Furthermore, ground stations can be destroyed during wartime. However, with the use of an autonomous navigation system, spacecraft can still work well when the ground communication is under interruption. Among many kinds of autonomous navigation systems, autonomous navigation using natural landmarks based on machine vision is a newly proposed navigation method and has potential applications for future space missions.

For the landmark-based autonomous navigation, landmarks are used as reference and measuring objects. At first, the landmarks combined with position information are gathered and stored onboard the satellite. For on-orbit satellite, the camera can capture ground targets that have been stored in the satellite. During the period when the satellite runs over the target, several images from different view angles can be captured. These images are used as the inputs of image matching algorithm. Once the matching succeeds, the location information corresponding to this landmark is used to determine the position of the satellite. Then the orbit of the satellite is estimated by different positions. Figure 1 shows the schematic of autonomous navigation using natural landmarks.

Image registration is a vital technology for landmark-based spacecraft autonomous navigation. However, autonomous navigation systems should be stable with real-time characteristics. In order to satisfy these stringent requirements, image registration progress must be fast and stable. This paper aims to propose an alternative method to improve the above performances.

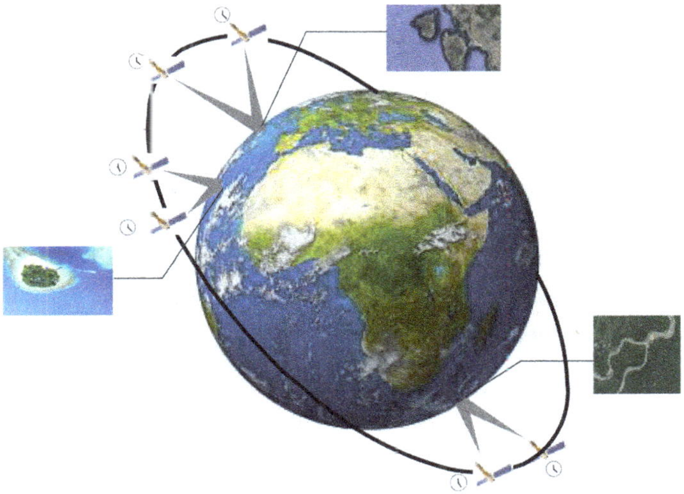

FIGURE 1: Sketch of landmark-based automatic navigation.

To improve the correctness and real-time performance in the process of image matching, Sha et al. proposed a fast matching algorithm based on image $K$-gray-degree clustering, which was robust and fast under the condition of nonlinear changing of local lighting, noise, target matching of irregular shape, and even complex background [2]. However, the problem of quickly obtaining a $K$-degree template reflecting main features of the matching object precisely still needs to be further researched. Xu et al. proposed a novel method called DFOB for the detection, orientation computation, and description of feature points. The method was computationally efficient as it was implemented by integral images. Compared with SIFT and SURF algorithms, the computational cost of this method was much lower [3]. However, this method does not work well under large affine and perspective deformations, making it unable to perform well in wide baseline matching.

Zhao et al. studied a method based on (principal component analysis) PCA [4] to speed up the image registration progress. The resulted computing time was reduced to 60% compared to single gray level normalized cross-correlation matching. He and Jiang proposed a fast image matching algorithm based on discrete Hartley transform (DHT) [5], which reduced data calculation and storage. Besides, it also improved image matching accuracy and efficiency. The PCA-based method proposed by Peng Zhao greatly improved the speed performance. However, its performance is poor when the rotation is larger than 20 deg. The image matching algorithm based on DHT also increased the image registration speed compared to traditional algorithm based on (fast Fourier transformation) FFT. But its computation time reached 18 s, which cannot satisfy the real-time requirement for natural landmark-based autonomous navigation. In order to overcome the above shortcomings, a fast image registration algorithm based on chessboard segmentation is proposed. Furthermore, RANSAC algorithm is applied to remove the error-matched key points for further repeatability improvement.

The reminder of this paper is organized as follows. Firstly, Section 1 introduces the purpose of this research. Then Section 2 reviews the basic theory of PCA-SURF algorithm. Later, Section 3 presents a method called chessboard segmentation algorithm to speed up image registration progress. In addition, random sample consensus algorithm is also presented in this section to obtain the registration statistical results about the number of matches and mismatches. In Section 4, the time consumptions of different methods are compared to verify the advantages of our algorithm. In Section 5, some metrics are defined to evaluate repeatability of image registration result. Besides, several tests are designed to evaluate the performance of the proposed method. Finally, Section 6 summarizes the contributions of this work.

## 2. Review of PCA-SURF

Principal component analysis (PCA) is a classical feature extraction and data representation technique widely used in the area of computer vision [6, 7]. PCA-SURF, which is a combination method of PCA and SURF [8], aims to reduce the computation by compressing the data dimension.

*2.1. PCA-SURF Description.* Suppose that $\mathbf{U}_i$ is a $64 \times 1$ description vector of a reference image $I_1$ and $\mathbf{V}_i$ is a $64 \times 1$ description vector of an object image $I_2$. Let $\mathbf{X}$ denote all SURF description vectors in the reference image and the object image:

$$\mathbf{X} = [\mathbf{V}_1 \ \mathbf{V}_2 \ \cdots \ \mathbf{V}_n \ \mathbf{U}_1 \ \mathbf{U}_2 \ \cdots \ \mathbf{U}_m], \quad (1)$$

where $n$ and $m$ are the number of SURF features in the reference and object images, respectively. Now, a group of orthogonal projection directions is required to be found to project $\mathbf{X}$ into a lower dimension space. A covariance matrix can be defined as follows:

$$\mathbf{R} = \mathbf{X}\mathbf{X}^T. \quad (2)$$

Then the eigenvalue and eigenvector corresponding to $\mathbf{R}$ are $\lambda_i$ and $\boldsymbol{\mu}_i$. Actually, the value of $\lambda_i$ is related to the amount of information in its corresponding vector $\boldsymbol{\mu}_i$. Larger eigenvalue means more amount of information. Then a matrix is defined

$$\mathbf{C} = [\mu_1 \quad \mu_2 \quad \cdots \quad \mu_{64}]. \tag{3}$$

Here, it is important to satisfy that **C** is an orthogonal matrix. Only with orthogonal transformation, the original vectors can be invariant to their Euclidean distance or angle with each other. What is more, in order to compress the dimension with least information loss, the following relation should be satisfied:

$$\lambda_1 > \lambda_2 > \cdots > \lambda_{64}. \tag{4}$$

Then with the orthogonal transformation, we have

$$\mathbf{Y} = \mathbf{C}^T\mathbf{X}, \tag{5}$$

where **Y** is the feature in the new space corresponding to the original SURF feature. A new equation is defined as

$$\mathbf{Y}^T = [\mathbf{y}_1 \quad \mathbf{y}_2 \quad \cdots \quad \mathbf{y}_{64}]. \tag{6}$$

Here, each column $\mathbf{y}_1\, \mathbf{y}_2 \ldots \mathbf{y}_{64}$ of **Y** is the principal component. Compressing the dimension of the descriptor, then $\mathbf{y}_{64}, \mathbf{y}_{63}, \ldots$ can be removed sequentially, because the smaller the eigenvalues are, the less the information is.

*2.2. Application of Image Registration.* In practical application, two conceptions called contribution rate and cumulative contribution rate should be defined:

$$\begin{aligned} \text{Contribution rate} &= \frac{\lambda_i}{\sum_{k=1}^{n}\lambda_k}, \\ \text{Cumulative contribution rate} &= \frac{\sum_{i=1}^{m}\lambda_i}{\sum_{k=1}^{m}\lambda_k}. \end{aligned} \tag{7}$$

Image registration task will be completed by using the first $m$ principal components which occupy a large proportion of the contribution rate. In order to find the best value of the cumulative contribution rate, some simulation results are presented in the following.

Figure 2 shows the repeatability of PCA-SURF with different cumulative contribution rates. Different lines in the figure represent the object image with different rotations. Figure 3 presents the remaining dimensions with different cumulative contribution rates. From the above comparison, it can be seen that PCA-SURF worked poorly when the rotation is larger than 20 degrees. When the rotation is less than 20 degree, the cumulative contribution rate can be chosen as 0.95; therefore, it can largely compress the dimension of the SURF descriptor with little accuracy loss.

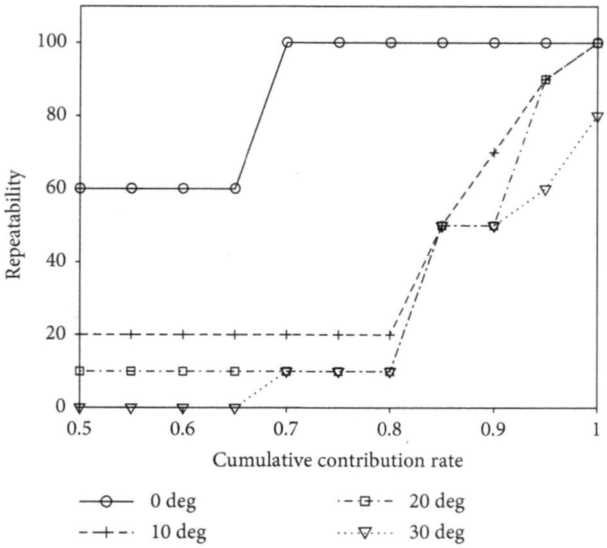


FIGURE 2: Repeatability for different cumulative contribution rates.

## 3. CSA and RANSAC

*3.1. Chessboard Segmentation Algorithm.* According to [9–11], there are three strategies that can be implemented to speed up the image registration progress. The first strategy is to decrease the dimension of the key points' descriptors. And the second one is to decompose the task and deal them with multithreads. The third is to decrease the number of key points. Here, in this paper, the latter two methods are used to improve the speed performance.

Chessboard segmentation algorithm, which combined these two ideas, can greatly improve the speed without losing performance of repeatability.

Figure 4 shows an island image that has been segmented into $N \times M$ parts. The number in each block represents the amount of SURF features. In order to speed up image registration progress, considering the second idea, the feature extraction tasks in each block will be allocated to different threads. Besides, by considering the third idea, only some representative blocks rather than all of them will be selected. In order to satisfy these requirements, several factors needed to be considered, for example, if the block with the most number of SURF features has been selected, then the weight of blocks around the chosen block should be decreased to make sure that regional distribution is more even. So, a block selection model should be established. The coordinate of each block is defined in Figure 4. And we suppose that $S_{i,j}$ is the amount of SURF features of block $D_{i,j}$.

Due to the diversity of each image, the SURF feature amounts for different images must be distinct so normalization is defined as

$$\widehat{S}_{i,j} = \frac{S_{i,j} - S_{\min}}{S_{\max} - S_{\min}}, \tag{8}$$

where $S_{\min}$ and $S_{\max}$ are the minimum and maximum values, respectively, of all $S_{i,j}$. And $\widehat{S}_{i,j}$ is the normalization of $S_{i,j}$. It is assumed that $\widehat{S}_{i,j}$ is the weight of importance for each block. Here is the weight matrix **W**

$$\mathbf{W} = \begin{bmatrix} \widehat{S}_{1,1} & \cdots & \widehat{S}_{1,M} \\ \vdots & \ddots & \vdots \\ \widehat{S}_{N,1} & \cdots & \widehat{S}_{N,M} \end{bmatrix} = \begin{bmatrix} w_{1,1} & \cdots & w_{1,M} \\ \vdots & \ddots & \vdots \\ w_{N,1} & \cdots & w_{N,M} \end{bmatrix}. \tag{9}$$

Local field attenuation effect simulates the phenomenon that the importance of blocks around the selected block needs

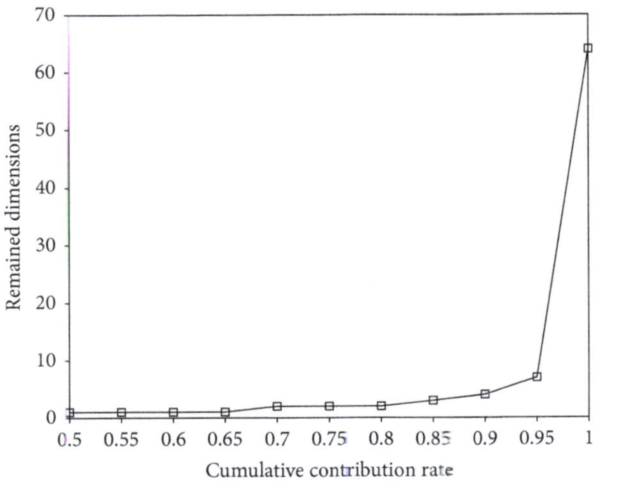


FIGURE 3: Remaining dimension for different cumulative contribution rates.

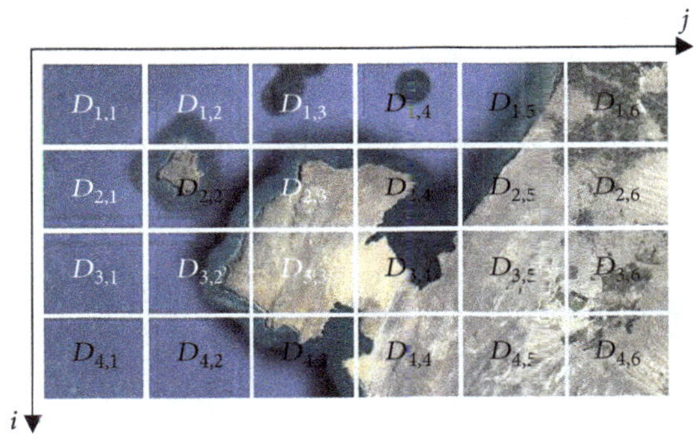

FIGURE 4: Illustration of the number of SURF features in each block.

to decrease. The diagram of local field attenuation effect is also demonstrated in Figure 4. It is supposed that block $D_{2,2}$ has been selected; then the normalized SURF feature amount $\widehat{S}_{i,j}$ of each local field with white color should be decayed. Assuming that the attenuation threshold is $T$, then $\widehat{S}_{i,j}$ is updated according to

$$\widehat{S}_{i,j} = \widehat{S}_{i,j} - T. \tag{10}$$

The entire diagram flow is presented in Figure 5.

The major steps of chessboard segmentation algorithm are as follows.

*(1) Image Segmentation.* In the first step, CSA splits the source image into $N \times M$ blocks. Simultaneously, the SURF features of each block will be extracted with the help of parallel computing technology.

*(2) Data Normalization.* Then the amount $S_{i,j}$ of SURF features of each block will be normalized to establish the weight matrix **W**.

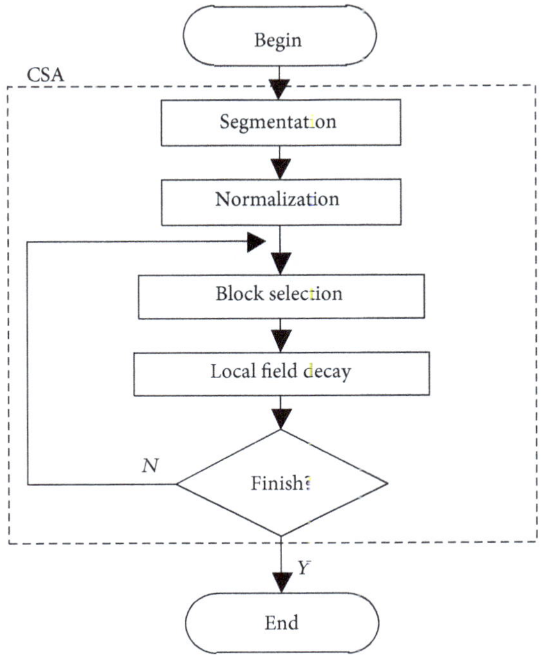


FIGURE 5: Diagram flow of CSA.

*(3) Block Selection.* In each step, the block with maximum weight among the candidate set will be selected and then be removed from the candidate set.

*(4) Local Attenuation Effect.* After the maximum weight has been selected, the weight coefficient around the selected block should minus a threshold $T$ to simulate the local attenuation effect.

Two parameters will affect the stability of CSA: the number of blocks in $x$ and $y$ directions. The experimental determination of the number of blocks that maximizes the stability of CSA is shown in Figure 6, which is based on an image registration task by using a collection of different natural landmarks. The terms $N_x$ and $N_y$ in this figure represent the number of blocks in $x$ direction and $y$ direction. Besides, the size of the object image that needs to be matched with the reference image is $800 \times 684$. The origin of an image coordinate system is located at the upper left corner of the corresponding image.

Figure 6 shows the repeatability for different combination of $N_x$ and $N_y$. The red marker is the best result corresponding to 100% repeatability. According to experimental results, the CSA works better when the following equation is satisfied:

$$\frac{N_y}{N_x} \approx \frac{\text{image width}}{\text{image height}}. \tag{11}$$

Besides, CSA threshold $T$ represents the extent of local attenuation effect and it is also an important parameter that will affect the CSA result. Here, a solution is proposed to

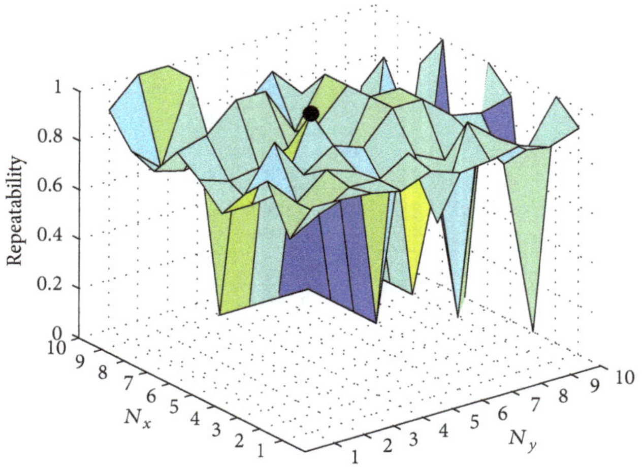


FIGURE 6: Repeatability for combined $N_x$ and $N_y$.

automatically calculate the threshold. A dataset is defined as follows:

$$\alpha = \{w_{i,j}, i = 1, \ldots, N, j = 1, \ldots, M\}, \tag{12}$$

where $w_{i,j}$ is the weight coefficient in weight matrix. Then the threshold $T$ is equal to standard derivation of weight coefficients.

$$T = \sqrt{\frac{1}{N \times M - 1} \sum_{j}^{M} \sum_{i}^{N} (w_{i,j} - \bar{w})^2}, \tag{13}$$

in which $\bar{w}$ is the mean of dataset $\alpha$. By this equation, a suitable threshold $T$ can be obtained.

Figures 7–9 present the results of CSA with different thresholds. In these figures, the selected blocks have been marked by green color. In Figure 7, the threshold $T = 0$ means that the local attenuation effect is invalid and illustrates that these blocks whose feature number is not equal to 0 are selected. Figure 8 shows the CSA result with the threshold $T = 0.1$ or 0.2. And Figure 9 shows the CSA result with automatic determined threshold $T = 0.3333$. It can be seen that Figure 9 gives the most typical and minimal number of blocks with the result of automatic threshold determination.

*3.2. Random Sample Consensus Algorithm.* Once the result of image registration is obtained, it is hard to analyze statistically the match ratio because it may have hundreds of matched key points in result. In this section, an algorithm called random sample consensus [12–14] is introduced to deal with the above problem. Besides, this algorithm also can be used to remove the wrong matched key points [15].

*3.2.1. Perspective Transformation.* Perspective transformation is a commonly used model to represent the relationship between two images from different views.

Figure 10 is a demonstration of perspective transformation. $P_1$ is a point in plane 1, and its coordinate is $(x, y)$. $Q_1$

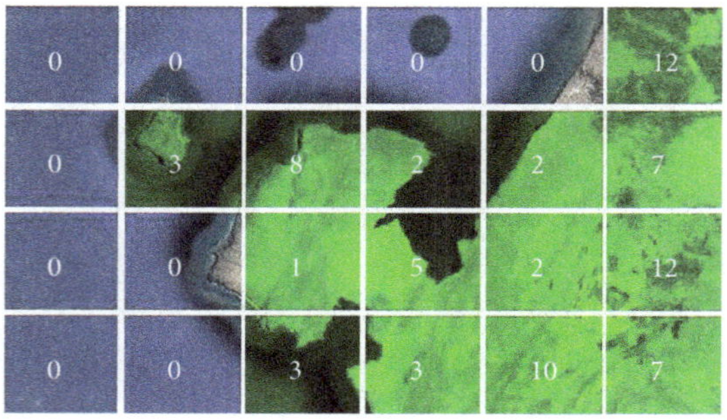

FIGURE 7: CSA result with threshold $T = 0$.

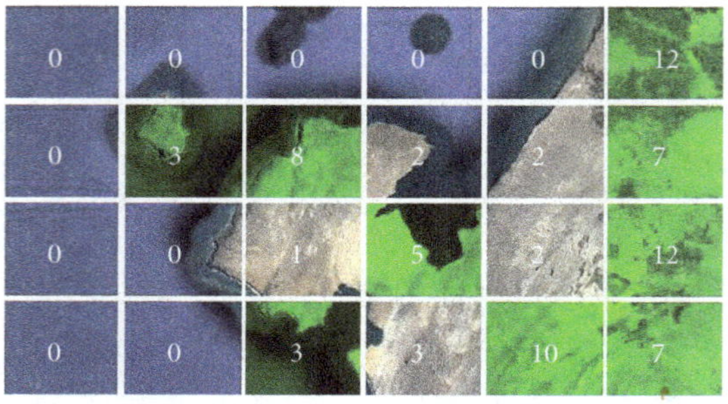

FIGURE 8: CSA result with threshold $T = 0.1$ or 0.2.

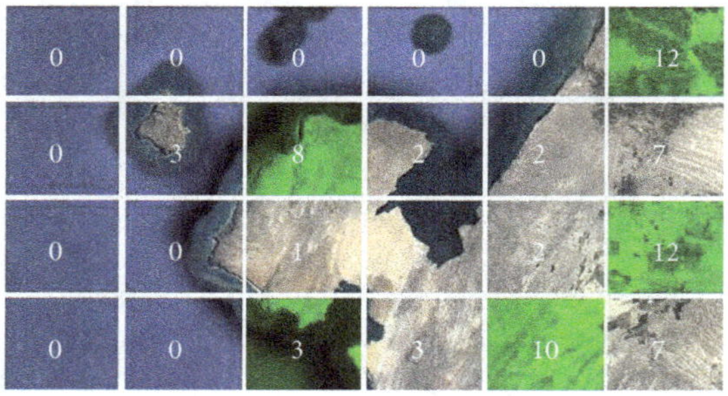

FIGURE 9: CSA result with automatic determined threshold $T = 0.3333$.

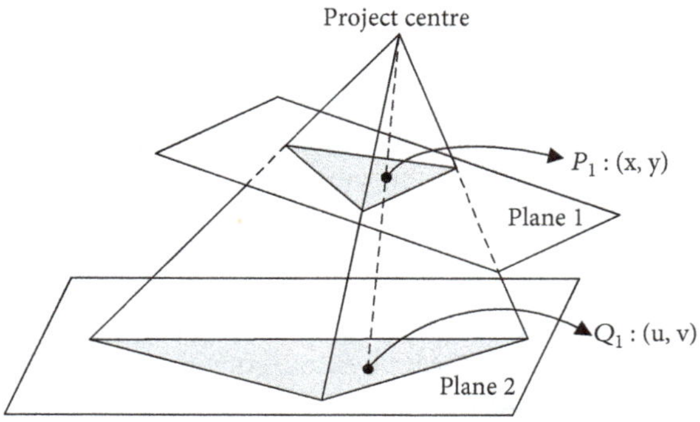


FIGURE 10: Perspective transformation.

is the corresponding point in plane 2 with coordinate $(u, v)$. In order to get the relationship between $P_1$ and $Q_1$, a perspective transformation matrix $\mathbf{H}$ is defined as

$$\mathbf{H} = \begin{bmatrix} h_1 & h_2 & h_3 \\ h_4 & h_5 & h_6 \\ h_7 & h_8 & 1 \end{bmatrix}. \quad (14)$$

Then the relationship between $P_1$ and $Q_1$ is

$$\lambda \begin{bmatrix} u \\ v \\ 1 \end{bmatrix} = \mathbf{H} \begin{bmatrix} x \\ y \\ 1 \end{bmatrix}, \quad (15)$$

where $\lambda$ is the scaling factor. By dividing the first and second rows with the third row, it will achieve

$$u = \frac{h_1 x + h_2 y + h_3}{h_7 x + h_8 y + 1}, \qquad v = \frac{h_4 x + h_5 y + h_6}{h_7 x + h_8 y + 1}. \quad (16)$$

Then we assume that $\widehat{\mathbf{H}} = [h_1 \ \ldots \ h_8]^T$; to transfer the above equation into the matrix form, we can obtain

$$\begin{bmatrix} x & y & 1 & 0 & 0 & 0 & -xu & -yu \\ 0 & 0 & 0 & x & y & 1 & -xv & -yv \end{bmatrix} \widehat{\mathbf{H}} = \begin{bmatrix} u \\ v \end{bmatrix}. \quad (17)$$

In order to solve $\widehat{\mathbf{H}}$, four or more corresponding pairs are required. So, we suppose that there are four known corresponding groups $P_1, P_2, P_3, P_4$ and $Q_1, Q_2, Q_3, Q_4$, and satisfying

$$\begin{bmatrix} x_1 & y_1 & 1 & 0 & 0 & 0 & -x_1u_1 & -y_1u_1 \\ 0 & 0 & 0 & x_1 & y_1 & 1 & -x_1v_1 & -y_1v_1 \\ & & & \cdots & & & & \\ x_4 & y_4 & 1 & 0 & 0 & 0 & -x_4u_4 & -y_4u_4 \\ 0 & 0 & 0 & x_4 & y_4 & 1 & -x_4v_4 & -y_4v_4 \end{bmatrix} \begin{bmatrix} h_1 \\ h_2 \\ \vdots \\ h_7 \\ h_8 \end{bmatrix} = \begin{bmatrix} u_1 \\ v_1 \\ \vdots \\ u_4 \\ v_4 \end{bmatrix}, \quad (18)$$

it can be simplified as

$$\mathbf{A}\widehat{\mathbf{H}} = \mathbf{b}. \quad (19)$$

Because of distortion, noise, or other reasons, the above equation may be a contradictory equation. Therefore, the least squares method is applied to find the satisfied result

$$\widehat{\mathbf{H}} = \left(\mathbf{A}^T\mathbf{A}\right)^{-1}\mathbf{A}^T\mathbf{b}. \quad (20)$$

Eventually, the perspective transformation matrix $\mathbf{H}$ can be recovered from $\widehat{\mathbf{H}}$.

Assuming that $P(x, y)$ is the key point in the first picture. $Q(u, v)$ is the corresponding matched key point in the second image. $\mathbf{H}_1$ is the perspective transformation matrix between the first and second image. Then we have

$$\lambda\left[x' \ \ y' \ \ 1\right]^T = \mathbf{H}_1\left[x \ \ y \ \ 1\right]^T, \quad (21)$$

where $(x', y')$ is the corresponding point after perspective projection transformation on key point $P$. Then the perspective projection error can be defined as follows:

$$\text{Err} = \sqrt{\left(x' - u\right)^2 + \left(y' - v\right)^2}. \quad (22)$$

*3.2.2. RANSAC Algorithm Progress.* Figure 11 shows the diagram flow of RANSAC algorithm. The main stages of RANSAC algorithm are described as follows:

*(1) Initialization.* The threshold of minimum perspective project error $T_{\text{ppe}}$ should be initialized at first. Besides, the maximum iterator time $M_{\text{iter}}$ should also be initialized here.

*(2) Perspective Transformation Solving.* The perspective transformation $\mathbf{H}$ can be solved by least squares method with four groups of key points, among which four key points are randomly selected from the dataset of source image key points. The other four key points are randomly selected from the dataset of target image key points.

*(3) Perspective Matrix Validation.* Once a new perspective projection error has been obtained in each iterator, a validation needs to be performed. The minimum of perspective projection error Err between the key points of the source image and the key points of the target image will be calculated. If Err is smaller than the smallest perspective projection error $E_{\min}$ which is stored before, the minimum perspective projection error $E_{\min}$ will be updated.

*(4) Loop Termination.* If $E_{\min} > T_{\text{ppe}}$ and iterator $> M_{\text{iter}}$, the loop will go back to step 2; otherwise, the loop will be terminated.

## 4. Time Consumption Comparison

In order to verify the advantage of the proposed CSA-SURF algorithm, the SURF and PCA-SURF algorithms are used for comparison. Feature extraction and matching progress are different for the above three algorithms, of which the time complexities are discussed in the following.

*4.1. Time Consumption for Feature Extraction.* It is well known that SIFT or SURF algorithms use the sliding window method to detect local extrema. PCA-SURF and CSA-SURF, having the same mechanism, are the improved algorithms based on SURF. However, there are some differences among them. In order to descript more clearly, the width and height of the image are defined as $I_w$ and $I_h$ and the step size is $\sigma$. The time for solving the eigenvalue and eigenvector is $T_{\text{eig}}$. The time for transferring the old feature space to the new feature space is $T_{\text{tran}}$. The time for computing the local

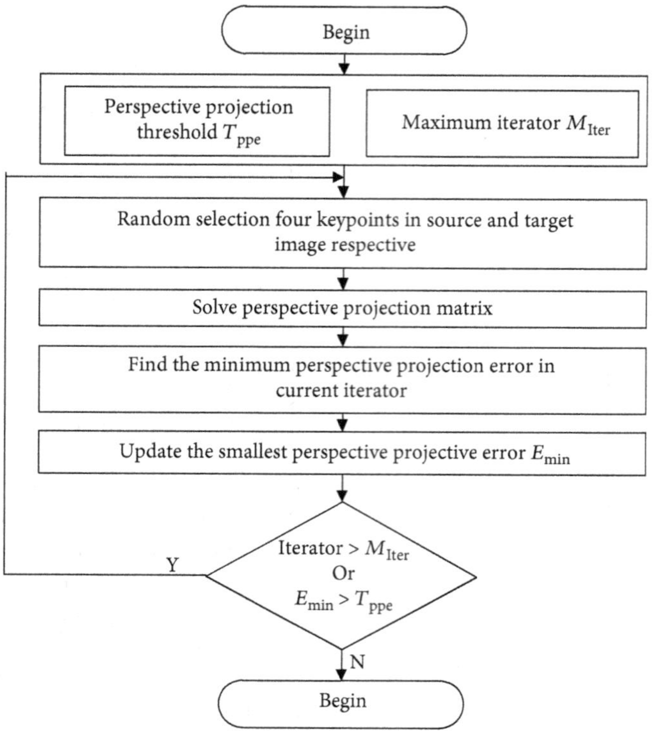


Figure 11: Diagram flow of RANSAC algorithm.

```
for i = 1 : σ : (I_w − 3)
  for j = 1 : σ : (I_h − 3)
      compute local feature for each pixel
  end
end
```

Pseudocode 1: Pseudocode for SURF feature extraction progress.

```
for i = 1 : σ : (I_w − 3)
  for j = 1 : σ : (I_h − 3)
      compute local feature for each pixel
  end
end
solve for eigenvalue and eigenvector
transfer feature space
```

Pseudocode 2: Pseudocode for PCA-SURF feature extraction progress.

feature is $T_{local}$. The extraction times of SURF, PCA-SURF, and CSA-SURF are $T^e_{surf}$, $T^e_{pca-surf}$, and $T^e_{csa-surf}$, respectively. The pseudocodes for the feature extraction progress of SURF, PCA-SURF, and CSA-SURF algorithms are presented as in Pseudocodes 1, 2, and 3.

The extraction time of SURF, PCA-SURF, and CSA-SURF can be calculated by

```
for i = 1 : n
  for j = 1 : m
      for k_1 = 1 : ⌊I_w/n − 3⌋
        for k_2 = ⌊I_h/m − 3⌋
            compute local feature for each pixel
        end
      end
  end
end
```

Pseudocode 3: Pseudocode for CSA-SURF feature extraction progress.

$$
\begin{aligned}
T^e_{surf} &= \left\lfloor \frac{(I_w - 3)}{\sigma} \right\rfloor \left\lfloor \frac{(I_h - 3)}{\sigma} \right\rfloor T_{local}, \\
T^e_{pca-surf} &= \left\lfloor \frac{(I_w - 3)}{\sigma} \right\rfloor \left\lfloor \frac{(I_h - 3)}{\sigma} \right\rfloor T_{local} + T_{eig} + T_{tran}, \\
T^e_{csa-surf} &= nm \left\lfloor \frac{I_w/n - 3}{\sigma} \right\rfloor \left\lfloor \frac{I_h/m - 3}{\sigma} \right\rfloor T_{local},
\end{aligned}
\tag{23}
$$

satisfying

$$
T^e_{csa-surf} < T^e_{surf} < T^e_{pca-surf}, \tag{24}
$$

in which the symbol $\lfloor \cdot \rfloor$ is a rounding down function.

The relationship of (24) can be proven as follows. Obviously, we have

$$nm\left\lfloor\frac{(I_w/n-3)}{\sigma}\right\rfloor\left\lfloor\frac{(I_h/m-3)}{\sigma}\right\rfloor < \frac{nm(I_w/n-3)(I_h/m-3)}{\sigma^2}, \tag{25}$$

$$nm\left\lfloor\frac{(I_w/n-3)(I_h/m-3)}{\sigma^2}\right\rfloor = \frac{(I_w-3n)(I_h-3m)}{\sigma^2}, \tag{26}$$

$$\frac{(I_w-3n)(I_h-3m)}{\sigma^2} < \left\lfloor\frac{I_w-3}{\sigma}\right\rfloor\left\lfloor\frac{I_h-3}{\sigma}\right\rfloor. \tag{27}$$

Combining (26) and (27), we obtain

$$nm\left\lfloor\frac{(I_w/n-3)}{\sigma}\right\rfloor\left\lfloor\frac{(I_h/m-3)}{\sigma}\right\rfloor < \left\lfloor\frac{(I_w-3)}{\sigma}\right\rfloor\left\lfloor\frac{(I_h-3)}{\sigma}\right\rfloor, \tag{28}$$

that is,

$$T^e_{\text{csa-surf}} < T^e_{\text{surf}}. \tag{29}$$

Obviously,

$$T^e_{\text{surf}} < T^e_{\text{pca-surf}}. \tag{30}$$

Therefore, (24) has been verified.

*4.2. Time Consumption of Matching.* In CSA-SURF, the image is divided into $n$ blocks in horizontal and $m$ blocks in vertical. The number of extracted features for SURF and PCA-SURF is $N$, while the corresponding number for CSA-SURF is $M$ and is much smaller than $N$. Besides, the dimension of PCA-SURF is $l$. The time consumption for two features matching is $T_{\text{match}}$. The matching time for SURF, PCA-SURF, and CSA-SURF are $T^m_{\text{surf}}$, $T^m_{\text{pca-surf}}$, and $T^m_{\text{csa-surf}}$ and can be calculated by

$$\begin{aligned} T^m_{\text{surf}} &= N^2 T_{\text{match}}, \\ T^m_{\text{pca-surf}} &= \frac{N^2 l}{64} T_{\text{match}}, \\ T^m_{\text{csa-surf}} &= M^2 T_{\text{match}}. \end{aligned} \tag{31}$$

As $M$ is much less than $N$, then we have

$$T^m_{\text{csa-surf}} < T^m_{\text{pca-surf}}, \quad \text{when } M \ll N, \tag{32}$$

which can be proven as follows.

As we can see,

$$\begin{aligned} M^2 &< N^2, \\ \frac{N^2 l}{64} &< N^2. \end{aligned} \tag{33}$$

Therefore, we have

$$\begin{aligned} T^m_{\text{pca-surf}} &< T^m_{\text{surf}}, \\ T^m_{\text{csa-surf}} &< T^m_{\text{surf}}. \end{aligned} \tag{34}$$

If $M \ll N$, then

$$\left(\frac{M}{N}\right)^2 \longrightarrow \text{small number}. \tag{35}$$

For the spacecraft autonomous navigation problem using natural landmarks, only 5 to 10 features satisfying certain relative distance constraints are required; therefore, it is easy to guarantee that $(M/N)^2 < N^2/l$ and then we have $T^m_{\text{csa-surf}} < T^m_{\text{pca-surf}}$.

From the above comparison, the proposed CSA-SURF method consumes the least time.

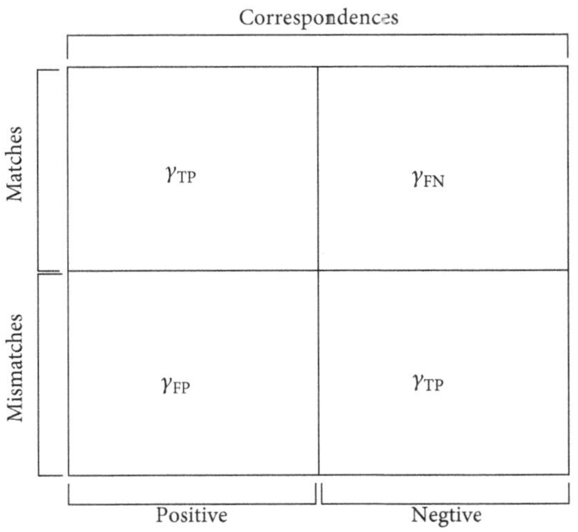


Figure 12: Relationship of different symbols.

## 5. Evaluation

*5.1. Evaluation Metrics.* In order to descript the repeatability performance, the following evaluation metrics are defined in this section. Positive in this paper means the key points that can be matched. Negative means the key points that cannot be matched.

Then we suppose that $\gamma_{TP}$ is the number of positive key points that are correct matches. $\gamma_{FP}$ is the number of negative key points that are false matches. $\gamma_{TN}$ is the number of negative key points that correct mismatches. And $\gamma_{FN}$ is the number of negative key points that are false mismatches. To explain it more clearly, Figure 12 shows the relationship of different symbols.

Then we have the following definitions:

$$\text{Precision} = \frac{\gamma_{TP}}{\gamma_{TP}+\gamma_{FP}} = \frac{\text{correct matches}}{\text{positive}}, \tag{36}$$

$$\text{Recall} = \frac{\gamma_{TP}}{\gamma_{TP}+\gamma_{FN}} = \frac{\text{correct matches}}{\text{positive}}, \tag{37}$$

Precision means the proportion of correct matches to positive. Recall means the proportion of correct matches to matches. Recall and precision are conflicting, so a recal

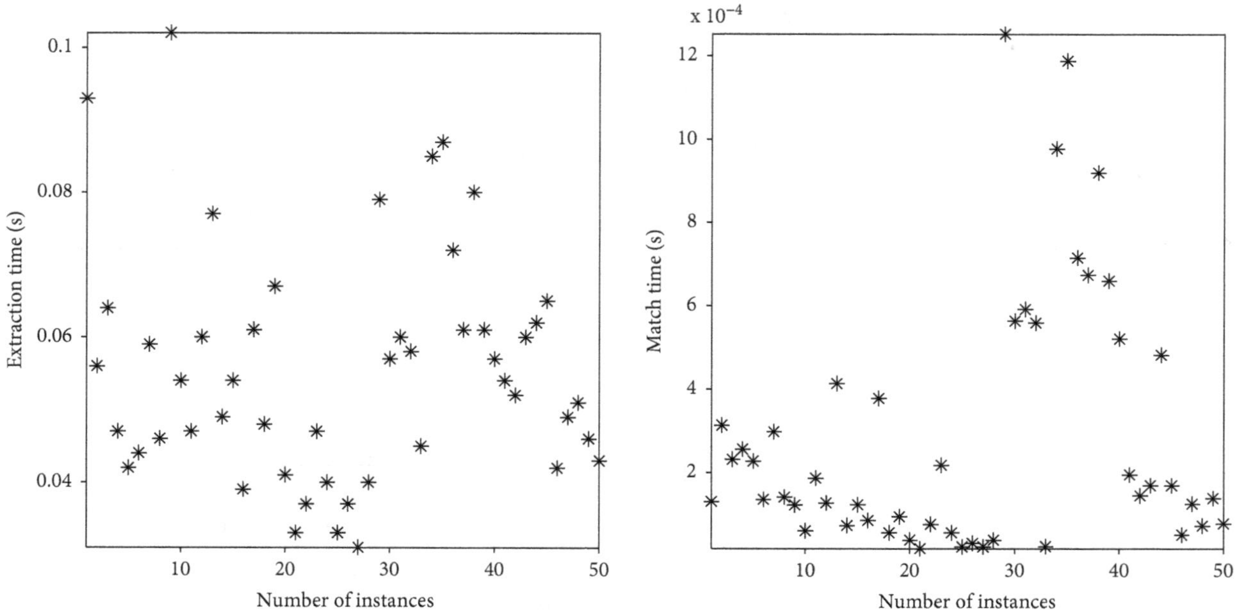


FIGURE 13: Time consumption for SURF.

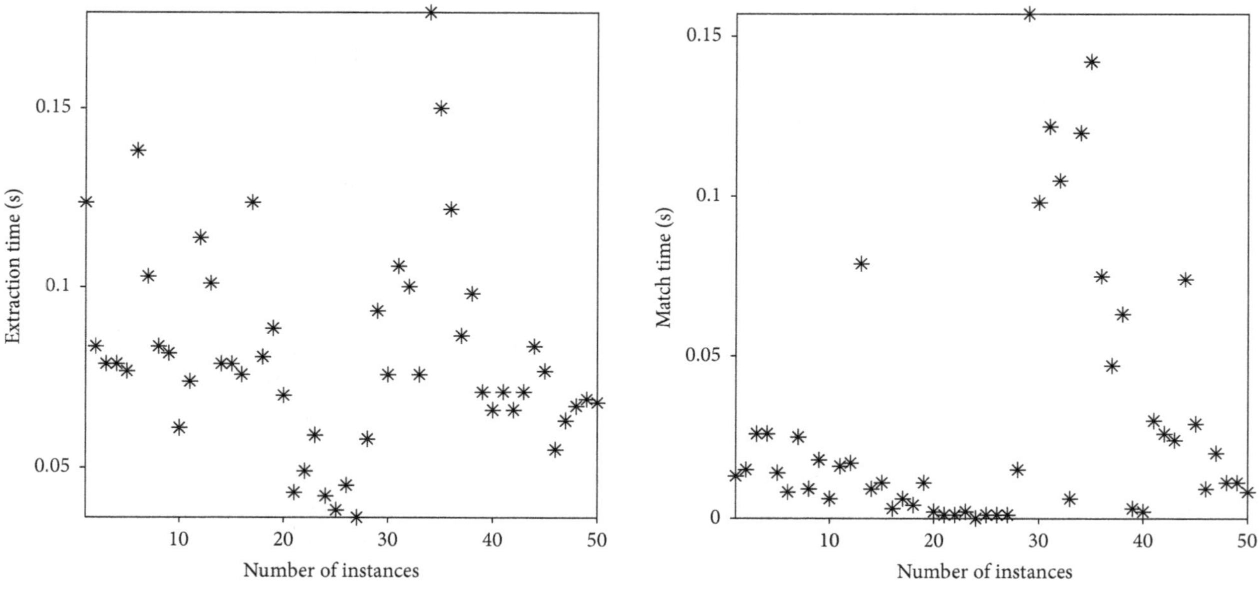


FIGURE 14: Time consumption for PCA-SURF.

l versus precision graph will help in analyzing the repeatability performance.

*5.2. Speed Test.* The following experiments are tested on Core I5 2.3 GHz CPU, 6GB RAM laptop with Windows 7 operation system. VS2013 and OpenCV2.4 are used to carry out all the experiments. And all datasets are from an open source database called UC Merced Land Use Dataset. And the resolution is $256 \times 256$.

Figures 13–15 are the matching comparison results of fifty images about SURF, PCA-SURF, and CSA-SURF. It can be seen that PCA-SURF consumes much more time than SURF and CSA-SURF during feature extraction process. The CSA-SURF algorithm works faster than SURF and PCA-SURF algorithms for both feature extraction and matching processes.

*5.3. Repeatability Test.* Figure 16 shows the repeatability of different methods for different conditions. But recall and precision may be conflict in some condition, so a recall versus precision graph has been established to help us analyze the repeatability performance. The red line is the test under

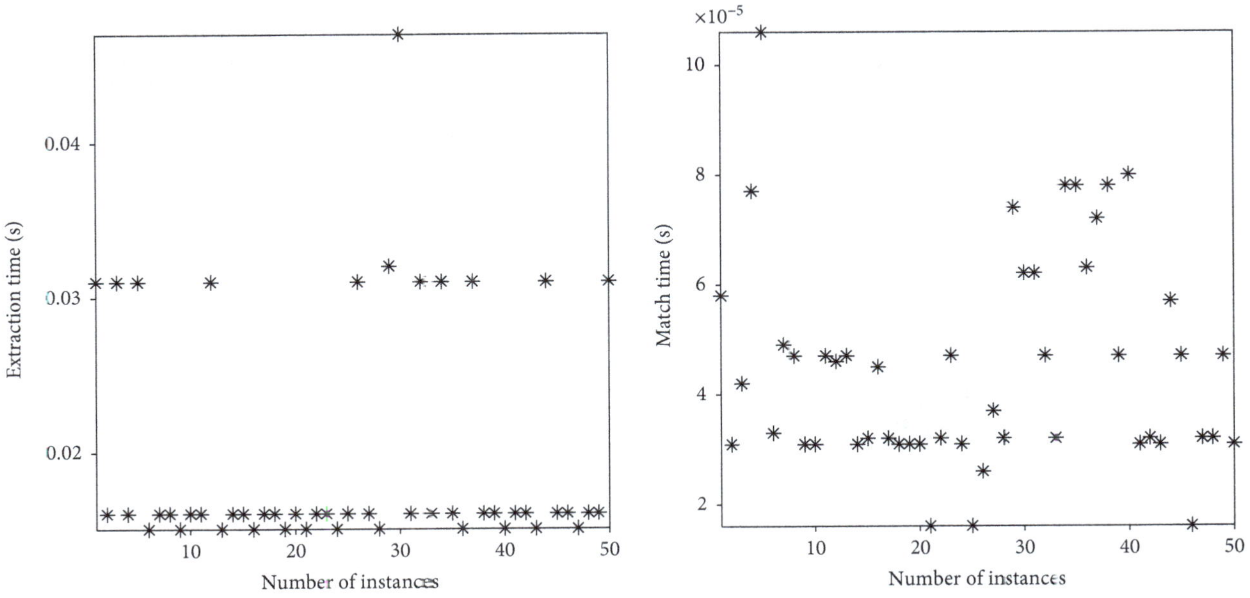


FIGURE 15: Time consumption for CSA-SURF.

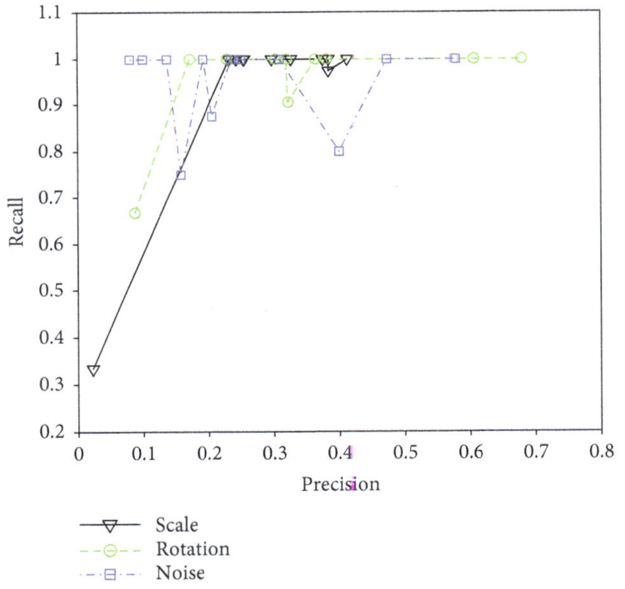


FIGURE 16: Recall versus precision graph of three methods under different conditions.

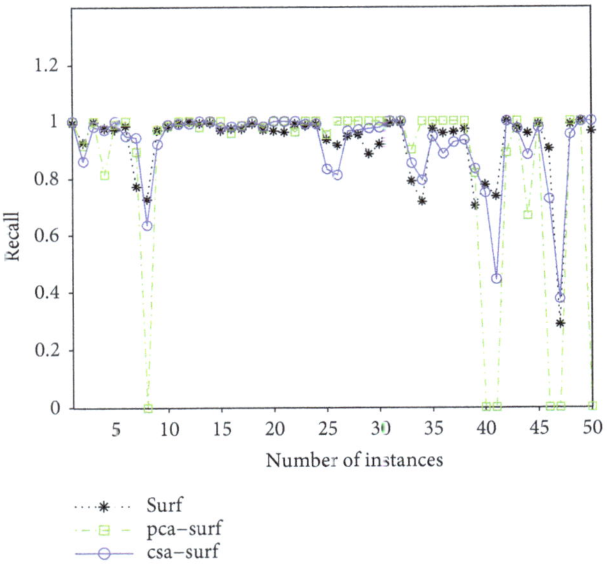


FIGURE 17: Recall of different methods.

scaling; it can be seen that CSA-SURF performs worse when precision $< 0.2$ and its performance is not stable when $0.2 <$ precision $< 0.45$. For precision $< 0.45$, it works well with recall $= 1$. The green line represents the test of CSA-SURF algorithm under rotation. Its performance is much the same as the solid line's case, while its unstable region is $0.2 <$ precision $< 0.4$. The blue line represents the test of CSA-SURF algorithm under salt and pepper noise [16] with a noise density equal to 0.35. It can be seen that CSA-SURF algorithm is not stable under salt and pepper noise, but recall still maintains its value above 0.7.

Furthermore, the above dataset with fifty images is also used to test the repeatability performance of the three methods. Figure 17 is the recall of different methods for these images. The results show that the repeatability of SURF and CSA-SURF is very close. Therefore, CSA-SURF can speed up the registration progress without losing its repeatability performance. However, PCA-SURF is unstable as it can get the correct matches in some conditions.

*5.4. Application on Natural Landmark Registration.* The major purpose of this paper is to apply the proposed method into natural landmark registration task. Five different kinds of natural landmarks, including island, airport,

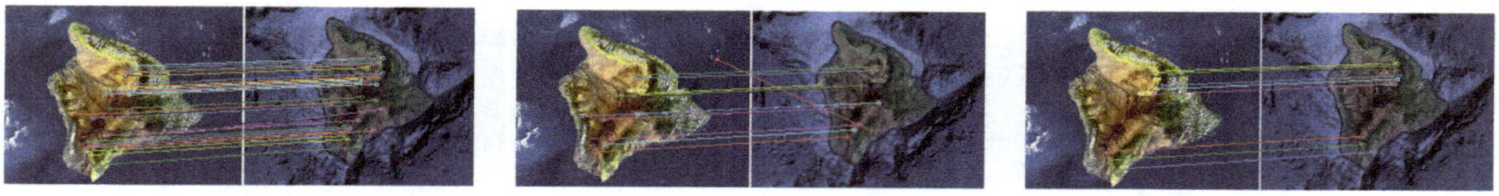

Figure 18: Hawaiian island registration result with SURF, PCA-SURF, and CSA-SURF.

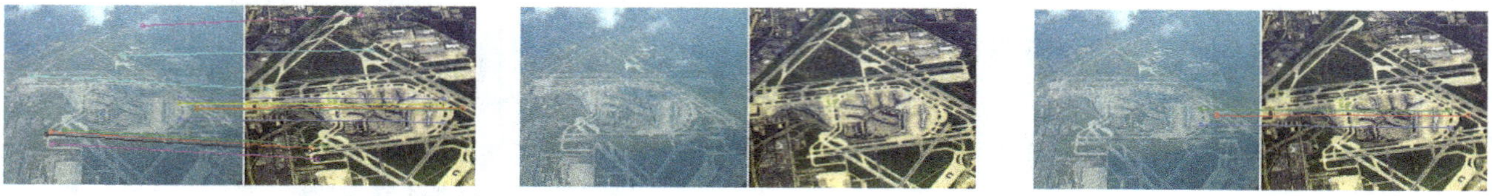

Figure 19: O'Hare International Airport registration result with SURF, PCA-SURF, and CSA-SURF.

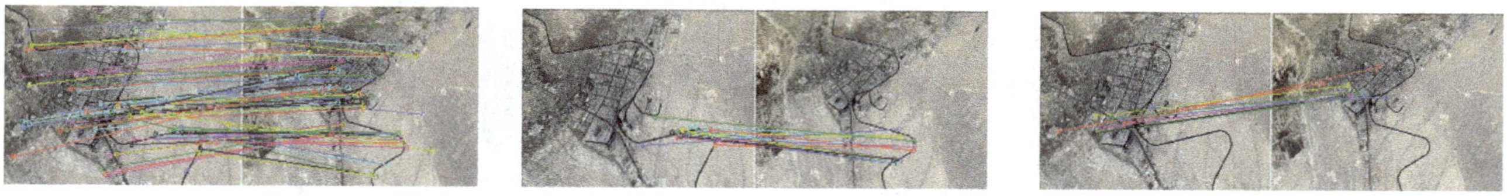

Figure 20: Qingzang railway registration result with SURF, PCA-SURF, and CSA-SURF.

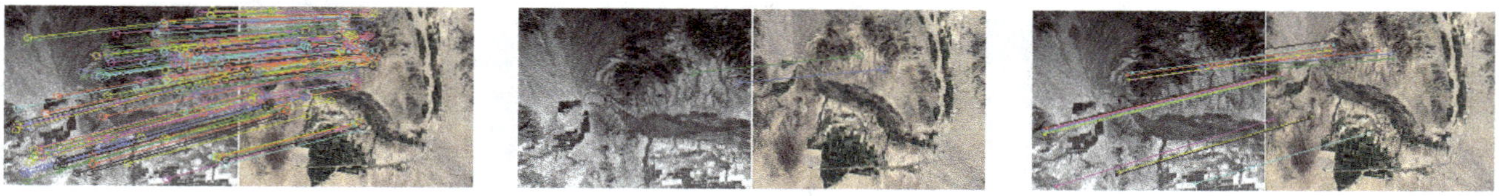

Figure 21: River landmark registration result with SURF, PCA-SURF, and CSA-SURF.

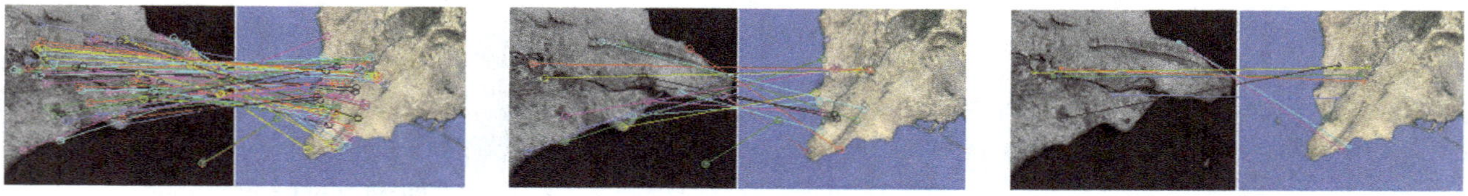

Figure 22: Coastline landmark registration result with SURF, PCA-SURF, and CSA-SURF.

railway, river, and coastline, are collected for the demonstration testing.

Figure 18 presents the registration results for Hawaiian islands with different algorithms, in which SURF and CSA-SURF work fine but PCA-SURF has a wrong match. Figure 19 shows the matching results of two O'Hare International Airport images with different algorithms, indicating that the PCA-SURF algorithm is unstable as it cannot find any matched key points in this example. Figures 20–22 are the matching results of Qingzang railway, river, and coastline images with different algorithms.

It can be seen from the above testing results that CSA-SURF algorithm works well for different images with significant contour features. The proposed CSA-SURF algorithm, which is improved from SURF, keeps the repeatability of SURF unchanged and improves the image matching speed. Many other pictures have also been tested; however, the results are not presented for the page limitation of the paper.

## 6. Conclusion

Because of the restricted computational resource and horrible environment for an autonomous navigation system, traditional image registration methods cannot satisfy practical requirements in speed. A novel algorithm called chessboard segmentation algorithm is proposed to solve the above problem. Because the new method is based on SURF features, it inherits lots of their advantages, such as scale and rotation invariant properties. To verify the improvement of the proposed algorithm, the PCA-SURF algorithm which was proposed in recent years is also presented in this paper for comparison. Besides, RANSAC algorithm is applied to remove the false negative key points to further improve the accuracy of the proposed algorithm. Thorough experiments have been carried out to demonstrate the performance of the proposed method; the corresponding simulation results show great improvement in image registration speed without losing repeatability. Finally, the CSA-SURF algorithm is applied to natural landmark registration task, showing that it works fine. The proposed method is a good candidate for image registration task for spacecraft autonomous navigation based on natural landmarks.

## Conflicts of Interest

The authors declare that they have no conflicts of interest.

## Acknowledgments

The authors would like to express their acknowledgment for the support provided by the National Natural Science Foundation of China (nos. 61403197 and 61673212), the Natural Science Foundation of Jiangsu Province (no. BK20140830), the National Key Research and Development Plan (no. 2016YFB0500901), and the Open Fund of National Defense Key Discipline Laboratory of Micro-Spacecraft Technology (no. HIT.KLOF.MST.201705).

## References

[1] A. C. Vigneron, A. H. J. de Ruiter, B. V. Burlton, and W. K. H. Soh, "Nonlinear filtering for autonomous navigation of spacecraft in highly elliptical orbit," *Acta Astronautica*, vol. 126, pp. 138–149, 2016.

[2] S. Sha, C. Jianer, and L. Sanding, "A fast matching algorithm based on K-degree template," in *2009 4th International Conference on Computer Science & Education*, pp. 1967–1971, Nanning, China, 2009.

[3] Z. Xu, Y. Liu, S. Du, P. Wu, and J. Li, "DFOB: detecting and describing features by octagon filter bank for fast image matching," *Signal Processing: Image Communication*, vol. 41, pp. 61–71, 2016.

[4] P. Zhao, Z. Bai, and W. Fan, "Research of fast image matching based on PCA," *Computer Technology and Its Applications*, vol. 4, pp. 132–134, 2010.

[5] D. He and P. Jiang, "Fast image matching algorithm based on discrete Hartley transform," *Modern Defense Technology*, vol. 44, no. 5, pp. 61–65, 2016.

[6] Y. Ke and R. Sukthankar, "PCA-SIFT: a more distinctive representation for local image descriptors," in *Proceedings of the 2004 IEEE Computer Society Conference on Computer Vision and Pattern Recognition, 2004. CVPR 2004*, pp. II-506–II-513, Washington, DC, USA, 2004.

[7] J. Yang, D. Zhang, A. F. Frangi, and J.-y. Yang, "Two-dimensional PCA: a new approach to appearance-based face representation and recognition," *IEEE Transactions on Pattern Analysis and Machine Intelligence*, vol. 26, no. 1, pp. 131–137, 2004.

[8] H. Bay, A. Ess, T. Tuytelaars, and L. Van Gool, "Speeded-up robust features (SURF)," *Computer Vision and Image Understanding*, vol. 110, no. 3, pp. 346–359, 2008.

[9] M. Bleyer and M. Gelautz, "Graph-cut-based stereo matching using image segmentation with symmetrical treatment of occlusions," *Signal Processing: Image Communication*, vol. 22, no. 2, pp. 127–143, 2007.

[10] A. Pancham, D. Withey, and G. Bright, "Tracking image features with PCA-SURF descriptors," in *14th IAPR International Conference on Machine Vision Applications (MVA)*, pp. 365–368, Tokyo, Japan, 2015.

[11] E. E. Maraş, M. Caniberk, and H. H. Maraş, "Automatic coastline detection using image enhancement and segmentation algorithms," *Polish Journal of Environmental Studies*, vol. 25, no. 6, pp. 2519–2525, 2016.

[12] Y. Wang, J. Zheng, Q. Z. Xu, B. Li, and H. M. Hu, "An improved RANSAC based on the scale variation homogeneity," *Journal of Visual Communication and Image Representation*, vol. 40, pp. 751–764, 2016.

[13] Y. Chen, Q. Sun, H. Xu, and L. Geng, "Matching method of remote sensing images based on SURF algorithm and RANSAC algorithm," *Jisuanji Kexue yu Tansuo*, vol. 6, no. 9, pp. 822–828, 2012.

[14] Y. Zhao, R. Hong, and J. Jiang, "Visual summarization of image collections by fast RANSAC," *Neurocomputing*, vol. 172, pp. 48–52, 2016.

[15] F. Yang, J. Guo, and J. Wang, "Image mismatching eliminating algorithm using structural similarity and geometric constraint," *Journal of Signal Processing*, vol 32, no. 1, pp. 83–90, 2016.

[16] R. H. Chan, Chung-Wa, and M. Nikolova, "Salt-and-pepper noise removal by median-type noise detectors and detail-preserving regularization," *IEEE Transactions on Image Processing*, vol. 14, no. 10, pp. 1479–1485, 2005.

# 3

# Design of Modular Power Management and Attitude Control Subsystems for a Microsatellite

**Anwar Ali**,[1] **Shoaib Ahmed Khan**,[2] **M. Usman Khan**,[2] **Haider Ali**,[1] **M. Rizwan Mughal**,[3,4] **and Jaan Praks**[4]

[1]*Electrical Technology Department, University of Technology (UoT), Nowshera, Pakistan*
[2]*National University of Computer & Emerging Sciences Peshawar Campus, Pakistan*
[3]*Department of Electrical Engineering, Institute of Space Technology, Islamabad, Pakistan*
[4]*Department of Electronics and Nanoengineering, School of Electrical Engineering, Aalto University, FI-00076 AALTO, 02150 Espoo, Finland*

Correspondence should be addressed to Anwar Ali; safi2000pk@gmail.com

Academic Editor: Jeremy Straub

The Electric Power System (EPS) and attitude control system (ACS) are the essential components of any satellite. EPS and ACS efficiency and compactness are substantial for the proper operation and performance of the satellite's entire mission life. So, realizing the significance of EPS and ACS subsystems for any satellite, they have been assimilated and developed in modular forms focusing on efficiency and compactness. The EPS is comprised of three modules called the solar panel module (SPM), power conditioning module (PCM), and power distribution module (PDM) while the ACS has an embedded magnetorquer coil. For compactness and miniaturization purposes, the magnetorquer coil is embedded inside the SPM. The components used are commercial off-the-shelf (COTS) components emphasizing on their power efficiency, small dimensions, and weight. Latch-up protection systems have been designed and analyzed for CMOS-based COTS components, in order to make them suitable for space radioactive environment. The main design features are modularity, redundancy, power efficiency, and to avoid single component failure. The modular development of the EPS and ACS helps to reuse them for future missions, and as a result, the overall budget, development, and testing time and cost are reduced. A specific satellite mission can be achieved by reassembling the required subsystems.

## 1. Introduction

Many universities around the world are working on various projects of small satellites [1–6]. These satellites are classified into various categories on the basis of their mass and dimensions, i.e., pico, nano, and micro. The main driving force for the development of small satellites is their low budget requirement, short development time, low mass and size, new technology, and because they provide more cost-effective and reliable access to space [7]. Small satellites are designed for low earth orbit (LEO); therefore, launching cost is also comparatively low. Small satellites can provide an ideal testbed to innovative technologies (e.g., hardware and software) which can be further used in large and expensive space missions. University satellite programs teach students the whole lifecycle of a satellite on a real situation and develop their capabilities in the space sector. Due to mentioned facts, small satellites have been considered a vital technology by many space mission experts.

The number of student-built small satellites is rapidly growing, specifically in Europe and in North America. The list of the number of universities working on satellite projects is too long, and it is impossible to mention all of them here. But some of the universities that are particularly moving the small satellite industry forward and working on very nice projects are mentioned here. From the United States of America, the significant contribution is from Cal Poly and Stanford University, University of Michigan, University of Texas, Air Force Academy, Massachusetts Institute of Technology, Colorado University Boulder, Santa Clara University,

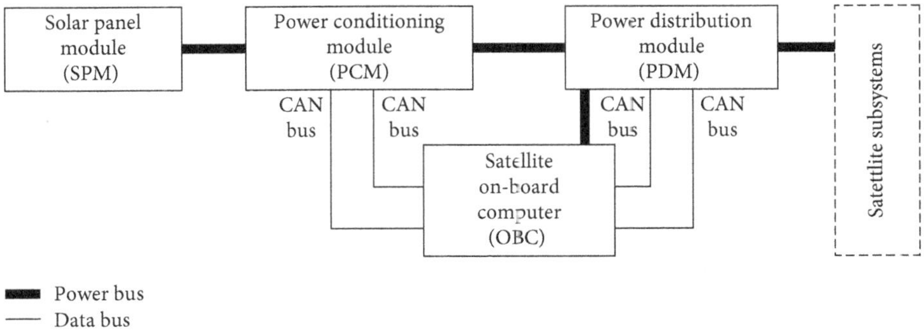


FIGURE 1: Block diagram of EPS subsystems.

Arizona State University, and many more [8, 9]. From Europe, the contribution worth mentioning is from the Swedish Institute of Space Physics, Aalborg University, Technical University of Denmark, Norwegian University of Technology and Science (NTNU), University of Wurzburg, Aachen University of Applied Sciences, Delft University of Technology, Swiss Federal Institute of Technology (EPFL), Budapest University of Technology and Economics, University of Montpellier II, Warsaw University of Technology, Sapienza University of Technology, University of Vigo, Polytechnic University of Turin, Berlin Technical University, the Space Research Centre of Polish Academy of Sciences, and a lot of other institutions. In Asia, the main contributing universities are from Japan, China, South Korea, Singapore, and India [9].

Keeping in view the significance of small satellites, the Pakistan National Space and Upper Atmospheric Research Corporation (SUPARCO) has framed the Pakistan National Student Satellite (PNSS) program [10]. This program will involve the local universities and industry in collaboration with SUPARCO for the development and launching of small satellites into space. In this program, PNSS-1 is the first small satellite that will be deployed into space. Universities are part of the designing process of the satellite. The main aim of this work is to design and analyze modular EPS and ACS for the PNSS-1 satellite which has a size of 50 $cm^2$ with 50 kg mass and one-year service life [11].

The EPS is responsible for fulfilling the power needs of the entire satellite. The paper proposes a novel approach to design and elegantly interconnect the EPS subsystems of a modular small satellite [12]. In the proposed modular design technique, the EPS is divided into small modules that are designed and developed independently and are attached together to attain the intended results. The major subsystems of the EPS are the solar panel subsystem, conversion of the solar panel power to the power distribution bus (PDB) level subsystem, and conversion of the unregulated PDB voltage level into various low voltage levels required for operation of the satellite subsystem components. The EPS also has a subsystem to store the excess generated power into onboard batteries and extract power from them during satellite night and solar eclipses. Different housekeeping sensors are embedded in the EPS subsystems to monitor and ensure the proper operation of the complete satellite. To perform all the abovementioned tasks, the whole EPS is subdivided into three modules called the solar panel module (SPM), power conditioning module (PCM), and power distribution module (PDM). This modular design approach is not only cost effective because the design, development, and testing costs are shared among these modules but also time saving because of the parallel development approach rather than the classical serial one [13–16]. In addition, the modular design extensively results in overall cost reduction because the cost is shared among many modules that can be used several times in other missions.

Each module offers standardized power and data interfaces. A block diagram of the EPS subsystems is depicted in Figure 1.

The SPM is responsible for the power generation of PNSS-1. Four solar panels are attached on the outside periphery of the satellite accountable for power generation. One solar panel generates sufficient power to meet the power needs of the complete satellite. The SPM output power is not constant and has variable voltage and current levels depending on solar power density and environment factors. The power conditioning module (PCM) converts this unstable SPM output voltage to a stable PDB voltage level. The PCM operates the solar panels at its maximum power point (MPP) to harvest maximum power. For this purpose, the PCM has four Maximum Power Point Tracking- (MPPT-) based converters. Each converter is connected to four solar panels on the input side and regulates the battery charge and discharge on the MPPT scheme on the output side. The functional requirements of the PCM are to generate the PDB for the spacecraft at $28\,V \pm 6\,V$ from raw power available from four body-mounted solar panels. The input voltage of the PCM may vary from 40 V to 80 V. The PCM is connected with the satellite onboard computer (OBC) through redundant CAN buses. It has an interface with all of the power sources (i.e., solar panels and battery) and outputs the PDB to the power distribution module (PDM). The PDM converts the PDB voltage level into lower specific voltage levels and distributes them to loads as per their requirements [17]. The PDM also has the EMI filtration as well as different protections on each distribution line in order to limit the electromagnetic noise on the main bus and prevent fault propagation. The ON/OFF switching of all switchable loads is also implemented inside the PDM. These switches are controllable through telecommands. The complete design of the specific module is discussed in the respective section of this paper, in detail.

The ACS of the PNSS-1 is responsible for attitude control as is obvious from its name. The ACS is responsible for orienting the satellite solar panels for harvesting maximum solar power and the satellite antenna toward ground station antennas. For the attitude control of small satellites, three types of systems are normally used, i.e., permanent magnet, reaction wheel, and magnetic rods. All the three systems have their pros and cons. For example, the permanent magnet is inexpensive, but its weight and pointing accuracy makes it incompatible with small satellites. Reaction wheels and magnetic rods are more precise, but their high price, greater weight, and large size are unacceptable for small satellites. In this paper, an innovative concept of an embedded magnetorquer coil is presented for the attitude control of small satellites. The magnetorquer coil comprises four subcoils embedded inside the four internal layers of an eight-layer PCB of the SPM module. This innovative design not only reduces cost and size of the magnetorquer unit but also results in an almost weightless system.

The paper is organized according to the following sequence. Section 1 provides the EPS introduction and Section 2 describes the design and analysis of the SPM subsystem. Section 3 discusses the design description of PCM subsystems according to general constraints and explains in detail the design and simulations of a wide range input MPPT-based buck converter. Section 4 describes the design details of the PDM, while the last, Section 5, discusses the results and conclusions.

## 2. Solar Panel Module

Selecting the correct power ratings for the SPM is the most important aspect of the photovoltaic (PV) system design for a microsatellite. The inappropriate string size will result in many problems. Choosing a small string size will result in small efficiency. Selecting a large string size will damage the converter and the associated equipment [18]. Solar cells can be connected in any fashion, i.e., series or parallel. Series connections increase the voltage with the current kept constant while parallel connections increase the current of the string with the voltage remaining constant. Calculating the string size, the relationship between ambient temperature and string voltage must be taken into consideration. Ambient temperature has adverse effect on PV array voltage output. Low temperatures result in increases to output voltage while high temperatures decrease string voltage. Proper and accurate power analysis is essential to compute the power ratings for efficient SPM design [19].

Triple-junction GaAs solar cells are selected for the solar panel design of PNSS-1. The single cell has a maximum power point voltage ($V_{mp}$) of 2396 mV and short circuit current density ($J_{SC}$) of 17.67 mA/cm$^2$ which gives 0.42 A current. According to the PNSS-1 solar panel design requirement, a single solar panel should generate at least 50 W with maximum output voltage of 80 V and minimum voltage should not be less than 40 V under normal sunlight conditions at sun incidence angle of 5°. It means that multiple

Table 1: GaAs triple-junction solar cell characteristics.

| Type | $GalnP_2/GaAs/Ge$ |
|---|---|
| Cell dimension | 39.8 mm × 60.4 mm |
| Short circuit current density ($J_{sc,mA/cm^2}$) | 17.67 |
| Open circuit voltage ($V_{oc,mV}$) | 2700 |
| Maximum power point current density ($J_{mp,mA/cm^2}$) | 17.13 |
| Maximum power point voltage ($v_{mp,mV}$) | 2396 |
| Fill factor (FF) | 0.86 |
| Efficiency ($\eta$) | 30% |
| Change in open circuit voltage per degree change in temperature ($\Delta V_{oc}/\Delta T$) | −5.589 mV/°C |
| Change in maximum power point voltage per degree change in temperature ($\Delta V_{mp}/\Delta T$) | −5.964 mV/°C |

string arrays will be used for the solar panel design. The efficiency of the solar cell array is always less than the sum of individual cells because of the manufacturing issues and environmental factors, collectively called as the degradation factor [20]. For the solar panel design calculations, it is essential to take into account the degradation factor. The data of the required triple-junction GaAs solar cell for this design is given in Table 1.

For calculation of the minimum voltage limit of a single solar cell, the maximum temperature limit is applied, because as the temperature increases above the standard test conditions (STC), i.e., 25°C, the voltage reduces per degree rise in temperature. Let us assume that the maximum temperature is 85°C. Difference between maximum temperature and STC is 60°C which will reduce the single cell minimum voltage to $V_{min} = 2.396 + 60 \times (-5.94 \times 10^{-3}) = 2.038\ \text{V}$ where $-5.94 \times 10^{-3}$ is the change in open circuit voltage per degree change in temperature ($\Delta V_{oc}/\Delta T$) as given in Table 1. After applying the degradation factor of 0.85, the resultant minimum voltage for the single cell becomes 1.732 V.

For upper voltage limit calculation, minimum operating temperature is considered. Let us assume that the minimum temperature is −45°C. Difference between minimum temperature and STC is −70°C which will increase the single cell maximum voltage to $V_{max} = 2.396 + (-70 \times (-5.589 \times 10^{-3})) = 3.09\ \text{V}$ where $-5.589 \times 10^{-3}$ is the change in MPP voltage per degree change in temperature ($\Delta V_{mp}/\Delta T$) as given in Table 1. After applying the degradation factor of 0.85, the resultant maximum voltage becomes 2.63 V.

From the degradation analysis given above, we can find the minimum and maximum number of cells required to achieve the upper and lower voltage limits. The maximum number of the solar cell limit is found by dividing the upper voltage limit with the single cell maximum voltage, i.e., $(80\ \text{V}/2.63\ \text{V}) = 31$, while the minimum number of the required cell limit is found by dividing the lower voltage limit with the single cell minimum voltage, i.e., $(40\ \text{V}/1.73\ \text{V}) = 23$.

After applying the degradation factor, the single cell output current is 0.36 A.

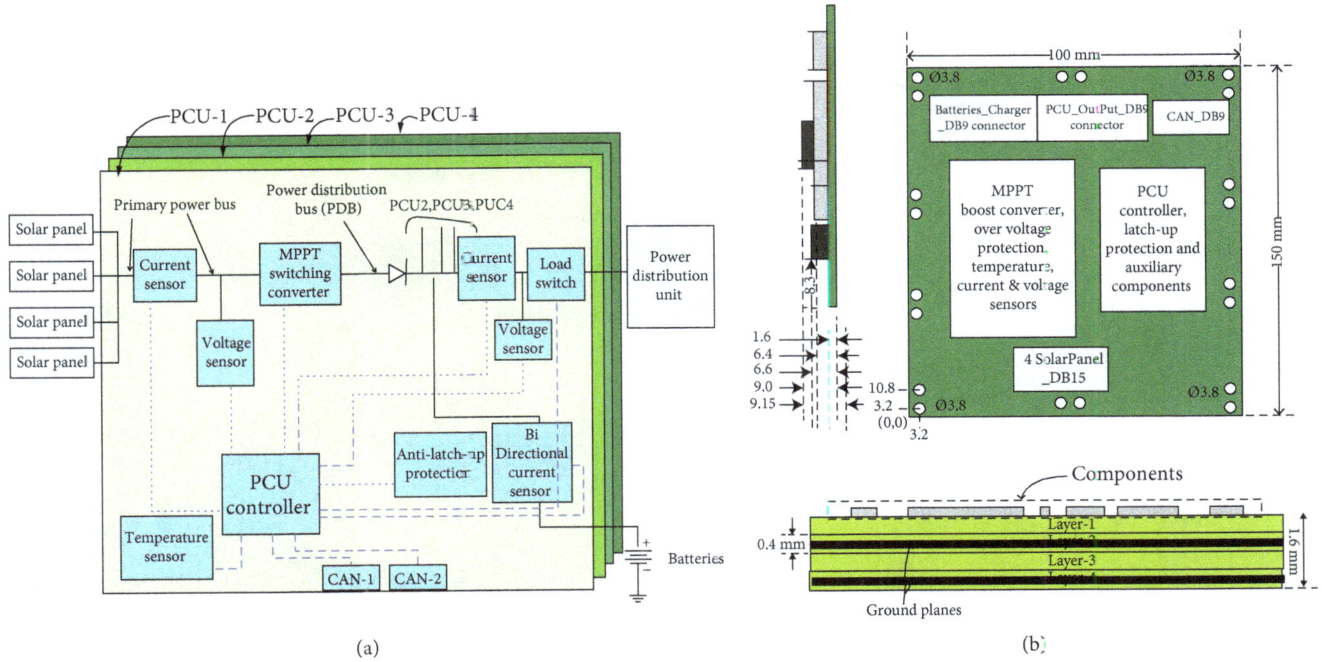


Figure 2: (a) PCM block diagram and (b) PCB dimensions.

According to the design requirements, the average output power from the solar panel should not be less than 50 W. The power analysis shows that minimum required power is 40 W and maximum required power is 60 W. Single cell average output power is 0.78 W. The solar panel has dimensions of 470 mm × 450 mm, and single cell dimensions are 39.8 mm × 60.4 mm. It means that a maximum of 87 solar cells can be accommodated on the solar panel. Excluding the area required for the sun sensor, temperature sensor, and bypass diodes, the total number of cells that can be accommodated on the solar panel is 64. These cells are divided into two strings; each has 32 solar cells connected in series. The average voltage of the single cell is 2.18 V and maximum numbers of solar cells are 32 which result in average output voltage and current of 69 V and 0.72A, respectively. The resultant output power after applying all the degradation factors is 50 W. A bypass diode is connected in parallel to each solar cell which ensures proper operation of the single cell if one of them is damaged. A protection diode is added at the end of each string of 32 solar cells to protect it from reverse current flow.

For the SPM to operate within its temperature limits, a glass-protected NTC (negative temperature coefficient) thermistor is mounted onto the solar panel for protection against transient voltage spikes. Output voltage of the sensor is linearly proportional to temperature. Its resistance range varies between 2.2 k and 100 k, which is highly accurate with 1% of tolerance.

## 3. Power Conditioning Module (PCM)

The PCM is responsible for converting solar panels' unstable voltage (40 V~80 V) into a stable voltage level delivered to the PDB (28 ± 6 V). It is also responsible for voltage regulation, battery charging, and health monitoring of different PCM subsystems. A MPPT-based buck converter is designed and simulated to operate the solar panel at its MPP. To achieve maximum power point, the MPPT buck converter can be designed using different MPPT algorithms. Most common algorithms for recognition of MPP are the constant voltage method [21], short-current pulse method [22], open voltage method [23], perturb and observe method [24], and incremental conductance method [25]. These seven MPPT algorithms are examined and evaluated on the basis of efficiency [26]. The perturb and observe method shows the best result and better efficiency as compared to others [27, 28]; therefore, this algorithm is chosen for the PCM MPPT design. Batteries are charged and discharged from the PDB through a battery charging system. The PCM controller is used to perform the power management and control operations of the entire unit. It communicates with the OBC of the satellite using redundant CAN buses. PCM casing dimensions and input/output power/signal have standardized connectors which result in a fully modular system that can be attached and detached as a separate unit to the rest of the satellite subsystems. Figure 2(a) shows the conceptual-level block diagram of the PCM subsystem while Figure 2(b) shows the PCB dimensions of the PCM. Casing dimensions of the PCM are shown in Figure 3.

*3.1. PCM Design Description.* The PCM consists of four modular and redundant MPPT units and CAN controller. Each MPPT unit has a local microcontroller (MSP430) to which different subsystems like MPPT switching converter, current, voltage, and temperature sensors are connected. These controllers are CMOS-based COTS components which are prone to radiation. In order to protect them from radiation damage, a latch-up protection system is designed for the

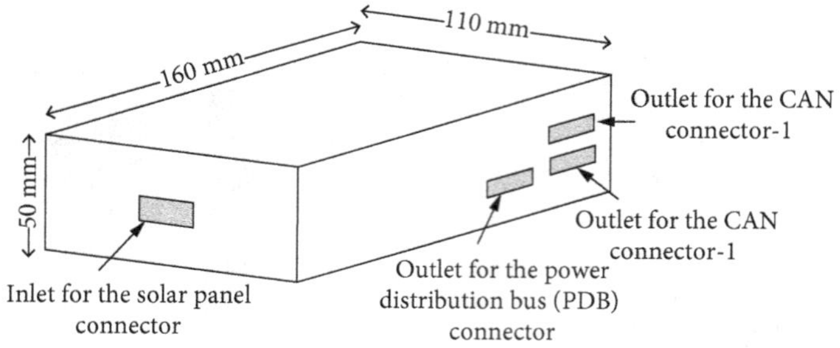


FIGURE 3: Casing dimensions of the PCM.

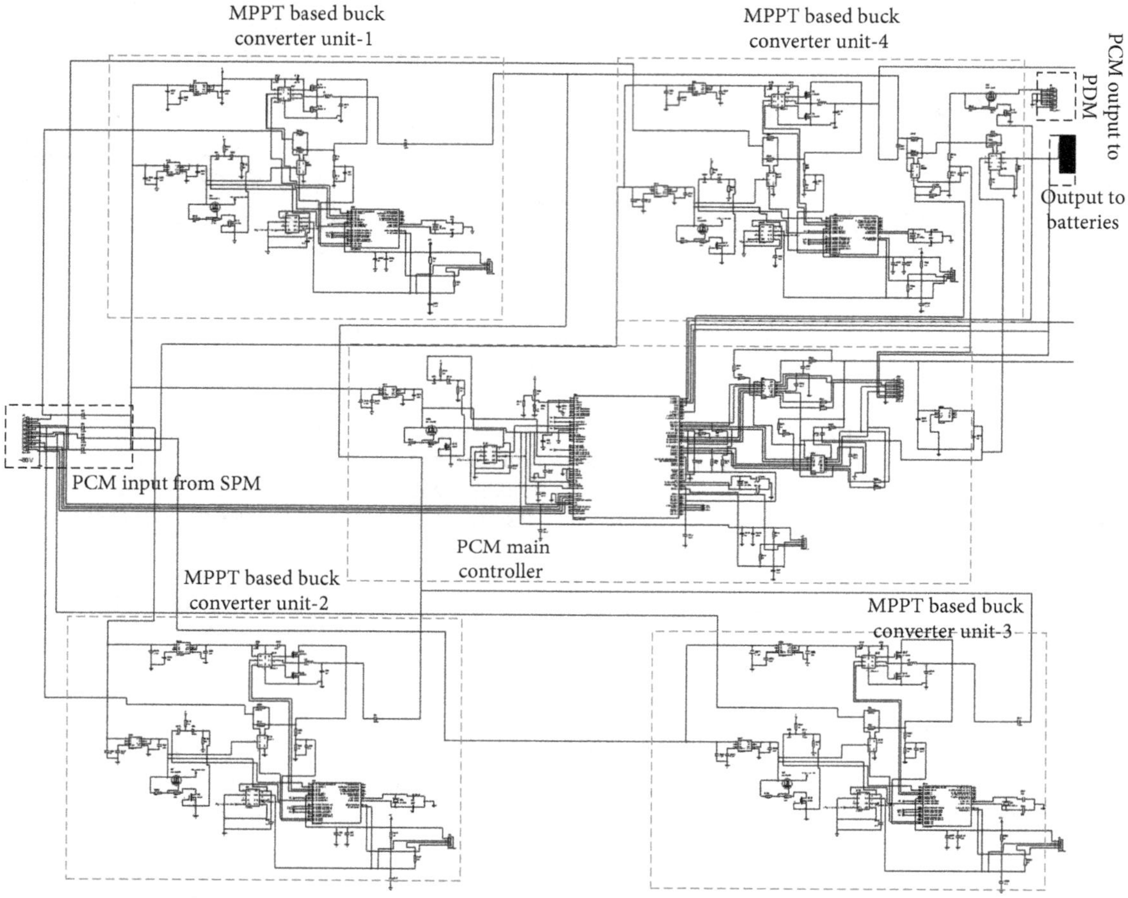


FIGURE 4: PCM complete schematic.

PCM unit. The four MPPT controllers (MSP430) and CAN controller (PIC24HJ256GP10) are communicating with each other through SPI protocol while the CAN microcontroller is communicating with the OBC of the satellite through an embedded CAN transceiver system having redundant CAN buses. A schematic of the PCM is shown in Figure 4.

In the PCM design, the perturb and observe algorithm is opted to continuously locate the maximum power point and extract maximum power from the solar panels connected with it. The P&O algorithm works by continuously calculating the power (current and voltage) from PV arrays and comparing it with preceding calculated power values. If there is a difference and the power calculated from the current cycle is greater than the previous cycle, the controller moves the MPP in that direction by increasing the duty cycle; otherwise, if the power levels are same, it is the maximum power point [27]. Current, voltage, and temperature sensors ensure proper voltage, current, and temperature limits of the solar panel output and on the PDB level. One set of voltage and current sensors is connected at the output of the solar panel (input of the buck converter) in all four redundant MPPT units, and a second set of current and voltage sensors is connected at the output of the buck converter. A bidirectional current sensor is employed at the input of batteries to monitor the battery charging and discharging currents. Latch-up protection and watchdog timer blocks protect the MPPT local controller of each unit from single-event latch-ups. A single MPPT buck converter unit is shown in Figure 5.

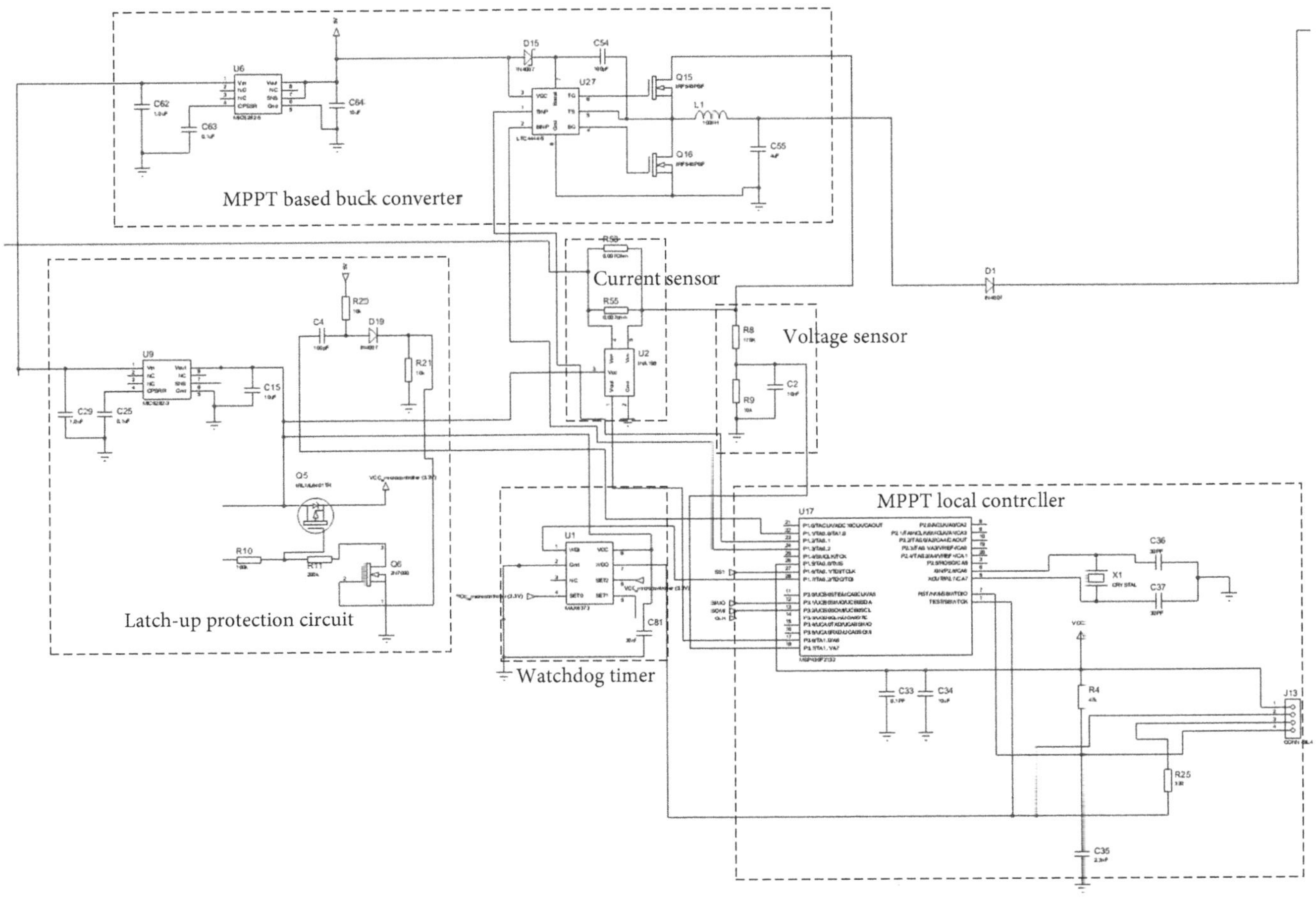


FIGURE 5: Single MPPT buck converter unit.

*3.2. Latch-Up Protection System.* MPPT local controller units of PCM are CMOS-based COTS devices which are sensitive to radiation [29]. A latch is a momentary effect in which the device gets short circuited and high current flows through the device from the power supply to ground which damages the device. The latch-up problem can be solved by using bipolar devices, which are immune to latch-up because they require an extremely high energy to trigger this event. But the microcontrollers are CMOS-based and require latch-up protection circuits. To avoid this condition, the microcontroller supply should be disconnected on a temporary basis. A latch-up protection circuit was designed and simulated to cut off the microcontroller supply during latch-up to protect the device from damage. The simulations of the latch-up protection circuit designed for the PCM are shown in Figure 6, where V1 is the microcontroller supply voltage connected with the microcontroller VCC pin, through MOSFET M2 and M1 switching. V3 is the voltage coming from the microcontroller pin. In normal operation condition, C1 is charged through R4 from the supply V2 and MOSFET M1 is ON, which further keeps M2 ON, and the voltage supply $V1 = 5\,V$ is connected with the microcontroller VCC pin. Figure 6(b) shows the waveforms of the latch-up protection circuit of Figure 6(a). The waveform V (R3:2) shows the microcontroller supply voltage, V (M1:g) represents the voltage across the C1 capacitor, and V (C1:1) indicates the microcontroller pin voltage. In waveforms, the normal operation state is shown from time 0 to 4 ms and 8 ms to 12 ms. During latch-up, the microcontroller pin, which is represented in the simulation circuit by V3, will be grounded and capacitor C1 will discharge. As a result of the C1 discharge, M1 and M2 will switch off and supply to the microcontroller VCC pin will cut off. The off state is shown in the simulation waveform from time 4 ms to 8 ms and 12 ms to 16 ms. The off state time duration is decided by the RC time constant of C1 and R4 components of the simulation circuit. At the moment, we have chosen it to be 4 ms as seen from the simulation waveform of Figure 6(b). The implemented latch-up protection circuit is shown in Figure 5.

*3.3. Data Interface.* The PCM has a dual redundant CAN interface with OBC for all telemetries and all of its telecommands using DB9 connectors. It provides sufficient telemetries (on CAN interface) in order to monitor the health of its own along with the health of all solar panels and battery. Telemetries of the PCM include the following at minimum:

(1) Primary bus voltage

(2) Primary bus current

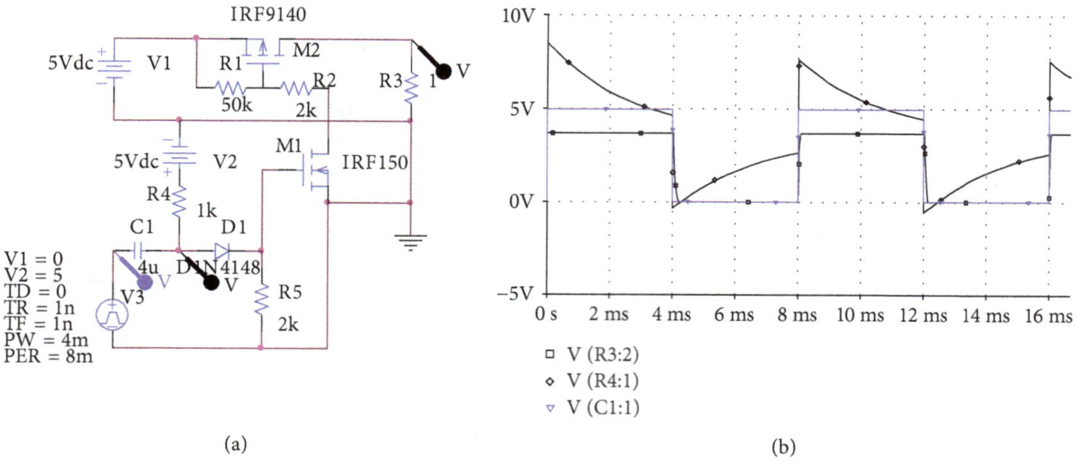


Figure 6: Latch-up protection simulation (a) circuit and (b) waveforms.

(3) Battery charge and discharge current

(4) Battery temperature

(5) Output current of each solar panel

(6) Voltage of each solar panel

(7) Temperature of each solar panel

The PCM collects analog temperature telemetries from each solar panel and battery, and it provides these telemetries (along with all other telemetries) to the OBC on the CAN interface.

*3.4. Power Interface.* PCM receives 2 : 1 redundant input power lines separately from four independent solar panels through the DB9 connector and provides 2 : 1 redundant output lines for battery charging as well as to PDM at bus voltage level using DB9 connectors. A gap of at least one unused pin is kept between positive power lines and their returns in all power interface connectors (for both input and output). To ensure modularity, all output connectors are female type and input connectors are male type.

*3.5. Simulations of MPPT Buck Converter.* The output range of the solar panel is 40 V~80 V as mentioned earlier. To cover this wide range input voltage to a stable output voltage, an MPPT buck converter is designed with wide range input. First, design considerations are made to select the appropriate values of filter components, i.e., inductor, capacitor, and resistor for an efficient and robust MPPT buck converter. The values selected according to the design consideration are assigned to all devices shown in Figure 7. The standard equations (1) and (2) are used for inductor ($L$) and capacitor ($C$) value selection [30], respectively.

$$L = \frac{(V_s - V_o) \times D}{(\Delta I_L \times f)}, \quad (1)$$

$$C = \frac{1 - D}{8L \times (\Delta V_o / V_o) \times f^2}. \quad (2)$$

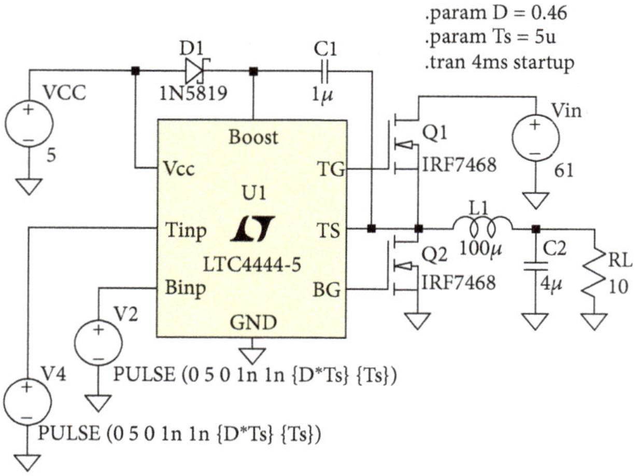


Figure 7: Simulation schematic of MPPT buck converter.

The LTspice simulation schematic of the buck converter with its driver circuitry is shown in Figure 7. The four MPPT buck converters are mounted on two PCBs, i.e., each PCB has two buck converter units. In order to avoid synchronization problem, 10 kHz frequency gap is kept between all four converters. The designed frequency for the 1st MPPT converter unit is 180 kHz, while the 2nd, 3rd, and 4th converters have 190 kHz, 200 kHz, and 210 kHz switching frequencies, respectively. All the components are selected on the basis of power loss analysis and small dimensions.

The two pulses V2 and V4 in Figure 7 represent the PWM signals coming from the microcontroller to increase or decrease the duty cycle according to the perturb and observe algorithm. The converter is simulated for a wide input voltage range, i.e., 80 V, 70 V, 61 V, 50 V, and 41 V, and the respective output voltage is measured which is the PDB voltage level (28 ± 6 V). The output current, input and output powers, and resultant efficiency are measured for the four designed buck converter units. The simulations of the MPPT buck converter are given in Figures 8–17, and simulation results are summarized in Table 2.

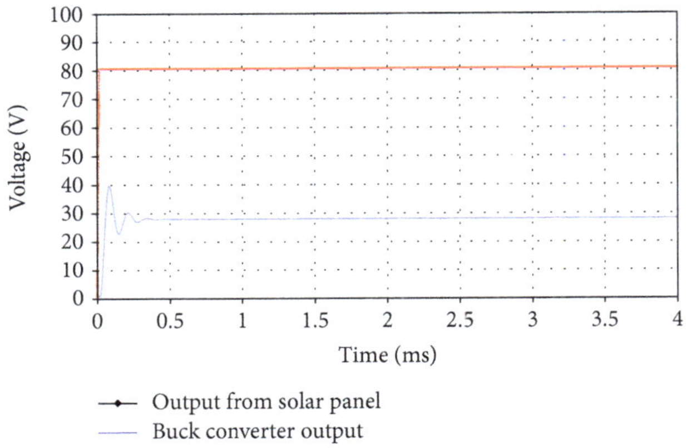


FIGURE 8: Input and output voltage waveforms of the MPPT buck converter for 80 V input.

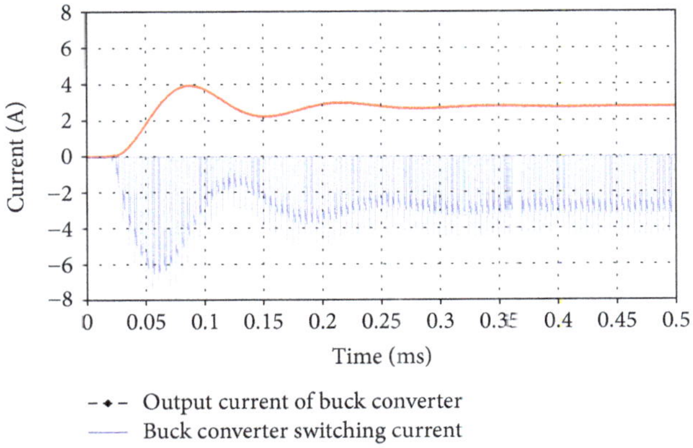


FIGURE 9: Input and output current waveforms for 80 V input.

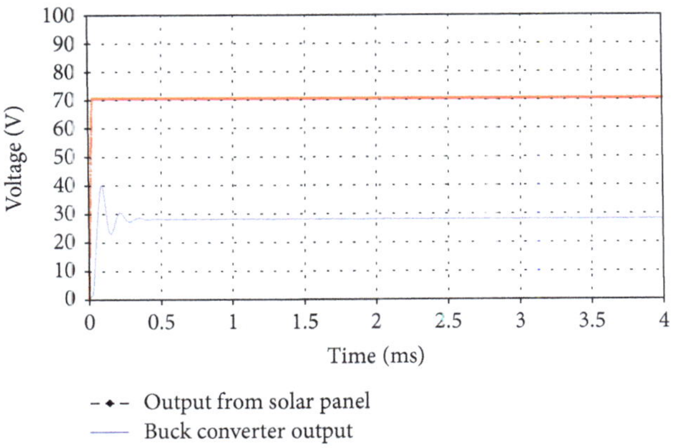


FIGURE 10: Input and output voltage waveforms for 70 V input.

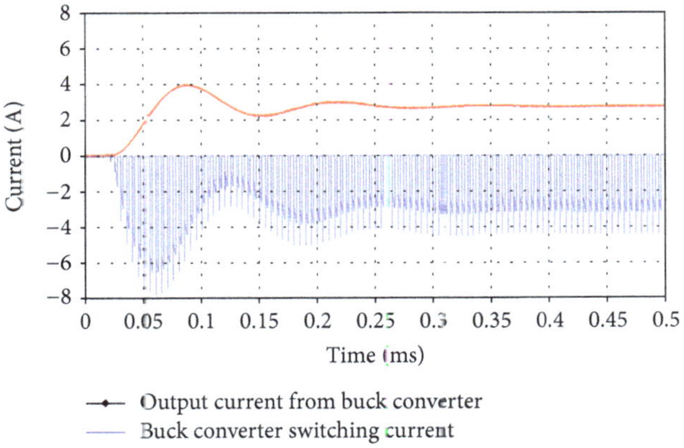


FIGURE 11: Input and output current waveforms for 70 V input.

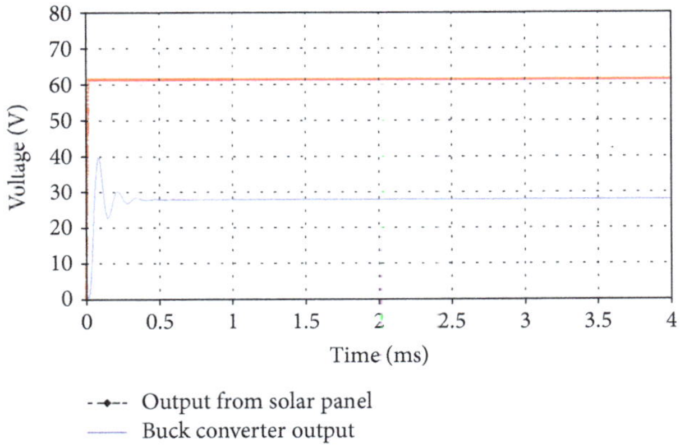


FIGURE 12: Waveforms of input and output voltage for 61 V input.

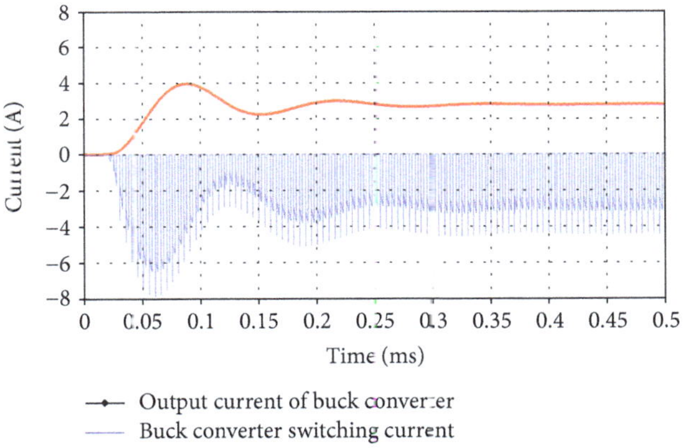


FIGURE 13: Input and output current waveforms for 61 V input.

Simulation results shown in Table 2 clearly demonstrate that the designed MPPT buck converter has a wide input range and possesses high efficiency (95.30% to 96.75%).

## 4. Power Distribution Module (PDM)

PDM is the submodule of the Electric Power System (EPS), responsible for the conversion of the PDB voltage level into the lower voltage levels and for distributing it to loads. EMI filtration is done by the PDM. There are certain protection systems installed on each distribution line. The main purpose of these protection systems is to limit the electromagnetic noise produced on the main bus and also to overcome the fault propagation. Inside the PDM, there is implementation of ON/OFF switching of all the switchable loads. With the help of telecommands, these switches are controlled from the OBC. To overcome failure and to improve the reliability,

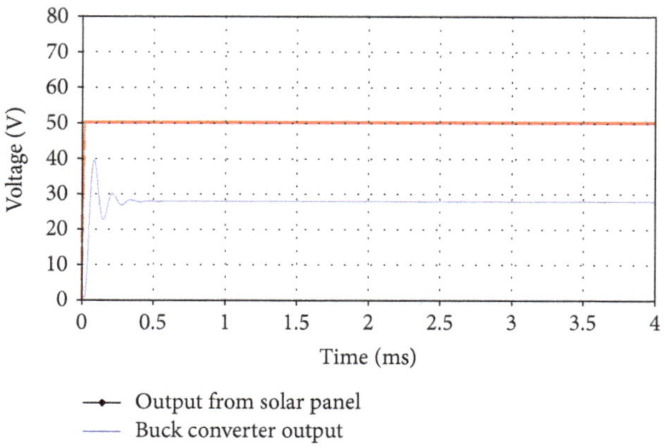


FIGURE 14: Input and output voltage waveforms for 50 V input.

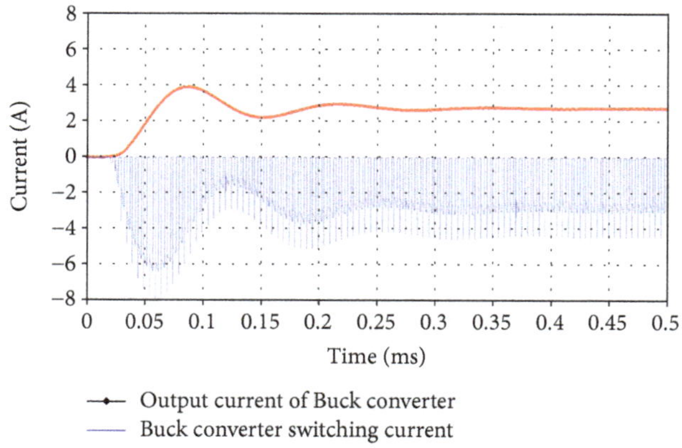


FIGURE 15: Input and output current waveforms for 50 V input.

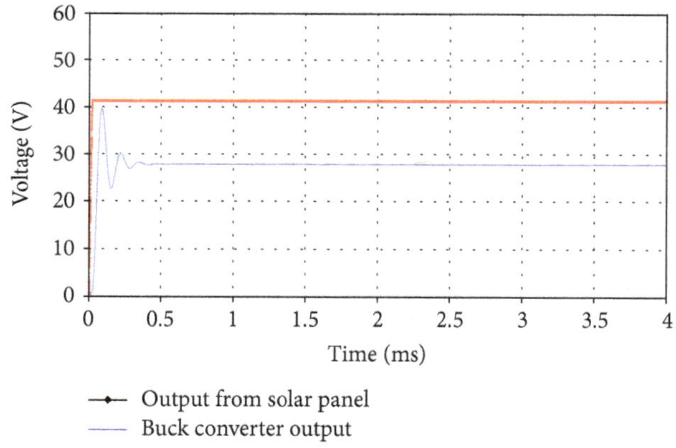


FIGURE 16: Input and output voltage waveforms for 41 V input.

redundancy is provided at all levels. No single electrical component failure in the PDU disables or degrades its required functions in a way that may lead to mission failure.

The power distribution unit receives primary power from the PCM at bus voltage 28 V ± 6 V (unregulated) and generates regulated secondary voltage levels at 12 V and 5 V. Current, voltage, and temperature sensors are mounted in the PDM unit to monitor and ensure proper operation of the module. The function of these sensors is to check current,

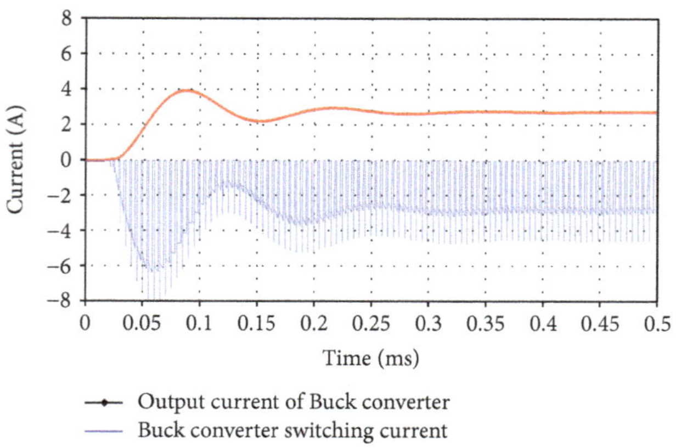


FIGURE 17: Input and output current waveforms for 41 V input.

voltage, and temperature limitations. The PDM controller further communicates with the OBC of the microsatellite through redundant CAN buses. The block diagram of the PDM is shown in Figure 18, and the design schematic is shown in Figure 19.

*4.1. Power Regulators.* The power regulators convert the PDB voltage level to low voltage levels (i.e., 5 V and 12 V) used by different subsystem components. For this purpose, the PDM unit requires two switching regulators. To meet the power and redundancy requirement for 12 V regulated output, the PDM has two redundant 12 V regulators with part number MGDD-20-R-E. For 5 V, four regulators with part number MGDB-35-H-C are connected in parallel to meet the power and redundancy requirements.

*4.2. Power Interface.* The PDM receives unregulated power through the PDB from the PCM and regulates it according to the requirement. For redundancy, the PDM receives 2 : 1 power lines from the PCM at 28 V ± 6 V. The PDM has a total of 40 output lines of different voltage and current levels. Some of these lines are pair switchable and some are non-switchable as shown in Table 3.

All output distribution lines of the PDM are protected against short circuit current and transient voltage spikes. A protection concept is implemented in the PDM to avoid fault and failure propagation between the loads. All PDM subsystems are connected and intercommunicate with the PDM controller (PIC24HJ256GP10) for monitoring and controlling purposes.

*4.3. Data Interface.* The PDM has a dual redundant CAN interface with the OBC for all telemetries and all of its telecommands. With the help of the CAN interface, the OBC receives certain telemetries from the PDM in order to check all the switchable distribution lines as well as to monitor the health of the primary and secondary voltages. In order to control the ON/OFF operation of switchable lines, the PDM has the capability of receiving telecommands through the CAN interface. Standard D-type connectors are used for CAN buses.

Table 2: Simulation results of MPPT buck converter designed for PCM.

| S. no | Input voltage ($V_{in}$) | Output voltage ($V_o$) | Duty cycle ($D$) | Input current ($I_{in}$) | Output current ($I_o$) | Input power ($P_{in}$) | Output power ($P_o$) | Efficiency ($\eta$) |
|---|---|---|---|---|---|---|---|---|
| 1 | 80 V | 27.33 V | 0.35 | 0.981A | 2.73A | 78.28 W | 74.61 W | 95.30 |
| 2 | 70 V | 27.36 V | 0.40 | 1.11A | 2.73A | 77.50 W | 74.69 W | 96.37 |
| 3 | 61 V | 27.45 V | 0.46 | 1.28A | 2.74A | 77.87 W | 75.21 W | 96.58 |
| 4 | 50V | 27.45 V | 0.56 | 1.56A | 2.74A | 77.79 W | 75.21 W | 96.68 |
| 5 | 41 V | 27.39 V | 0.68 | 1.89A | 2.73A | 77.28 W | 74.77 W | 96.75 |

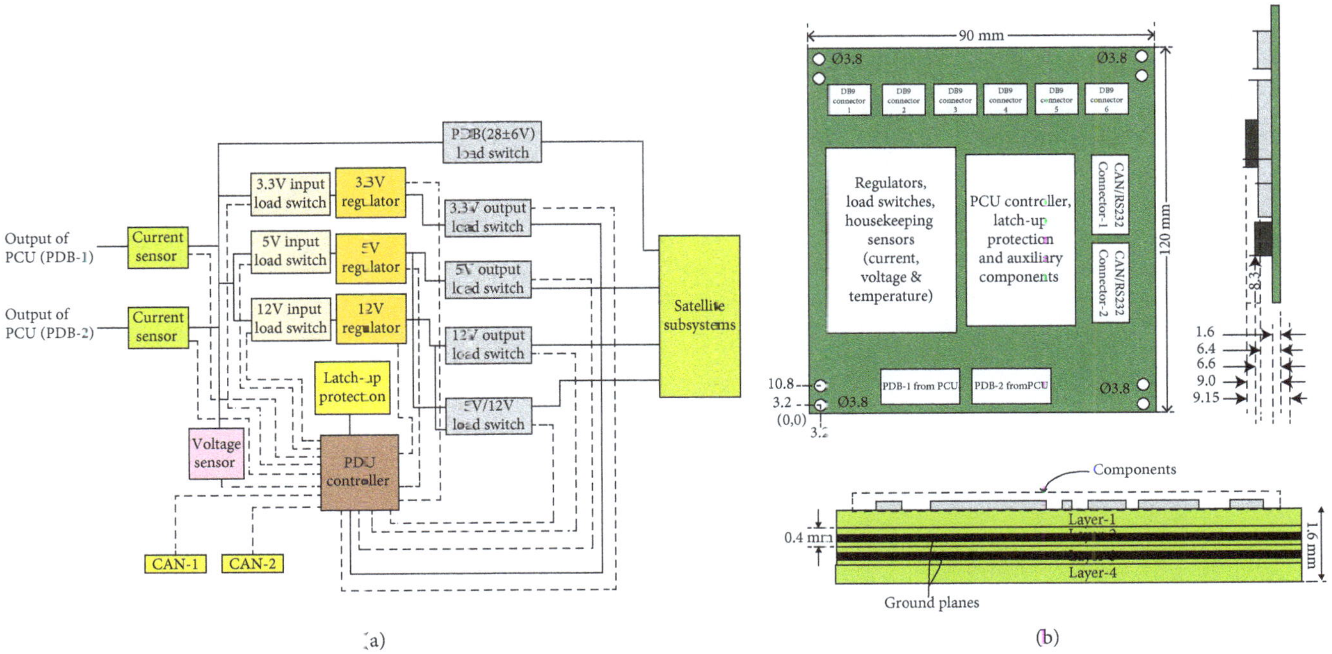


Figure 18: (a) PDM block diagram and (b) PCB dimensions.

## 5. Attitude Control System (ACS)

Due to dimension and weight constraints of small satellites, designers are focusing on the miniaturization of subsystems such as power, attitude determination and control, telecommunication, and payload. In the PNSS-1 attitude control subsystem design, the major aim was to achieve a system with small dimensions and weight. The available options for the attitude control were permanent magnets, reaction wheels, and magnetic rods [31, 32]. Permanent magnets are cheaper, simpler, and lightweight, but they have inadequate pointing accuracy. The other options were magnetic rods and reaction wheels which have better pointing accuracy, but the problem is their price, weight, and size which make them unsuitable for small satellites. To achieve the goal of a miniaturized compatible system with PNSS-1, an innovative concept of an embedded magnetorquer is introduced. The magnetorquer coil is copper traces embedded inside the internal layers of the solar panel PCB. Solar panel PCB is an eight-layer PCB with copper traces in five internal layers. These copper traces are just like the magnetorquer rod producing a magnetic field when current flows through it.

*5.1. PNSS-1 Embedded Magnetorquer Coil.* The magnetorquer coil works on the motor action concept that when a current carrying the coil is placed in a magnetic field, a torque is exerted on the coil. The current-carrying coil generates a magnetic moment which interacts with the already existing magnetic field and produces torque. According to Fleming's left-hand rule, the middle finger gives the direction of the magnetic moment, the index finger shows the direction of the existing magnetic field, and the thumb indicates the direction of exerted torque. Let us suppose the current-carrying coil shown in Figure 20 with the magnetic moment ($D$) generated is toward the page and is given by

$$\vec{D} = N \cdot S \cdot I, \tag{3}$$

where $N$ is the number of turns of the coil, $S$ is the area of the coil, and $I$ is the current flowing through the coil.

If the magnetic field ($\vec{B}$) direction is toward the right side of the paper, then the torque generated is toward the bottom of the paper and can be found by

$$\vec{\tau} = \vec{D} \times \vec{B} = DB \sin \theta \vec{n}, \tag{4}$$

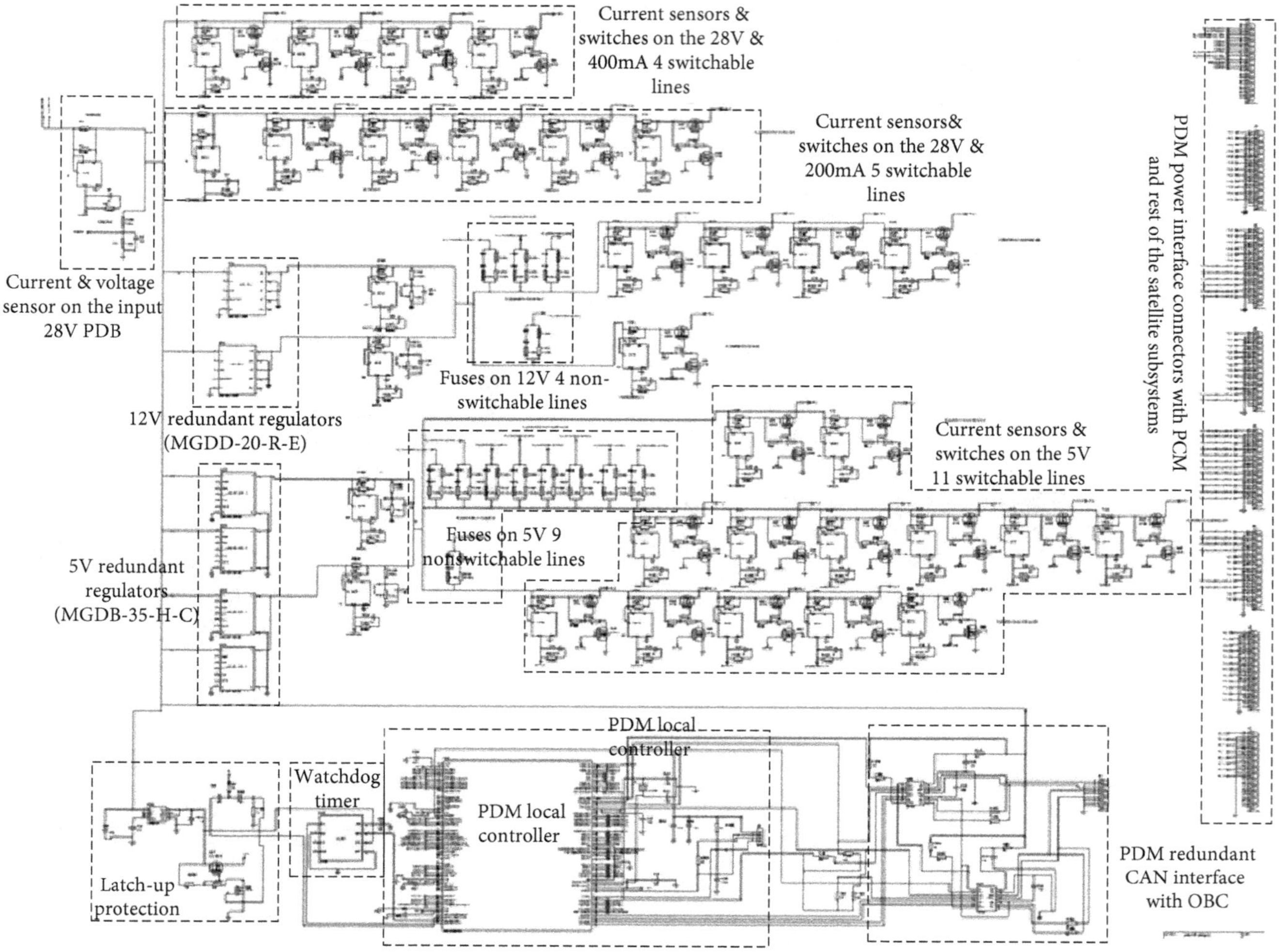


FIGURE 19: Design schematic of PDM.

TABLE 3: Voltage distribution of switchable/nonswitchable lines.

| S. no | Voltage level | No. of power lines and current rating | Switchable/ nonswitchable |
|---|---|---|---|
| 1 | 28 V | 4 lines at 400 mA each | Switchable |
| 2 | 28 V | 5 lines at 200 mA each | Switchable |
| 3 | 12 V | 3 lines at 400 mA each | Nonswitchable |
| 4 | 12 V | 4 lines at 400 mA each | Switchable |
| 5 | 12 V | 1 line at 100 mA each | Switchable |
| 6 | 12 V | 1 line at 100 mA each | Nonswitchable |
| 7 | 5 V | 2 lines at 2000 mA each | Switchable |
| 8 | 5 V | 8 lines at 800 mA each | Nonswitchable |
| 9 | 5 V | 6 lines at 800 mA each | Switchable |
| 10 | 5 V | 1 line at 200 mA each | Nonswitchable |
| 11 | 5 V | 5 lines at 200 mA each | Switchable |

where $\vec{n}$ is the direction of torque $\vec{\tau}$ and $\theta$ is the angle between $\vec{D}$ and $\vec{B}$.

Earth is a big magnet, and in low earth orbit (LEO), from 400 km to 2000 km altitude, the earth magnetic field varies between 0.15 gauss and 0.45 gauss [33]. Four solar panel modules with embedded magnetorquer coils are mounted on the four external faces of the PNSS-1 satellite as shown in Figure 21. When current flows through the coils, a magnetic moment is generated which interacts with the earth magnetic field and a torque is exerted on the satellite to rotate it in the desired direction.

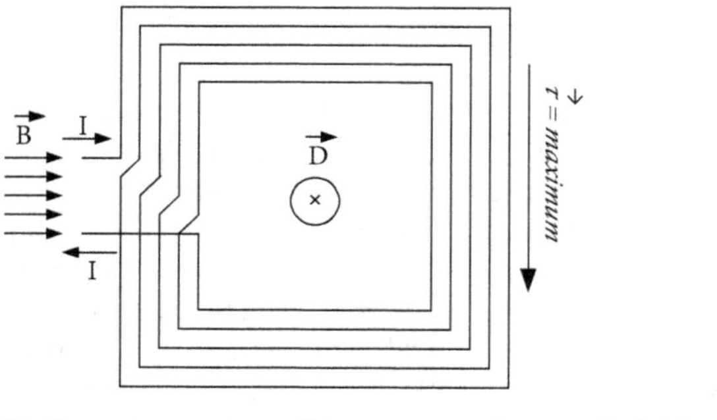


FIGURE 20: Current-carrying coil in presence of magnetic field.

The coils are embedded in the five internal layers of the solar panel PCB, i.e., the 2nd, 3rd, 4th, 5th, and 6th layers. These five coils are treated as individual coils and attached through switches. By changing the arrangement

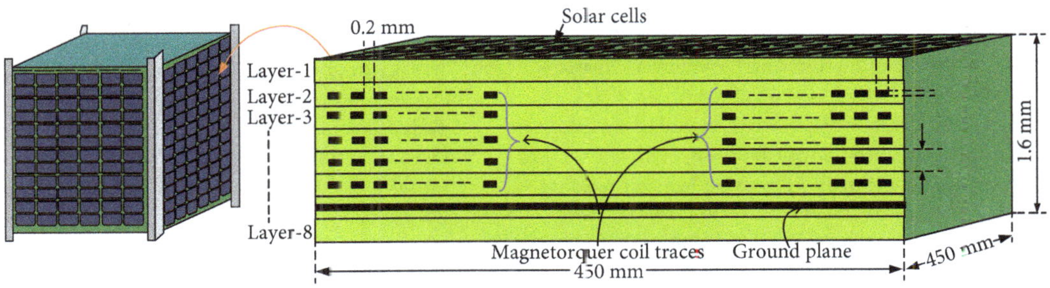


FIGURE 21: Solar panel module (PNSS-1) cross-sectional view with embedded magnetorquer.

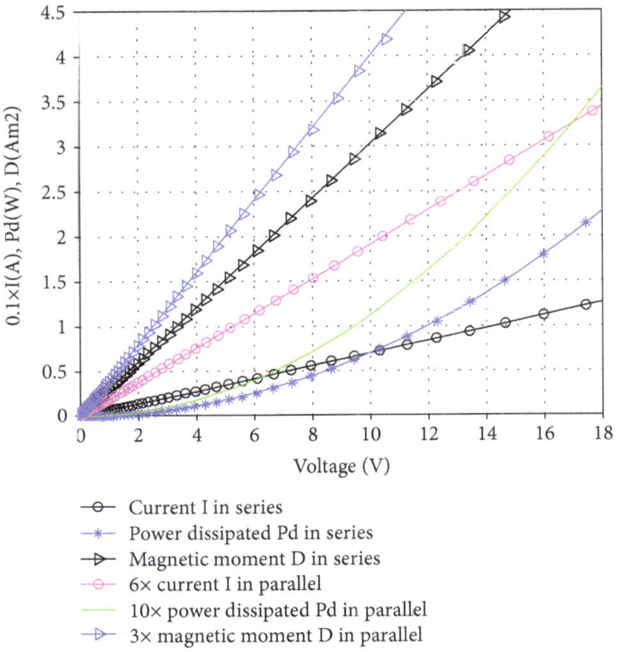


FIGURE 22: Voltage versus current, magnetic moment, and power dissipated for four-coil combination (series/parallel).

of these switches, the five coils can be attached/detached in different configurations such as a single coil, five in series, and five coils in parallel. According to the satellite torque requirements, the magnetorquer coil should be able to generate a specific magnetic field. To generate the required magnetic moment of PNSS-1, the magnetorquer coil composed of copper traces with 60 turns in each layer and a total of 300 turns is embedded in five internal layers. Each coil trace has 1.8 mm width and 18 $\mu$m thickness. Space between two adjacent traces is 0.2 mm.

Current, magnetic moment, and power dissipated are plotted against voltage in Figure 22, for four coils connected in series and parallel. At a particular voltage, one can observe and compare all the other parameters.

## 6. Conclusion

The paper proposed a modular design for the EPS and ACS of a microsatellite. The complete EPS is divided into three submodules, i.e., the SPM, PCM, and PDM. Every module has data and power interfaces which are standardized with specific number of input/output lines that meet the power and data requirement of microsatellite subsystems. COTS components were used for the subsystem implementation. The COTS components were selected on the basis of small dimension, minimum weight, and low power consumption. To avoid single component failure and increase the reliability, certain subsystems and components were made redundant in the design. To operate the solar panels at MPP and extract maximum power from them, a wide range highly efficient MPPT-based buck converter is designed and analyzed. Each solar panel has a separate MPPT unit with a 10 kHz frequency gap to avoid synchronization between them. To protect the CMOS-based controllers from radiation, a latch-up protection system is designed and simulated. Different sensors such as current, temperature, and voltage are mounted at different locations on the respective modules to monitor and ensure proper operation of these units. For miniaturization purposes, an embedded magnetorquer is designed for the microsatellite which consumes reasonable power and generates the required magnetic moment. In the end, an EPS for the microsatellite is achieved which is fully modular, compact, redundant, and power efficient.

## Conflicts of Interest

The authors announce that there is no conflict of interest concerning the publication of this article.

## References

[1] C. Passerone, M. Tranchero, S. Speretta, L. Reyneri, C. Sansoe, and D. Del Corso, "Design solutions for a university nano-satellite," in *2008 IEEE Aerospace Conference*, pp. 1–13, Big Sky, MT, USA, March 2008.

[2] J. C. de los Rios, D. Roascio, L. Reyneri et al., "ARAMIS: a fine-grained modular architecture for reconfigurable space missions," in *1st Conference on University Satellite Missions*, Rome, January 2011.

[3] A. Ali, L. M. Reyneri, and M. Rizwan Mughal, *Innovative Electric Power Supply System for Nano-Satellites*, 64TH IAC, Beijing China, 2013.

[4] D. Del Corso, C. Passerone, L. M. Reyneri et al., "Architecture of a small low-cost satellite," in *10th Euromicro Conference on Digital System Design Architectures, Methods and Tools (DSD 2007)*, Lubeck, Germany, August 2007.

[5] D. De Bruyn, "Power distribution and conditioning for a small student satellite, design of the NUTS backplane & EPS module," in *Master of Science in Engineering Cybernetics*, Norwegian University of Science and Technology (NTNU), 2011.

[6] "Satellite mass categories," http://www.daviddarling.info/encyclopedia/S/satellite_mass_categories.html.

[7] S. Dahbi, A. Aziz, S. Zouggar et al., "Power budget analysis for a LEO polar orbiting nano-satellite," in *2017 International Conference on Advanced Technologies for Signal and Image Processing (ATSIP)*, pp. 1–6, Fez, Morocco, May 2017.

[8] "Universities in space: seriously higher education," http://spaceflight.com/universities-in-space-seriously-higher-education/.

[9] "World's largest database of nanosatellites, over 2150 nanosats and Cube Sats," https://www.nanosats.eu/.

[10] "Introduction to PNSS-P (1)," January 2014, http://suparco.gov.pk/pages/intro.asp,suparco.gov.pk/pages/presentations-pdf/day-1/session-1/12B-III/4.pdf.

[11] C.-S. Sun and J.-C. Juang, "Design and implementation of a microsatellite electric power subsystem," *Journal of Aeronautics, Astronautics and Aviation Series A*, vol. 44, no. 2, pp. 67–74, 2012.

[12] S. Greenland, P. Mendham, M. Macdonald, C. McInnes, and C. Clark, "Implementation of a product-centric space system approach to nanosatellite systems design and deployment," in *The 4S Symposium 2016*, Valletta, Malta, May-June 2016.

[13] M. R. Mughal, J. C. De Los Rios, L. M. Reyneri, and A. Ali, "Scalable plug and play tiles for modular nano satellites," in *63rd International Astronautical Congress*, Naples, Italy, October 2012.

[14] J. Esper, "Modular, adaptive, reconfigurable systems: technology for sustainable, reliable, effective, and affordable space exploration," *AIP Conference Proceedings*, vol. 746, pp. 1033–1043, 2005.

[15] K. B. Clark and C. Y. Baldwin, *Design Rules. Vol. 1: The Power of Modularity*, MIT Press, Cambridge, Massachusetts, 2000.

[16] P. M. Wegner, P. Blower, and J. Wilkenfeld, "Standard buses, modular buses, and plug-and-play buses; what is the difference and why does it matter?," Report No. A553345, 2008.

[17] J. S. O. Alvarado, J. S. R. Mora, and L. E. A. Pico, "Designs and implementations for CubeSat Colombia 1 satellite power module," *International Journal of Applied Engineering Research*, vol. 12, no. 18, pp. 7360–7371, 2017.

[18] B. Burger and R. Rüther, "Inverter sizing of grid-connected photovoltaic systems in the light of local solar resource distribution characteristics and temperature," *Solar Energy*, vol. 80, no. 1, pp. 32–45, 2006.

[19] M. Green, *General Temperature Dependence of Solar Cell Performance and Implications for Device Modeling*, John Wiley & Sons, Ltd., 2003.

[20] W. J. Larson, *Space Mission Analysis and Design, Third Edition*, p. 217, 1999.

[21] P. C. M. De Carvalho, R. S. T. Pontes, D. S. Oliveira, D. B. Riffel, R. G. V. de Oliveira, and S. B. Mesquita, "Control method of a photovoltaic powered reverse osmosis plant without batteries based on maximum power point tracking," in *2004 IEEE/PES Transmision and Distribution Conference and Exposition: Latin America (IEEE Cat. No. 04EX956)*, pp. 137–142, Sao Paulo, Brazil, November 2004.

[22] M. Park and I. K. Yu, "A study on optimal voltage for MPPT obtained by surface temperature of solar cell," in *30th Annual Conference of IEEE Industrial Electronics Society, 2004. IECON 2004*, pp. 2040–2045, Busan, South Korea, November 2004.

[23] T. Takashima, T. Tanaka, M. Amano, and Y. Ando, "Maximum output control of photovoltaic (PV) array," in *Collection of Technical Papers. 35th Intersociety Energy Conversion Engineering Conference and Exhibit (IECEC) (Cat. No.00CH37022)*, pp. 380–383, Las Vegas, NV, USA, July 2000.

[24] N. Femia, D. Granozio, G. Petrone, G. Spagnuolo, and M. Vitelli, "Optimized one-cycle control in photovoltaic grid connected applications," *IEEE Transactions on Aerospace and Electronic Systems*, vol. 42, no. 3, pp. 954–972, 2006.

[25] W. Wu, N. Pongratananukul, W. Qiu, K. Rustom, T. Kasparis, and I. Batarseh, "DSP-based multiple peak power tracking for expandable power system," in *Eighteenth Annual IEEE Applied Power Electronics Conference and Exposition, 2003. APEC '03*, pp. 525–530, Miami Beach, FL, USA, February 2003.

[26] C. Hua and C. Shen, "Comparative study of peak power tracking techniques for solar storage system," in *APEC '98 Thirteenth Annual Applied Power Electronics Conference and Exposition*, pp. 679–685, Anaheim, CA, USA, February 1998.

[27] D. P. Hohm and M. E. Ropp, "Comparative study of maximum power point tracking algorithms using an experimental, programmable, maximum power point tracking test bed," *Conference Record of the Twenty-Eighth IEEE Photovoltaic Specialists Conference - 2000 (Cat. No.00CH37036)*, 1699–1702, Anchorage, AK, USA, September 2000, 1699–1702.

[28] A. Dolara, R. Faranda, and S. Leva, "Energy comparison of seven MPPT techniques for PV systems," *Journal of Electromagnetic Analysis and Applications*, vol. 1, no. 3, pp. 152–162, 2009.

[29] P. Ehrenfreund, B. H. Foing, and F. Salama, "Organics in space: results from space exposure platforms and nanosatellites," in *IAU General Assembly, Meeting #29, id. 2257551*, American Astronomical Society, 2015.

[30] D. W. Hart, *Power Electronics*, mcgraw-Hill, New York, NY, USA, 2011, Print.

[31] N. Shams, F. Tanveer, and S. Ahmad, "Design and development of attitude control system (ACS) using COTS based components for small satellites," in *2008 2nd International Conference on Advances in Space Technologies*, pp. 6–11, Islamabad, Pakistan, November 2008.

[32] V. Francois-Lavet, *Study of Passive and Active Attitude Control Systems for the OUFTI Nanosatellites*, University of Liege Faculty of Applied Sciences, 2009–2010.

[33] "SPENVIS: The Space Environment Information System," http://www.spenvis.oma.be.

# 4

# A Novel Technique to Compute the Revisit Time of Satellites and Its Application in Remote Sensing Satellite Optimization Design

**Xin Luo,[1] Maocai Wang,[1,2] Guangming Dai,[1,2] and Xiaoyu Chen[1,3]**

[1]*School of Computer, China University of Geosciences, Wuhan 430074, China*
[2]*Hubei Key Laboratory of Intelligent Geo-Information Processing, China University of Geosciences, Wuhan 430074, China*
[3]*Institute of Computer Sciences, Heidelberg University, 69120 Heidelberg, Germany*

Correspondence should be addressed to Maocai Wang; mcwang@cug.edu.cn

Academic Editor: Christian Circi

This paper proposes a novel technique to compute the revisit time of satellites within repeat ground tracks. Different from the repeat cycle which only depends on the orbit, the revisit time is relevant to the payload of the satellite as well, such as the tilt angle and swath width. The technique is discussed using the Bezout equation and takes the gravitational second zonal harmonic into consideration. The concept of subcycles is defined in a general way and the general concept of "small" offset is replaced by a multiple of the minimum interval on equator when analyzing the revisit time of remote sensing satellites. This technique requires simple calculations with high efficiency. At last, this technique is used to design remote sensing satellites with desired revisit time and minimum tilt angle. When the side-lap, the range of altitude, and desired revisit time are determined, a lot of orbit solutions which meet the mission requirements will be obtained fast. Among all solutions, designers can quickly find out the optimal orbits. Through various case studies, the calculation technique is successfully demonstrated.

## 1. Introduction

Satellite missions devoted to the observation of the Earth as well as navigation satellites commonly use repeat ground track orbits [1]. With the development of civilian satellites technology, constellation composed of a number of satellites plays an important role in remote sensing [2]. The Earth observation missions often require the constant solar illumination, the same ground resolution, and small repeat cycles, which often results in the design of Repeat Sun-Synchronous Orbit (RSSO) satellites as the most suitable one. This kind of satellites allows the observation of a given region of the Earth at the same local time after a time interval [3, 4]. The approach to design repeat ground track orbit for the Earth observations (EO) is quite mature [5, 6]. As well, the RSSO satellites are also used for Mars observations [7–10].

Repeat ground track (RGT) orbits allows a satellite to reobserve the same area after a repeat cycle. Some articles have shown the various uses of RGT orbits. Fu et al. [11] presented a strategy for design and maintenance of low RGT successive-coverage orbits and their analysis is based on the drift over the entire ground track. Li et al. [12] introduced a special repeat coverage orbit which is a special class of RGT orbit, such orbits can visit a target site at both the ascending and descending stages in one revisit cycle. Circi et al. [13] showed the concepts of sliding ground track pattern, which allows one RGT orbit to transfer to another RGT orbit using a low-$\Delta v$ technique. This technology guarantees the fulfillment of several objectives in the course of the same mission. Recent studies have shown the possibility of using the Periodic Multi-Sun-Synchronous orbits (PMSSOs) for Earth and Mars observation; these orbits allow the observation of the same area under different solar illumination conditions and have a repetition period of the solar illumination conditions which is multiple of the repeat cycle [9, 10]. Wang et al. [14] divided the region by latitude stripes. The relationship between the cumulative coverage and the altitude can be quickly got, which is helpful for orbit designer to select optimal orbits for Earth observation.

Revisit time (RT) of a single satellite is the time elapsed between two successive observations of the same ground point on the surface of the Earth [15]. Different from the repeat cycle which is only relevant to the satellite orbits, the revisit time is relative to both of the orbit and the payload of satellite, such as tilt angle and swath width [16]. Most of the previous papers which aim to design orbits for remote sensing concentrate on the repeat cycle and near-repeat cycle [4, 16]. However, they have not considered the revisit time when designing satellites. In general, the orbit design should be treated as a multidisciplinary process, which needs to consider the satellite system such as payload properties. Saboori et al. [17] proposed a multiobjective optimization tool to design repeat sun-synchronous orbits for remote sensing satellites; they considered the revisit time as a function of the tilt angle and side-lap of the satellite. However, the calculation in their work does not consider the subcycles of the orbit. Pie and Schutz [18] introduced the subcycles of repeat ground track orbits and the charts of subcycles are used to decompose the repeat cycle into three main subcycles.

Nadoushan and Assadian [19] presented a novel technique to design RGT orbit with desired revisit time and optimal tilt angle; their calculation is based on the analyses of subcycles. However, their approach could be used only when the repeat cycle is prime relative to the revisit time. When the repeat cycle is not prime relative to the revisit time, the design approach could not be used. As a result, a lot of feasible orbits which meet the mission requirements are ignored. As a design tool, it is preferable to conduct an ergodic performance on the altitude.

This paper proposes a novel technique to compute the revisit time and minimum revisit time of remote sensing satellites. The relationship between the revisit time and the tilt angle of a satellite can be computed fast by this technique. Compared to the approach proposed by Nadoushan and Assadian [19], when the swath width, side-lap, the range of altitude, and desired revisit time are given, a lot of orbit solutions which meet the mission requirements will be obtained fast. It will be helpful for the orbit designer to select the best orbits in all solutions.

Section 2 illustrates the orbital relationships that have to be satisfied to obtain regular cycles of observation of the Earth with a uniform ground track pattern. And the subcycles of repeat ground track orbit are considered and developed using Bezout's lemma. In the third section, the procedure of the proposed technique is raised to calculate the RT of satellite. At last, the technique is used to compute and select the best orbits according to the mission requirements. Finally, some cases are investigated for evaluation of the technique.

## 2. Repeating Sun-Synchronous Orbits and Subcycles

*2.1. Repeating Sun-Synchronous Orbits.* The repeating orbits are also known as the repeat ground track orbits, which are defined as orbits with periodic repeating ground tracks. Their ground track will repeat after a whole number of revolutions $R$ in $D$ nodal days. These orbits have good appearances for the Earth's coverage, which is good for remote sensing. For optical observations, it is important to ensure that illumination conditions remain the same or vary as little as possible when observing the same ground area; this is known as sun-synchronous orbit. We consider an intersection of the equator and a satellite's descending (ascending) ground track. For a satellite, the interval between two successive equatorial crossings of the ground track on the equator is

$$\Delta\lambda = T_N\left(\omega_E - \dot{\Omega}\right), \tag{1}$$

where $\omega_E$ is the Earth's rotation rate with respect to the vernal equinox. $\dot{\omega}$ is the variation rates of argument of perigee and $\dot{M}$ is the variation rates of mean anomaly. $T_N$ is the nodal period of the motion of the satellite, which is expressed as

$$T_N = \frac{2\pi}{\dot{M} + \dot{\omega}}. \tag{2}$$

The condition for repeating ground track orbits can be written as

$$R \cdot T_N\left(\omega_E - \dot{\Omega}\right) = R \cdot \Delta\lambda = D \cdot 2\pi \tag{3}$$

or

$$R \cdot T_N = D \cdot D_N, \tag{4}$$

where $R$ and $D$ are positive integers and they are prime to one another. $D_N$ is nodal day; it is expressed as

$$D_N = \frac{2\pi}{\omega_E - \dot{\Omega}}. \tag{5}$$

If the orbit is sun-synchronous orbit as well, the nodal precession rate $\dot{\Omega}$ equals the rotational angular speed of the Earth $\omega_S$. A nodal day of the repeating sun-synchronous orbit is a solar day which is equal to 86400 s. This is a useful relationship unique to RSSOs when evaluating how fast the ground track advances in longitude as a function of time [4].

In engineering practice, the repeating factor Q represents the number of orbits completed per day. $Q$ determines the location and sequence of all ground traces, which is defined as

$$Q = \frac{R}{D} = I + \frac{K}{D}. \tag{6}$$

In (6), $Q$ can be written as an integer number $I$ plus a fractional part $K/D$, where $K$ is an integer number which is prime to $D$ and $0 \le K < D$.

This paper only considers the descending node passes or ascending node passes. The terms related to this article are defined as follows.

*Definition 1.* Fundamental interval $S_Q$: $S_Q$ is the interval on equator between two successive ground tracks. $S_Q = 2\pi/Q = \Delta\lambda$.

*Definition 2.* Minimum interval $S_D$: $S_D$ is the minimum interval on equator between two ground tracks after a repeat cycle. $S_D = S_Q/D$.

Table 1 shows the parameters of some repeating sun-synchronous satellite orbits.

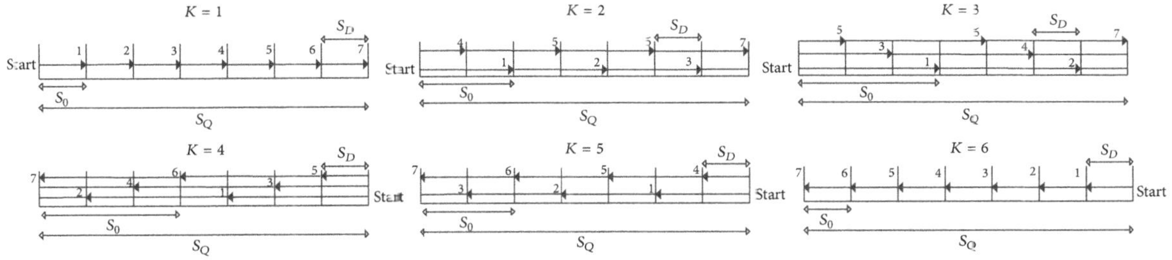


FIGURE 1: Ground track sequences for $D = 7$.

TABLE 1: Parameters of some repeating sun-synchronous orbits.

| | a (km) | i (°) | I | K | D | $S_Q$ (°) | $S_Q$ (km) | $S_D$ (°) | $S_D$ (km) |
|---|---|---|---|---|---|---|---|---|---|
| A | 605.512 | 97.81 | 14 | 6 | 7 | 24.23 | 2697.36 | 3.46 | 385.33 |
| B | 650.737 | 97.99 | 14 | 5 | 7 | 24.47 | 2723.54 | 3.50 | 389.08 |
| C | 696.701 | 98.18 | 14 | 4 | 7 | 24.71 | 2750.25 | 3.53 | 392.89 |
| D | 743.421 | 98.37 | 14 | 3 | 7 | 24.95 | 2777.48 | 3.56 | 396.78 |
| E | 790.919 | 98.57 | 14 | 2 | 7 | 25.20 | 2805.25 | 3.60 | 400.75 |
| F | 839.216 | 98.78 | 14 | 1 | 7 | 25.45 | 2833.59 | 3.64 | 404.80 |

Ground track analyses play an important role in designing the repeating sun-synchronous orbits. Every fundamental interval $S_Q$ is divided into $D$ minimum fundamental intervals by $(D - 1)$ time-successive ground tracks. The parameter $K$ determines the way where these subdivisions are carried out. The minimum ground track distance $S_0$ on equator in consecutive nodal days can be expressed as follows:

$$S_0 = S_D \cdot K \quad \text{for } K < \frac{D}{2}, \tag{7a}$$

$$S_0 = S_D \cdot (K - D) \quad \text{for } K > \frac{D}{2}. \tag{7b}$$

Depending on the parameters $K$, the ground traces can be considered direct or skipping. If $K$ is 1 or $D - 1$, then each successive trace falls next to the one which is before it. Take into account orbits named E, F in Table 1 as an example. The fundamental interval $S_Q$ is divided into 5 equally spaced intervals. For the case of orbit F, the parameter $K$ equals 1. The daily interval $S_0$ is equal to $S_D$ and the ground tracks related to the successive nodal days are eastward shifted of $S_D$. For the case of orbit E, the parameter $K$ is equal to 2. The daily interval $S_0 = 2S_D$ and the ground tracks related to the successive nodal days are eastward shifted of $2S_D$. Figure 1 shows the ground track sequences of the satellite when the parameter $D = 7$. In fact, if $S_0 > 0$, the progress of ground trace is eastward and if $S_0 < 0$, the progress is westward.

*2.2. Subcycles in a Repeat Cycle.* According to the paper [20, 21], a subcycle is an integer value of days after which the ground track of the satellite nearly repeats itself within a small offset. A subcycle of a satellite can be viewed as a near-repeat cycle which is equal to an integer number of nodal days.

From Figure 1, we can get a conclusion: for any repeat ground track orbits whose values of $K$ and $D$ are decided, there exists a time of interval $d_1$ nodal days after which the ground track of the satellite passes at a minimum interval $S_D$ of the original node to the east. The ground track will pass at a minimum interval $S_D$ of the original node to the west in $(D - d_1)$ days.

In Lim and Schutz's definition, there are no specifics given for the offset, other than being "small" [21]. The offset is always considered to be equal to a minimum interval $S_D$ in some literatures [16, 18].

However, for different purpose, it may be required to obtain the subcycle of a RGT orbit with a specific offset. The analyses for subcycles have a significant role in designing RGT orbits; in this paper, the small offset is redefined and replaced by a multiple of minimum intervals $S_D$. For a specific value $D$, there exists a large range of different subcycle patterns due to the value $K$. The approach to calculate subcycles of a satellite is presented in the paper [18]. Basically, the definition of subcycles is based on *Bezout* lemma, which can be expressed as [22]

$$d \cdot K - m \cdot D = k, \tag{8}$$

where $d$, $m$, and $k \neq 0$ are integer numbers. An integer solution $d$ exists if and only if the parameter $k$ is a multiple of the great common divisor of $K$ and $D$. Since the value $K$ and value $D$ are coprime to one another, so if $k$ is an integer number, the *Bezout* lemma ensures the existence of the solution. In (8), a subcycle $d_k$ is the number of days after which the ground track will repeat itself with an offset equal to $k \cdot S_D$. The subcycles can be, respectively, labeled $(d_k, k \cdot S_D)$. For example, the first subcycle $d_1$ is the number of days required for the ground track to pass an offset $S_D$ of the original descending node. Since the offset can be east or west, so the first subcycle $d_1$ has two values $d_{1\min}$, $d_{1\max}$ and $d_{1\max} > d_{1\min}$. And two values are relative to each other by $d_{1\min} = D - d_{1\max}$.

Orbit data of HJ-1A is used in this section, HJ-1A is a Chinese satellite of satellite constellation which aims to the environment monitoring and forecasting, and the design goal is to cover the Earth within four days [6]. Choose the orbit of HJ-1A as an example of which the parameter $Q = 14 + 23/31$. In Table 2, we present the subcycles of HJ-1A when setting different values of $k$. As illustrated in Table 2, when the required offset is twice as long as $S_D$, the corresponding subcycles are 23 days or 8 days. The 23 days of subcycle is relative to east offset which is labeled $(23, 2S_D)$, while the 8

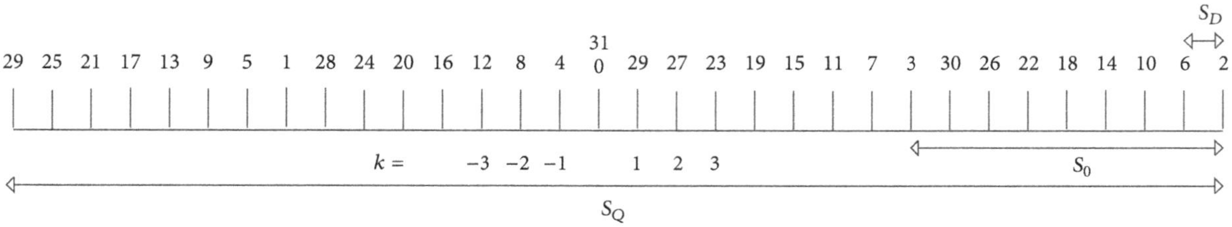


FIGURE 2: Ground track sequences when the parameter $Q = 14 + 23/31$ and subcycles corresponding to different offsets.

TABLE 2: Subcycles for a RGT orbit of which the parameter $Q = 14 + 23/31$.

| $K$ | $D$ | $k$ | Subcycles $d$ (days) | Labeled |
|---|---|---|---|---|
| 23 | 31 | 1 | 27 | $(27, S_D)$ |
| 23 | 31 | −1 | 4 | $(4, -S_D)$ |
| 23 | 31 | 2 | 23 | $(23, 2S_D)$ |
| 23 | 31 | −2 | 8 | $(8, -2S_D)$ |
| 23 | 31 | 3 | 19 | $(19, 3S_D)$ |
| 23 | 31 | −3 | 12 | $(12, -3S_D)$ |

days is relative to a west offset which is labeled $(8, -2S_D)$. The temporal order of subsatellite trackes is illustrated in Figure 2.

## 3. Revisit Time and Optimization in the Satellite Design

In this section, the revisit time of a satellite is discussed. Revisit time (RT) of a single satellite is the time elapsed between two successive observations of the same ground point using off-nadir pointing of the payload or the attitude of the satellite [15]. The revisit time is relevant to both of the orbits and the payload of the satellite, such as tilt angle and swath width [16]. In practice, a satellite's optical instrument can view an area (using off-nadir observation) before and after the orbit passes over it. Off-nadir observation of a ground area is possible in two ways. The first one is the swath width of a sensor which is wide enough to cover the adjacent ground tracks, which is illustrated in Figure 3, where $\alpha$ is the half of field of view (FOV) angle. Another one is using the attitude maneuvers (tilt angles) $\pm\theta$ of the satellite with a small swath width, which is illustrated in Figure 4, and where $\theta$ is the tilt angle of the satellite. In either case, the satellite in adjacent orbits can observe the ground target, thus making the revisit time less than the repeat cycle.

*3.1. Maximum Revisit Time.* It should be noted that, in the case of a satellite with a swath width less than the minimum interval $S_D$ and with no tilting capacity, the revisit time (or equivalently maximum revisit time) is equal to the repeat cycle. For these orbits, the revisit time of ground point at equator covered by the field of regard (FOR) of the satellite payload is equal to the repeat cycle. In addition, complete coverage at equator cannot be provided by the satellite and the revisit time is only related to the points which can be covered by the sensor. FOR is the area covered by a detector or sensor

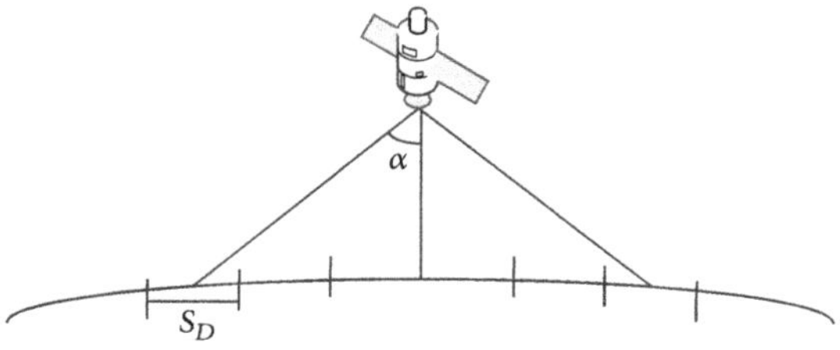


FIGURE 3: Observe the ground target by a sensor with a wide swath width.

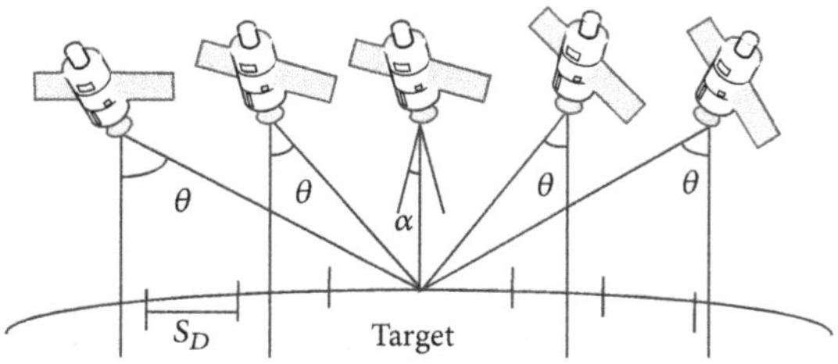


FIGURE 4: Observe the ground target from adjacent orbit using tilt capacity.

when pointing to all mechanically possible positions which is different from the FOV. The FOV is the area covered by the sensor or detector when pointing to one position [17].

The maximum cone angle $\phi$ where the surface of the Earth can be observed by a satellite is calculated as follows [19]:

$$\phi = \sin^{-1}\frac{R_E}{R_E + a}, \tag{9}$$

where $a$ is the altitude of orbit and $R_E$ is the mean radius of the Earth.

*3.2. Calculation of Revisit Time.* The calculation of the revisit time is divided into two kinds of cases. The first case is where the sensor has a wide swath width with no tilting capacity; another case is that the sensor with tilting capacity has a small swath width. In either case, the maximum revisit time is smaller than the repeat cycle of the orbit. In both two cases, we can get the swath width $\omega$ corresponding to FOV, which can be calculated as follows [4]:

$$\omega = R\left\{2\sin^{-1}\left(\frac{R+a}{R}\sin\frac{\text{FOV}}{2} - \text{FOV}\right)\right\}. \tag{10}$$

When calculating the swath width corresponding to the FOR, the FOV in (11) should be replaced by the FOR. In order

TABLE 3: Different revisit time of P6 satellite with different tilt capacity and side-lap.

| $n$ | Side-lap (%) | Tilt angle range (°) | $S'$ | RT (days) | $RT_{min}$ (days) |
|---|---|---|---|---|---|
| 1 | 0 | [7.98, 15.60) | {0, 5, 19} | 14 | 5 |
| 2 | 0 | [15.60, 22.60) | {0, 5, 10, 14, 19} | 5 | 4 |
| 3 | 0 | [22.60, 28.81) | {0, 5, 9, 10, 14, 15, 19} | 5 | 1 |
| 4 | 0 | [28.81, 34.20) | {0, 4, 5, 9, 10, 14, 15, 19, 20} | 4 | 1 |
| 1 | 5 | [8.39, 16.37) | {0, 5, 19} | 14 | 5 |
| 2 | 5 | [16.37, 23.63) | {0, 5, 10, 14, 19} | 5 | 4 |
| 3 | 5 | [23.63, 30.01) | {0, 5, 9, 10, 14, 15, 19} | 5 | 1 |
| 4 | 5 | [30.01, 35.48) | {0, 4, 5, 9, 10, 14, 15, 19, 20} | 4 | 1 |

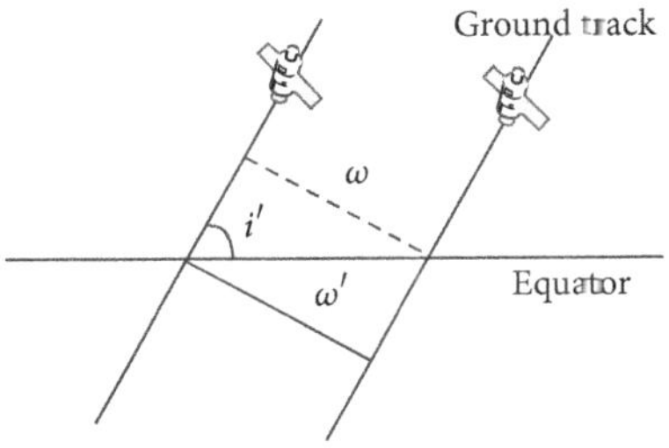


FIGURE 5: Apparent inclination $i'$ and the swath on the equator $\omega'$.

to get a more accurate result, the apparent inclination $i'$ is introduced in this paper. Apparent inclination is the angle between the equator and the ground track of the satellite in Earth-centered Earth-Fixed Coordinate (ECEF) system. And for circular orbits, $i'$ is defined as [22]

$$\tan i' = \frac{\sin i}{\cos i - 1/Q}. \tag{11}$$

In some cases, in order to get a better spatial resolution, a margin of safety must be taken into account. Assuming a 10% overlap is required, so the effective swath width $\omega$ of the satellite should be replaced by $0.9\omega$.

The swath of the satellite on the equator $\omega'$ is depicted in Figure 5 and it can be calculated as follows [22]:

$$\omega' = \frac{\omega}{\sin i'}. \tag{12}$$

In order to achieve a complete longitude coverage on the equator in a repeat cycle $D$, $\omega'$ must be greater than the minimum interval $S_D$. When the swath width of the satellite is wide enough to cover several minimum intervals, the half of number of minimum separations $n$ which can be covered is calculated as follows:

$$n = \text{int}\left(\frac{\omega'}{2 \cdot S_D}\right), \tag{13}$$

where function $\text{int}(x)$ is to get the integer part of $x$. According to (13), the number of minimum separation intervals which can be covered by the satellite payload is obtained. In this case, the satellite can observe the ground target with an offset which is smaller than $n \cdot S_D$, thus making the satellite in adjacent tracks able to observe the ground target as well. The feasible number $n$ of minimum intervals $S_D$ which can be covered by the satellite forms a set $R$, and $R = \{-n, -n+1, \ldots, -1, 0, 1, \ldots, n-1, n\}$.

Choose an element $k$ in the set $R$, and bring it into (8); then the subcycle $d_k$ can be obtained. Through this step, a set of subcycles $S$ corresponding to the set $R$ can be derived as

$$S = \{d_{-n}, d_{-n+1}, \ldots, d_{-1}, d_0, d_1, \ldots, d_n\}. \tag{14}$$

Next, the set of subcycles $S$ should be sorted from small to large; the sorted set is named as $S'$. Assume that $S' = \{d_0, d_1, d_2, \ldots, d_{2n-1}, d_{2n}\}$. The revisit time (equivalently maximum revisit time $RT_{max}$) and the minimum revisit time $RT_{min}$ can be computed as

$$RT_{max} = \max\{d_{k+1} - d_k\}, \quad 0 \le k < 2n, \tag{15a}$$

$$RT_{min} = \min\{d_{k+1} - d_k\}, \quad 0 \le k < 2n. \tag{15b}$$

Choose the orbit of HJ-1A as an example of which $Q = 14 + 23/31$. The swath on the equator of the satellite is 6 times as long as minimum interval $S_D$ and $n = 3$ according to (13). So the set $R = \{-3, -2, -1, 0, 1, 2, 3\}$. The set of corresponding subcycles $S = \{12, 8, 4, 0, 27, 23, 19\}$ which is depicted in Table 2. According to (15a) and (15b), the revisit time of HJ-1A is 7 nodal days and the minimum revisit time is 4 nodal days.

Actually, different revisit time corresponds to the different tilt capacity and side-lap; Table 3 shows the different revisit time of P6 satellite [19] with different tilt capacity and required side-lap.

There are some analyses for Table 3 as follows:

(1) The revisit time and minimum revisit time of a satellite vary with different tilt angle and required side-lap.

(2) In some cases, an increase of tilt angle could not result in reduction of the RT. For example, when required side-lap is 5% and the tilt angle ranges from 8.39° to 23.63°, the revisit time remains 5 days.

The procedure to calculate the revisit time of a satellite with specific parameters is shown in Figure 6.

*3.3. Design for Remote Sensing Satellites.* Orbital parameters design is a very important task during the mission analysis and design phase of a satellite. For remote sensing satellites,

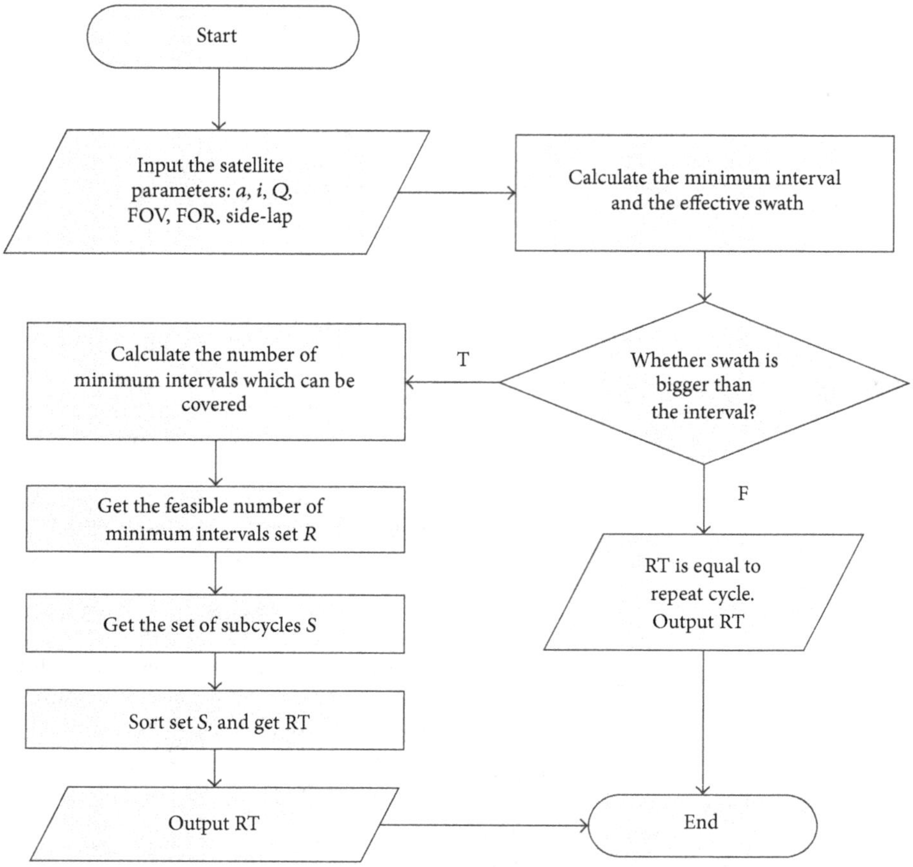


FIGURE 6: The procedure to compute the revisit time of a satellite.

the performance of a satellite can be represented in terms of resolution, revisit time, repeat cycle, tilt angle, and overlapping width [16]. Those characteristics should be taken into account in the orbit design of remote sensing satellites.

According to the approach shown in the last section, when the orbit and payload parameters are determined, the relationships between the tilt angle and the revisit time can be quickly obtained. This approach needs simple calculations and has high efficiency. In this section, the proposed approach is used to design and select the optimal orbits for remote sensing satellites.

When designing satellites for the Earth observation, the basic inputs are the orbital altitude range, the required side-lap, desired revisit time, and the maximum repeat cycle. The output of this approach is the minimum tilt angle in the premise of the desired revisit time and available side-lap constraint. This approach makes an ergodic search on the range of altitude and generates the orbital parameters which meet all of the operation constraints.

The calculation procedure of designing a satellite with required revisit time and optimal tilt angle are given as follows.

*Step 1.* Input the range of altitude, required side-lap, and desired revisit time $RT_{des}$.

*Step 2.* Input the maximum repeat cycle $D_{max}$.

*Step 3.* Generate the set $P$ composed of all the feasible orbits which is in the altitude range (the approach can be seen in [3]).

*Step 4.* Calculate the repeating factor $Q$ and inclination $i$ of each orbit. Since the orbit is sun-synchronous, so the inclination of orbit is determined. In this type of orbits, $Q$ is only a function of the orbital altitude (the approach can be seen in [4]).

*Step 5.* Generate the maximum cone angle $\phi$ corresponding to each orbital altitude using (9).

*Step 6.* Choose an orbit of which the revisit time and tilt angle are not calculated in the solution set $P$.

*Step 7.* Search all the tilt angles which are smaller than maximum cone angle $\phi$, and get the corresponding revisit rime (RT).

*Step 8.* If the RT is equal to $RT_{des}$, then record minimum tilt angle and the orbit in the set $P'$.

*Step 9.* Turn to Step 6 when an orbit of which the revisit time is not computed in the set $P$ exists.

*Step 10.* Output all the orbits in the solution set $P'$.

*Step 11.* Compare the solutions and choose the optimal orbits in set $P'$.

TABLE 4: The orbital parameters and mission characteristics of two satellites.

| Parameters | HJ-1A | P6 satellite |
|---|---|---|
| Orbit | Repeating Sun-synchronous | Repeating Sun-synchronous |
| Altitude (km) | 649.093 | 816.964 |
| Inclination (°) | 97.9486 | 98.6799 |
| Repeat cycle | 457 orbits within 31 days | 341 orbits within 24 days |
| Q value | 14 + 23/31 | 14 + 5/24 |
| Swath (km) | 720 | 141 |
| Tilting capacity (°) | 0 | 26 |

TABLE 5: The revisit time of two satellites.

| Parameters | HJ-1A | P6 satellite |
|---|---|---|
| $S_D$ (km) | 87.6915 | 117.522 |
| Apparent inclination $i'$ (°) | 101.756 | 102.654 |
| Effective swath on the equator $\omega'$ (km) | 735.426 | 830.166 |
| $n$ | 4 | 3 |
| Feasible offset multiple $R$ | $\{-4, -3, -2, -1, 0, 1, 2, 3, 4\}$ | $\{-3, -2, -1, 0, 1, 2, 3\}$ |
| Set of subcycles $S$ | $\{16, 12, 8, 4, 0, 27, 23, 19, 15\}$ | $\{9, 14, 19, 0, 5, 10, 15\}$ |
| Sorted set of subcycles $S'$ | $\{0, 4, 8, 12, 15, 16, 19, 23\}$ | $\{0, 5, 9, 10, 14, 15, 19\}$ |
| RT (days) | 4 | 5 |
| $\text{RT}_{\min}$ (days) | 1 | 1 |

## 4. Case Studies

*4.1. Calculation of Revisit Time.* The proposed technique is applied to compute the revisit time of some remote sensing satellites. HJ-1A [2] and P6 satellite [19] are chosen to verify the method. The orbital parameters and mission characteristics are presented in Table 4.

It could be found that the former satellite has a wide swath width while the latter satellite has a tilting capacity with a narrow swath. Utilizing the proposed technique, the revisit time of two satellites are calculated and presented in Table 5.

From Table 5, the maximum time gap between two successive observations of the HJ-1A satellite is 5 days. Actually, the revisit time of the HJ-1A is designed as 4 days [2], which can be seen in Table 4. At the same time, from Table 5, the revisit time of the P6 satellite is 5 days, which equals the values given in Table 4. These two cases demonstrate the accuracy of the proposed technique.

*4.2. Optimization Design for Remote Sensing Satellites.* In this section, some optimizations will be implemented on the P6 satellite. Since the altitude of the P6 satellite is 816.964 km, so we search all the feasible orbits in the altitude range from 810 km to 820 km. In Step 1, the swath width of satellite is 141 km, the required side-lap is set to 0%, and the required revisit time is equal to 5 nodal days. Table 6 illustrates the number of the orbits in set $P$ and $P'$ with different maximum repeat cycle $D_{\max}$.

In Step 2, the maximum repeat cycle is set to 100 days. Compute all the orbits in set $P$ and all the orbit solutions which meet the mission requirements will be obtained. Figure 7 illustrates the feasible orbit solutions within $\text{RT}_{\text{des}} = 5$ when $D_{\max} = 100$; Figure 8 illustrates the feasible orbit solutions within $\text{RT}_{\text{des}} = 5$ when $D_{\max} = 200$.

TABLE 6: The number of the orbits in sets $P$ and $P'$ with different $D_{\max}$.

| $D_{\max}$ (days) | Number of the orbits In the set $P$ | Number of the orbits In the set $P'$ |
|---|---|---|
| 20 | 4 | 2 |
| 40 | 16 | 14 |
| 60 | 34 | 32 |
| 80 | 58 | 56 |
| 100 | 88 | 86 |
| 200 | 356 | 354 |

TABLE 7: Parameters and characteristics of some orbit solutions within $\text{RT}_{\text{des}} = 5$.

| Solution parameters | Sol.1 | P6 satellite | Sol.2 |
|---|---|---|---|
| Altitude (km) | 814.967 | 816.96 | 813.917 |
| Inclination (°) | 98.6716 | 98.6799 | 98.6671 |
| Q value | 14 + 3/14 | 14 + 5/24 | 14 + 5/23 |
| Revisit time (days) | 5 | 5 | 5 |
| Minimum tilt angle (°) | 13.5074 | 15.604 | 16.2889 |

In Figures 7 and 8, some orbit solutions termed, respectively, as Sol.1, Sol.2 in all feasible orbit solutions are shown in Table 7.

The most remarkable solutions are Sol.1, Sol.2, and the case of P6 satellite. Of course, Sol.1 could be the best choice since its tilt angle is the smallest among three solutions.

Figures 9 and 10 illustrate all the orbits which can meet the conditions where revisit time is 4 days in the altitude range of 810 km–820 km.

TABLE 8: Parameters and characteristics of some orbit solutions within $RT_{des} = 4$.

| Solution parameters | Sol.3 | Sol.4 | Sol.5 | P6 satellite |
|---|---|---|---|---|
| Altitude (km) | 812.285 | 810.579 | 814.967 | 816.96 |
| Inclination (°) | 98.6602 | 98.653 | 98.6716 | 98.6799 |
| Q value | 14 + 2/9 | 14 + 5/22 | 14 + 3/14 | 14 + 5/24 |
| Revisit time (days) | 4 | 4 | 4 | 4 |
| Minimum tilt angle (°) | 20.4385 | 24.5137 | 25.4045 | 28.81 |

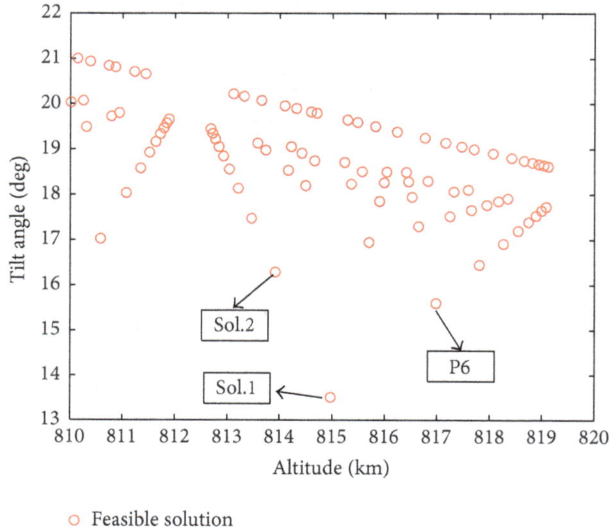


FIGURE 7: Feasible orbit solutions within $RT_{des} = 5$ when repeat cycle $D_{max} = 100$.

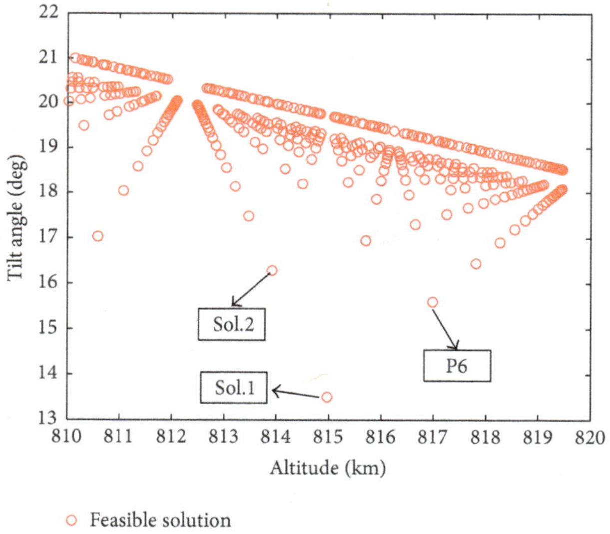


FIGURE 8: Feasible orbit solutions within $RT_{des} = 5$ when repeat cycle $D_{max} = 200$.

In Figures 7 and 8, some orbit solutions termed, respectively, as Sol.3, Sol.4, and Sol.5 in all feasible orbit solutions are shown in Table 8.

The most remarkable solutions are Sol.3, Sol.4, and Sol.5 and the case of p6 satellite. Of course, Sol.3 could be the best choice since its tilt angle is the smallest among three solutions.

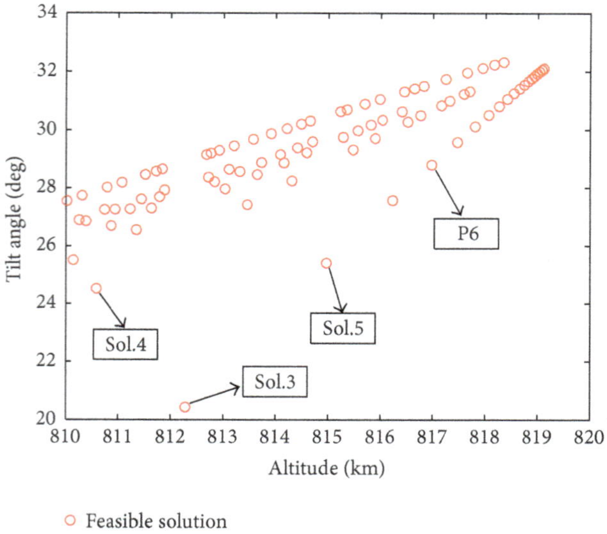


FIGURE 9: Feasible orbit solutions within $RT_{des} = 4$ when repeat cycle $D_{max} = 100$.

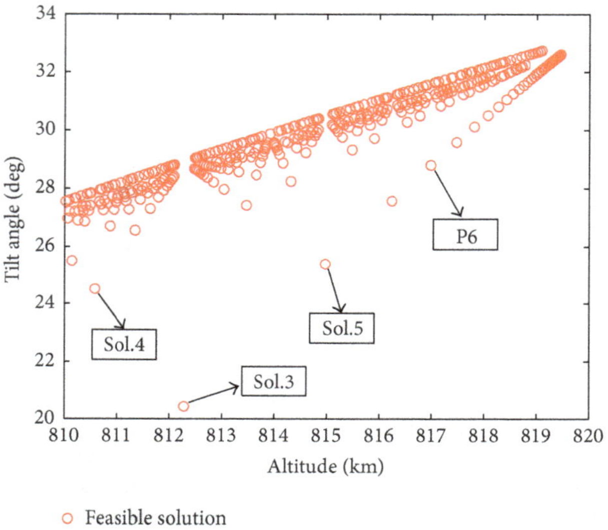


FIGURE 10: Feasible orbit solutions within $RT_{des} = 4$ when repeat cycle $D_{max} = 200$.

Comparing the feasible orbit solutions in Figures 7 and 10, there are some analyses as follows:

(1) With the increase of maximum repeat cycle, the number of feasible orbit solutions will increase as well. At the same time, the best orbit of which the desired revisit time is 5 days is not the same as the best orbit of which the desired revisit time is 4 days.

(2) Sol.3, Sol.4, and Sol.5 have different repeat cycles and ground tracks, but all of them can observe the same ground target within 4 days. According to the paper [13], when the orbital altitude is approximately the same, one orbit can transfer to another orbit using a low $\Delta v$ technique. This technology guarantees the fulfillment of several objectives in the course of the same mission, which is useful for different task.

## 5. Conclusion

A novel technique which takes into account the orbits and the payload is proposed to calculate the revisit time of a remote sensing satellite. The technique is discussed using the Bezout equation and takes the gravitational second zonal harmonic into consideration. The relationships between altitude and minimum tilt angle can be quickly obtained when the revisit time is determined. So, the proposed technique can make an ergodic search on the altitude and a lot of orbit solutions which meet the mission requirements will be got fast as well. Therefore, the approach can be used in the optimization design for remote sensing satellites. It will be helpful for orbit designer to calculate and select the best orbits in all solutions.

## Competing Interests

The authors declare that there is no conflict of interests regarding the publication of this paper.

## Acknowledgments

This work is supported by Natural Science Foundation of China under Grant no. 41571403, Joint Funds of Equipment Pre-Research and Ministry of Education of China, and Fundamental Research Funds for the Central Universities, China University of Geosciences (Wuhan), under Grants nos. CUG2017G01, CUG160207, and 1610491B21. The first author also acknowledges the encouragement and care by Qin Luo (father) and E. Liu (mother).

## References

[1] R. Sandau, K. Brieß, and M. D'Errico, "Small satellites for global coverage: potential and limits," *ISPRS Journal of Photogrammetry and Remote Sensing*, vol. 65, no. 6, pp. 492–504, 2010.

[2] T. Tan and F. Yang, "Hj-1a/1b constellation orbit design," *Spacecraft Engineering*, vol. 18, no. 6, pp. 27–30, 2009.

[3] R. J. Boain, *A-B-Cs of Sun-Synchronous Orbit Mission Design*, Jet Propulsion Laboratory, National Aeronautics and Space Administration, Pasadena, Calif, USA, 2005.

[4] D. Casey and J. Way, "Orbit selection for the EOS mission and its synergism implications," *IEEE Transactions on Geoscience and Remote Sensing*, vol. 29, no. 6, pp. 822–835, 1991.

[5] Q. U. Hong-Song, Z. Ye, and G. Jin, "Repeat sun-synchronous orbit design method based on q value selection," *Optics and Precision Engineering*, vol. 16, no. 9, pp. 1688–1694, 2008.

[6] M. Xu and L. Huang, "An analytic algorithm for global coverage of the revisiting orbit and its application to the CFOSAT satellite," *Astrophysics and Space Science*, vol. 352, no. 2, pp. 497–502, 2014.

[7] C. Circi, E. Ortore, F. Bunkheila, and C. Ulivieri, "Elliptical multi-sun-synchronous orbits for Mars exploration," *Celestial Mechanics and Dynamical Astronomy*, vol. 114, no. 3, pp. 215–227, 2012.

[8] X. Liu, H. Baoyin, and X. Ma, "Five special types of orbits around mars," *Journal of Guidance, Control, and Dynamics*, vol. 33, no. 4, pp. 1294–1301, 2010.

[9] C. Ulivieri and L. Anselmo, "Multi-sun-synchronous (MSS) orbits for earth observation," in *Proceedings of the AAS/AIAA Astrodynamics Conference*, pp. 123–133, San Diego, Calif, USA, August 1991.

[10] E. Ortore, C. Circi, F. Bunkheila, and C. Ulivieri, "Earth and Mars observation using periodic orbits," *Advances in Space Research*, vol. 49, no. 1, pp. 185–195, 2012.

[11] X. Fu, M. Wu, and Y. Tang, "Design and maintenance of low-earth repeat-ground-track successive-coverage orbits," *Journal of Guidance, Control, and Dynamics*, vol. 35, no. 2, pp. 686–691, 2012.

[12] T. Li, J. Xiang, Z. Wang, and Y. Zhang, "Circular revisit orbits design for responsive mission over a single target," *Acta Astronautica*, vol. 127, pp. 219–225, 2016.

[13] C. Circi, E. Ortore, and F. Bunkheila, "Satellite constellations in sliding ground track orbits," *Aerospace Science and Technology*, vol. 39, pp. 395–402, 2014.

[14] M. Wang, X. Luo, G. Dai, and X. Chen, "Application of latitude stripe division in satellite constellation coverage to ground," *International Journal of Aerospace Engineering*, vol. 2016, Article ID 4315026, 9 pages, 2016.

[15] H.-D. Kim, O.-C. Jung, and H. Bang, "A computational approach to reduce the revisit time using a genetic algorithm," in *Proceedings of the International Conference on Control, Automation and Systems (ICCAS '07)*, pp. 184–189, IEEE, Seoul, Republic of Korea, October 2007.

[16] G. Taini, A. Pietropaolo, and A. Notarantonio, "Criteria and trade-offs for leo orbit design," in *Proceedings of the IEEE Aerospace Conference Proceedings*, pp. 1–11, March 2008.

[17] B. Saboori, A. M. Bidgoli, and B. Saboori, "Multiobjective optimization in repeating sun-synchronous orbits design for remote-sensing satellites," *Journal of Aerospace Engineering*, vol. 27, no. 5, 2014.

[18] N. Pie and B. E. Schutz, "Subcycle analysis for ICESat's repeat groundtrack orbits and application to phasing maneuvers," *Journal of the Astronautical Sciences*, vol. 56, no. 3, pp. 325–340, 2008.

[19] M. J. Nadoushan and N. Assadian, "Repeat ground track orbit design with desired revisit time and optimal tilt," *Aerospace Science and Technology*, vol. 40, pp. 200–208, 2015.

[20] W. G. Rees, "Orbital subcycles for Earth remote sensing satellites," *International Journal of Remote Sensing*, vol. 13, no. 5, pp. 825–833, 1992.

[21] S. Lim and B. E. Schutz, "Repeat ground track analysis for the geoscience laser altimeter system," *Advances in the Astronautical Sciences*, vol. 93, pp. 1615–1622, 1996.

[22] M. Capderou, *Satellites: Orbits and Missions*, Springer, Paris, France, 2005.

# Error Analysis and Error Allocation for Turntable Systems Used in GyroWheel Calibration Tests

**Yuyu Zhao, Hui Zhao, Xin Huo, and Yu Yao**

*Control and Simulation Center, Harbin Institute of Technology, Harbin 150001, China*

Correspondence should be addressed to Yu Yao; yaoyu_academic@163.com

Academic Editor: Paul Williams

Calibration tests are of great importance to ensure rate-sensing accuracy of GyroWheel, an innovative attitude determination and control device. In the process of calibration tests, turntable errors are inevitable, which hinder the calibration accuracy and rate-sensing capability. Hence, error analysis for GyroWheel calibration tests is conducted, and the relationship between the calibration accuracy and the orientation error is established based on analytical derivation and numerical simulations. Subsequently, an error model of the turntable system is derived using rigid body kinematics, by which the relationship between the orientation error and turntable errors is described. According to sensitivity analysis and manufacturing capability, an error allocation method is proposed to determine the accuracy requirement of the test turntable, and the effectiveness of the proposed method is verified by repeated simulation tests. Based on the presented analysis and proposed method in this paper, the effects of various turntable errors on the calibration accuracy can be obtained quantitatively, and a theoretical basis for the determination of the turntable accuracy is provided, which are of great significance to guide the calibration tests and improve the calibration accuracy of GyroWheel.

## 1. Introduction

The development of small spacecrafts has received a lot of attention in recent years [1, 2]. As GyroWheel is an innovative attitude determination and control device, it offers the potential to meet the performance, mass, and cost requirements of small spacecrafts. It provides control torques about three axes while also measuring the spacecraft angular rates about the two axes perpendicular to the spin direction, which improves the integration and efficiency of attitude control system in small spacecrafts [3, 4].

The conception of GyroWheel is inspired by a dynamically tuned gyroscope (DTG). However, it has a larger rotor and tilt angles, as well as a time-varying spinning rate due to its multifunction capability. When the GyroWheel is used to measure angular rates, erroneous torques, which are often caused by design limitation and constructional deficiencies, act on the rotor of the GyroWheel. These imperfections give rise to precession of the rotor, resulting in measurement errors of angular rates [5–7].

To maintain high-accuracy measurement, calibration tests are of great importance in the application of the GyroWheel. The goal of calibration tests is to fully characterize the outputs of the inertial instruments so that a nonideal behavior can be modelled and compensated [5, 7–10]. Generally, calibration tests are carried out using a turntable to provide the intended orientation of the GyroWheel. Turntable errors, including position error, wobble error, orthogonality error, and intersection error, will inevitably affect the GyroWheel system in calibration tests [11]. Although the effects of turntable errors can be reduced by using a high-accuracy turntable, the costs of calibration tests will increase significantly. To reconcile the requirement of calibration accuracy and test costs, the effects of turntable errors on calibration accuracy should be analyzed, and error allocation methods for turntable system should be investigated in consideration of the sensitivities of these errors and the manufacturing capability in practical engineering before calibration tests are carried out.

Generally, error models of multi-DOF motion systems, such as machine tools and turntables, can be derived using rigid body kinematics [11–14]. Additionally, a few studies about error effect analysis for calibration tests have been conducted. An error analysis of precision centrifuge, an equipment used in calibration tests, was conducted, and its effect on the calibration accuracy of gyro accelerometers was discussed in [15, 16]. The effect of position error on calibration accuracy of inertial instruments has been analyzed in [17, 18]. As mentioned in [19], the models of a turntable's orthogonality error and horizontality error were established, and quantitative analyses of their effects on calibration accuracy were presented. However, the existing studies have mainly analyzed a single error component's effect on calibration accuracy, in which other error factors are not considered; obviously, this is not consistent with the actual situation. To the authors' knowledge, little attention has been focused on the error allocation problem for the multi-DOF motion systems. By assuming that each of the errors contributes equally to the overall error, the error allocation problem for a 6-DOF parallel manipulator was discussed in [20, 21]. However, the assumption of equivalent effects of the errors is inappropriate for the turntable system. Although orthogonal experimental design has been used to determine the significant level of each error factor for a 3-DOF robotic mechanism [22], it is difficult to obtain a quantitative error allocation result. There were no relevant studies on the error allocation method for the turntable system used in calibration tests.

Motivated by these facts, the error analysis and error allocation problems of the turntable used in GyroWheel's calibration tests are investigated in this paper. According to the multiposition calibration theory and rate-sensing model of the GyroWheel, error analysis for GyroWheel calibration tests is conducted quantitatively. With numerical simulations, the relationships between the rate-sensing errors of the GyroWheel and the orientation errors are obtained. The orientation error due to manufacture imperfections and design limitations of the turntable system is modelled based on rigid body kinematics, with consideration of the interaction of these turntable errors. Then an error allocation method for the turntable system based on sensitivity analysis is proposed, and the effectiveness of the proposed method is verified by repeated simulation tests.

The remainder of this paper is as follows. In Section 2, the GyroWheel calibration theory is described, and the effect of orientation error on the calibration accuracy is analyzed. In Section 3, the error model of a two-axis turntable system is established based on rigid body kinematics, which develops a relationship between the turntable error components and the orientation error of the GyroWheel. In Section 4, sensitivities of the orientation error to the individual error components and the manufacturing capability are discussed. On this basis, an error allocation method is given to determine the accuracy requirement of the test turntable. And several concluding remarks are given in Section 5.

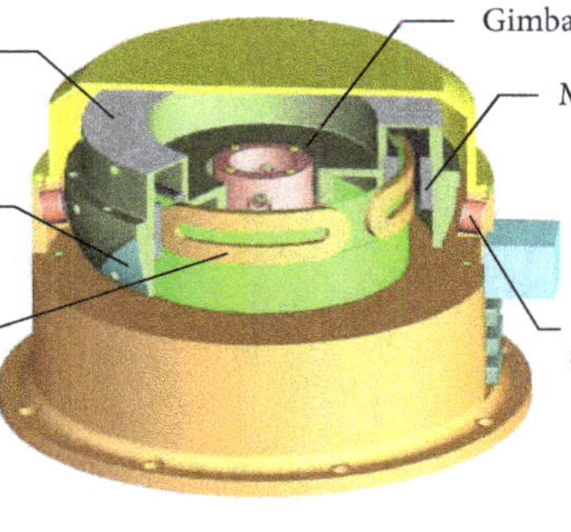


FIGURE 1: Cutaway view of a GyroWheel.

## 2. GyroWheel Calibration Theory and Error Analysis for Calibration Tests

*2.1. Overview of GyroWheel Calibration Theory.* The GyroWheel is an innovative attitude determination and control instrument. A cutaway isometric view of the GyroWheel is shown in Figure 1. When the GyroWheel is used to measure angular rates, erroneous torques give rise to precession of the rotor resulting in drift errors. In the null tilt condition, the rate-sensing equations of the GyroWheel are given in the following equation, with consideration of the drift errors [5]:

$$\begin{aligned}\omega_x &= k_y i_y - D(x)_F - D(x)_x a_x - D(x)_y a_y - D(x)_z a_z \\ &\quad - D(x)_{xx} a_x^2 - D(x)_{zz} a_z^2 - D(x)_{xy} a_x a_y, \\ \omega_y &= k_x i_x - D(y)_F - D(y)_x a_x - D(y)_y a_y - D(y)_z a_z \\ &\quad - D(y)_{yy} a_y^2 - D(y)_{zz} a_z^2 - D(y)_{yx} a_x a_y,\end{aligned} \tag{1}$$

where $D(x), D(y)$ represent the drift errors of $x, y$ axes, $D(x)_F, D(y)_F$ are $g$ insensitive terms, $D(x)_i, D(y)_i, i = x, y, z$, are $g$ sensitive error coefficients, and $D(x)_{ij}, ij = xx, zz, xy$, $D(y)_{ij}, ij = yy, zz, yx$, are $g^2$ sensitive error coefficients. $a_x$, $a_y$, $a_z$ represent acceleration components of the gravity vector, $\omega_x, \omega_y$ are external angular velocities, $k_x, k_y$ are torque scale factors of the GyroWheel, and $i_x, i_y$ are the currents in the torque coils.

In an effort to improve the rate-sensing accuracy of the GyroWheel, multiposition tests are performed to calibrate the GyroWheel. Multiposition tests make use of a two-axis turntable to provide the intended orientation for the GyroWheel. A schematic representation of a turntable is shown in Figure 2. The earth's rotation rate and gravitational acceleration are regarded as the nominal inputs

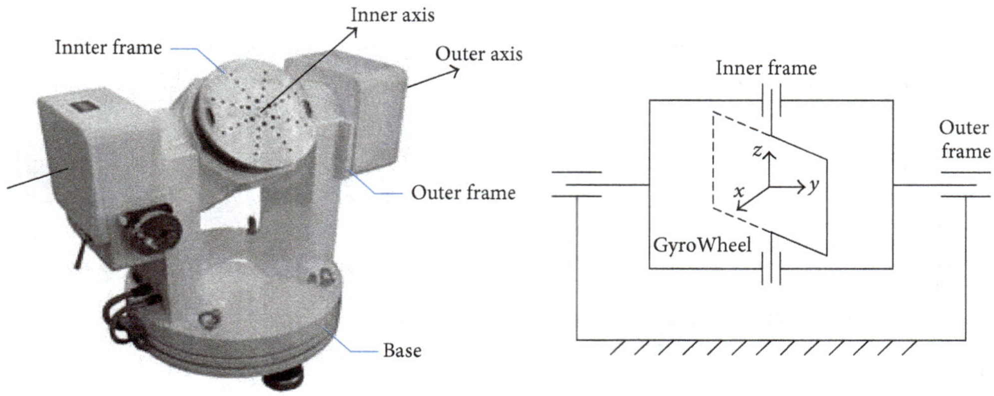


FIGURE 2: Schematic representation of a two-axis turntable.

of the GyroWheel. At each position, an equation expressed as (1) can be obtained using the inputs and outputs of the GyroWheel. Therefore, the error coefficients can be calculated when enough tests of different positions are conducted.

Take the $x$-axis model of the GyroWheel as an example. According to the rate-sensing model of the GyroWheel and multiposition calibration theory [10], we have

$$\begin{bmatrix} k_y i_{y1} \\ k_y i_{y2} \\ \vdots \\ k_y i_{yn} \end{bmatrix} - \begin{bmatrix} \omega_{x1} \\ \omega_{x2} \\ \vdots \\ \omega_{xn} \end{bmatrix} = \begin{bmatrix} 1 & a_{x1} & a_{y1} & a_{z1} & a_{x1}^2 & a_{z1}^2 & a_{x1}a_{y1} \\ 1 & a_{x2} & a_{y2} & a_{z2} & a_{x2}^2 & a_{z2}^2 & a_{x2}a_{y2} \\ \vdots & \vdots & \vdots & \vdots & \vdots & \vdots & \vdots \\ 1 & a_{xn} & a_{yn} & a_{zn} & a_{xn}^2 & a_{zn}^2 & a_{xn}a_{yn} \end{bmatrix} \begin{bmatrix} D(x)_F \\ D(x)_x \\ D(x)_y \\ D(x)_z \\ D(x)_{xx} \\ D(x)_{zz} \\ D(x)_{xy} \end{bmatrix}, \quad (2)$$

where $n$ is the number of test positions, $i_{yj}$, $j = 1, 2, \ldots, n$, represent the currents in the torque coils, $\omega_{xi}$, $i = 1, 2, \ldots, n$, represent the earth rate components of each test position, and $a_{xi}, a_{yi}, a_{zi}$, $i = 1, 2, \ldots, n$, represent the gravitational acceleration components of each test position.

The initial orientation of the GyroWheel axes is north-west-up. Define that $\omega_e$ is the earth's rotation rate, $\lambda$ is the latitude, $\phi_{1i}, \phi_{2i}$ are the rotation angles of the turntable, and the subscript $i$ means the $i$th test position. The gravitational acceleration components and the earth rate components can be expressed as follows:

$$\begin{aligned} a_{xi} &= -\sin\phi_{1i}\cos\phi_{2i}, \\ a_{yi} &= \sin\phi_{1i}\sin\phi_{2i}, \\ a_{zi} &= \cos\phi_{1i}, \\ \omega_{xi} &= \omega_e\cos\phi_{2i}(\cos\phi_{1i}\cos\lambda - \sin\phi_{1i}\sin\lambda), \\ \omega_{yi} &= \omega_e\sin\phi_{2i}(-\cos\phi_{1i}\cos\lambda + \sin\phi_{1i}\sin\lambda), \\ \omega_{zi} &= \omega_e(\sin\phi_{1i}\cos\lambda + \cos\phi_{1i}\sin\lambda). \end{aligned} \quad (3)$$

Denote

$$\mathbf{b}_1 = \begin{bmatrix} k_y i_{y1} \\ k_y i_{y2} \\ \vdots \\ k_y i_{yn} \end{bmatrix}, \quad \mathbf{b}_2 = \begin{bmatrix} \omega_{x1} \\ \omega_{x2} \\ \vdots \\ \omega_{xn} \end{bmatrix}, \quad \mathbf{A} = \begin{bmatrix} 1 & a_{x1} & a_{y1} & a_{z1} & a_{x1}^2 & a_{z1}^2 & a_{x1}a_{y1} \\ 1 & a_{x2} & a_{y2} & a_{z2} & a_{x2}^2 & a_{z2}^2 & a_{x2}a_{y2} \\ \vdots & \vdots & \vdots & \vdots & \vdots & \vdots & \vdots \\ 1 & a_{xn} & a_{yn} & a_{zn} & a_{xn}^2 & a_{zn}^2 & a_{xn}a_{yn} \end{bmatrix}, \quad \boldsymbol{\beta} = \begin{bmatrix} D(x)_F \\ D(x)_x \\ D(x)_y \\ D(x)_z \\ D(x)_{xx} \\ D(x)_{zz} \\ D(x)_{xy} \end{bmatrix}. \quad (4)$$

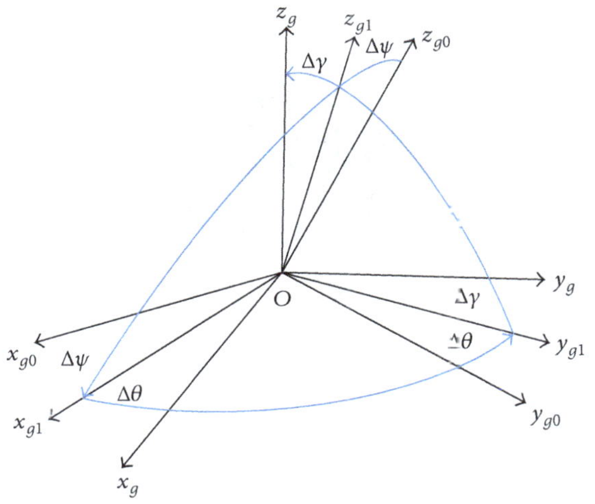


FIGURE 3: Orientation error of GyroWheel axes. $Ox_{g0}y_{g0}z_{g0}$: the intended orientation of the GyroWheel axes. $Oz_g y_g z_g$: the actual orientation of the GyroWheel axes.

Then (2) can be rewritten as

$$\mathbf{b}_1 - \mathbf{b}_2 = \mathbf{A}\boldsymbol{\beta}. \tag{5}$$

The calibration coefficient vector $\boldsymbol{\beta}$ can be solved utilizing the method of least squares:

$$\widehat{\boldsymbol{\beta}}_1 = \left(\mathbf{A}^T\mathbf{A}\right)^{-1}\mathbf{A}^T\left(\mathbf{b}_1 - \mathbf{b}_2\right) \tag{6}$$

*2.2. Error Analysis for GyroWheel Calibration Tests.* Due to manufacture imperfections and design limitations of the turntable system, the orientation of the GyroWheel axes will be deviated from the intended one. As a result, there are deviations between the actual earth rate components and the ideal earth rate components, as well as the gravitational acceleration components. Use $\Delta\psi, \Delta\theta$, $\Delta\gamma$ to denote rotational errors about $y$-, $z$-, and $x$-axes, respectively (Figure 3).

With the assumption of small angular errors, the transformation matrix describing the orientation error of the GyroWheel with respect to its ideal orientation is given below:

$$\Delta\mathbf{R} = \begin{bmatrix} 1 & -\Delta\theta & \Delta\psi \\ \Delta\theta & 1 & -\Delta\gamma \\ -\Delta\psi & \Delta\gamma & 1 \end{bmatrix}. \tag{7}$$

The actual earth rate components and gravitational acceleration components can be calculated as follows:

$$\begin{aligned} \left[\omega'_{xi}\ \ \omega'_{yi}\ \ \omega'_{zi}\right] &= \left[\omega_{xi}\ \ \omega_{yi}\ \ \omega_{zi}\right]\Delta\mathbf{R}, \\ \left[a'_{xi}\ \ a'_{yi}\ \ a'_{zi}\right] &= \left[a_{xi}\ \ a_{yi}\ \ a_{zi}\right]\Delta\mathbf{R}. \end{aligned} \tag{8}$$

In the presence of the orientation error, the regression model of multiposition calibration is given by

$$\mathbf{b}_1 - \mathbf{b}'_2 = \mathbf{A}'\boldsymbol{\beta}, \tag{9}$$

where

$$\mathbf{b}'_2 = \begin{bmatrix} \omega'_{x1} \\ \omega'_{x2} \\ \vdots \\ \omega'_{xn} \end{bmatrix},$$

$$\mathbf{A}' = \begin{bmatrix} 1 & a'_{x1} & a'_{y1} & a'_{z1} & a'^2_{x1} & a'^2_{z1} & a'_{x1}a'_{y1} \\ 1 & a'_{x2} & a'_{y2} & a'_{z2} & a'^2_{x2} & a'^2_{z2} & a'_{x2}a'_{y2} \\ \vdots & \vdots & \vdots & \vdots & \vdots & \vdots & \vdots \\ 1 & a'_{xn} & a'_{yn} & a'_{zn} & a'^2_{xn} & a'^2_{zn} & a'_{xn}a'_{yn} \end{bmatrix}. \tag{10}$$

Then the true value of the calibration coefficient vector can be calculated as follows:

$$\widehat{\boldsymbol{\beta}}_2 = \left(\mathbf{A}'^T\mathbf{A}'\right)^{-1}\mathbf{A}'^T\left(\mathbf{b}_1 - \mathbf{b}'_2\right). \tag{11}$$

The actual calibration results are given by (6), and the deviation of the calibration results caused by the orientation error can be expressed as

$$\Delta\boldsymbol{\beta} = \widehat{\boldsymbol{\beta}}_1 - \widehat{\boldsymbol{\beta}}_2. \tag{12}$$

The deviation $\Delta\boldsymbol{\beta}$ results in GyroWheel's rate-sensing error $\Delta\omega_x$; the rate-sensing error vector at the test positions can be calculated:

$$\Delta\widehat{\boldsymbol{\omega}}_x = \left(\mathbf{b}_1 - \mathbf{A}\widehat{\boldsymbol{\beta}}_2\right) - \mathbf{b}'_2. \tag{13}$$

Define

$$\Delta\omega_x = \left\|\Delta\widehat{\boldsymbol{\omega}}_x\right\|_\infty. \tag{14}$$

$\Delta\omega_x$ represents the rate-sensing error of $x$-axis. The rate-sensing error of $y$-axis, denoted by $\Delta\omega_y$, can be calculated in the same way. Obviously, the rate-sensing errors $\Delta\omega_x$ and $\Delta\omega_y$ are affected by the orientation error described by $\Delta\psi, \Delta\theta, \Delta\gamma$. If the values of $\Delta\psi, \Delta\theta, \Delta\gamma$ are given, the corresponding values of the rate-sensing errors $\Delta\omega_x$ and $\Delta\omega_y$ can be calculated by several matrix operations. To facilitate an error analysis, a numerical method is utilized. A series of rate-sensing errors are obtained with numerical simulations when a series of $\Delta\psi, \Delta\theta, \Delta\gamma$ are given. The relationships between the rate-sensing errors $\Delta\omega_x, \Delta\omega_y$ and the orientation error described by $\Delta\psi, \Delta\theta, \Delta\gamma$ are expressed in a visualized way, as shown

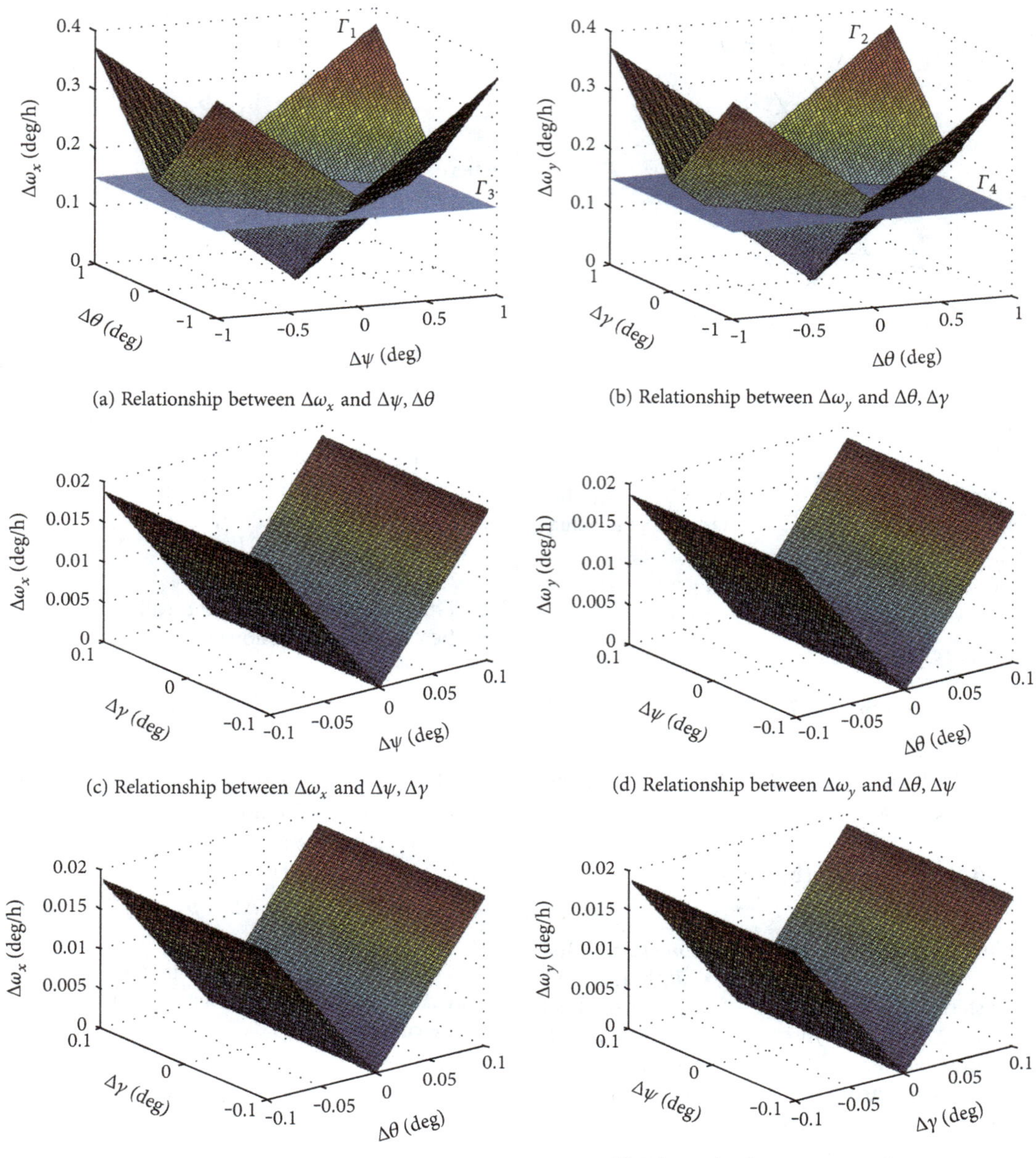


(a) Relationship between $\Delta\omega_x$ and $\Delta\psi, \Delta\theta$

(b) Relationship between $\Delta\omega_y$ and $\Delta\theta, \Delta\gamma$

(c) Relationship between $\Delta\omega_x$ and $\Delta\psi, \Delta\gamma$

(d) Relationship between $\Delta\omega_y$ and $\Delta\theta, \Delta\psi$

(e) Relationship between $\Delta\omega_x$ and $\Delta\theta, \Delta\gamma$

(f) Relationship between $\Delta\omega_y$ and $\Delta\gamma, \Delta\psi$

FIGURE 4: Relationships between rate-sensing errors and orientation error.

in Figure 4. The two surfaces describing the relationships in Figures 4(a) and 4(b) are denoted by $\Gamma_1, \Gamma_2$. From the simulation results, the following conclusions can be made:

(1) As seen in Figures 4(a)–4(f), the rate-sensing error of $x$-axis$\Delta\omega_x$ is caused by $\Delta\psi$, $\Delta\theta$, and it is not affected by $\Delta\gamma$. The rate-sensing error of $y$-axis$\Delta\omega_y$ is caused by $\Delta\theta$, $\Delta\gamma$, and it is not affected by $\Delta\psi$.

(2) As seen in Figure 4(a), the surface $\Gamma_1$ is symmetric about the planes $\Delta\psi = 0$ and $\Delta\theta = 0$. As seen in Figure 4(b), the surface $\Gamma_2$ is symmetric about the planes $\Delta\theta = 0$ and $\Delta\gamma = 0$. Therefore, $\Delta\omega_x$ is an even function of $\Delta\psi$, $\Delta\theta$, and $\Delta\omega_y$ is an even function of $\Delta\theta$, $\Delta\gamma$.

(3) As seen in Figures 4(a)–4(f), $\Delta\omega_x$ rises with the increasing of $|\Delta\psi|, |\Delta\theta|$, and $\Delta\omega_y$ rises with the increasing of $|\Delta\theta|, |\Delta\gamma|$.

Use $\omega_m$ to denote the required calibration accuracy of the GyroWheel; that is, $\Delta\omega_x \leq \omega_m, \Delta\omega_y \leq \omega_m$. By drawing two planes defined by the implicit equations $\Gamma_3 : \Delta\omega_x = \omega_m$

and $\Gamma_4 : \Delta\omega_y = \omega_m$ in Figures 4(a) and 4(b), respectively, two lines of intersection are obtained, and the requirement of the orientation error can be determined easily and is denoted by

$$\begin{aligned} |\Delta\psi| &\le I_\psi, \\ |\Delta\theta| &\le I_\theta, \\ |\Delta\gamma| &\le I_\gamma. \end{aligned} \tag{15}$$

*2.3. Example.* A GyroWheel with the required calibration accuracy of 0.01°/h is taken as an example. The nominal values of the calibration coefficients are listed in Table 1. According to the above analysis, the line of intersection between $\Gamma_1$ and $\Gamma_3$, and the line of intersection between $\Gamma_2$ and $\Gamma_4$ are given in Figure 5.

As seen in Figure 5, the rectangular areas in black represent the ranges of $\Delta\psi, \Delta\theta, \Delta\gamma$, in which the requirement of the calibration accuracy can be met. To maximize the areas of the two rectangles in Figure 5, the requirement of the orientation error is determined as follows:

$$\begin{aligned} |\Delta\psi| &\le 0.027^\circ, \\ |\Delta\theta| &\le 0.027^\circ, \\ |\Delta\gamma| &\le 0.027^\circ. \end{aligned} \tag{16}$$

## 3. Error Analysis of Turntable System

*3.1. Overview of Turntable Errors.* There are four main sources of errors in a turntable system; they are position error, wobble error, orthogonality error, and intersection error.

Position error is defined as the difference between the actual and the intended rotation angles of the turntable's rotating shaft, which is influenced by servo precision and measurement accuracy of the turntable system.

Wobble error can be divided into axial wobble error, radial wobble error, and angular wobble error, as illustrated in Figure 6, where $\Delta s_a$ is the axial wobble error, $\Delta s_r$ is the radial wobble error, and $\Delta\alpha$ is the angular wobble error. In practical engineering, $\Delta s_a$ and $\Delta s_r$ are small enough to be ignored; thus, wobble error refers specifically to angular wobble error $\Delta\alpha$ in this paper [11, 23, 24].

Orthogonality error and intersection error are induced by mechanical imperfections of the turntable structure and misalignment of turntable elements.

For the two-axis turntable given in Figure 2, a summary of the error components involved is shown in Table 2.

*3.2. Error Modeling for Turntable System.* The error model of the turntable system can be developed based on rigid-body kinematics. The two-axis turntable system, as shown in Figure 2, consists of three bodies: the base, the outer frame, and the inner frame. In addition, the GyroWheel being attached to the inner frame should be considered. In an effort to derive the error model of the turntable, five coordinate systems are introduced. They are the base coordinate system $\mathcal{F}_0$: $O_0X_0Y_0Z_0$, the outer axis coordinate system $\mathcal{F}_1$: $O_1X_1Y_1Z_1$, the reference coordinate system

TABLE 1: Nominal values of calibration coefficients.

| | | | |
|---|---|---|---|
| $D(z)_F$ (deg/h) | −93.3 | $D(y)_F$ (deg/h) | 105.3 |
| $D(z)_x$ (deg/h/g) | −43.2 | $D(y)_x$ (deg/h/g) | 2.2 |
| $D(z)_y$ (deg/h/g) | 3.5 | $D(y)_y$ (deg/h/g) | −33.2 |
| $D(x)_z$ (deg/h/g) | −1.2 | $D(y)_z$ (deg/h/g) | −0.1 |
| $D(x)_{xx}$ (deg/h/$g^2$) | 2.1 | $D(y)_{yy}$ (deg/h/$g^2$) | −10.6 |
| $D(x)_{zz}$ (deg/h/$g^2$) | −0.089 | $D(y)_{zz}$ (deg/h/$g^2$) | −10.38 |
| $D(x)_{xy}$ (deg/h/$g^2$) | 17.4 | $D(y)_{xy}$ (deg/h/$g^2$) | 47.8 |

of the inner axis $\mathcal{F}_2$: $O_2X_2Y_2Z_2$, the inner axis coordinate system $\mathcal{F}_3$: $O_3X_3Y_3Z_3$ and the GyroWheel coordinate system $\mathcal{F}_4$: $O_gX_gY_gZ_g$. $\mathcal{F}_0$, $\mathcal{F}_1$, $\mathcal{F}_3$, and $\mathcal{F}_4$ are four body-fixed coordinate systems, which are attached to the base, the outer frame, the inner frame, and the GyroWheel, respectively. $\mathcal{F}_2$ is an interim coordinate system; the relation between $\mathcal{F}_2$ and $\mathcal{F}_1$ is determined by the orthogonality error and intersection error of the turntable system. An illustration of these coordinate systems is shown in Figure 7.

Actually, the base coordinate system $\mathcal{F}_0$ is fixed with respect to the earth. The outer axis of the turntable is aligned along the $y$-axis of the outer axis coordinate system $\mathcal{F}_1$. The inner axis of the turntable is aligned along the $z$-axis of the inner axis coordinate system $\mathcal{F}_3$. Since the GyroWheel is attached to the inner frame of the turntable, the GyroWheel coordinate system $\mathcal{F}_4$ is coincident with $\mathcal{F}_3$. The relations between these coordinate systems and the error components of the turntable system are expressed in Table 3. $\phi_1, \phi_2$ are the rotation angles about the outer axis and inner axis, respectively.

The relative position and orientation of one coordinate system with respect to another coordinate system can be modeled using a homogeneous transformation matrix (HTM), denoted as ${}^iT_j$, $i, j = 0, 1, 2, 3, 4$. The presuperscript represents the coordinate system we are transferring from, and the post-subscript represents the coordinate system we want the results to be represented in.

Hence, the transform relation between $\mathcal{F}_1$ and $\mathcal{F}_0$ can be calculated with the assumption of small angular errors:

$${}^0\mathrm{T}_1 = \begin{bmatrix} \cos\phi_1 & 0 & \sin\phi_1 & 0 \\ 0 & 1 & 0 & 0 \\ -\sin\phi_1 & 0 & \cos\phi_1 & 0 \\ 0 & 0 & 0 & 1 \end{bmatrix} \cdot \begin{bmatrix} 1 & -\Delta\alpha_{z1} & \Delta\phi_{y1} & 0 \\ \Delta\alpha_{z1} & 1 & -\Delta\alpha_{x1} & 0 \\ -\Delta\phi_{y1} & \Delta\alpha_{x1} & 1 & 0 \\ 0 & 0 & 0 & 1 \end{bmatrix}. \tag{17}$$

The transform relation between $\mathcal{F}_2$ and $\mathcal{F}_1$ can be calculated as follows:

$${}^1\mathrm{T}_2 = \begin{bmatrix} 1 & 0 & \Delta\varepsilon_y & 0 \\ 0 & 1 & -\Delta\varepsilon_x & 0 \\ -\Delta\varepsilon_y & \Delta\varepsilon_x & 1 & 0 \\ 0 & 0 & 0 & 1 \end{bmatrix} \cdot \begin{bmatrix} 1 & 0 & 0 & \Delta\eta \\ 0 & 1 & 0 & 0 \\ 0 & 0 & 1 & 0 \\ 0 & 0 & 0 & 1 \end{bmatrix}. \tag{18}$$

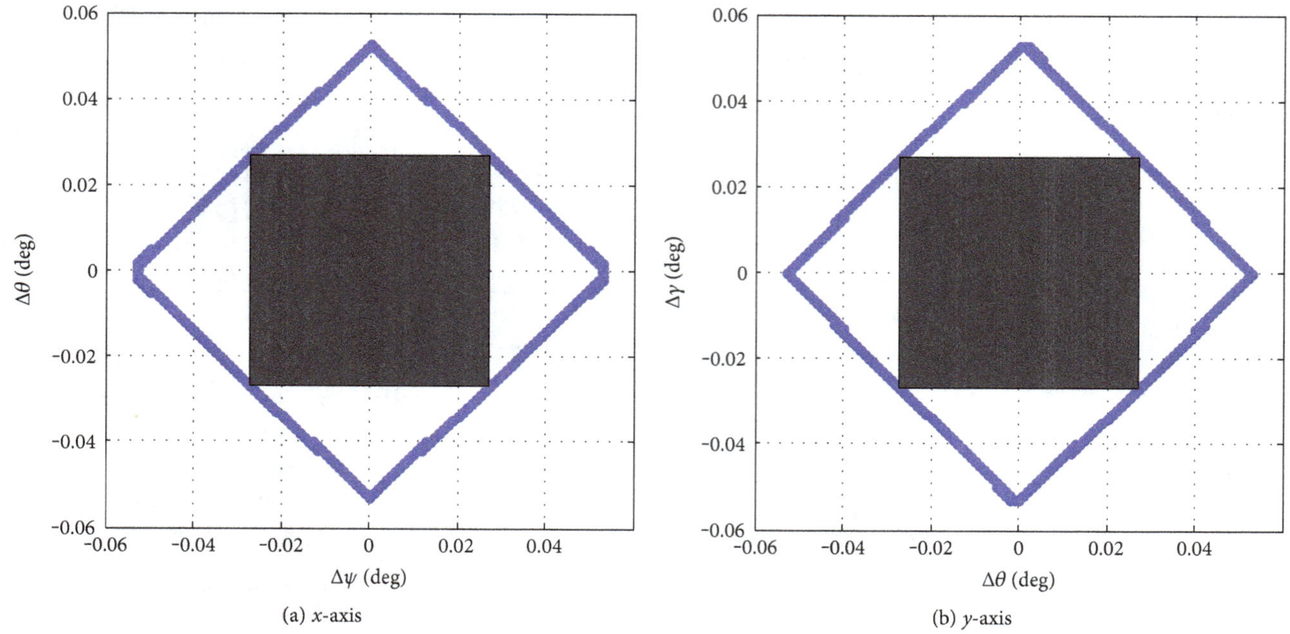


FIGURE 5: Lines of intersection.

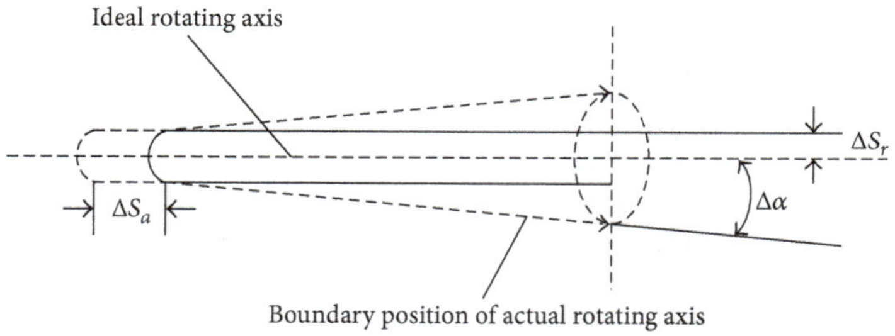


FIGURE 6: Schematic of a wobble error.

TABLE 2: Error components of a two-axis turntable system.

| Location of error components | Symbols | Definitions |
|---|---|---|
| Outer axis | $\Delta\phi_{y1}$ | Position error of outer axis |
| | $\Delta\alpha_{x1}$ | Wobble error component of outer axis about $x$-axis |
| | $\Delta\alpha_{z1}$ | Wobble error component of outer axis about $z$-axis |
| Inner axis | $\Delta\phi_{z2}$ | Position error of inner axis |
| | $\Delta\alpha_{x2}$ | Wobble error component of inner axis about $x$-axis |
| | $\Delta\alpha_{y2}$ | Wobble error component of inner axis about $y$-axis |
| Between outer and inner axes | $\Delta\varepsilon_x$ | Orthogonality error component about $x$-axis |
| | $\Delta\varepsilon_y$ | Orthogonality error component about $y$-axis |
| | $\Delta\eta$ | Intersection error |

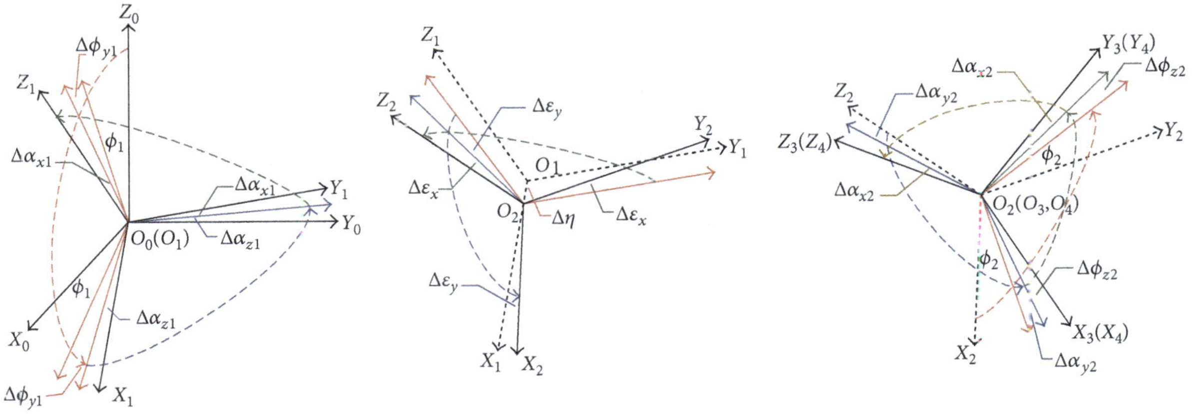


Figure 7: An illustration of the coordinate systems.

Table 3: Rotational and translational motion relations between adjacent coordinate systems.

| $i$ | $\mathscr{F}_i \rightarrow \mathscr{F}_{i+1}$ |
|---|---|
| 0 | A rotation $\phi_1$ about $y$-axis, followed by a rotation $\Delta\phi_{y1}$ about $y$-axis<br>A rotation $\Delta\alpha_{z1}$ about $z$-axis and then a rotation $\Delta\alpha_{x1}$ about $x$-axis |
| 1 | A translation $\Delta\eta$ along $x$-axis, followed by a rotation $\Delta\varepsilon_y$ about $y$-axis, and then a rotation $\Delta\varepsilon_x$ about $x$-axis |
| 2 | A rotation $\phi_2$ about $z$-axis, followed by a rotation $\Delta\alpha_{y2}$ about $y$-axis; a rotation $\Delta\phi_{z2}$ about $z$-axis, and then a rotation $\Delta\alpha_{x2}$ about $x$-axis |

And the transform relation between $\mathscr{F}_3$ and $\mathscr{F}_2$ is given below:

$$^2\mathbf{T}_3 = \begin{bmatrix} \cos\phi_2 & -\sin\phi_2 & 0 & 0 \\ \sin\phi_2 & \cos\phi_2 & 0 & 0 \\ 0_1 & 0 & 1 & 0 \\ 0 & 0 & 0 & 1 \end{bmatrix} \cdot \begin{bmatrix} 1 & -\Delta\phi_{z2} & \Delta\alpha_{y2} & 0 \\ \Delta\phi_{z2} & 1 & -\Delta\alpha_{x2} & 0 \\ -\Delta\alpha_{y2} & \Delta\alpha_{x2} & 1 & 0 \\ 0 & 0 & 0 & 1 \end{bmatrix}. \tag{19}$$

As stated above, the transform relation between $\mathscr{F}_4$ and $\mathscr{F}_3$ is written in the following form:

$$^3\mathbf{T}_4 = \mathbf{I}_{4\times 4}. \tag{20}$$

We introduce homogeneous transformation matrixes $\mathbf{T}_d$, $\mathbf{T}_t$ to denote the ideal and actual transform relations between the GyroWheel and the base coordinate system. $\mathbf{T}_d$ is obtained through a series of perfect rotations $(\phi_1, \phi_2)$ and is given by the following:

$$\mathbf{T}_d = \begin{bmatrix} \cos\phi_1 & 0 & \sin\phi_1 & 0 \\ 0 & 1 & 0 & 0 \\ -\sin\phi_1 & 0 & \cos\phi_1 & 0 \\ 0 & 0 & 0 & 1 \end{bmatrix} \cdot \begin{bmatrix} \cos\phi_2 & -\sin\phi_2 & 0 & 0 \\ \sin\phi_2 & \cos\phi_2 & 0 & 0 \\ 0_1 & 0 & 1 & 0 \\ 0 & 0 & 0 & 1 \end{bmatrix}. \tag{21}$$

In presence of the turntable errors, $\mathbf{T}_t$ can be calculated by matrix multiplication:

$$\mathbf{T}_t = \prod_{i=0}^{3} {}^i\mathbf{T}_{i+1}. \tag{22}$$

The top left corner $3 \times 3$ matrixes of $\mathbf{T}_d$, $\mathbf{T}_t$ denoted by $\mathbf{R}_d$, $\mathbf{R}_t$ represent the ideal and actual direction cosine matrixes. Since $\mathbf{R}_d$, $\mathbf{R}_t$ are orthogonal matrixes, the orientation error of the GyroWheel caused by the turntable errors can be expressed as follows:

$$\Delta\mathbf{R} = \mathbf{R}_d^T \mathbf{R}_t. \tag{23}$$

According to (23) and (7), we have the following:

$$\begin{aligned} \Delta\psi &= \Delta R_{13}, \\ \Delta\theta &= \Delta R_{21}, \\ \Delta\gamma &= \Delta R_{32}, \end{aligned} \tag{24}$$

where $\Delta R_{ij}$ represents the element in the $i$th row and $j$th column of matrix $\Delta\mathbf{R}$. Substituting (15)–(21) into (22) yields the following expression:

$$\begin{aligned} \Delta\psi &= f_\psi(\phi, \chi), \\ \Delta\theta &= f_\theta(\phi, \chi), \\ \Delta\gamma &= f_\gamma(\phi, \chi), \end{aligned} \tag{25}$$

where $f_\psi, f_\theta, f_\gamma$ are nonlinear functions of the rotation angles and the turntable error components, $\phi = [\phi_1 \phi_2]^T$ represents the rotation angle vector, and $\chi = [\Delta\phi_{y1} \quad \Delta\phi_{z2} \quad \Delta\alpha_{x1} \; \Delta\alpha_{z1} \; \Delta\alpha_{x2} \; \Delta\alpha_{y2} \; \Delta\varepsilon_x \; \Delta\varepsilon_y]^T$ represents the turntable error vector. According to (25), the intersection error has no effect on the orientation error of the GyroWheel.

To simplify the analysis, (25) is rewritten into the following form with small angle approximation of the errors:

$$\begin{aligned}
\Delta\psi &= f_\psi(\phi_2, \chi_\psi) = \Delta\phi_{y1}\cos\phi_2 - \Delta\alpha_{x1}\sin\phi_2 + \Delta\alpha_{y2} \\
&\quad - \Delta\varepsilon_x\sin\phi_2 + \Delta\varepsilon_y\cos\phi_2, \\
\Delta\theta &= f_\theta(\chi_\theta) = \Delta\phi_{z2} + \Delta\alpha_{z1}, \\
\Delta\gamma &= f_\gamma(\phi_2, \chi_\gamma) = \Delta\phi_{y1}\sin\phi_2 + \Delta\alpha_{x1}\cos\phi_2 + \Delta\alpha_{x2} \\
&\quad + \Delta\varepsilon_x\cos\phi_2 + \Delta\varepsilon_y\sin\phi_2,
\end{aligned} \tag{26}$$

where

$$\begin{aligned}
\chi_\gamma &= [\Delta\phi_{y1} \quad \Delta\alpha_{x1} \quad \Delta\alpha_{x2} \quad \Delta\varepsilon_x \quad \Delta\varepsilon_y]^T, \\
\chi_\theta &= [\Delta\phi_{z2} \quad \Delta\alpha_{z1}]^T, \\
\chi_\psi &= [\Delta\phi_{y1} \quad \Delta\alpha_{x1} \quad \Delta\alpha_{y2} \quad \Delta\varepsilon_x \quad \Delta\varepsilon_y]^T.
\end{aligned} \tag{27}$$

## 4. Error Allocation of the Turntable System

*4.1. Sensitivity Analysis and Constraints Analysis.* To ensure the orientation accuracy and perform error allocation of the turntable system, sensitivities of the orientation error to the individual errors and the manufacturing capability should be considered.

According to (23), sensitivities of the orientation error can be analyzed:

$$\begin{aligned}
C_{\psi i} &= \left|\frac{\partial f_\psi}{\partial \chi_i}\right|, \\
C_{\theta i} &= \left|\frac{\partial f_\theta}{\partial \chi_i}\right|, \\
C_{\gamma i} &= \left|\frac{\partial f_\gamma}{\partial \chi_i}\right|, \\
i &= 1, 2, \ldots, 8.
\end{aligned} \tag{28}$$

$\chi_i$ represents the $i$th element of the turntable error vector $\chi$, and $C_{\psi i}, C_{\theta i}, C_{\gamma i}$ represent the sensitivities of the orientation error to the individual error components. Given that $C_{\psi i}$, $C_{\theta i}, C_{\gamma i}$ vary with the rotation angles, we here consider the average values, which are given in Figure 8. The values represent how sensitive the orientation error is to the individual error components. For example, if the orientation error is more sensitive to error component $\chi_i$ than to error component $\chi_j$, $\chi_i$ is required to be smaller than $\chi_j$ to achieve a reasonable error allocation.

Additionally, with the limitation of technological level and manufacturing costs in practical engineering, the manufacturing capability determines the maximum accuracy that the turntable can reach. Therefore, it gives the lower bounds of the error components of the turntable system, namely, the error components should satisfy the constraints:

$$|\chi_i| \geq \mathrm{LB}_i, \quad i = 1, 2, \ldots, 8. \tag{29}$$

$\mathrm{LB}_i$ are the lower bounds of the error components.

*4.2. Error Allocation Method for Turntable Systems.* For the error allocation problem of the turntable system, we expect to find the upper bounds $\mathrm{UB}_i$, $i = 1, 2, \cdots, 8$. The turntable system should reach the required orientation accuracy at any rotation angles when the error components satisfy the constraints, namely,

$$\begin{aligned}
&\left|f_\gamma(\phi, \chi)\right| \leq I_\gamma, \\
&\left|f_\theta(\phi, \chi)\right| \leq I_\theta, \\
&\left|f_\psi(\phi, \chi)\right| \leq I_\psi, \\
&\forall \phi \in \Omega_\phi, \forall \chi \in \Omega_\chi,
\end{aligned} \tag{30}$$

where $\Omega_\phi = \{\phi | 0 \leq \phi_i \leq 2\pi, i = 1, 2\}$ and $\Omega_\chi = \{\chi | \mathrm{LB}_i \leq |\chi_i| \leq \mathrm{UB}_i, i = 1, 2, \ldots, 8\}$.

With the simplification of the orientation error model given by (26), the error allocation of the turntable system can be described by the following optimization problems:

$$\begin{aligned}
\min_{\chi_\gamma} \quad & J_\gamma = \sum_{i=1}^{5} W_{\gamma i}^2 \frac{1}{\chi_{\gamma i}^2}, \\
\text{s.t.} \quad & f_\gamma(\phi_2, \chi_\gamma) \leq I_\gamma, \quad \forall \phi_2 \in \left[0, \frac{\pi}{2}\right], \\
& \chi_{\gamma i} \geq \mathrm{LB}_{\gamma i}, \quad i = 1, 2, \ldots, 5,
\end{aligned} \tag{31}$$

$$\begin{aligned}
\min_{\chi_\theta} \quad & J_\theta = \sum_{i=1}^{2} W_{\theta i}^2 \frac{1}{\chi_{\theta i}^2}, \\
\text{s.t.} \quad & f_\theta(\chi_\theta) \leq I_\theta, \\
& \chi_{\theta i} \geq \mathrm{LB}_{\theta i}, \quad i = 1, 2,
\end{aligned} \tag{32}$$

$$\begin{aligned}
\min_{\chi_\psi} \quad & J_\psi = \sum_{i=1}^{5} W_{\psi i}^2 \frac{1}{\chi_{\psi i}^2}, \\
\text{s.t.} \quad & f_\psi(\phi_2, \chi_\psi) \leq I_\psi, \quad \forall \phi_2 \in \left[\frac{3\pi}{2}, 2\pi\right], \\
& \chi_{\psi i} \geq \mathrm{LB}_{\psi i}, \quad i = 1, 2, \ldots, 5,
\end{aligned} \tag{33}$$

where $\chi_{\gamma i}, \chi_{\theta i}, \chi_{\psi i}$ are the $i$th elements of the vectors $\chi_\gamma$, $\chi_\theta, \chi_\psi$, respectively. $W_\gamma, W_\theta, W_\psi$ are weighted coefficients; the values of which are the multiplicative inverse of the corresponding sensitivities given in Figure 8. By solving the optimization problems shown in (31)–(35), the optimal

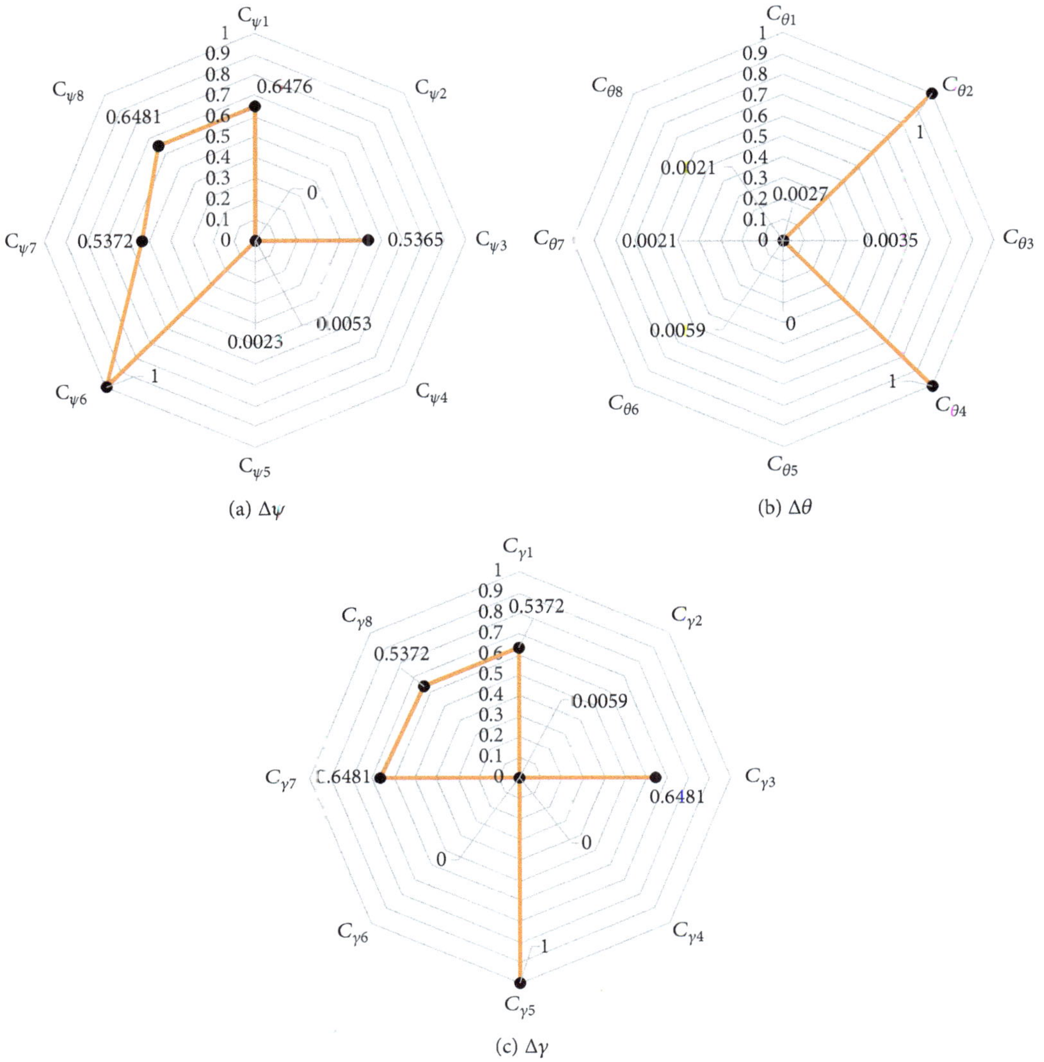


FIGURE 8: Sensitivities of orientation error to individual error components.

solutions $\chi_\gamma^*, \chi_\theta^*, \chi_\psi^*$ are obtained respectively. Then, the upper bounds of the turntable errors can be given by

$$\begin{aligned} UB_1 &= \min\left(\chi_{\gamma 1}^*, \chi_{\psi 1}^*\right), \\ UB_2 &= \chi_{\theta 1}^*, \\ UB_3 &= \min\left(\chi_{\gamma 2}^*, \chi_{\psi 2}^*\right), \\ UB_4 &= \chi_{\theta 2}^*, \\ UB_5 &= \chi_{\gamma 3}^*, \\ UB_6 &= \chi_{\psi 3}^*, \\ UB_7 &= \min\left(\chi_{\gamma 4}^*, \chi_{\psi 4}^*\right), \\ UB_8 &= \min\left(\chi_{\gamma 5}^*, \chi_{\psi 5}^*\right). \end{aligned} \tag{34}$$

According to (31)–(36), the error allocation of the turntable system can be accomplished. The error allocation results $UB_i$, $i = 1, 2, \ldots, 8$, are proven below.

**Lemma 1.** *Consider a turntable system with the orientation error model shown in (26). The error allocation results* $UB_i$, $i = 1, 2, \ldots, 8$, *are obtained from (31)–(36).* $\forall \phi \in \Omega_\phi, \forall \chi \in \Omega_\chi$, *where* $\Omega_\phi = \{\phi | 0 \le \phi_i \le 2\pi i = 12\}$ *and* $\Omega_\chi = \{\chi | LB_i \le |\chi_i| \le UB_i i = 12 \cdots 8\}$, *it is always true that* $|f_\gamma(\phi, \chi)| \le I_\gamma$, $|f_\theta(\phi, \chi)| \le I_\theta$, $|f_\psi(\phi, \chi)| \le I_\psi$.

*Proof*

*Case 1.* $\forall \phi_2 \in [0, (\pi/2)]$, then

$$\begin{aligned} \left| f_\gamma(\phi, \chi) \right| &= \left| \Delta\phi_{y1} \sin\phi_2 + \Delta\alpha_{x1} \cos\phi_2 + \Delta\alpha_{x2} + \Delta\varepsilon_x \cos\phi_2 \right. \\ &\quad \left. + \Delta\varepsilon_y \sin\phi_2 \right| \le \left| \Delta\phi_{y1} \right| \sin\phi_2 + \left| \Delta\alpha_{x1} \right| \cos\phi_2 \\ &\quad + \left| \Delta\alpha_{x2} \right| + \left| \Delta\varepsilon_x \right| \cos\phi_2 + \left| \Delta\varepsilon_y \right| \sin\phi_2 \le \chi_{\gamma 1}^* \sin\phi_2 \\ &\quad + \chi_{\gamma 2}^* \cos\phi_2 + \chi_{\gamma 3}^* + \chi_{\gamma 4}^* \cos\phi_2 + \chi_{\gamma 5}^* \sin\phi_2 \\ &= f_\gamma\left(\phi_2, \chi_\gamma^*\right) \le I_\gamma. \end{aligned} \tag{35}$$

*Case 2.* $\forall\phi_2 \in [(\pi/2), \pi]$, denote $\widehat{\phi}_2 = \pi - \phi_2$; then $\forall\widehat{\phi}_2 \in [0, (\pi/2)]$,

$$\begin{aligned}\left|f_\gamma(\phi, \chi)\right| &= \left|\Delta\phi_{y1}\sin\phi_2 + \Delta\alpha_{x1}\cos\phi_2 + \Delta\alpha_{x2} + \Delta\varepsilon_x\cos\phi_2 \right.\\ &\left.+ \Delta\varepsilon_y\sin\phi_2\right| \le \left|\Delta\phi_{y1}\right|\sin\widehat{\phi}_2 + |\Delta\alpha_{x1}|\cos\widehat{\phi}_2 \\ &+ |\Delta\alpha_{x2}| + |\Delta\varepsilon_x|\cos\widehat{\phi}_2 + \left|\Delta\varepsilon_y\right|\sin\widehat{\phi}_2 \le \chi^*_{\gamma1}\sin\widehat{\phi}_2 \\ &+ \chi^*_{\gamma2}\cos\widehat{\phi}_2 + \chi^*_{\gamma3} + \chi^*_{\gamma4}\cos\widehat{\phi}_2 + \chi^*_{\gamma5}\sin\widehat{\phi}_2 \\ &= f_\gamma\left(\widehat{\phi}_2, \chi^*_\gamma\right) \le I_\gamma.\end{aligned} \tag{36}$$

*Case 3.* $\forall\phi_2 \in [\pi, ((3\pi)/2)]$, denote $\widehat{\phi}_2 = \phi_2 - \pi$; then $\forall\widehat{\phi}_2 \in [0, (\pi/2)]$,

$$\begin{aligned}\left|f_\gamma(\phi, \chi)\right| &= \left|\Delta\phi_{y1}\sin\phi_2 + \Delta\alpha_{x1}\cos\phi_2 + \Delta\alpha_{x2} + \Delta\varepsilon_x\cos\phi_2 \right.\\ &\left.+ \Delta\varepsilon_y\sin\phi_2\right| \le \left|\Delta\phi_{y1}\right|\sin\widehat{\phi}_2 + |\Delta\alpha_{x1}|\cos\widehat{\phi}_2 \\ &+ |\Delta\alpha_{x2}| + |\Delta\varepsilon_x|\cos\widehat{\phi}_2 + \left|\Delta\varepsilon_y\right|\sin\widehat{\phi}_2 \le \chi^*_{\gamma1}\sin\widehat{\phi}_2 \\ &+ \chi^*_{\gamma2}\cos\widehat{\phi}_2 + \chi^*_{\gamma3} + \chi^*_{\gamma4}\cos\widehat{\phi}_2 + \chi^*_{\gamma5}\sin\widehat{\phi}_2 \\ &= f_\gamma\left(\widehat{\phi}_2, \chi^*_\gamma\right) \le I_\gamma.\end{aligned} \tag{37}$$

*Case 4.* $\forall\phi_2 \in [((3\pi)/2), 2\pi]$, denote $\widehat{\phi}_2 = 2\pi - \phi_2$; then $\forall\widehat{\phi}_2 \in [0, (\pi/2)]$,

$$\begin{aligned}\left|f_\gamma(\phi, \chi)\right| &= \left|\Delta\phi_{y1}\sin\phi_2 + \Delta\alpha_{x1}\cos\phi_2 + \Delta\alpha_{x2} + \Delta\varepsilon_x\cos\phi_2 \right.\\ &\left.+ \Delta\varepsilon_y\sin\phi_2\right| \le \left|\Delta\phi_{y1}\right|\sin\widehat{\phi}_2 + |\Delta\alpha_{x1}|\cos\widehat{\phi}_2 \\ &+ |\Delta\alpha_{x2}| + |\Delta\varepsilon_x|\cos\widehat{\phi}_2 + \left|\Delta\varepsilon_y\right|\sin\widehat{\phi}_2 \le \chi^*_{\gamma1}\sin\widehat{\phi}_2 \\ &+ \chi^*_{\gamma2}\cos\widehat{\phi}_2 + \chi^*_{\gamma3} + \chi^*_{\gamma4}\cos\widehat{\phi}_2 + \chi^*_{\gamma5}\sin\widehat{\phi}_2 \\ &= f_\gamma\left(\widehat{\phi}_2, \chi^*_\gamma\right) \le I_\gamma.\end{aligned} \tag{38}$$

In summary, $\forall\phi \in \Omega_\phi, \forall\chi \in \Omega_\chi$, we have $|f_\gamma(\phi, \chi)| \le I_\gamma$, where $\Omega_\phi = \{\phi | 0 \le \phi_i \le 2\pi, i = 1, 2\}$ and $\Omega_\chi = \{\chi | \text{LB}_i \le |\chi_i| \le \text{UB}_i, i = 1, 2, \ldots, 8\}$.

Similarly, $|f_\theta(\phi, \chi)| \le I_\theta$, $|f_\psi(\phi, \chi)| \le I_\psi$ can be demonstrated as well.

*4.3. Example.* Take the GyroWheel with the required calibration accuracy of $0.01°/h$ as an example. According to the analysis in Section 2, the requirement of the orientation accuracy is given in (16). Using the error allocation method proposed in this section, as shown in (31)–(36), the allocation results are given in Table 4.

TABLE 4: Error allocation results of the turntable system.

| Error sources | Symbols | Accuracy requirements (arc minute) |
|---|---|---|
| Position error | $\Delta\phi_{y1}$ | 0.4235 |
| | $\Delta\phi_{z2}$ | 0.7800 |
| Wobble error | $\Delta\alpha_{x1}$ | 0.4234 |
| | $\Delta\alpha_{z1}$ | 0.7800 |
| | $\Delta\alpha_{x2}$ | 0.2825 |
| | $\Delta\alpha_{y2}$ | 0.2824 |
| Orthogonality error | $\Delta\varepsilon_x$ | 0.4234 |
| | $\Delta\varepsilon_y$ | 0.4233 |

As seen in Table 4, the $x$-axis wobble error component of the inner axis $\Delta\alpha_{x2}$ and the $y$-axis wobble error component of the inner axis $\Delta\alpha_{y2}$ have higher accuracy requirements than the other error components. This is due to the fact that the orientation error is more sensitive to $\Delta\alpha_{x2}$ and $\Delta\alpha_{y2}$, as shown in Figure 8. According to Figure 8, only $\Delta\theta$ is sensitive to the position error of the inner axis $\Delta\phi_{z2}$ and the $z$-axis wobble error component of the outer axis $\Delta\alpha_{z1}$, and $\Delta\psi$ and $\Delta\gamma$ are insensitive to $\Delta\phi_{z2}$ and $\Delta\alpha_{z1}$. Meanwhile, $\Delta\theta$ is affected only by $\Delta\phi_{z2}$ and $\Delta\alpha_{z1}$; thus, $\Delta\phi_{z2}$ and $\Delta\alpha_{z1}$ have lower accuracy requirements. The error allocation results provide a basis for the selection of the suitable turntable and are of great significance to guide the calibration tests and improve the calibration accuracy of the GyroWheel.

Given that the error allocation of the turntable system is accomplished based on certain simplifications, Monte Carlo simulations are performed to verify the effectiveness of the error allocation results for the original orientation error model as shown in (25). According to the allocation results given in Table 4, values of the error components satisfying the accuracy requirements $\text{LB}_i \le |\chi_i| \le \text{UB}_i$, $i = 1, 2, \ldots, 8$ are generated randomly by utilizing MATLAB. 10,000 times of numerical simulations are performed, and the orientation error is calculated for each simulation test. The distribution of $\Delta\psi, \Delta\theta, \Delta\gamma$ describing the orientation error for the 10,000 times of simulations is shown in Figure 9.

It is seen in Figure 9 that the angular errors $\Delta\psi, \Delta\theta, \Delta\gamma$ always satisfy the corresponding accuracy requirements in the repeated random simulation tests, which verifies the validity of the proposed error allocation method in this section.

## 5. Conclusions

Error analysis and error allocation problems in calibration tests are studied for GyroWheel in this paper. The relationships between the rate-sensing errors of GyroWheel and the orientation error are obtained, and the orientation error induced by the turntable errors is modelled based on rigid

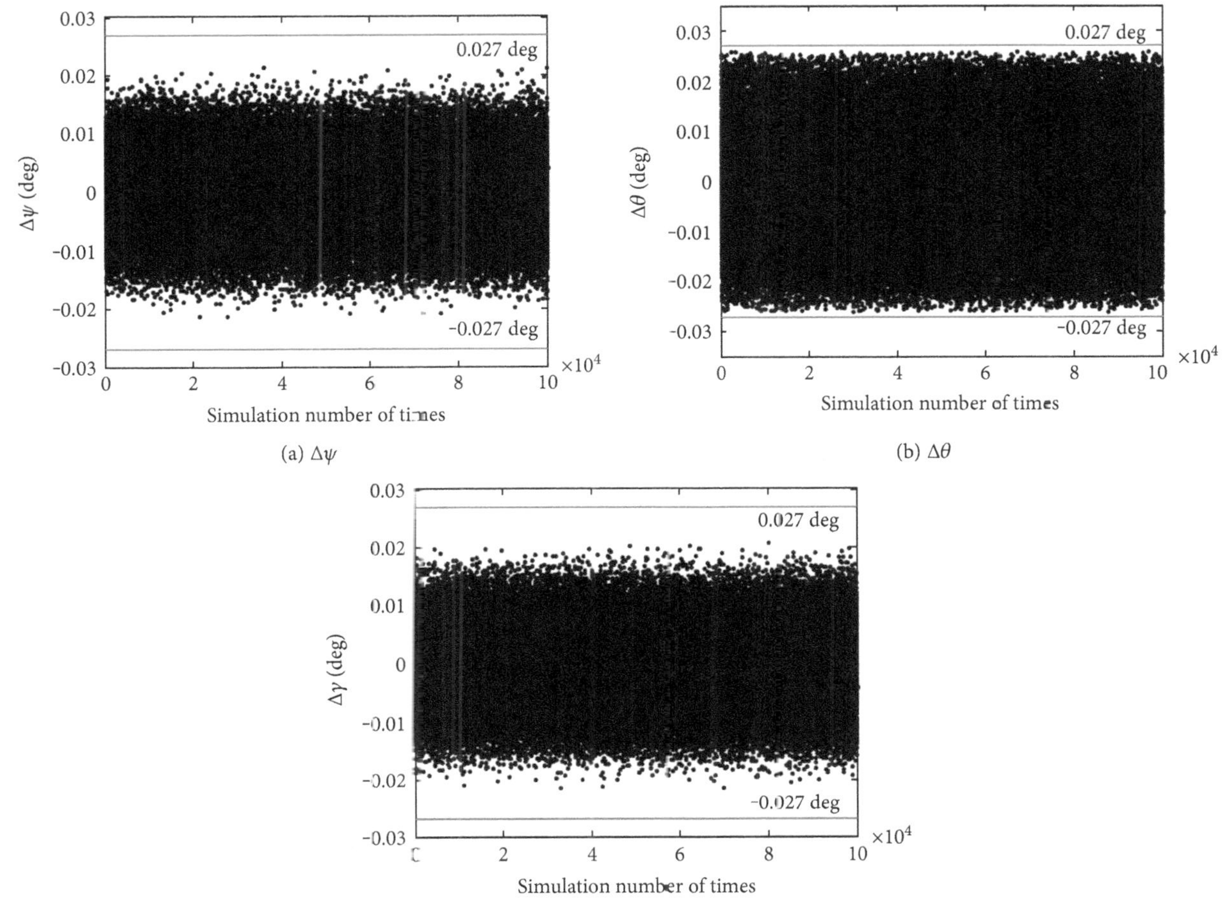


(a) $\Delta\psi$

(b) $\Delta\theta$

(c) $\Delta\gamma$

FIGURE 9: Distribution of orientation error.

body kinematics. A practical error allocation method is developed to determine the accuracy requirement of the test turntable. The salient features and contributions of this work are as follows:

(1) The error analysis for the GyroWheel calibration tests and the turntable system provides a way to express the effects of various turntable errors on the calibration accuracy quantitatively. Of all the error components, the wobble error of the inner axis contributes most to the calibration error.

(2) The proposed error allocation method provides a theoretical basis for the selection of the suitable turntable used in the calibration tests. With the considerations of the sensitivities of the orientation error to the individual error components, the requirements of these error components are determined according to the different sensitivities. Compared with the traditional allocation method assuming the equivalent effects of the errors, the proposed method is more reasonable and is beneficial to reducing the costs of calibration tests.

Although the study is focused on a two-axis turntable for GyroWheel calibration, it can be easily applied to other inertial instruments and other types of turntables, and it holds true for the cases of other calibration equipment such as centrifuges as well.

To further enhance the calibration accuracy of the GyroWheel, future research will focus on the compensation of the turntable errors.

## Conflicts of Interest

The authors declare that there is no conflict of interests regarding the publication of this paper.

## Acknowledgments

The research presented in this document is supported by the National Natural Science Foundation of China (NNSF) under Grant nos. 61427809 and 61773138, the China Postdoctoral Science Foundation under Grant 2015M571415, and the Heilongjiang Postdoctoral Foundation under Grant LBH-Z14088.

## References


[1] W. A. Bezouska, M. R. Aherne, J. T. Barrett, and S. J. Schultz, *Demonstration of Technologies for Autonomous Micro-Satellite Assembly*, AIAA SPACE 2009 Conference & Exposition, Pasadena, California, 2009.

[2] C. Balty and J. D. Gayrard, *Flexible Satellites: A New Challenge for the Communication Satellite Industry*, AIAA International Communications Satellite Systems Conference, 25th edition, 2007.

[3] G. Tyc, D. A. Staley, W. R. Whitehead et al., *GyroWheel™-an Innovative New Actuator/Sensor for 3-Axis Spacecraft Attitude Control*, Annual AIAA/USU Conference on Small Satellites, 13th edition, 1999.

[4] T. H. Paul, *Development and Testing of a GyroWheel Based Control System for the SCISAT-1 Scientific Satellite*, Carleton University, Canada, Master Dissertation, 2003.

[5] J. M. Hall, *Calibration of an Innovative Rate Sensing /Momentum Management Instrument for de-Tuned Operation and Temperature Effects*, Carleton University, Canada, Master Dissertation, 2008.

[6] J. C. Ower, *Analysis and Control System Design of an Innovative Tuned-Rotor Instrument*, Carleton University, Canada, Doctor Dissertation, 2000.

[7] D. H. Titterton and J. L. Weston, *Strapdown Inertial Navigation Technology*, Institution of Electrical Engineers, 2nd edition, 2004.

[8] L. Fu, X. Yang, and L. L. Wang, "A novel calibration procedure for dynamically tuned gyroscope designed by D-optimal approach," *Measurement*, vol. 46, no. 9, pp. 3173–3180, 2013.

[9] Z. F. Syed, P. Aggarwal, C. Goodall, X. Niu, and N. El-Sheimy, "A new multi-position calibration method for MEMS inertial navigation systems," *Measurement Science and Technology*, vol. 18, no. 7, p. 1897, 2007.

[10] Y. Y. Zhao, H. Zhao, and X. Huo, "A new multi-position calibration method for the Gyrowheel using multiple objective particle swarm optimization algorithm," *AIAA Guidance, Navigation, and Control Conference*, pp. 1906–1916, 2017.

[11] L. A. DeMore, R. A. Peterson, L. B. Conley, H. Havliscek, and N. P. Andrianos, "Design study for a high-accuracy three-axis test table," *Journal of Guidance, Control, and Dynamics*, vol. 10, no. 1, pp. 104–114, 1987.

[12] A. C. Okafor and Y. M. Ertekin, "Derivation of machine tool error models and error compensation procedure for three axes vertical machining center using rigid body kinematics," *International Journal of Machine Tools and Manufacture*, vol. 40, no. 8, pp. 1199–1213, 2000.

[13] Y. Yao and Z. Y. Qu, "Derivation of generic error models procedure for 6-DOF motion system using multi-body kinematics," in *2006 1st International Symposium on Systems and Control in Aerospace and Astronautics*, pp. 1267–1272, Harbin, China, January 2006.

[14] Z. Y. Qu and Y. Yao, "Derivation of error models and error compensation procedure for simulation turntable using multi-body kinematics," in *IEEE International Conference Mechatronics and Automation*, pp. 1408–1411, Niagara Falls, Ont., Canada, August 2005.

[15] S. M. Wang and S. Q. Ren, "Calibration of cross quadratic term of gyro accelerometer on centrifuge and error analysis," *Aerospace Science and Technology*, vol. 43, pp. 30–36, 2015.

[16] S. M. Wang and S. Q. Ren, "Relationship between calibration accuracy of error model coefficients of accelerometer and errors of precision centrifuge," *Journal of Astronautics*, vol. 33, no. 4, pp. 520–526, 2012.

[17] H. L. Zhang, Y. X. Wu, J. X. Lian, and W. Q. Wu, "Improved calibration scheme for high precision IMUs based on turntable error analysis," *Journal of Chinese Inertial Technology*, vol. 18, no. 1, pp. 129–134, 2010.

[18] Y. B. Li and M. Zeng, "Influence of three-axis turntable error source on pendulous integrating gyro accelerometer testing," *Aviation Precision Manufacturing Technology*, vol. 44, no. 2, pp. 28–31, 2008.

[19] G. P. Xiao, H. L. Zhang, M. Lv, W. Q. Wu, and Y. X. Wu, "Effect of turntable's orthogonal error and horizontal error on gyro calibration accuracy," *Journal of System Simulation*, vol. 20, pp. 370–373, 2009.

[20] Y. J. Zhao, X. H. Zhao, and W. M. Ge, "An algorithm for the accuracy synthesis of a 6-SPS parallel manipulator," *Mechanical Science and Technology*, vol. 23, no. 4, pp. 392–395, 2004.

[21] F. W. Pan, Z. S. Duan, L. L. He, and P. Wang, "Precision synthesis on novel 6DOF parallel robot based on fuzzy genetic algorithm," *Journal of Machine Design*, vol. 23, no. 9, pp. 45–48, 2006.

[22] W. Wang and C. Yun, "Orthogonal experimental design to synthesize the accuracy of robotic mechanism," *Journal of Mechanical Engineering*, vol. 45, no. 11, pp. 18–24, 2009.

[23] W. Li, S. Q. Ren, and H. B. Zhao, "Influence of three-axis turntable error on gyro calibration accuracy," *Electric Machines and Control*, vol. 15, no. 10, pp. 101–106, 2011.

[24] S. Q. Ren and J. Z. Wang, "Data processing method of calculating wobble error with level instrument," *Journal of Harbin Institute of Technology*, vol. 38, no. 6, pp. 837–847, 2006.

6

# Design Methodology and Performance Evaluation of New Generation Sounding Rockets

**Marco Pallone, Mauro Pontani, Paolo Teofilatto, and Angelo Minotti**

*University of Rome "La Sapienza", Via Salaria 851/831, 00138 Rome, Italy*

Correspondence should be addressed to Marco Pallone; pallone.1420138@studenti.uniroma1.it

Academic Editor: Corin Segal

Sounding rockets are currently deployed for the purpose of providing experimental data of the upper atmosphere, as well as for microgravity experiments. This work provides a methodology in order to design, model, and evaluate the performance of new sounding rockets. A general configuration composed of a rocket with four canards and four tail wings is sized and optimized, assuming different payload masses and microgravity durations. The aerodynamic forces are modeled with high fidelity using the interpolation of available data. Three different guidance algorithms are used for the trajectory integration: constant attitude, near radial, and sun-pointing. The sun-pointing guidance is used to obtain the best microgravity performance while maintaining a specified attitude with respect to the sun, allowing for experiments which are temperature sensitive. Near radial guidance has instead the main purpose of reaching high altitudes, thus maximizing the microgravity duration. The results prove that the methodology at hand is straightforward to implement and capable of providing satisfactory performance in term of microgravity duration.

## 1. Introduction

Sounding rockets are specialized missiles generally used to investigate the region of space between 50 and 700 km where it is difficult to enter with the traditional atmospheric balloon or low orbit satellites. They can be used in several missions, such as detection of the solar activity and anomalies, analysis of the constituents of the upper atmosphere, thermal analysis on new materials, and generally, measurements of the space surrounding the Earth [1–3]. The most important characteristic of a sounding rocket is the capability of achieving microgravity conditions for the payload, without the need of a manned mission in the ISS to perform experiments and then reducing the costs of the mission. The quality of microgravity depends on the absence of drag and other gyroscopic forces, so the guidance of the rocket assumes the utmost importance. Another good characteristic of the sounding rocket in contrast to the satellite missions is the sounding rocket's ability to retrieve the payload with the help of a heat shield and a parachute. For all of these reasons, a vast number of sounding rocket families were developed, such as the American "Black Brant" and the European "Maxus." The Black Brant family includes sounding rockets that range from one through three stages. The Black Brant VC [4] (Figure 1) can be launched from a steerable launch tower with different kick angles to achieve different apogee altitudes. It is controlled via spin-up motors and canard wings, which provide a fast response at the cost of small instability. Four rear tail surfaces guarantee an adequate static margin. The Maxus [4, 5] is a very powerful European single-stage sounding rocket launched in Kiruna (Sweden) and can reach a maximum altitude of 700 km, while the range does not exceed 100 km. It has a microgravity duration of about 12 min, and the microgravity level $\Delta g$ is near $10^{-4}$m/sec$^2$. TVC control system is highly sophisticated in order to guarantee strict margins for the range and to avoid damages to both people and goods. Table 1 portrays some useful data to compare the performance of the two rockets.

Both of these sounding rockets are propelled with a solid state motor, due to their simplicity and light weights with respect to liquid propellant. A crucial point is that a sounding rocket has to be cheaper than a high-tech rocket, and in order to achieve this objective, new design methodologies are needed. Many authors (e.g., Chowdhury et al. [6, 7] and

FIGURE 1: Black Brant VC.

TABLE 1: "Maxus" and "Black Brant VC" data.

| | BBVC | Maxus |
|---|---|---|
| Length (m) | 8.3 | 16.2 |
| Max diameter (m) | 0.44 | 1 |
| Mass liftoff (kg) | 1544 | 12298 |
| Propellant mass (kg) | 1017 | 10042 |
| Av. thrust (N) | 69374 | 372500 |
| Burnout time (sec) | 32 | 63 |

Casalino and Pastrone [8, 9]) are interested in designing new sounding rockets to explore the potentiality of new hybrid motors for hypersonic flight. Nonaka et al. [10] and Woo et al. [11] investigated the reusability of the sounding rocket, and Cremaschi et al. [12] investigated the optimization of a trajectory with the use of a commercial trajectory simulation tool (ASTOS). However, a very limited number of studies are concerned with fast sizing and aerodynamic losses during the ascent path. The purpose of this work is to find a novel and simple methodology to size and design for future sounding rockets, considering accurate aerodynamics and three guidance algorithms. In particular, the technique explained in this work is applied to a single-stage rocket whose three-dimensional trajectory is simulated using the following guidance schemes:

(i) Constant attitude guidance

(ii) Near radial guidance

(iii) Sun-pointing guidance

This work purpose is to find a procedure that can provide the minimal volume rocket that can guarantee a fixed time of microgravity with different scientific payloads inside. Definitely, the methodology at hand yields configurations that can provide a well-developed preliminary design for future sounding rockets, with an accurate aerodynamic model and using three distinct guidance schemes.

## 2. Rocket Geometry and Sizing

*2.1. Basic Geometry.* With the purpose of maintaining the configuration as simple as possible, the sounding rocket is composed of one stage and a module hosting the avionics, experiments, and recovery system (parachute and the heat shield).

The basic geometry of the rocket is composed of the following:

(i) A nose cone of parabolic shape

(ii) A fuselage of cylindrical shape

(iii) Four canard wings in cruciform configuration

(iv) Four tail surfaces of trapezoidal shape

This configuration is chosen with the purpose of creating an algorithm as general as possible. Incidentally, this geometry is adopted by a lot of sounding rockets in the market, like the "Black Brant VC" in Figure 1.

Every rocket configuration in this work is defined by the total length $l$ and the maximum diameter $d$, while all the other parameters like the propulsive ones are obtained through scaling factors. The grid used for the computation is

(i) $l$ from 3 m to 15 m with step of 1 m

(ii) $d$ from 0.2 m to 1.5 m with step of 0.05 m.

Thus, a total of 243 different configurations were analyzed. The design methodology that is being presented relates geometry, weight, aerodynamics, and propulsive performances.

*2.2. Main Body.* For each configuration, the length of the ogive is proportional to the total length of the sounding rocket

$$l_{\text{ogive}} = C_{\text{ogive}} l, \tag{1}$$

where $C_{\text{ogive}} = 0.17$ is a dimensionless coefficient. The dimensional ogive point distribution is then obtained by multiplying the nondimensional points in the $(\xi, \eta)$ frame in Figure 2 with $l_{\text{ogive}}$ and $d$. The main body, which contains the propulsive system and the avionics, is cylindrical with diameter $d$, while its length is found through

$$l_{\text{body}} = l - l_{\text{ogive}}. \tag{2}$$

*2.3. Aerodynamic Surfaces.* The aerodynamic surfaces are sized according to the following ratios:

$$\begin{aligned} A_{\text{P}} &= \left(l - l_{\text{ogive}}\right) d, \\ A_{\text{C}} &= \frac{1}{45} A_{\text{P}}, \\ A_{\text{T}} &= \frac{1}{12} A_{\text{P}}. \end{aligned} \tag{3}$$

Then the following quantities are computed as follows:

$$\begin{aligned} h_{\text{T}} &= \sqrt{\frac{4}{3} A_{\text{C}}}, \\ h_{\text{C}} &= \sqrt{2 A_{\text{T}}}, \end{aligned} \tag{4}$$

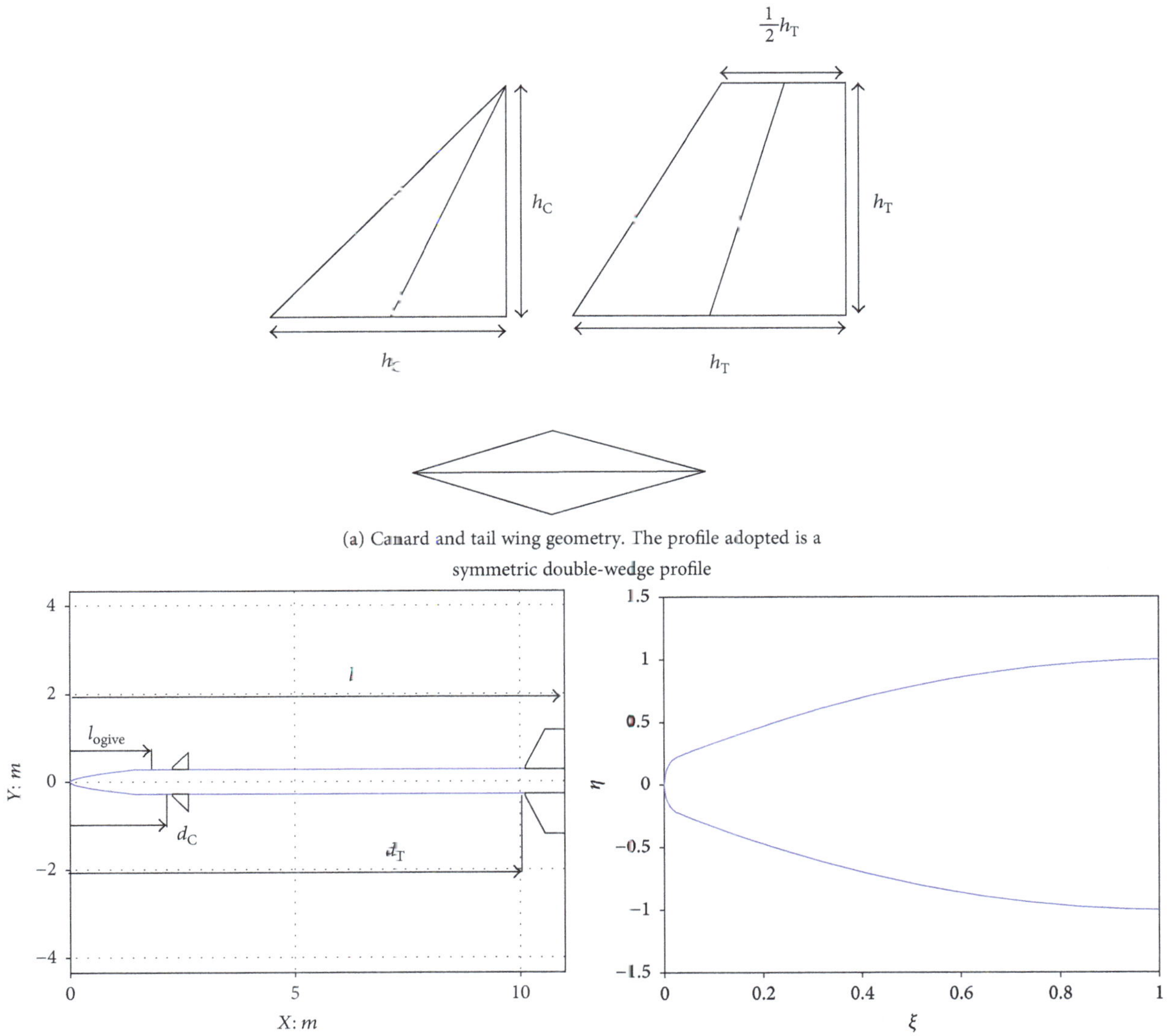


(a) Canard and tail wing geometry. The profile adopted is a symmetric double-wedge profile

(b) Canard and tail wing disposition along the longitudinal axis

(c) Nondimensional ogive

FIGURE 2: Main surfaces and reference quantities.

while the distances with respect to the nose tip are as follows:

$$
\begin{aligned}
d_C &= l_{ogive} + 0.4\text{m}, \\
d_T &= l - h_T.
\end{aligned} \tag{5}
$$

The main concept under this procedure is to maintain the shape of the wings, while selecting their size in relation to the overall size of the main body. This is in order to avoid wings with nonconventional shapes that would result in undesired aerodynamic effects. Figure 2 portrays the wing geometry with the profile adopted for both wing systems. The nonvariable distance in (5) 0.4 m is meant to provide a fixed volume needed for the canard avionic and the recovery system. Figure 2 portrays an example of a configuration with the disposition of the leading edges along the longitudinal axis of the rocket.

TABLE 2: Auxiliary mass $m_{aux}$ of the ogive.

| | |
|---|---|
| Parachute mass (kg) | 10 |
| Heat shield mass (kg) | 35 |
| Metal case mass (kg) | 5 |
| Total $m_{aux}$ (kg) | 50 |

*2.4. Mass Scaling.* In order to describe properly a real sounding rocket mission, a fixed mass including parachute, heat shield, and metal case is used for every configuration. Table 2 portrays in detail these auxiliary masses, which form $m_{aux}$.

With regard to the initial mass at launch, a common initial density is used in order to find a reasonable and scaled mass distribution for every rocket configuration. This initial density is $C_{m_0} = 1092.3\text{kg/m}^3$, which is taken from the Black

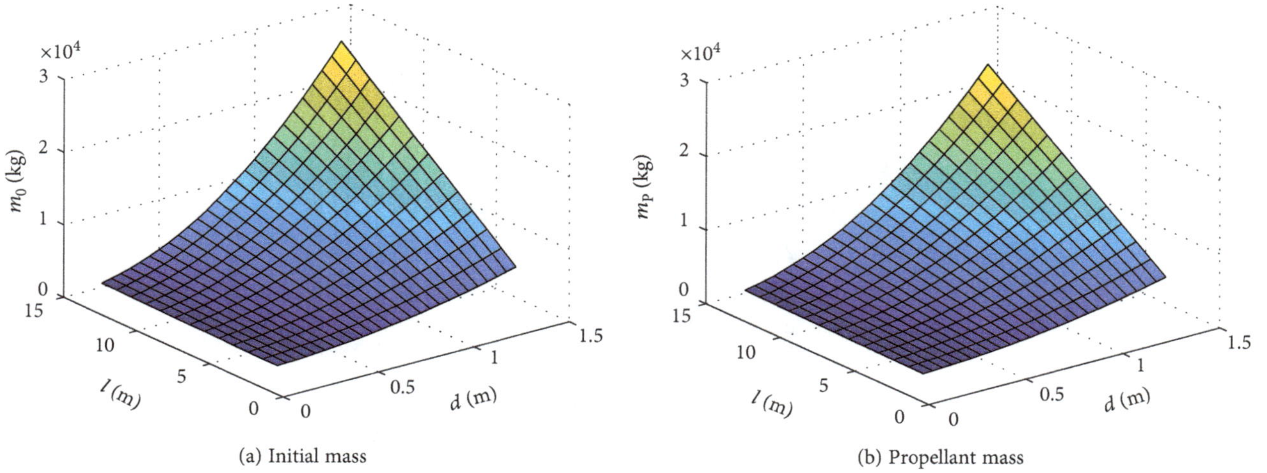


FIGURE 3: Example of scaled $m_0$ and $m_P$ with $m_U = 50$kg.

Brant VC and is considered to have a reasonable value. This value includes propellant, structural, and auxiliary masses. So the initial mass for every configuration without the ogive mass is

$$m_{0\text{body}} = C_{m_0} V_{\text{body}}. \tag{6}$$

For every configuration. 50, 100, and 200 kg of scientific payload are used, so the total initial mass at launch is

$$m_0 = m_{0\text{body}} + m_U + m_{\text{aux}}. \tag{7}$$

The propulsive technology of this single-stage rocket is a solid propellant one, allowing for the following structural coefficient:

$$\epsilon = \frac{m_S}{m_P + m_S} = 0.11. \tag{8}$$

The structural and propellant masses are obtained through the following expressions:

$$\begin{aligned} u &= \frac{m_U + \epsilon m_0 - \epsilon m_U}{m_0}, \\ m_P &= \frac{m_0}{1-u}, \\ m_S &= \frac{\epsilon m_0 (1-u)}{1-\epsilon}. \end{aligned} \tag{9}$$

In the end, (6), (7), (8), and (9) yield the overall rocket mass at launch, portrayed in Figure 3 as a function of $l$ and $d$, together with the corresponding propellant mass.

*2.5. Propulsion System Scaling.* To scale the thrust for every sounding rocket size, the initial launch acceleration and the specific impulse are maintained constant:

$$T = n_0 g m_0, \tag{10}$$

where $n_0 = 3g$, while the propellant mass consumption is

$$\dot{m} = -\frac{T}{I_{sp} g}. \tag{11}$$

The value for $n_0$ is chosen in order to constrain the rocket dynamical stress at launch. Equation (11) leads to the burnout time:

$$t_B = \frac{m_P}{\dot{m}}. \tag{12}$$

Figure 4 portrays the thrust and the burnout time scaling with the rocket size.

## 3. Rocket Aerodynamic Modeling

Aerodynamic forces affect the rocket motion, which partially occurs in the atmosphere. Thus, the rocket aerodynamics is to be modeled appropriately. In this work, four quantities are assumed to determine the aerodynamic forces: Mach number $M$, Reynolds number Re, angle of attack $\alpha$, and sideslip angle $\beta$. To provide an accurate model for each configuration, two fundamental steps are needed:

(1) Derivation of the aerodynamic coefficients $C_D$, $C_L$, and $C_Q$ from DATCOM [13] at a relevant number of Mach, Reynolds, angles of attack, and sideslip angles

(2) Cubic interpolation of the aerodynamic coefficients during the integration of the dynamics equations

Following the approach presented by Mangiacasale [14], Pallone et al. [15], and Pontani [16], the aerodynamics of the rocket was modeled through the Missile DATCOM software [13]. This software provides an accurate esteem of the aerodynamic forces without using a lot of computational resources; thus, it can provide a complete aerodynamic database for each configuration in a short time. The following discrete values of $M$, Re, $\alpha$, and $\beta$ are used:

(i) $M = [0.2, 0.5, 0.6, 0.8, 0.9, 1.1, 1.2, 1.5, 2, 2.5, 3, 3.5, 4, 4.5, 5, 5.5, 6, 6.5, 7, 7.5, 8, 10]$

(ii) $\text{Re} = [0.5, 1, 1.5, 2] * 10^5$

(iii) $\alpha = [0, 1, 2, 3, 4, 5]$ deg

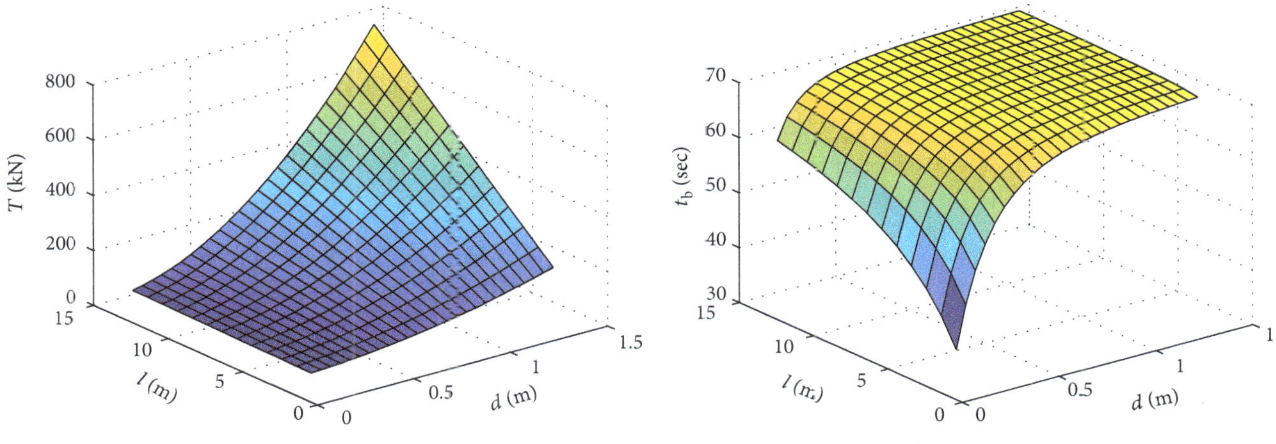


(a) Thrust

(b) Burnout time

FIGURE 4: Example of scaled $T$ and $t_B$ with $m_U = 50$ kg.

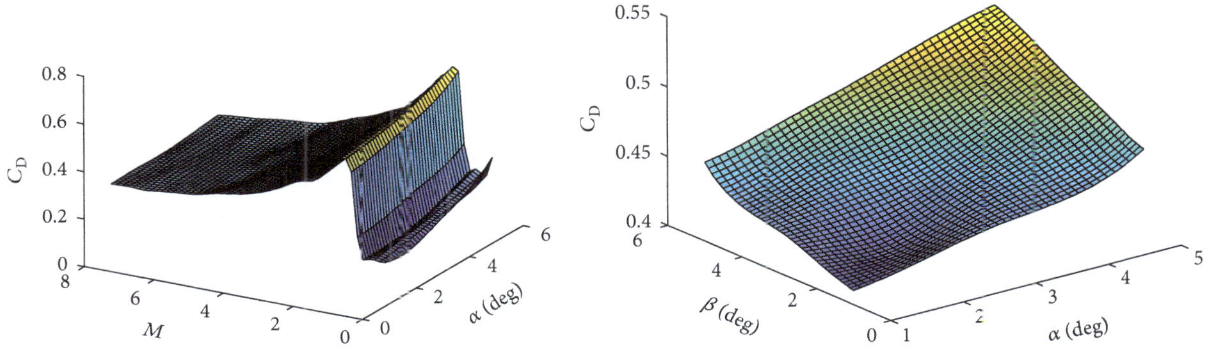


(a) Drag coefficient for $\beta = 0$ deg and $\mathrm{Re} = 4 * 10^7$

(b) Drag coefficient for $M = 3$ and $\mathrm{Re} = 4 * 10^7$

FIGURE 5: Example of the drag force coefficients provided by the DATCOM software for the configuration $l = 6$ m and $d = 0.8$ m.

(iv) $\beta = [0, 1, 2, 3, 4, 5]$ deg

This means that DATCOM was run 3024 times for each $(l, d)$ combination. The aerodynamic surfaces used in the computation are the cross surfaces of each configuration and correspond to the circular section, that is,

$$S^{(i)} = \pi \left(\frac{d}{2}\right)^2. \tag{13}$$

Figure 5 portrays an example of the aerodynamic coefficients for a single configuration.

## 4. Rocket Dynamics

The problem investigated in this work consists in finding the sounding rocket with the minimum total volume that can achieve microgravity durations not less than 10 or 15 minutes. In the absence of significant attitude angular velocities, a rocket is considered in microgravity conditions when it is flying at an altitude above 100 km [5].

This approximation is based on real evidence and calculations, allowing for it to be used as a practical rule to analyze the rocket performance. Thus, the problem at hand reduces to finding the minimal rocket volume $(l, d)$ that can guarantee $\Delta T_{mg}$ of 10 and 15 min, fixing the propulsive technology $(I_{sp}, \epsilon)$ and the geometrical ratios between its parts $(C_{m_0}, C_{ogive}, A_C/A_P, A_T/A_P)$.

*4.1. Reference Frames.* A description of the reference frames used in this work is useful to understand the guidance implementation. In the following, the notation $R_j(\mp\xi)$ $(j = 1, 2, 3)$ denotes the elementary rotation about axis $j$ by angle $\xi$ while the $\mp$ sign means, respectively, clockwise and counterclockwise.

(1) The *Earth-centered* frame (ECI) $(\hat{\mathbf{c}}_1, \hat{\mathbf{c}}_2, \hat{\mathbf{c}}_3)$ is a Cartesian inertial reference frame defined as follows. Its origin O is the center of the Earth. The unit vector $\hat{\mathbf{c}}_3$ is aligned with the Earth axis of rotation and is positive northward, whereas $\hat{\mathbf{c}}_1$ is aligned with the vernal axis, which corresponds to the intersection of the Earth equatorial plane with the ecliptic plane.

(2) The *Earth-centered Earth-fixed* frame (ECEF) $(\hat{\mathbf{i}}, \hat{\mathbf{j}}, \hat{\mathbf{k}})$ represents a reference system that is rigidly attached to the Earth. Under the assumption that

the planet rotates with a constant angular rate $\omega_E$, also the ECEF frame rotates with angular rate $\omega_E$. The origin of the coordinate system is the center of the Earth, while the unit vector $\hat{\mathbf{i}}$ intersects the Greenwich reference meridian at all times, and $\hat{\mathbf{k}}$ is aligned with the planet axis of rotation and is positive northward. $\omega_E\hat{\mathbf{k}}$ is the vector rotation rate of the ECEF frame with respect to the ECI frame. The reference meridian is the Greenwich side real-time $\theta_G$, given by $\theta_G(t) = \theta_G(\bar{t}) + \omega_E(t - \bar{t})$ where $\bar{t}$ refers to a generic time instant. The ECEF frame is obtained through a counterclockwise rotation by angle $\theta_G$ around the third axis, that is, $[\hat{\mathbf{i}}, \hat{\mathbf{j}}, \hat{\mathbf{k}}]^T = R_3(\theta_G)[\hat{\mathbf{c}}_1, \hat{\mathbf{c}}_2, \hat{\mathbf{c}}_3]^T$.

(3) The *local horizontal* frame (LH) $(\hat{\mathbf{r}}, \hat{\mathbf{E}}, \hat{\mathbf{N}})$ is a reference system with the origin in the rocket center of mass. Specifically, $\hat{\mathbf{r}}$ is the unit vector aligned with the instantaneous position $\hat{\mathbf{r}}$, $\hat{\mathbf{E}}$ is directed along the local east direction, and $\hat{\mathbf{N}}$ is aligned with the local north direction. LH frame is obtained from the ECEF frame through the following rotations: $[\hat{\mathbf{r}}, \hat{\mathbf{E}}, \hat{\mathbf{N}}]^T = R_2(-\mathrm{La})R_3(\lambda_g)[\hat{\mathbf{i}}, \hat{\mathbf{j}}, \hat{\mathbf{k}}]^T$.

(4) The *auxiliary orbital* frame (AO) $(\hat{\mathbf{r}}, \hat{\boldsymbol{\theta}}_r, \hat{\mathbf{h}}_r)$ is defined making reference to the rocket relative velocity $v_R = v_I - \omega_E \times r$. The unit vector $\hat{\boldsymbol{\theta}}_R$ is aligned with the projection of the velocity $v_r$ onto the horizontal plane $(\hat{\mathbf{E}}, \hat{\mathbf{N}})$. The AO frame is obtained from the LH frame with the simple rotation $[\hat{\mathbf{r}}, \hat{\boldsymbol{\theta}}_R, \hat{\mathbf{h}}_R]^T = R_1(\zeta_R)[\hat{\mathbf{r}}, \hat{\mathbf{E}}, \hat{\mathbf{N}}]^T$.

(5) The *relative velocity* frame (RV) $(\hat{\mathbf{n}}_R, \hat{\mathbf{v}}_R, \hat{\mathbf{h}}_R)$ is obtained from the AO frame through the following clockwise rotation: $[\hat{\mathbf{n}}_R, \hat{\mathbf{v}}_R, \hat{\mathbf{h}}_R]^T = R_3(-\gamma_R)[\hat{\mathbf{r}}, \hat{\boldsymbol{\theta}}_R, \hat{\mathbf{h}}_R]^T$.

(6) The *wind axis* frame (WA) $(\hat{\mathbf{n}}_W, \hat{\mathbf{v}}_W, \hat{\mathbf{h}}_W)$ is defined with reference to the rocket velocity with respect to the local velocity of the atmosphere $v_a = \omega_E \times r$ so $v_a = v_R$ and just a counterclockwise rotation about axis 2 by the bank angle $\sigma$: $[\hat{\mathbf{n}}_W, \hat{\mathbf{v}}_W, \hat{\mathbf{h}}_W]^T = R_2(\sigma)[\hat{\mathbf{n}}_R, \hat{\mathbf{v}}_R, \hat{\mathbf{h}}_R]^T$.

(7) The *auxiliary body axis* frame (ABA) $(\hat{\mathbf{i}}_B, \hat{\mathbf{j}}_B, \hat{\mathbf{k}}_B)$ is a reference system aligned with the rocket inertia axes. The unit vector $\hat{\mathbf{k}}_B$ is orthogonal to the rocket plane of symmetry, while the two remaining unit vectors lie on it. The ABA frame is obtained from the WA frame through a sequence of two rotations: $\sigma[\hat{\mathbf{i}}_B, \hat{\mathbf{j}}_B, \hat{\mathbf{k}}_B]^T = R_3(-\alpha)R_1(\beta)[\hat{\mathbf{n}}_W, \hat{\mathbf{v}}_W, \hat{\mathbf{h}}_W]^T$.

(8) The *body axis* (BA) $(\hat{\mathbf{x}}_B, \hat{\mathbf{y}}_B, \hat{\mathbf{z}}_B)$ is, like the ABA frame, also attached to the rocket inertial axes only with a different orientation. Specifically, $\hat{\mathbf{x}}_B$ is aligned with the rocket longitudinal axis $(\hat{\mathbf{x}}_B = \hat{\mathbf{j}}_B)$; $\hat{\mathbf{z}}_B$ is directed downward so $\hat{\mathbf{z}}_B = -\hat{\mathbf{i}}_B$ and as consequence $\hat{\mathbf{y}}_B = -\hat{\mathbf{k}}_B$. These relations lead to the following elementary rotation:

$$\begin{bmatrix} \hat{\mathbf{x}}_B \\ \hat{\mathbf{y}}_B \\ \hat{\mathbf{z}}_B \end{bmatrix} = \begin{bmatrix} 0 & 1 & 0 \\ 0 & 0 & -1 \\ -1 & 0 & 0 \end{bmatrix} \begin{bmatrix} \hat{\mathbf{i}}_B \\ \hat{\mathbf{j}}_B \\ \hat{\mathbf{k}}_B \end{bmatrix} = R_A \begin{bmatrix} \hat{\mathbf{i}}_B \\ \hat{\mathbf{j}}_B \\ \hat{\mathbf{k}}_B \end{bmatrix}. \tag{14}$$

The knowledge of the instantaneous orientation of $(\hat{\mathbf{x}}_B, \hat{\mathbf{y}}_B, \hat{\mathbf{z}}_B)$ is equivalent to identifying the instantaneous attitude of the rocket.

(9) The *launch inertial* frame (LI) $(\hat{\mathbf{i}}_0, \hat{\mathbf{j}}_0, \hat{\mathbf{k}}_0)$ is an inertial system centered in the launch location at the launch instant. The unit vector $\hat{\mathbf{i}}_0$ is aligned with the local north direction, $\hat{\mathbf{j}}_0$ is directed toward the local east direction, and $\hat{\mathbf{k}}_0$ is aligned with the downward direction. The LI frame is obtained from the ECI frame with a sequence of three rotations: the first two of them take the ECI frame to an intermediary reference frame, named *auxiliary launch inertial* frame (ALI) $(\hat{\mathbf{r}}_0, \hat{\mathbf{E}}_0, \hat{\mathbf{n}}_0)$ using the geographical latitude and longitude of the launch base at the launch time, while the third one relates the axes of the ALI frame to the LI frame. The complete sequence is as follows:

$$\begin{bmatrix} \hat{\mathbf{i}}_0 \\ \hat{\mathbf{j}}_0 \\ \hat{\mathbf{k}}_0 \end{bmatrix} = R_2(-L_L)R_3(\lambda_{aL}) \begin{bmatrix} 0 & 0 & 1 \\ 0 & 1 & 0 \\ -1 & 0 & 0 \end{bmatrix} \begin{bmatrix} \hat{\mathbf{c}}_1 \\ \hat{\mathbf{c}}_2 \\ \hat{\mathbf{c}}_3 \end{bmatrix} = R_2(-L_L)R_3(\lambda_{aL})R_B \begin{bmatrix} \hat{\mathbf{c}}_1 \\ \hat{\mathbf{c}}_2 \\ \hat{\mathbf{c}}_3 \end{bmatrix}. \tag{15}$$

The BA frame is also related to the LI frame through a sequence of three rotations using the attitude angles $(\phi, \theta, \psi)$. The complete sequence is $[\hat{\mathbf{x}}_B, \hat{\mathbf{y}}_B, \hat{\mathbf{z}}_B]^T = R_1(\phi)R_2(\theta)R_3(\psi)[\hat{\mathbf{i}}_0, \hat{\mathbf{j}}_0, \hat{\mathbf{k}}_0]^T$.

In order to avoid ambiguities in the definition of the vectors of interest, the angles introduced in this section have to be constrained. Table 3 shows the limitation for the angles used in this work.

Any vector can be obtained with respect to any reference frame using the rotations defined in this section. These rotations can be used to define the rocket kinematics and dynamics through the motion, and Figures 6 and 7 portray all the reference frames used in this work, together with the related rotations.

TABLE 3: Angle constraints.

| | |
|---|---|
| $-\pi < \theta_G \leq \pi$ | $-\pi < \lambda_G \leq \pi$ |
| $-\frac{\pi}{2} < \mathrm{La} \leq \frac{\pi}{2}$ | $-\pi < \zeta_R \leq \pi$ |
| $-\frac{\pi}{2} < \gamma_R \leq \frac{\pi}{2}$ | $-\pi < \sigma \leq \pi$ |
| $-\frac{\pi}{2} < \beta \leq \frac{\pi}{2}$ | $-\pi < \alpha \leq \pi$ |
| $-\pi < \psi \leq \pi$ | $-\frac{\pi}{2} < \theta \leq \frac{\pi}{2}$ |
| $-\pi < \phi \leq \pi$ | — |

*4.2. Forces.* The main term of the gravitational force $G$ is given in the RV frame simply by

$$\frac{G}{m} = -\frac{\mu_E}{r^3} = -\frac{\mu_E}{r^2}\begin{bmatrix}\cos\gamma_R\\ \sin\gamma_R\\ 0\end{bmatrix}^T\begin{bmatrix}\hat{\mathbf{n}}_E\\ \hat{\mathbf{v}}_E\\ \hat{\mathbf{h}}_E\end{bmatrix}. \quad (16)$$

The components of the aerodynamic force can be referred to the WA frame and then projected in the RV frame with a simple rotation. The three components are the lift force, the drag force, and the side force.

$$\frac{A}{m} = \frac{L+D+Q}{m} = \frac{L\hat{\mathbf{n}}_W - D\hat{\mathbf{v}}_W + Q\hat{\mathbf{h}}_W}{m}, \quad (17)$$

where

$$\begin{aligned} L &= L\hat{\mathbf{n}}_W = \frac{1}{2}\rho v_W^2 C_L(\alpha,\beta,\mathrm{Re},M)\hat{\mathbf{n}}_W,\\ D &= -D\hat{\mathbf{n}}_W = -\frac{1}{2}\rho v_W^2 C_D(\alpha,\beta,\mathrm{Re},M)\hat{\mathbf{n}}_W,\\ Q &= Q\hat{\mathbf{n}}_W = \frac{1}{2}\rho v_W^2 C_Q(\alpha,\beta,\mathrm{Re},M)\hat{\mathbf{n}}_W \end{aligned} \quad (18)$$

The aerodynamic forces have to be projected in the RV frame with a rotation about the bank angle around the second axis:

$$A = [A_n A_v A_h]\begin{bmatrix}\hat{\mathbf{n}}_R\\ \hat{\mathbf{v}}_R\\ \hat{\mathbf{h}}_R\end{bmatrix} = [L - DQ]R_2(\sigma)\begin{bmatrix}\hat{\mathbf{n}}_R\\ \hat{\mathbf{v}}_R\\ \hat{\mathbf{h}}_R\end{bmatrix}. \quad (19)$$

The aerodynamic forces are taken into account up to 100 km of altitude where the atmospheric density becomes negligible.

The thrust can be projected along the RV frame with the in-plane $\alpha_T$ and the out-of-plane $\beta_T$:

$$\frac{\mathbf{T}}{m} = \left[\frac{\mathbf{T}}{m}\cos\beta_T\sin\alpha_T \quad \frac{\mathbf{T}}{m}\cos\beta_T\cos\alpha_T \quad \frac{\mathbf{T}}{m}\sin\beta_T\right] \cdot \left[\hat{\mathbf{n}}_R \quad \hat{\mathbf{v}}_R \quad \hat{\mathbf{h}}_R\right]^T, \quad (20)$$

and to avoid ambiguities, the following constraints must hold:

$$\begin{aligned} -\pi < \alpha_T \leq \pi,\\ -\frac{\pi}{2} \leq \beta_T \leq \frac{\pi}{2}. \end{aligned} \quad (21)$$

The thrust vector $\mathbf{T}$ is considered aligned with the rocket longitudinal axis so two angles, $\alpha_T$ and $\beta_T$ suffices to describe its direction.

*4.3. Equations of Motion.* The equations of motion that govern the three-dimensional rocket dynamics can be conveniently written in terms of $\{r, \lambda_G, \mathrm{La}, \gamma_R, v_R, \zeta_R\}$. These variables refer to the relative motion in the RV frame. They form the state vector $\mathbf{x}_R$ of the launch vehicle (in rotating coordinates).

$$\begin{aligned} \dot{r} &= v_R\sin\gamma_R,\\ \dot{\lambda}_G &= \frac{v_R\cos\gamma_R\cos\zeta_R}{r\cos\mathrm{La}},\\ \dot{\mathrm{La}} &= \frac{v_R\cos\gamma_R\sin\zeta_R}{r},\\ \dot{\gamma}_R &= \left(-\frac{\mu_E}{v_R r^2} + \frac{v_R}{r}\right)\cos\gamma_R + \frac{(T/m)\cos\beta_T\sin\alpha_T}{v_R} + \frac{A_n}{v_R}\\ &\quad + \frac{\omega_E^2 r}{v_R}(\cos\mathrm{La}(\cos\mathrm{La}\cos\gamma_R + \sin\mathrm{La}\sin\gamma_R\sin\zeta_R))\\ &\quad + 2\omega_E\cos\mathrm{La}\cos\zeta_R,\\ \dot{v}_R &= -\frac{\mu_E}{r^2}\sin\gamma_R + \frac{T}{m}\cos\beta_T\cos\alpha_T + A_v\\ &\quad + \omega_E^2 r\cos\mathrm{La}(\cos\mathrm{La}\sin\gamma_R - \sin\mathrm{La}\cos\gamma_R\sin\zeta_R),\\ \dot{\zeta}_R &= -\frac{v_R}{r}\tan\mathrm{La}\cos\gamma_R\cos\zeta_R + \frac{T/m}{v_R\cos\gamma_R}\sin\beta_T\\ &\quad + 2\omega_E\cos\mathrm{La}\tan\gamma_R\sin\zeta_R + \frac{A_h}{v_R\cos\gamma_R}\\ &\quad - \frac{\omega_E^2 r}{v_R\cos\gamma_R}\sin\mathrm{La}\cos\mathrm{La}\cos\zeta_R - 2\omega_E\sin\mathrm{La}. \end{aligned} \quad (22)$$

The initial values for (22) are reported in Table 4.

Equation (22) is valid for both the propelled arc and the ballistic flight of the ogive. The values for La and $\lambda_G$ in Table 4 are referred to the Malindi launch base, while the launch is in the east direction ($\zeta_R$). The control laws for $(\alpha_T, \beta_T)$, are determined using three distinct guidance scheme, namely, the constant attitude guidance, the near radial guidance, and the sun-pointing guidance, which are described in the following section. The state $\mathbf{x}_R$ is continuos across the first stage separation, which occurs at time $t_B$. After that, the payload continues its motion in a ballistic trajectory.

## 5. Guidance Algorithms

*5.1. Constant Attitude Guidance.* With this guidance, the rocket orientation does not change during the flight and

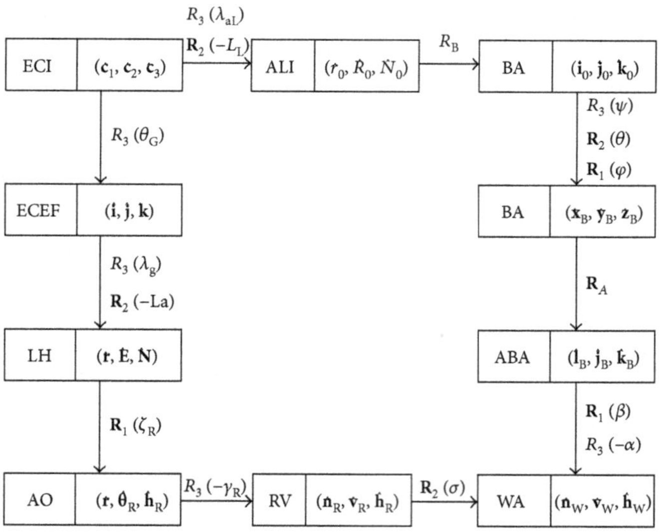


FIGURE 6: Reference frames and related rotation matrices.

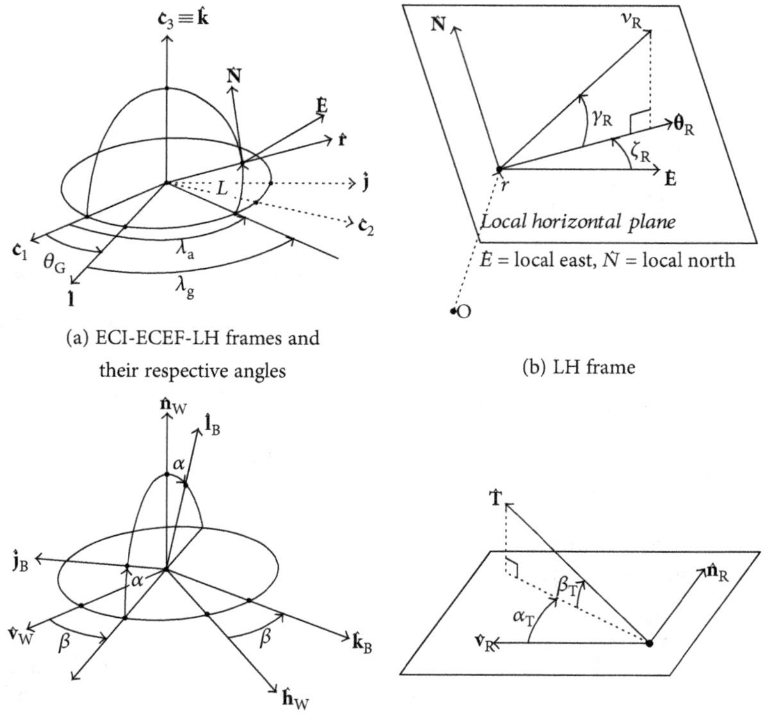


(a) ECI-ECEF-LH frames and their respective angles

(b) LH frame

(c) WA-ABA frames and their relative angles

(d) Thrust vector direction with respect to the RV frame

FIGURE 7: Main reference frames and related angles.

was actually implemented in the Black Brant X (cf. [17]). This is ideal to achieve nominal (perfect) microgravity conditions ($\Delta g = 0$), due to rotation rates equal to 0. The attitude angles $\{\phi, \theta, \psi\}$ remain constant; therefore, in this type of guidance, the following expressions can be assumed as follows:

$$\phi = \phi_0 \theta = \theta_0 \psi = \psi_0. \tag{23}$$

TABLE 4: Initial condition for (22).

| | |
|---|---|
| $r$ (km) | $R_E$+0.001 |
| $v_R$ (km/s) | 0.008 |
| $\gamma_R$ (deg) | 86 |
| La (deg) | −2.938 |
| $\zeta_R$ (deg) | 0 |
| $\lambda_G$ (deg) | 40.213 |

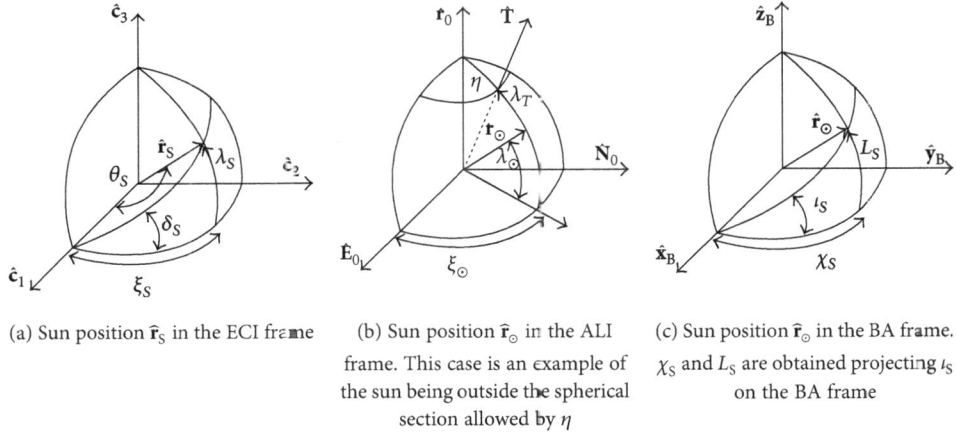

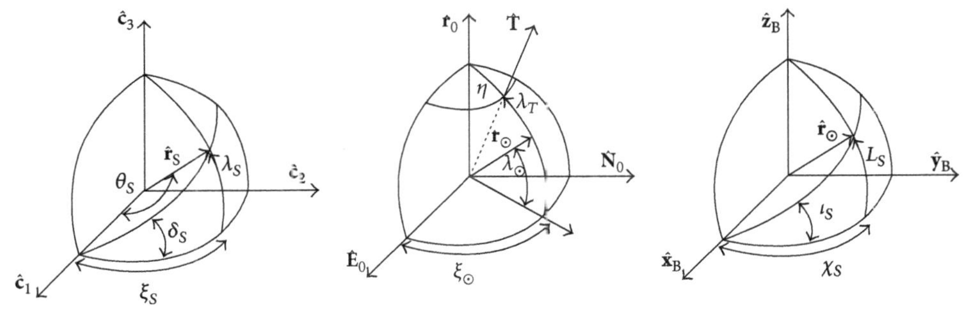


(a) Sun position $\hat{\mathbf{r}}_S$ in the ECI frame

(b) Sun position $\hat{\mathbf{r}}_\odot$ in the ALI frame. This case is an example of the sun being outside the spherical section allowed by $\eta$

(c) Sun position $\hat{\mathbf{r}}_\odot$ in the BA frame. $\chi_S$ and $L_S$ are obtained projecting $\iota_S$ on the BA frame

FIGURE 8: Sun-pointing guidance principal angles and axes.

The remaining angles $\{\alpha_T, \beta_T, \alpha, \beta, \sigma\}$ can be found using analytical formulas at any times using the rotation matrices portrayed in Figure 6.

*5.2. Near Radial Guidance.* This guidance is called near radial because it is defined in the AO reference frame and the thrust direction is defined by

$$\alpha_T = \gamma_{R0} + \frac{t}{t_B}\left(\frac{\pi}{2} - \gamma_{R0}\right) - \gamma_R, \qquad \beta_T = 0 \tag{24}$$

whereas $\gamma_{R0} = \theta_0$ is the pitch angle at the initial time and is equal to the initial flight path angle. With $\beta_T = 0$, the angle that the rocket forms with the local horizontal axis is given by $(\alpha_T + \gamma_R)$. At the initial time, this angle corresponds to the initial flight path angle, while at the burnout time, it corresponds to $(\alpha_T + \gamma_R) = \pi/2$. The roll angle is can be chosen arbitrarily and is set to 0 deg. For the remaining angles $\{\theta, \psi, \alpha, \beta, \sigma\}$, they are obtained through analytical expressions through the rotation matrices in Figure 6. The thrust direction tends to be aligned with the radial direction. Thus, in principle, this guidance leads to obtain the mean maximal altitude for the rocket at hand. Moreover, with this guidance, the usual values of $\Delta g$ are in the order of $10^{-4}$m/sec$^2$.

*5.3. Sun-Pointing Guidance.* The sun-pointing guidance is based on maintaining constant attitude of the rocket with respect to the sun position. The attitude angles $\{\phi, \theta, \psi\}$ are computed in an analytical fashion using the rotations in Figure 6. The first step to obtain the attitude of the sounding rocket is to locate the sun position with respect to the ECI frame:

$$\hat{r}_S = [\cos\lambda_S \cos\xi_S \quad \cos\lambda_S \sin\xi_S \quad \sin\lambda_S][\hat{c}_1 \quad \hat{c}_2 \quad \hat{c}_3]^T, \tag{25}$$

where $\lambda_S$ and $\xi_S$ are the sun declination and the right ascension, respectively. Then, projection of this unit vector into the ALI frame yields as follows:

$$\begin{aligned} \hat{\mathbf{r}}_\odot &= \hat{\mathbf{r}}_S R_3(-\lambda_{aL}) R_2(L_L) \begin{bmatrix} \hat{\mathbf{r}}_0 \\ \hat{\mathbf{E}}_0 \\ \hat{\mathbf{N}}_0 \end{bmatrix} \\ &= [\cos\lambda_\odot \cos\xi_\odot \quad \cos\lambda_\odot \sin\xi_\odot \quad \sin\lambda_\odot] \begin{bmatrix} \hat{\mathbf{r}}_0 \\ \hat{\mathbf{E}}_0 \\ \hat{\mathbf{N}}_0 \end{bmatrix} \\ &= [X_\odot \quad Y_\odot \quad Z_\odot] \begin{bmatrix} \hat{\mathbf{r}}_0 \\ \hat{\mathbf{E}}_0 \\ \hat{\mathbf{N}}_0 \end{bmatrix}. \end{aligned} \tag{26}$$

So the thrust direction is obtained by aligning the body axis of the launcher with the sun direction in the ALI frame, as shown in Figure 8. To avoid that the rocket will fly with a $\gamma_r$ near to zero while trying to catch the sun slightly above the local horizon, the thrust direction is constrained to have a displacement from $\hat{\mathbf{r}}_\odot$ not exceeding the maximal value $\eta = 10$deg (cf. Figure 8). So the thrust direction is defined imposing $\xi_T = \xi_\odot$ and $\lambda_T = \max(\lambda_\odot, \pi/2 - \eta)$. The angles $\alpha_T$ and $\beta_T$ are then obtained through rotation of $\hat{\mathbf{T}}$ from the ALI frame to the RV frame using the rotations reported in Figure 6 and considering that the thrust is aligned with the longitudinal axis of the rocket $\hat{\mathbf{x}}_B$. The pitch and yaw angles are then obtained with the following expressions:

$$\begin{aligned} \sin\theta &= \cos\lambda_T \cos\xi_T \rightarrow \theta, \\ \cos\psi &= \sin\lambda_T, \\ \sin\psi &= \cos\lambda_T \sin\xi_T \rightarrow \psi. \end{aligned} \tag{27}$$

As roll angle is arbitrary, it can be selected after choosing the angle $\iota_S$ related to the sun position in the body frame. The roll angle $\phi$ is obtained by comparing $\hat{\mathbf{r}}_\odot$ in the ALI frame with its projection in the RV frame.

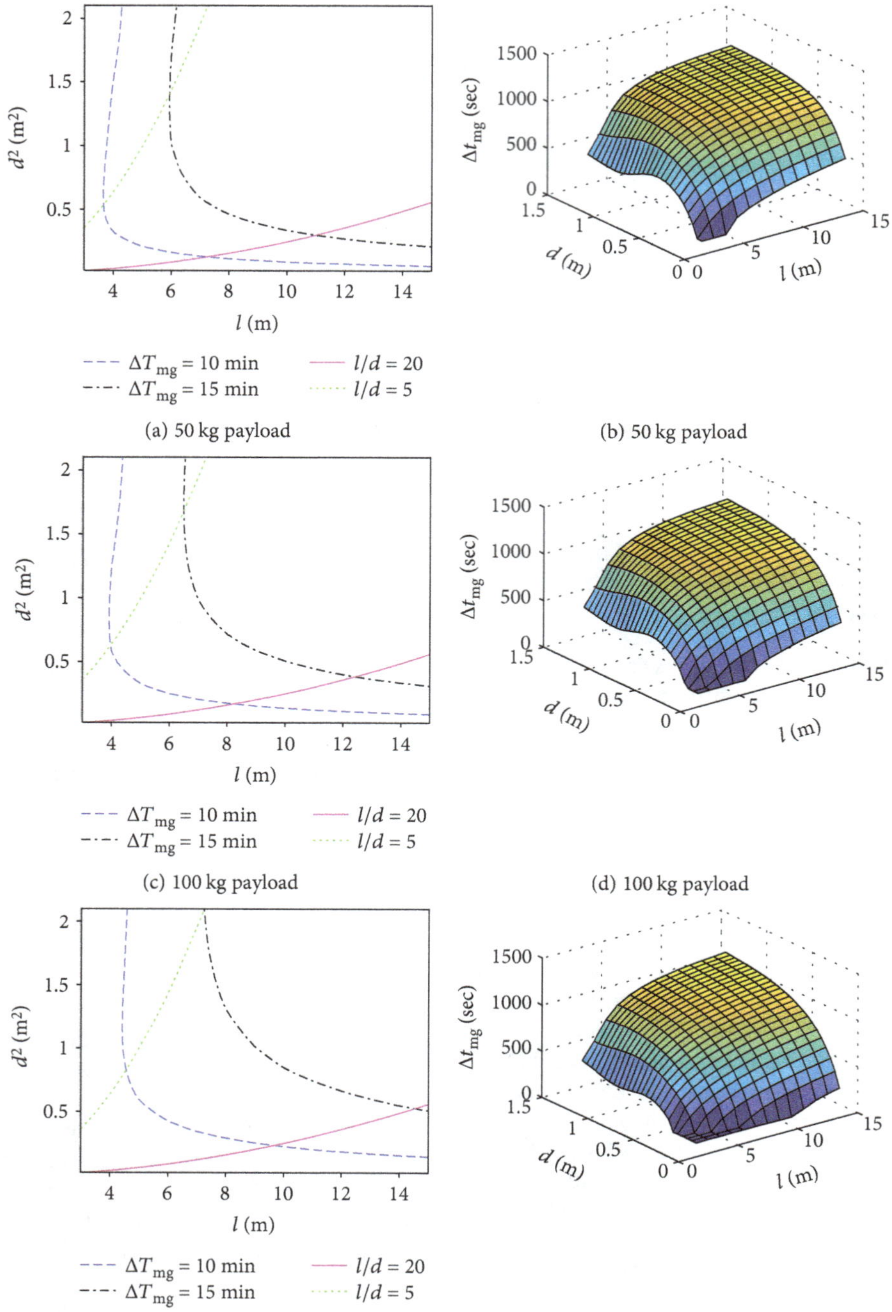


(a) 50 kg payload (b) 50 kg payload

(c) 100 kg payload (d) 100 kg payload

(e) 200 kg payload (f) 200 kg payload

Figure 9: Optimal results for the constant attitude guidance.

From Figure 8(c), it is straightforward to obtain the following angles:

$$\frac{\sin L_S}{\sin \iota_S} = \sin \lambda_T \longrightarrow L_S,$$
$$\frac{\tan L_S}{\tan \iota_S} = \sin \xi_S, \tag{28}$$
$$\frac{\cos \lambda_T}{\cos L_S} = \cos \xi_S \longrightarrow \xi_S.$$

They univocally define the vector $\hat{\mathbf{r}}_{\odot} = [X_{B\odot} Y_{B\odot} Z_{B\odot}, \hat{\mathbf{x}}_B \hat{\mathbf{y}}_B \hat{\mathbf{z}}_B]^T$ in the body frame. Comparing the coordinates of $\hat{\mathbf{r}}_{\odot}$ in the ALI frame to the ones in the BA frame leads the following expression:

$$\begin{aligned}[X_{\odot} \quad Y_{\odot} \quad Z_{\odot}] R_B{}^T R_3{}^T(\psi) R_2{}^T(\theta) R_1{}^T(\phi) \\ = [X_{B\odot} \quad Y_{B\odot} \quad Z_{B\odot}].\end{aligned} \tag{29}$$

Equation (29) leads to the roll angle $\phi$.

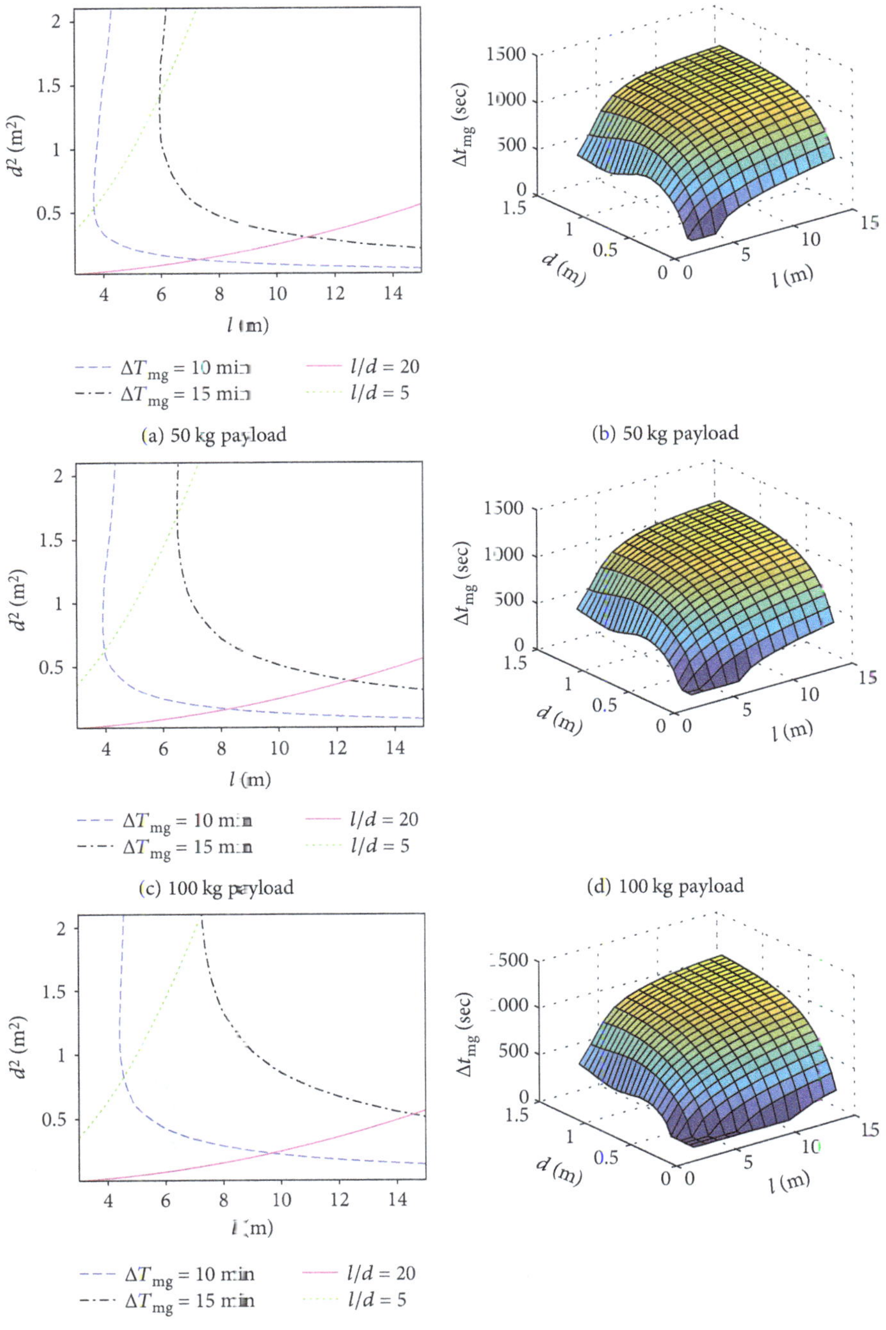


(a) 50 kg payload (b) 50 kg payload

(c) 100 kg payload (d) 100 kg payload

(e) 200 kg payload (f) 200 kg payload

FIGURE 10: Optimal results for the near radial guidance.

Then, the aerodynamic angles $\alpha, \beta, \sigma$ can be obtained through the rotation matrices in Figure 6, which are not reported for the sake of brevity. In short, the sun-pointing guidance requires specifying the following parameters:

(i) $\theta_S$ which is the sun position in the ECI frame

(ii) $\eta$ which constraints the possible thrust direction

(iii) $\theta_{G0}$, defined by the date and hour of the launch

(iv) $\iota_S$, needed to obtain the roll angle $\phi$

As the attitude is also in this case constant, the sun-pointing guidance is a particular case of the constant attitude guidance. Hence, it maintains its main advantages like the ideally perfect microgravity condition ($\Delta g = 0$) and can provide the perfect conditions for thermal experiments in the space.

## 6. Numerical Results

Figures 9–11 portray the results of the computation, which are, respectively, for the constant attitude guidance

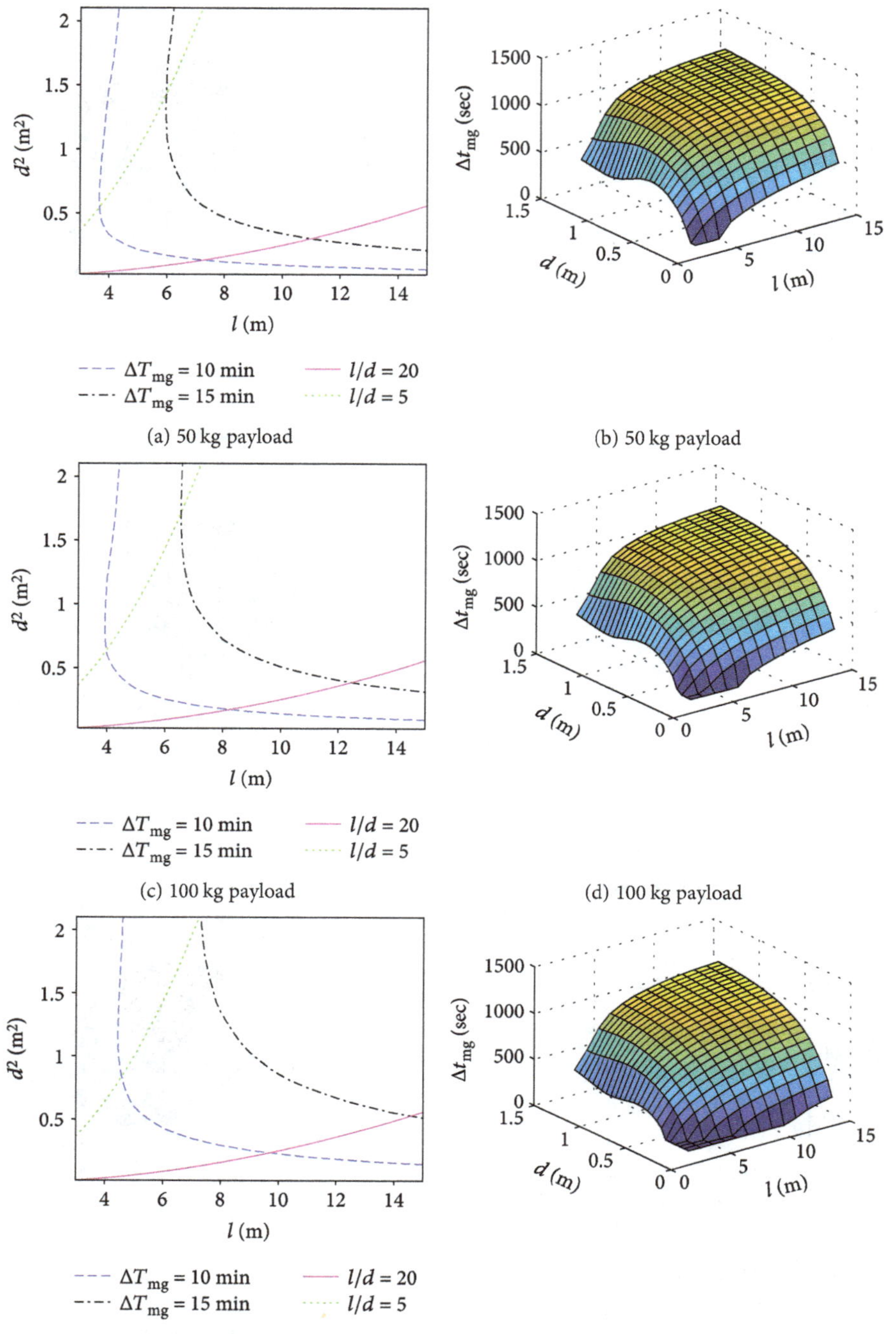


(a) 50 kg payload (b) 50 kg payload

(c) 100 kg payload (d) 100 kg payload

(e) 200 kg payload (f) 200 kg payload

FIGURE 11: Optimal results for the sun-pointing guidance.

TABLE 5: Optimal configurations for each payload mass, $\Delta T_{mg}$ = 10 min.

| $m_U$ (kg) | 50 | 100 | 200 |
|---|---|---|---|
| $l$ (m) | 7.29 | 8.25 | 9.69 |
| $d$ (m) | 0.36 | 0.41 | 0.48 |

TABLE 6: Optimal configurations for each payload mass, $\Delta T_{mg}$ = 15 min.

| $m_U$ (kg) | 50 | 100 | 200 |
|---|---|---|---|
| $l$ (m) | 10.98 | 12.42 | 14.53 |
| $d$ (m) | 0.55 | 0.62 | 0.73 |

(Figure 9), the near radial guidance (Figure 10), and the sun-pointing guidance (Figure 11). From Figure 9 to Figures 11(b), 11(d), and 11(f)), if the desired $\Delta T_{mg}$ is not chosen, it is possible to obtain a maximum $\Delta T_{mg}$ for each length. This result is particularly visible for the shorter lengths while is less evident for the bigger configurations.

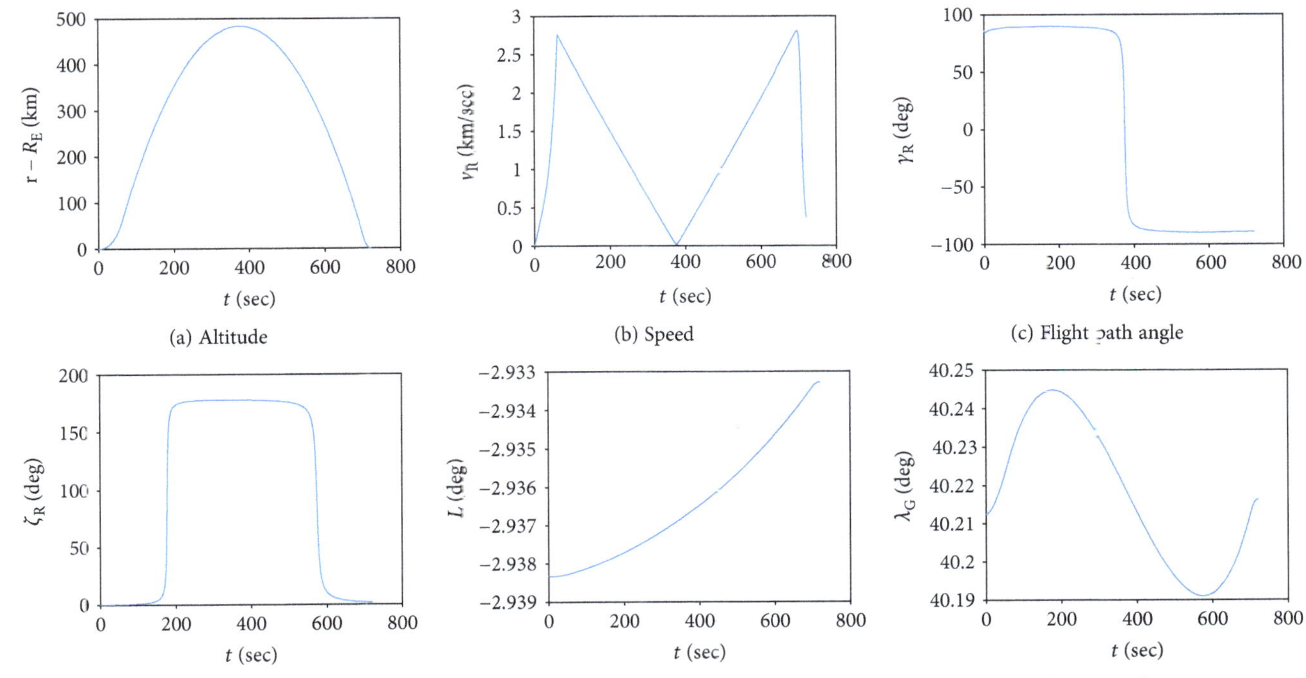


(a) Altitude (b) Speed (c) Flight path angle

(d) Heading angle (e) Latitude (f) Longitude

FIGURE 12: State variables for the configuration with $l = 7.29$ m and $d = 0.36$ m (near radial guidance).

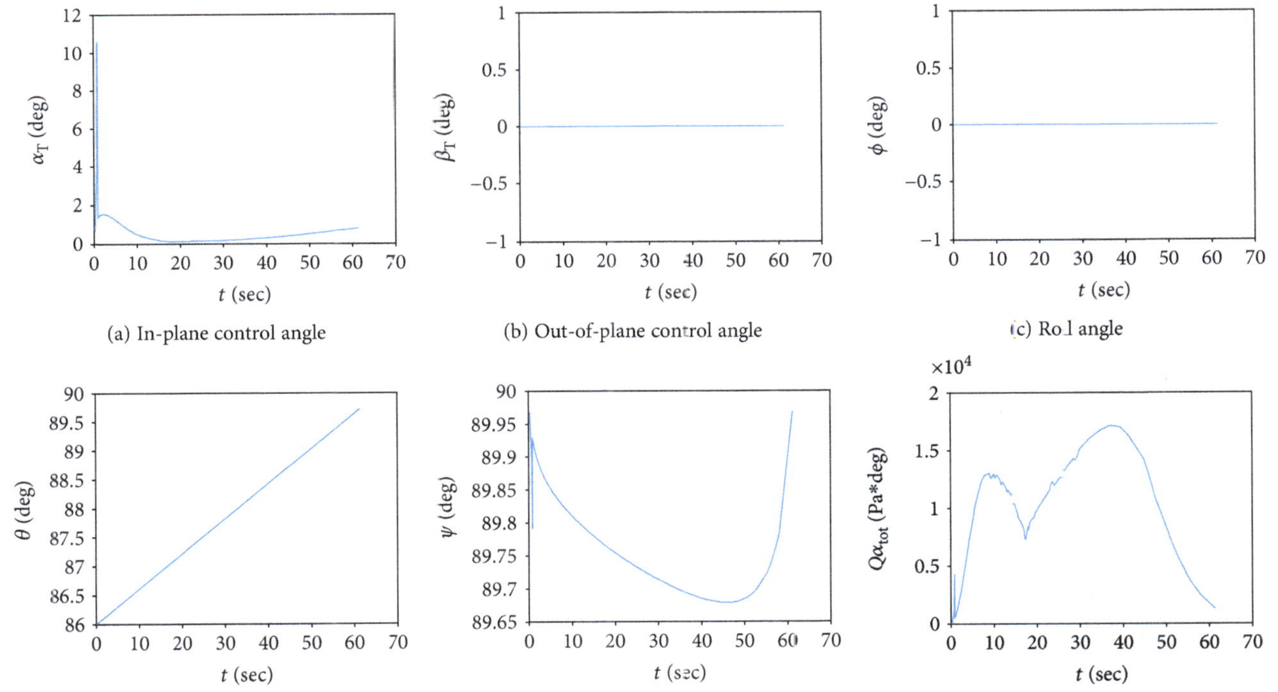


(a) In-plane control angle (b) Out-of-plane control angle (c) Roll angle

(d) Pitch angle (e) Yaw angle (f) Dynamic pressure constraint

FIGURE 13: Control law, attitude angles, and dynamic pressure in the thrust arc for the configuration with $l = 7.29$ m and $d = 0.36$ m (near radial guidance).

Panels (a), (c), and (e) of Figures 9–11 show every $(l, d)$ combination which guarantee 600 and 900 sec of $\Delta T_{mg}$. The optimal configuration corresponds to the point in which each curve is tangent to an equilateral isovolume hyperbola in the $(l, d^2)$ graph. As the tangence point always occurs in the lowest right part of the graph, two constraints were

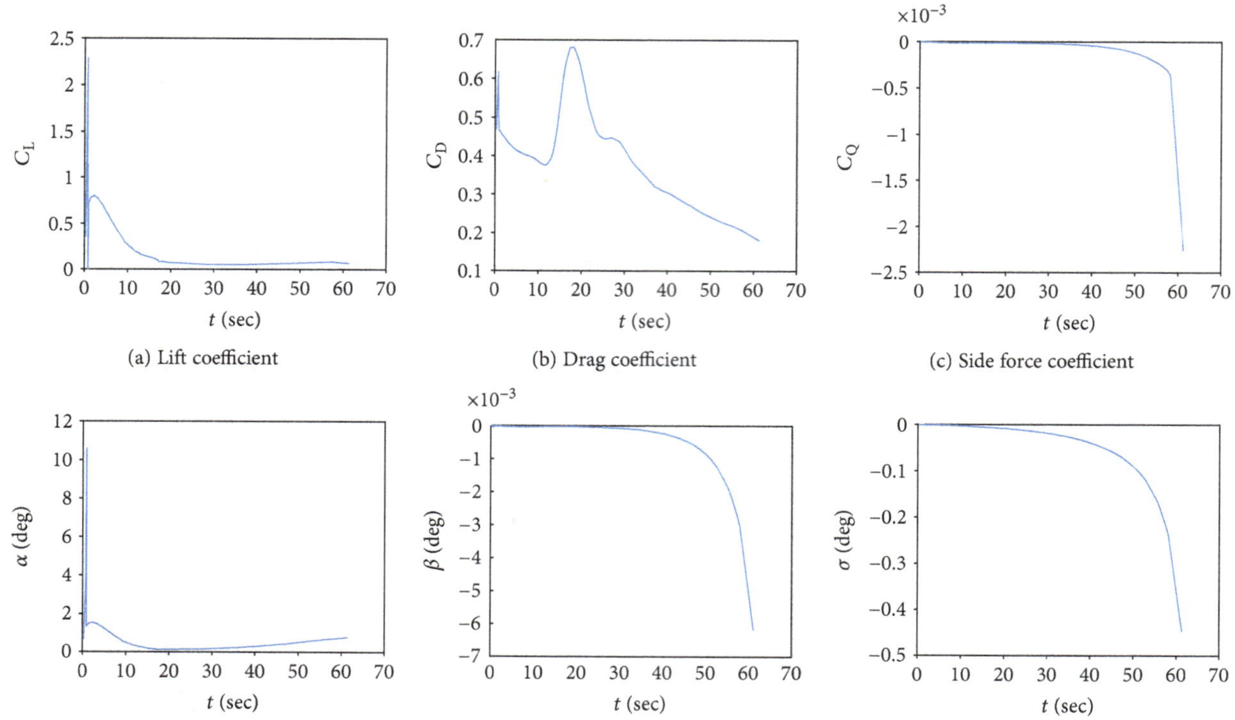


(a) Lift coefficient (b) Drag coefficient (c) Side force coefficient

(d) Angle of attack (e) Sideslip angle (f) Bank angle

FIGURE 14: Aerodynamic angles and coefficients in the thrust arc for the configuration with $l = 7.29$ m and $d = 0.36$ m (near radial guidance).

inserted, for the sake of avoiding too slender or bulky configurations. In general, the constraint for the $l/d$ is

$$5 \leq \frac{l}{d} \leq 20. \tag{30}$$

It should be noticed that that the optimal configurations for all the three guidance are nearly the same, and this means that the preliminary design of the sounding rocket can be done regardless of the guidance type. Tables 5 and 6 show the optimal rocket configuration for 600 and 900 sec of $\Delta T_{mg}$ and for all the payload masses. The configuration with $l = 7.29$m and $d = 0.36$m in Table 5 is particularly similar to the Black Brant VC, demonstrating that with similar $m_U$ and $\Delta T_{mg}$, the algorithm can be effective to optimize the size of the rocket.

Figures 12–14 portray the complete state vector elements and other variables for the optimal configuration with $m_U = 50$ kg and $\Delta T_{mg} = 10$ min, using the near radial guidance.

For the rocket structural integrity, the constraint $Q\alpha_{tot} < 150000\text{Pa} * \text{deg}$ must be satisfied, where

$$Q = \frac{1}{2}\rho v_R^2 \tag{31}$$

and

$$\alpha_{tot} = \arccos(\cos\alpha \cos\beta). \tag{32}$$

From Figure 13(f), it is apparent that the constraint is satisfied, and the same occurs for all the cases.

## 7. Conclusions

Designing a sounding rocket is a difficult task, and this work proposes a method to simplify the early stage modeling process and performance evaluation. The new methodology at hand has four principal phases: (i) mass and geometry sizing based on reasonable assumptions, (ii) scaled propulsion modeling, (iii) accurate aerodynamic modeling, and (iv) the performance evaluation using different guidance schemes. With regard to the guidance algorithms, three schemes where presented which are the constant attitude, near radial, and sun-pointing guidances. While the constant attitude guidance is a well-known scheme, the other two are novel guidances. In particular, the benefits of the sun-pointing guidance are presented, which are the nearly perfect microgravity condition and the possibility to orient the rocket with a defined attitude with respect to the sun. The results clearly show that the algorithm is capable of providing a fast and reasonable sizing in accordance with the mission requirements. The results obtained can be used as a reference point for a montecarlo campaign to analyze the impact points in the presence of local winds or as a first guess solution for more sophisticated optimization algorithms.

## Nomenclature

$g$: Earth gravitational acceleration, m/sec$^2$
$\mu_E$: Earth gravitational parameter, km$^2$/sec$^3$
$d$: Rocket diameter, m
$l_{ogive}$: Ogive length, m
$C_{m_0}$: Dimensionless total initial mass

$m_0$: Initial mass, kg
$m_P$: Propellant mass, kg
$V_{body}$: Rocket volume, $m^3$
$u$: Mass ratio
$T$: Thrust, N
$I_{SP}$: Specific impulse, sec
$t_B$: Burnout time, sec
$C_L$: Lift coefficient
$C_Q$: Side force coefficient
$D$: Drag force, N
$M$: Mach number
$\alpha$: Angle of attack, deg
$S$: Reference surface, $m^2$
$\Delta T_{mg}$: Microgravity duration, sec
$\theta_G$: Greenwich sidereal time, deg
$L_L$: Launch base latitude, deg
$\lambda_G$: Longitude, deg
$v_R$: Relative velocity m/sec$^2$
$\sigma$: Bank angle, deg
$\theta$: Pitch angle, deg
$\alpha_T$: In-plane control angle, deg
$m$: Mass, kg
$\lambda_{\odot}$: Sun declination ALI frame, deg
$\lambda_S$: Sun declination ECI frame, deg
$\lambda_T$: Thrust declination ALI frame, deg
$L_S$: Sun declination BA frame, deg
$\omega_E$: Earth rotation rate, deg/sec$^2$
$l$: Rocket total length, m
$t$: Time, sec
$l_{body}$: Rocket body length, m
$C_{ogive}$: Dimensionless ogive length
$m_{aux}$: Auxiliary mass, kg
$m_S$: Structural mass, kg
$\epsilon$: Structural coefficient
$m_U$: Payload mass, kg
$n_0$: Acceleration at launch, g
$\dot{m}$: Mass rate, kg/sec
$m_{0body}$: Cylinder initial mass, kg
$C_D$: Drag coefficient
$L$: Lift force, N
$Q$: Side force, N
Re: Reynolds number
$\beta$: Sideslip angle, deg
$\rho$: Atmospheric density, kg/$m^3$
$r$: Radius, m
La: Latitude, deg
$\lambda_{aL}$: Launch base longitude, deg
$\gamma_R$: Relative flight path angle, deg
$\zeta_R$: Heading angle, deg
$\phi$: Roll angle, deg
$\psi$: Yaw angle, deg
$\beta_T$: Out-of-plane control angle, deg
$Q$: Dynamic pressure, Pa
$\xi_{\odot}$: Sun right ascension ALI frame, deg
$\xi_S$: Sun right ascension ECI frame, deg
$\xi_T$: Thrust right ascension ALI frame, deg
$\chi_S$: Sun right ascension BA frame, deg.

## Conflicts of Interest

The authors declare that they have no conflicts of interest.

## References

[1] A. Asensio Ramos and R. Manso Sainz, "Signal detection for spectroscopy and polarimetry," *Astronomy and Astrophysics*, vol. 547, p. A113, 2012.

[2] E. Luvsandamdin, S. Spießberger, M. Schiemangk et al., "Development of narrow linewidth, micro-integrated extended cavity diode lasers for quantum optics experiments in space," *Applied Physics B*, vol. 111, no. 2, pp. 255–260, 2013.

[3] S. Wieman, L. Didkovsky, T. Woods, A. Jones, and C. Moore, "Sounding rocket observations of active region soft X-ray spectra between 0.5 and 2.5 nm using a modified SDO/EVE instrument," *Solar Physics*, vol. 291, no. 12, pp. 3567–3582, 2016.

[4] Sounding Rocket Program Handbook, *National Aeronautics and Space Administration, Goddard Space Flight Center*, 2001.

[5] G. Seibert, *The History of Sounding Rockets and Their Controibution to European Space Research*, ESA Publications Division, 2006.

[6] S. M. Chowdhury, J. P. de la Beaujardiere, M. Brooks, and L. Roberts, "An integrated six degree-of-freedom trajectory simulator for hybrid sounding rockets," in *49th AIAA Aerospace Sciences Meeting including the New Horizons Forum and Aerospace Exposition*, Orlando, Florida, January 2011.

[7] S. M. Chowdhury, *Design and Performance Simulation of a Hybrid Sounding Rocket*, Master's thesis, University of Kwa-Zulu-Natal, 2012.

[8] L. Casalino and D. Pastrone, "Optimization of hybrid sounding rockets for hypersonic testing," in *45th AIAA/ASME/SAE/ASEE Joint Propulsion Conference & Exhibit*, Denver, Colorado, August 2009.

[9] L. Casalino and D. Pastrone, "Optimization of hybrid sounding rockets for hypersonic testing," *Journal of Propulsion and Power*, vol. 28, no. 2, pp. 405–411, 2012.

[10] S. Nonaka, H. Ogawa, Y. Naruo, and Y. Inatani, "System design and tecnical demonstrations for reusable sounding rocket," in *20th ESA Symposium on European Rocket and Balloon Programmes and Related Research*, Hyére, France, October 2011.

[11] D. L. Y. Woo and J. A. Martin, "Reusable sounding-rocket design," *Journal of Spacecraft and Rockets*, vol. 32, no. 2, pp. 376–378, 1995.

[12] F. Cremaschi, S. Weikert, A. Wiegand, W. Jung, and F. Scheuerpflug, "Sounding rocket trajectory simulation and optimization with astos," in *19th ESA Symposium on European Rocket and Balloon Programmes and Related Research*, 2009.

[13] W. B. Blake, *Missile DATCOM User's Manual – 1997 Fortran 90 Revision*, United States Air Force, 1998.

[14] L. Mangiacasale, *Meccanica del volo atmosferico. Terne di riferimento, equazioni di moto, linearizzazione, stabilità*, Ingegneria, 2000.

[15] M. Pallone, M. Pontani, and P. Teofilatto, "Accurate modeling and near optimal ascent trajectory of microsatellite launch vehicles via firework algorithm," in *67th International Astronautical Congress*, 2016.

[16] M. Pontani, "Particle swarm optimization of ascent trajectories of multistage launch vehicles," *Acta Astronautica*, vol. 94, no. 2, pp. 852–864, 2014.

[17] L. Ljunge and L. Hall, "S19 guidance of the Black Brant X sounding rocket," *Journal of Guidance, Control, and Dynamics*, vol. 7, no. 2, pp. 156–160, 1984.

# 7

# Mission Overview and Initial Observation Results of the X-Ray Pulsar Navigation-I Satellite

**Xinyuan Zhang,[1] Ping Shuai,[1] Liangwei Huang,[1] Shaolong Chen,[1] and Lihong Xu[2]**

[1]*Qian Xuesen Laboratory of Space Technology, Beijing 100094, China*
[2]*China Academy of Space Technology, Beijing 100094, China*

Correspondence should be addressed to Liangwei Huang; huangliangwei@qxslab.cn

Academic Editor: Linda L. Vahala

The newly launched X-ray pulsar navigation-I (XPNAV-1) is an experimental satellite of China that is designed for X-ray pulsar observation. This paper presents the initial observation results and aims to recover the Crab pulsar's pulse profile to verify the X-ray instrument's capability of observing pulsars in space. With the grazing-incidence focusing type instrument working at the soft X-ray band (0.5–10 keV), up to 162 segments of observations of the Crab pulsar are fulfilled, and more than 5 million X-ray events are recorded. Arrival times of photons are corrected to the solar system barycentre, and the 33 ms pulse period is sought out for Crab. Epoch folding of all the corrected photon times generates the refined pulse profile of Crab. The characteristic two-peak profile proves that the Crab pulsar has been clearly seen, so that the conclusion is made that XPNAV-1's goal of being capable of observing pulsars is achieved.

## 1. Introduction

Pulsars [1], which are recognized as rotating neutron stars, can emit regular signals from the radio to the high energy band [2]. Navigation using X-ray pulsars [3–5] is regarded as a revolutionary technology providing autonomous spacecraft navigation capability in the whole solar system.

To demonstrate this technology, China Academy of Space Technology (CAST) has brought forward a three-step space demonstration scheme. The first step is the X-ray pulsar navigation-I (XPNAV-1) satellite, the core goal of which is to validate the capability of observing X-ray pulsars. The following step is to launch a medium-sized satellite in about 2~3 years to accumulate more X-ray data for pulsar parameter database construction and to test pulsar navigation algorithm onboard after 3~5 pulsars are timed enough accurately. The third step is to build a constellation system to demonstrate navigation application and time service using X-ray pulsars.

XPNAV-1 was launched at Beijing time Nov. 10, 2016, by the Long March 11 rocket at the Jiuquan Satellite Launch Centre. It is a small satellite weighing about 270 kg and works on the 500 km LEO orbit. The purpose of the XPNAV-1 mission is to test the technology of pulsar observation in the soft X-ray band through the X-ray instruments developed by CAST. Three objectives are outlined: (1) test function and performance for the X-ray instruments in the outer space; (2) detect typical X-ray pulsars' radiation photons and acquire the pulse profiles to verify the ability of X-ray pulsar observation; (3) accumulate X-ray data for a long time to measure the pulsars' parameters via X-ray timing.

The XPNAV-1 satellite works at the three-axis stabilization attitude mode with the ability to quickly point to any inertial position according to demand as accurate as 2 arcminutes and to provide up to 90 minutes' sustained observation limited by the power supply. The satellite operates two X-ray devices. One is the Time-resolved soft X-ray spectrometer (TSXS), and the other is the high time-resolution photon counter (HTPC). The TSXS uses a Wolter-I type lens of 4 nested mirror shells with the collecting area of 30 $cm^2$ to focus X-ray photons within 15 arcminutes field-of-view onto a silicon drift detector (SDD). A GPS calibrated Rb clock is included to provide accurate time and a quasi-parallelly optical star is used to assist the inertial pointing. The TSXS device provides the 1.5 $\mu s$ time resolution and the

TABLE 1: Default targets' angular positions and pulse periods.

| Number | Pulsar name | J2000 right ascension (deg) | J2000 declination (deg) | Pulse period (ms) |
|---|---|---|---|---|
| 1 | B0531+21 (Crab) | 83.63303 | 22.01449 | 33.085 |
| 2 | B1617-155 (Sco X-1) | 244.979 | −15.640 | 3.200 |
| 3 | B1758-250 | 270.284 | −25.079 | 3.000 |
| 4 | B1813-140 | 274.006 | −14.036 | 3.300 |
| 5 | GROJ1744-28 | 266.138 | −28.741 | 467.000 |
| 6 | B0540-69 | 85.04668 | −69.33171 | 50.499 |
| 7 | B1509-58 | 228.48175 | −59.13583 | 150.658 |
| 8 | J1846-0258 | 281.60392 | −2.97503 | 325.684 |

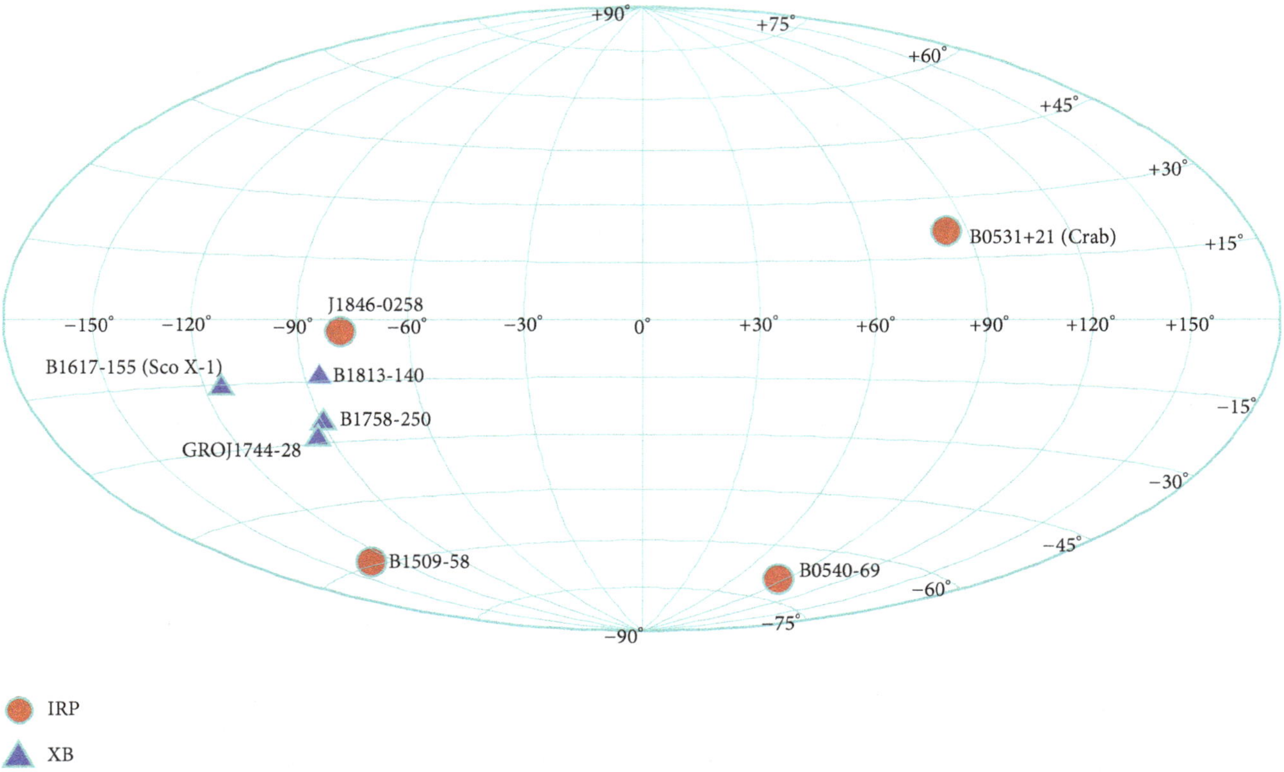


FIGURE 1: Default targets' angular positions plotted in the J2000 frame.

180 eV@5.9 keV energy resolution in the 0.5~10 keV energy band. The HTPC device uses the collimator to confine the field-of-view to 2 degrees and the microchannel plate (MCP) X-ray detector to count the X-ray photons in the 1~10 keV energy band from the pulsar. Compared to the TSXS, the MCP of the HTPC has a higher time resolution of 100 ns and a bigger collecting area of 1200 cm$^2$.

In order to examine the satellite-borne X-ray instruments, we choose the Crab pulsar [6] as the calibration target because this pulsar is believed to be one of the best studied objects in the sky and one of the brightest X-ray sources regularly studied. Current data show that TSXS exhibits a good and steady performance in X-ray photon collection. In this paper, we will present the data analysis results of the TSXS observation on Crab as well as the mission overview. We organize this paper as follows. The mission flight system is described in Section 2. The XPNAV-1 X-ray data analysis software system is introduced in Section 3. The observations and the data analysis results are provided in Section 4. The conclusions and discussions are made in Section 5.

## 2. XPNAV-1 Mission Overview

The core objective of the XPNAV-1 mission is to validate the ability of observing X-ray pulsars. To achieve this, eight X-ray sources are chosen as default targets. Not only 4 isolated rotation-powered (IRP) pulsars but also 4 X-ray binaries (XB) are included because of their brightness. The eight sources' angular positions and pulse periods [7–9] are shown in Table 1 and Figure 1. XPNAV-1 provides four observation modes: the self-test mode, the scanning mode, the default target

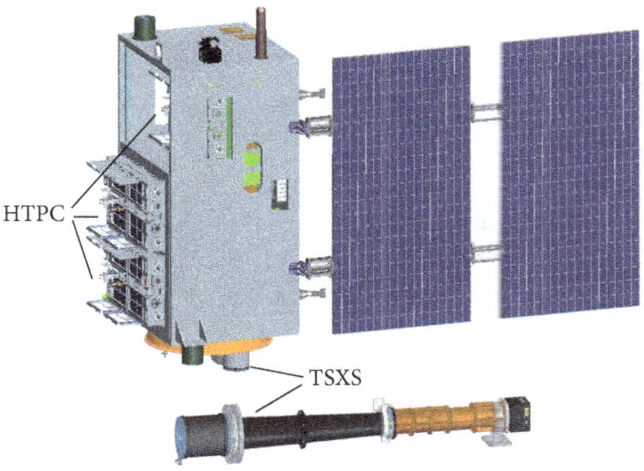


FIGURE 2: XPNAV-1 satellite structure overview.

observation mode, and the arbitrary target pointing mode. The self-test mode works when the lens is covered with the hood in order that the instrument noise can be measured. The scanning mode is to scan a belt of the space via rotating the satellite in order to evaluate the background noise. The default target observation mode makes the X-ray instrument point at one of the 8 default targets. The arbitrary target pointing mode is similar to the former mode except that the angular position parameters should be uploaded from ground.

Figure 2 provides an overview of the XPNAV-1 satellite structure. The satellite applies integrated electronics design, with which the functions of onboard data handling (OBDH), tracking telemetry & command (TTC), attitude determination and control, GNSS navigation, and power control are integrated in one electronic component. A zero momentum three-axis attitude control system is exploited with high-precision inertial pointing capability.

The XPNAV-1 satellite operates on a sun-synchronous orbit, of which the orbit semimajor axis is 6878.137 km, the inclination is 97.4 degrees, and the local time of descending node (LTDN) is 6:00 AM. Two instruments, TSXS and HTPC, are mounted along different directions. When the observation carries on, the satellite spins to make the X-ray instrument inertially pointing at the target, and when the observation is finished, the satellite spins to make the solar array point to the sun. The two instruments cannot work simultaneously because of the insufficient capability of the power supply.

In Beijing time Nov. 17, 2016, XPNAV-1 fulfilled the satellite test and entered the observation phase. Until Feb., 2017, XPNAV-1 has observed three sources, that is, PSR B0531+21 (Crab), PSR B0540-69, and PSR B1509-58. During each observation, X-ray devices are shut down in the region of south Atlantic anomaly (SAA) to avoid the potentially harmful impact on the working detector. During the Earth occultation, no observation is planned. Besides, when the angle between the target and the sun or the moon is less than 45 degrees, observation is ceased to avoid the interference signal.

The preliminary data show that TSXS has steady performance in the signal-to-noise ratio. Thus, we implement most observations on the Crab pulsar using TSXS. Totally, 162 TSXS observations of Crab are performed, each of which lasts 10 to 90 minutes. In the following sections, we will introduce the data analysis software system and present the analysis results.

## 3. Data Analysis Software System

The data analysis software system (DASS) is constructed to process XPNAV-1 data. The main purpose of DASS is to acquire the pulsar's pulse profile from the observed X-ray data in order to testify to the X-ray instrument's capability of observing pulsars. The system consists of nine parts, each of which is designed as a functional software. The nine pieces of software are separately named as *ORIDATA*, *XSELECT*, *XFLUX*, *XENERGY*, *ORBITPROP*, *XCORR*, *NOMDB*, *XFSEARCH*, and *XFOLD*, which are organized as Figure 3 shows.

*ORIDATA* deals with the original data unpacking. The telemetry, platform, and payload data transmitted from the satellite are unpacked and regrouped into the GPS data, the onboard orbit data, the satellite status flag data, the satellite attitude data, and the X-ray photon data. *XSELECT* performs X-ray photon selection according to the satellite status flag and the satellite attitude. The photons at the time when the satellite works well and the X-ray instrument points accurately to the source remained. *XFLUX* counts the photons to calculate the source or the background flux. *XENERGY* analyzes the photon energy data for the spectrum and radiation model researches. *ORBITPROP* propagates the orbit to determine the satellite's inertial position at every single photon's arrival time from the GPS data or the onboard orbit data. *XCORR* corrects the photon arrival time at the satellite to the arrival time at the solar system barycentre (SSB) or the emission time at the pulsar [10–14]. *NOMDB* [15] is the nominal pulsar parameter database, which provides the current known pulsar parameters for the data handling. *XFSEARCH* searches for the pulse period and its derivatives from the corrected photon arrival times in the time domain as well as the frequency domain [16, 17], and it will update the nominal database's pulsar parameters if necessary. *XFOLD* generates the pulsar's profile through epoch folding [18] and analyze its timing property.

## 4. Observations and Results

From coordinate universal time (UTC) 57709.0 in modified Julian date (MJD) to 57872.0 MJD, TSXS carried out 162 segments of observations upon PSR B0531+21. These TSXS observation data are made public and can be downloaded at the Beidou official website [19]. The data analysis work in this section is performed using this set of public data. The mean duration of the observations is about 39 minutes. From these observations, 5824511 photons in the 0.5~10 keV band

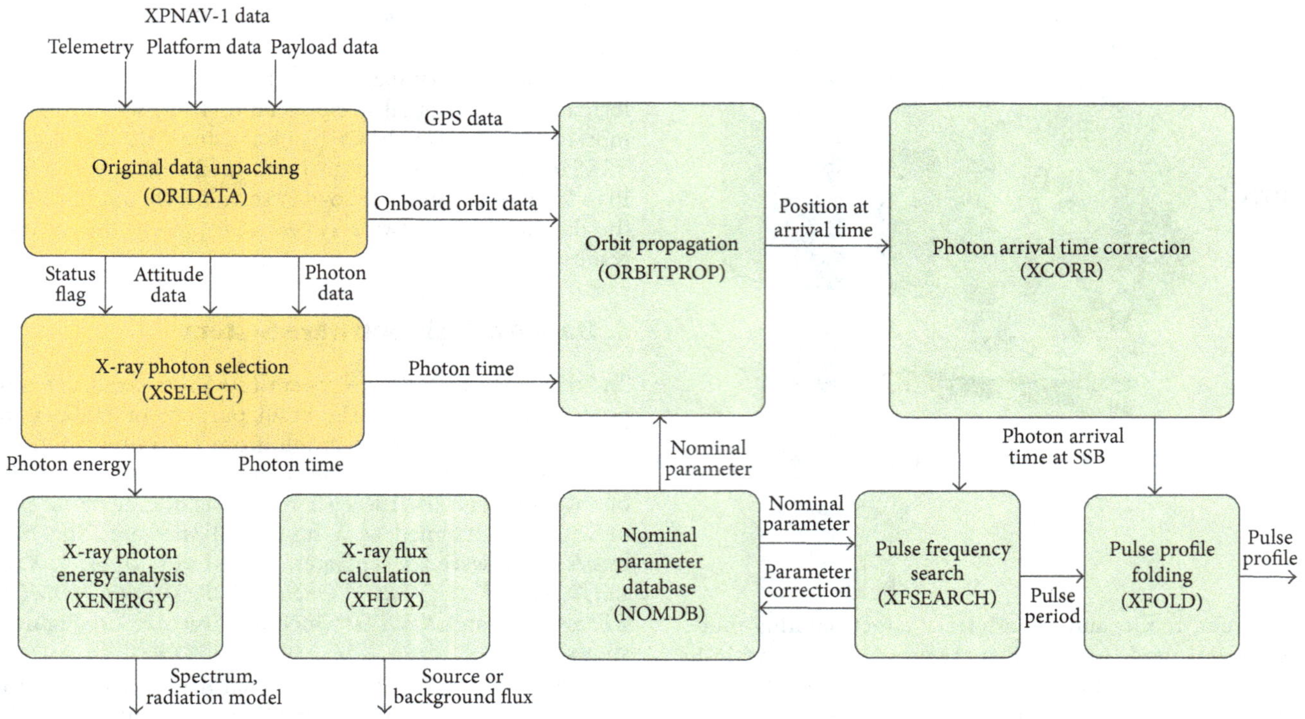


FIGURE 3: Structure and data flow of the data analysis software system.

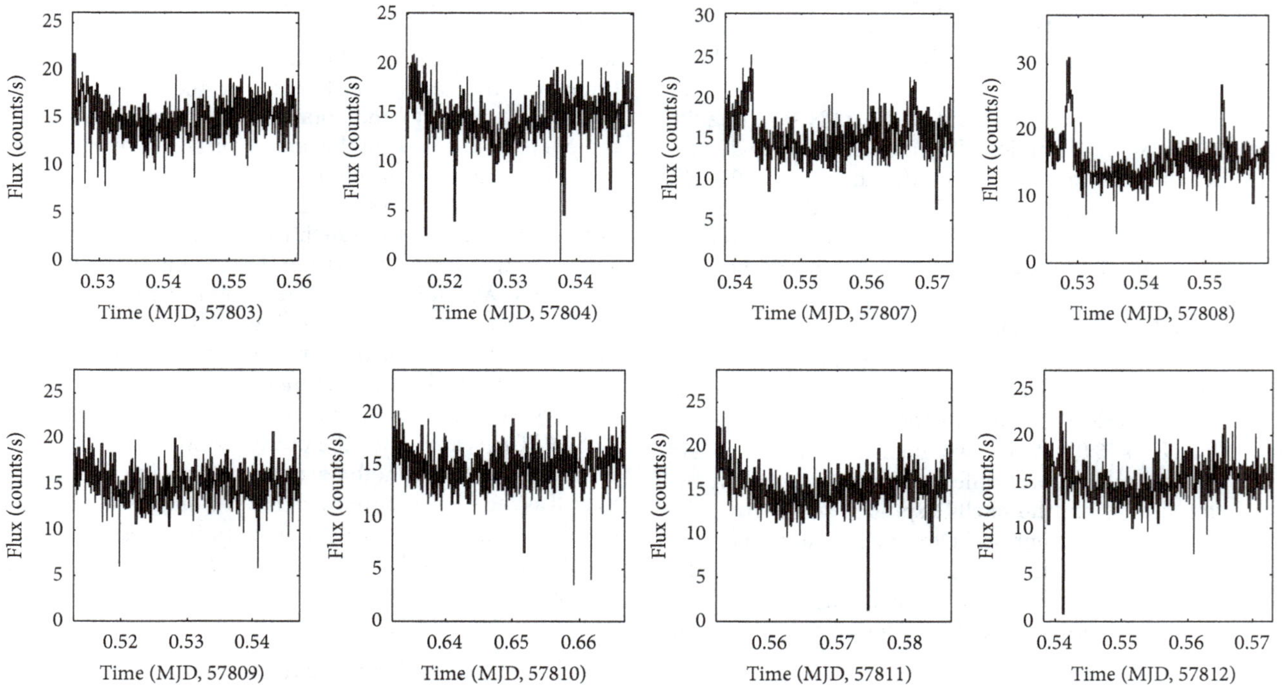


FIGURE 4: TSXS flux data of eight observations on Crab.

are collected. The mean flux in this band is 15.4 counts/s. The flux data of eight observations from UTC 57803.5 MJD to UTC 57812.6 MJD are illustrated in Figure 4.

The derived spectrum of Crab in the 0.5~7 keV band is shown in Figure 5. The vertical line in the figure represents the 3-sigma error. The spectrum analysis and the calibration of TSXS's energy property with respect to the radiation model will be our future work.

The photon arrival times at the X-ray device are corrected to the solar system barycentre (SSB) arrival times via the

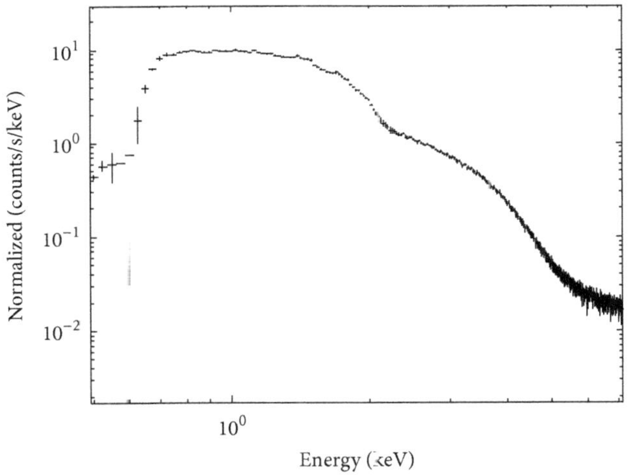


FIGURE 5: Spectrum of Crab in the 0.5~7 keV band from TSXS.

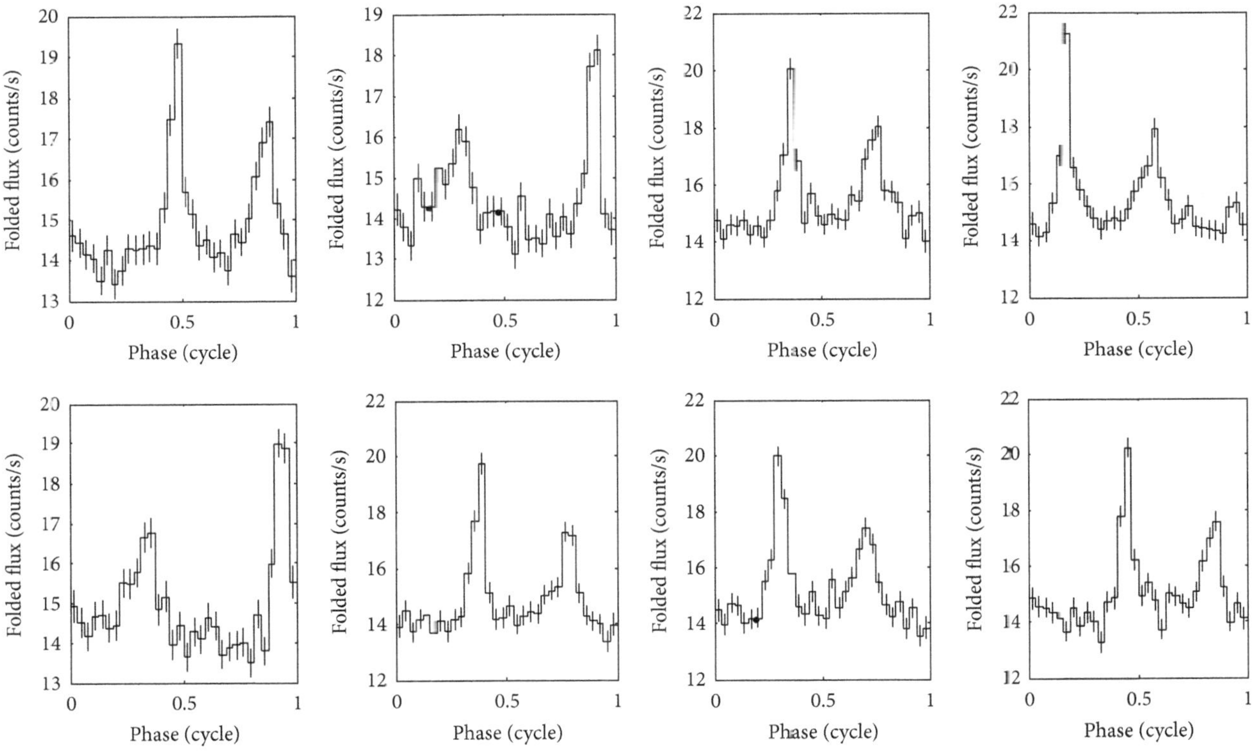


FIGURE 6: Crab profiles in the 0.5~9 keV band from eight observations from UTC 57803.5 MJD to UTC 57812.6 MJD.

orbit data. The periodicity is found from the corrected times in the frequency domain, and then the accurate pulse period and its derivatives are searched for in the time domain. With the period parameters, the pulse profile can be acquired by epoch folding from each observation. Furthermore, all observations can be combined to form a much finer profile.

In Figure 6, we provide eight folded Crab profiles in the 0.5~9 keV band derived from the corresponding observations in Figure 4. The profiles use 32 bins, the vertical lines of which indicate the folded 1 sigma folding flux error. The zero phase time in Figure 6 is taken as the observation beginning time. Therefore, by shifting the profile [20], the pulse time of arrival (TOA) corresponding to the peak time can be accurately figured out so that 162 pulse TOAs are acquired.

The combined Crab profile in the 0.5~9 keV band is plotted in Figure 7. This profile uses 512 phase bins, the characteristic two peaks [21] of which kept apart at about 0.4 cycles are clearly identified. The zero phase of the plot is aligned to the main peak. From this profile, the pulsed fraction of the Crab radiation in the 0.5~9 keV band can be estimated to be 5.3%.

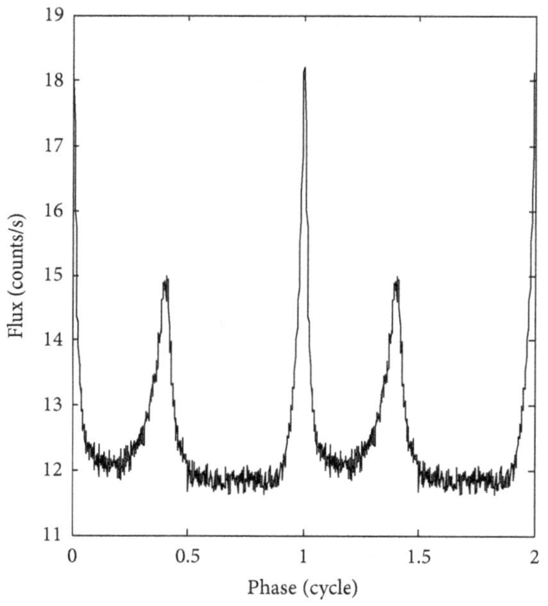


FIGURE 7: Crab profiles in the 0.5∼9 keV band from all 162 observations.

## 5. Conclusions and Discussion

In this paper, we give an overview of the XPNAV-1 mission, the properties of the X-ray instruments, and the data analysis software system. The TSXS observations on Crab are described in detail. Up to 162 segments of observations are collected containing over 5 million X-ray photons with a coming rate at 15.4 counts/s on average. Arrival times of photons within 0.5∼9 keV are corrected to SSB and epoch-folded, which produces folded flux curves corresponding to 162 pulse TOAs. Combination of the observations generates the refined profile for Crab, which shows a 5.3% pulsed fraction in the 0.5∼9 keV band. The characteristic profile of Crab has validated XPNAV-1's capability of observing pulsars in the soft X-ray band, which implies that the aim of the XPNAV-1 mission has been achieved. Next, we will research on how to use the Crab observation data to assist the orbit determination. Although three-dimensional positioning cannot be realized by observing only one pulsar, the divergence of the orbit propagation could be slowed down by incorporating the radial range measurement along Crab's line-of-sight. We will also spend more time observing the weaker sources and explore the possibility of navigation if more pulsars are observed. XPNAV-1 satellite is the first step of the consecutive missions designed by CAST for the demonstration of X-ray pulsar navigation. Next mission will bring to space an X-ray detector with a much larger collection area with the hope to accurately time 3∼5 pulsars and fulfill the onboard navigation.

## Conflicts of Interest

The authors declare that there are no conflicts of interest regarding the publication of this paper.

## Acknowledgments

This work was supported by the National Natural Science Foundation of China (Grants nos. 61403391, 61601463, and 11405265).

## References

[1] A. Hewish, S. J. Bell, J. D. H. Pilkington, P. F. Scott, and R. A. Collins, "Observation of a rapidly pulsating radio source," *Nature*, vol. 217, no. 5130, pp. 709–713, 1968.

[2] R. D. Lorimer and M. Kramer, *Handbook of Pulsar Astronomy*, Cambridge Univ Press, Cambridge, UK, 2005.

[3] E. Hanson J, *Principles of X-ray navigation. Ph.D Dissertation, [Ph.D. thesis]*, Dept of Aeronautics and Astronautics, Stanford Univ, Stanford, Calif, USA, 1996.

[4] S. I. Sheikh, D. J. Pines, P. S. Ray, K. S. Wood, M. N. Lovellette, and M. T. Wolff, "Spacecraft navigation using X-ray pulsars," *Journal of Guidance, Control, and Dynamics*, vol. 29, no. 1, pp. 49–63, 2006.

[5] P. Shuai, M. Li, and L. S. Chen, *Principles and Methods for X-ray Pulsar-based navigation System*, China Astronautic Publishing House, Beijing, China, 2009.

[6] M. G. F. Kirsch, U. G. Briel, D. Burrows et al., "Crab: The standard X-ray candle with all (modern) X-ray satellites," in *Proceedings of the UV, X-Ray, and Gamma-Ray Space Instrumentation for Astronomy XIV, SPIE*, pp. 1–12, August 2005.

[7] R. N. Manchester, G. B. Hobbs, A. Teoh, and M. Hobbs, "The Australia telescope national facility pulsar catalogue," *Astronomical Journal*, vol. 129, no. 4, pp. 1993–2006, 2005.

[8] The Australia Telescope National Facility (ANTF) Pulsar Catalogue. 2011, available: http://www.atnf.csiro.au/research/pulsar/psrcat.

[9] S. I. Sheikh, *The use of variable celestial X-ray sources for spacecraft navigation [Ph.D. thesis]*, Dept Aero Eng, Maryland Univ, College Park, MD, USA, 2005.

[10] T. Damour and N. Deruelle, "General relativistic celestial mechanics of binary systems. I. The post-Newtonian motion," *Annales de l'Institut Henri Poincaré (Physicque théorique)*, vol. 43, no. 1, pp. 107–132, 1985.

[11] T. Damour and N. Deruelle, "General relativistic celestial mechanics of binary systems II. The post-Newtonian timing formula," *Annales de l'Institut Henri Poincaré (Physicque théorique)*, vol. 44, no. 3, pp. 263–292, 1986.

[12] G. B. Hobbs, R. T. Edwards, and R. N. Manchester, "TEMPO2, a new pulsar-timing package - I. An overview," *Monthly Notices of the Royal Astronomical Society*, vol. 369, no. 2, pp. 655–672, 2006.

[13] R. T. Edwards, G. B. Hobbs, and R. N. Manchester, "TEMPO2, a new pulsar timing package - II. The timing model and precision estimates," *Monthly Notices of the Royal Astronomical Society*, vol. 372, no. 4, pp. 1549–1574, 2006.

[14] L. Huang, B. Liang, T. Zhang, and C. Zhang, "Navigation using binary pulsars," *Science China: Physics, Mechanics and Astronomy*, vol. 55, no. 3, pp. 527–539, 2012.

[15] L. Huang, P. Shuai, Q. Lin, and X. Zhang, "Research on nominal database for X-ray pulsar navigation," *Chinese Space Science and Technology*, vol. 35, no. 3, pp. 66–74, 2015.

[16] X.-Y. Zhang, P. Shuai, and L.-W. Huang, "Profile folding distortion and period estimation for pulsar navigation," *Journal of Astronautics*, vol. 36, no. 9, pp. 1056–1060, 2015.

[17] Y. Q. Zhou, F. J. Ji, and H. F. Ren, "Quick search algorithm of X-ray pulsar period based on unevenly spaced timing data," *Acta Physica Sinica*, vol. 62, no. 1, pp. 1–8, 2013.

[18] A. A. Emadzadeh and J. L. Speyer, "X-ray pulsar-based relative navigation using epoch folding," *IEEE Transactions on Aerospace and Electronic Systems*, vol. 47, no. 4, pp. 2317–2328, 2011.

[19] XPNAV-1 Observation Data, 2017, http://www.beidou.gov.cn/xpnavdata.rar.

[20] A. A. Emadzadeh and J. L. Speyer, "On modeling and pulse phase estimation of X-ray pulsars," *IEEE Transactions on Signal Processing*, vol. 58, no. 9, pp. 4484–4495, 2010.

[21] MPIfR EPN Pulsar Profiles Database, 2017, http://rian.kharkov.ua/decameter/EPN/browser.html.

# 8

# Investigation of Dual-Vortical-Flow Hybrid Rocket Engine without Flame Holding Mechanism

A. Lai,[1] T.-H. Chou,[1] S.-S. Wei,[1] J.-W. Lin,[1] J.-S. Wu,[1] and Y.-S. Chen[2]

[1]Department of Mechanical Engineering, National Chiao Tung University, Hsinchu, Taiwan
[2]National Space Organization, National Applied Research Laboratories, Hsinchu, Taiwan

Correspondence should be addressed to J.-S. Wu; chongsin@faculty.nctu.edu.tw

Academic Editor: Angel Velazquez

A 250 kgf thrust hybrid rocket engine was designed, tested, and verified in this work. Due to the injection and flow pattern of this engine, this engine was named dual-vortical-flow engine. This propulsion system uses $N_2O$ as oxidizer and HDPE as fuel. This engine was numerically investigated using a CFD tool that can handle reacting flow with finite-rate chemistry and coupled with the real-fluid model. The engine was further verified via a hot-fire test for 12 s. The ground $I_{sp}$ of the engine was 232 s and 221 s for numerical and hot-fire tests, respectively. An oscillation frequency with an order of 100 Hz was observed in both numerical and hot-fire tests with less than 5% of pressure oscillation. Swirling pattern on the fuel surface was also observed in both numerical and hot-fire test, which proves that this swirling dual-vortical-flow engine works exactly as designed. The averaged regression rate of the fuel surface was found to be 0.6~0.8 mm/s at the surface of disk walls and 1.5~1.7 mm/s at the surface of central core of the fuel grain.

## 1. Introduction

Hybrid rocket propulsion has attracted tremendous attention in the past decade due to its distinct characteristics and performance as compared to liquid and/or solid rocket propulsion. The advantages of hybrid rocket propulsion include the following [1, 2]: (1) extremely high safety because of separation of oxidizer and fuel storage and solid form of fuel minimizing fuel explosion; (2) good cost-effectiveness attributing to reduced complex plumbing and valve system; (3) good throttle capability similar to liquid rocket engines; and (4) highly green and environmental friendly combustion technology which allows various choices of fuel and oxidizer. Despite the benefits mentioned above, there are some issues that a hybrid rocket engine design must take into account before it becomes a useful rocket propulsion technology. These well-known issues are mostly due to the inherent characteristics of hybrid rocket engine (HRE), which are described next.

These issues include (1) $O/F$ ratio shift during combustion due to varying burning area and regression rate [3]; (2) limited total operating duration, which affects the size (diameter or length) of the hybrid rocket engine; (3) low combustion efficiency due to the nature of diffusion flame as compared to premixed flame; and (4) various choices of fuel and oxidizer [2, 4]. When designing a new type of hybrid rocket engine, one needs to take the above issues into account based on the specific mission requirements.

There are many different types of designs of HRE nowadays. For example, Nagata et al. [5] proposed an HRE called CAMUI (cascaded multistage impinging-jet). Fuel grain of this HRE has small ports along the engine axis, similar to the multiport design that increases the fuel burning area. The innovative part of this engine was the cascading feature of the ports. This creates large turbulence intensity as the flow impinges at the surface of the next fuel grain which boosts mixing efficiency while increasing the reaction area. A second example was proposed by Knuth et al. [6]. This HRE came up with the idea of coaxial, corotating vortex flow field engine called Orbitec Hybrid. This HRE injects the oxidizer at the rear end of the engine in the tangential direction. After injection, the oxidizer flows to the front part of the

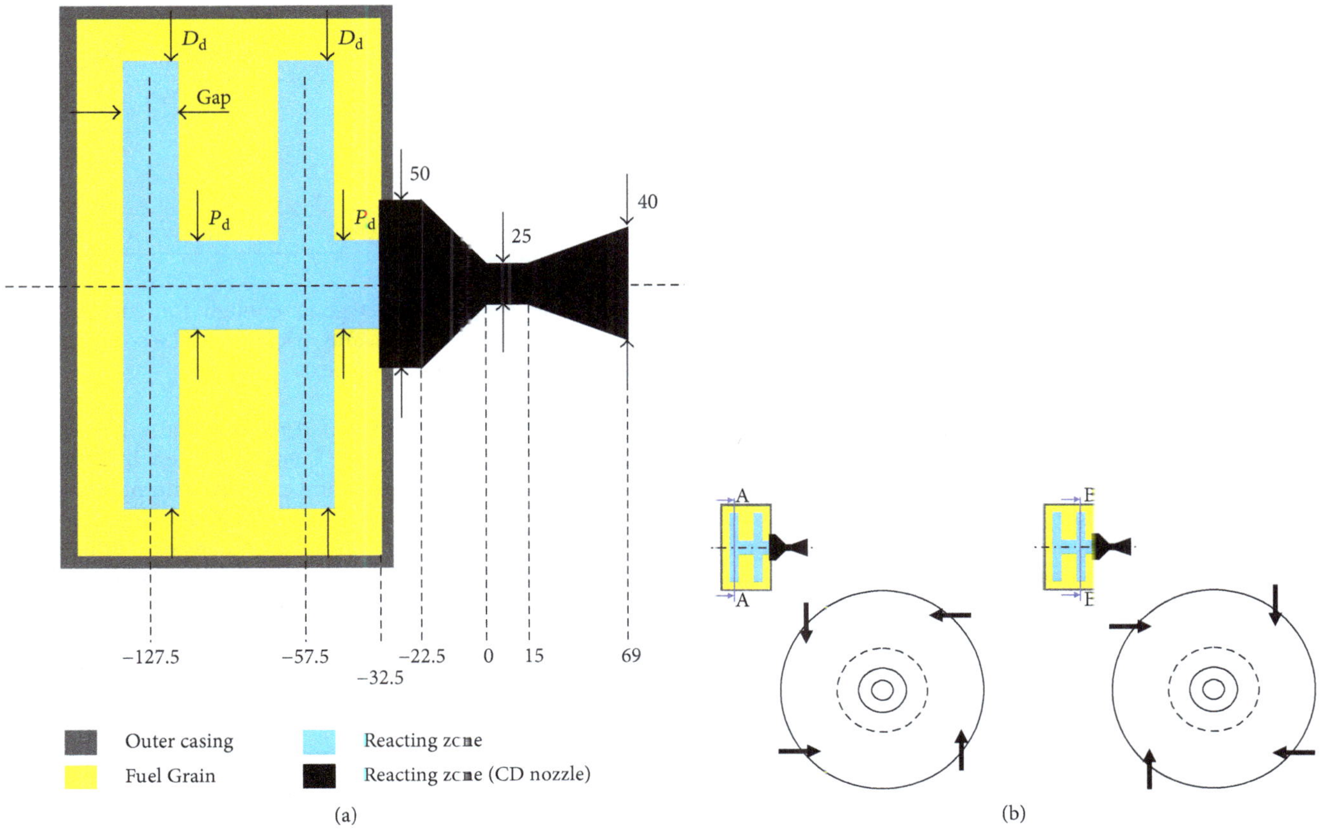


FIGURE 1: Schematic diagram of the 250 kgf class DVF engine. (a) *XY* plane cross section. (b) *YZ* plane cross section.

engine along the outer surface then back to the nozzle with swirling motion. A similar engine was also studied by Wall [7]. This engine increases the fuel regression rate by applying swirl injection [8], in which this design also enhances mixing by introducing bidirectional axial flow. More information on alternative designs on HRE could be found in Haag's study [9]. All these studies are designed to meet specific requirement missions, such as large thrust for short period of time (boosters) and small thrust for long operation (cruising). Lai et al. [10] recently proposed a highly efficient dual-vortical-flow (DVF) engine. The overall performance of the proposed configuration may give a very stable and high efficiency combustion and thrust. That specific configuration includes a flame holding mechanism implemented inside the high-temperature combustion chamber, which is very technically challenging or even impractical for application purposes. In this study, we would like to present a similar but easy to fabricate HRE without the flame holding mechanism for sounding rocket flight mission based on numerical and experimental investigation.

This engine consists of two counter-rotating flow reacting zones perpendicular to the engine axis with four tangential oxidizer injectors each. The fuel grain disks are connected by a central port to the nozzle, as shown in Figures 1(a) and 1(b). The main objective of this design is to provide a relatively stable thrust throughout the entire operating period. With this design, the burning surface could maintain a nearly constant combustion area throughout the operation. The counter-rotating flow maximizes the mixing and combustion efficiency with possible roll control of the rocket. Furthermore, this DVF HRE has a very small aspect ratio ($L/D$~1), which is favorable for gimbal-based thrust vector control (TVC) if needed.

TABLE 1: Dimensions of the 250 kgf class DVF engine.

| | Cases A1~A3 |
|---|---|
| $P_d$ | 40 |
| $D_c$ | 210 |
| Gap | 10 |
| | Unit: mm |

## 2. Research Methods

This study initially analyzed the described engine numerically using the well-known computational fluid dynamics (CFD) technique [11] and was verified via hot-fire tests. The numerical tool used in this work was UNIC-UNS [12], which is to be described later. Figure 1(a) shows the schematic diagram of the design, in which the engine was designed to fit into a casing of 266 mm in diameter and 148 mm in length (not including the convergent-divergent nozzle). In this design, the reacting zone (blue) is surrounded by fuel grain (yellow) which regresses while combustion

Table 2: Boundary settings for numerical analysis.

| Boundaries | | Flow condition | Thermal condition |
|---|---|---|---|
| Fuel | — | Flow rate (pyrolysis), normal to the wall | $T > T_{\text{pyrolysis}}$ (820 K) |
| | — | No slip | $T < T_{\text{pyrolysis}}$ |
| Inlet | — | Fixed total pressure (40 atm) | 300 K |
| Outlet | — | Supersonic-free expansion | N/A |
| Wall | — | No slip | Adiabatic |

occurs. The arrangement of fuel grain automatically prevents the casing from the flame during operation before it is burned out. Due to the pyrolysis of the fuel grain during combustion, some of the dimensions of the engine change over time, namely, $P_d$, $D_d$, and Gap in Figure 1(a). The dimensions for the numerical investigation are summarized in column "Cases $A_1$~$A_3$" of Table 1. The pyrolysis rate of the fuel grain is a function of many physical properties such as temperature, flow pattern, turbulence intensity, and oxidizer mass flux, to name a few.

*2.1. Numerical Analysis.* The engine model was simplified and solved numerically by computational fluid dynamics (CFD) method according to the governing equations, namely, continuity equation (mass conservation equation), species conservation equations, Navier-Stokes equations (momentum conservation equation), and energy conservation equation with various well-known physical models such as HBMS (Hirschfelder, Beuler, McGee, and Sutton) real-fluid model [13] and extended $k$-$\varepsilon$ turbulence model [14]. These equations and models were discretized using the cell-centered finite-volume method, parallelized using MPI protocol and had been applied successfully to similar problems [10, 12]. For the reacting species, the finite rate reaction model (a.k.a. Arrhenius reaction model) was implemented. Together with a simplified set of species and reaction path [10], the UNIC-UNS code handles the species conservation equation. In general, the simulations were performed using a time step of $5 \times 10^{-6}$ s with boundary conditions summarized in Table 2. These computations are performed using 64 cores of processors on the IBM 1350 PC cluster at the National Center for High-Performance Computing (NCHC) of Taiwan, in which a single node consists of 4 cores with 3.0 GHz of CPU for each core and 16 GB of RAM for each node.

For the CFD model, we performed a series of grid convergence tests using 5.03, 2.34, and 1.63 million cells which were labeled as Case $A_1$, $A_2$, and $A_3$, respectively. We have compared three simulated quasi-steady-state data which include thrust, mass flow rate ($\dot{m}$), and $I_{sp}$. Figure 2 illustrates the corresponding convergence history, in which all the cases were calculated long enough to reach a quasi-steady-state where the abovementioned values are nearly constant. The thrust oscillation caused by the pressure oscillation will be explained later in Results and Discussion. For all three cases, the simulated mass flow rates are essentially the same as 1.05 kg/s. But the thrust (or $I_{sp}$) of Case $A_3$ (coarse mesh) is slightly higher than the $A_1$ (fine mesh) and $A_2$ (medium mesh) which are ~254 kgf (or 234 s) and ~246 kgf (or 232 s), respectively. With these results, one can summarize that the resolution of Case $A_2$ is good enough for engine design purpose, further analysis, and discussion. The corresponding hot-fire tests are described next.

*2.2. Hot-Fire Test Setup and Analysis.* Figure 3 shows the schematic sketch of the DVF engine design in detail. For hot-fire test purpose, the chamber body, external plumbing, and both bulkheads were made of 304 stainless steel. The nozzle was fabricated using high-density antioxidation graphite. The insulators are made of silicone rubber, EPDM, or ceramic depending on the location. The fuel grains were manufactured using off-the-shelf high-density polyethylene (HDPE) [4] with a density of 0.945 g/cm$^3$. Photographs of all manufactured components are shown in Figure 4(a) and are assembled into a DVF engine as shown in Figure 4(b).

Figure 5 shows the test setup of this hot-fire test. For this system, the $N_2$ cylinder was filled to exceed 120 atm and was connected to the top of the specially prepared oxidizer ($N_2O$) running tank with a regulator set as 57 atm. A pressure transducer was also mounted to the top of the running tank to monitor the tank pressure during operation. The bottom opening of the running tank was connected to each of the injectors of the engine with a plumbing system including a main valve followed by a flow distributor and some pressure transducers. The DVF engine was then mounted on a horizontal thrust stand. The pipe size used before the flow distribution system was 3/4 inch (~19 mm) and was split into eight 1/4-inch (~6.35 mm) tube before injection into the combustion chamber. The valve used in this system was a traditional ball valve with nominal diameter 19 mm controlled by a pneumatic valve.

A reliable electrical ground support equipment (EGSE) is also required to accomplish the task. To acquire the experimental data, we mounted a series of sensors on the engine system: these sensors were managed on a data acquisition system (DAQ) based on National Instruments Corporation's (NI) products. The sensors used in this test were pressure transducers and load cells. These sensors output 0~10 VDC according to the physical quantity measured. The pressure transducers were JPT-131 series provided by Jetek Electronics Co. Ltd. [14]. The pressure transducer can measure 0~100 bar gauge pressure with the accuracy of +/−0.5%. For the thrust of the engine and mass difference of the supply tank, S-type load cell with the desired load range provided by Sensolink [15] was used. These sensors were wired to the cRIO 9074 of NI using ordinary signal cables. Then the cRIO 9074 is connected to a PC using Ethernet cable, which is suitable for long-distance monitoring and controlling as shown in Figure 6. The programming was done using the

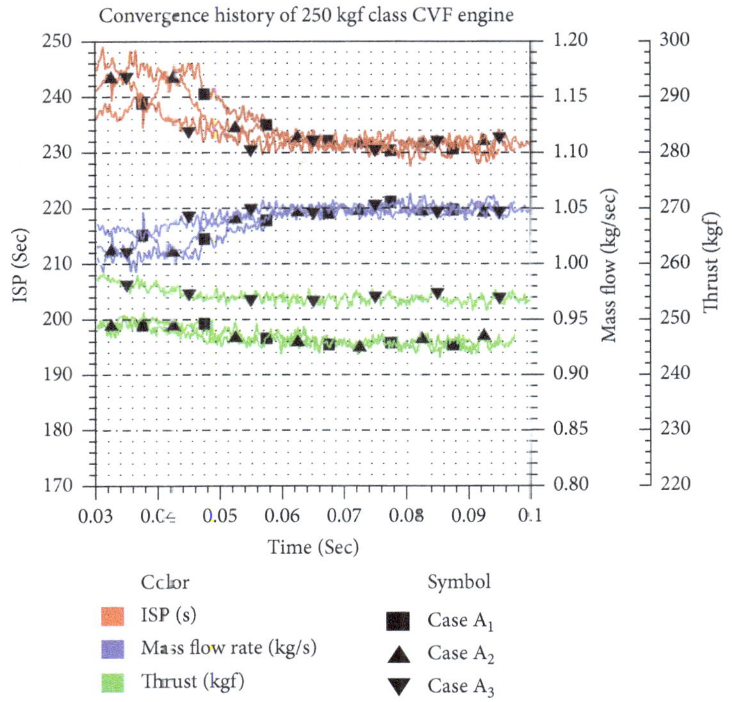


FIGURE 2: Convergence history of the numerical calculation of 250 kgf class DVF engine.

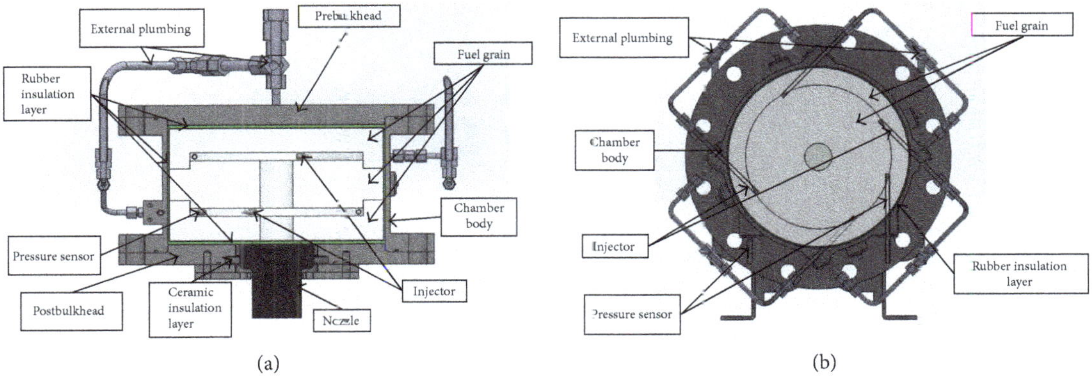


FIGURE 3: Detailed design of 250 kgf class DVF engine.

software called LabVIEW, which is also provided by NI. This platform makes the programming process simple and easy. The obtained data is stored in the PC at the rate of 1 kHz. These data are then postprocessed and analyzed after the hot-fire test.

## 3. Results and Discussion

*3.1. Simulation Results.* We have performed the calculation of thermodynamic equilibrium reactions for $N_2O$ [16] (oxidizer) and $C_2H_4$ (fuel) using NASA CEA online [17–19]. The chamber pressure was set to 38 atm with an $O/F$ ratio of 4.2, and the area expansion ratio of the nozzle was 2.56. For this case, the resulting outlet pressure 2.85 atm was underexpanded for sea level operation. The optimal (theoretical) $I_{sp}$, $C^*$, and $C_f$ value for this case was 236 s, 1759 m/s, and 1.32, respectively. Figure 7(a) shows the summary of the simulated equilibrium mass fractions of all species. The major species include $N_2$, CO, $H_2O$, $CO_2$, and $H_2$ with the mass fraction of 0.5136, 0.3572, 0.0664, 0.0417, and 0.0200, respectively. These five species sum up to 0.9989 of the composition. Figure 7(b) shows the simulated mass fractions by UNIC-UNS CFD code. The major species are identical to those of CEA, but the mass fractions are slightly different. The mass fraction of $N_2$, CO, $H_2O$, $CO_2$, and $H_2$ is 0.4715, 0.2924, 0.0713, 0.0554, and 0.0121, respectively. In addition to the above species, there are other species that sum up to 0.0973, which are mostly radical species of combustion. The main reason for this difference was that for CEA the species were calculated assuming thermodynamic equilibrium while, for UNIC, they were obtained using the finite-rate chemistry. The majority of the mixture is $N_2$, which comes from the direct decomposition of $N_2O$ oxidizer under high combustion temperature exceeding 1640°C. Due to the low $O/F$ ratio

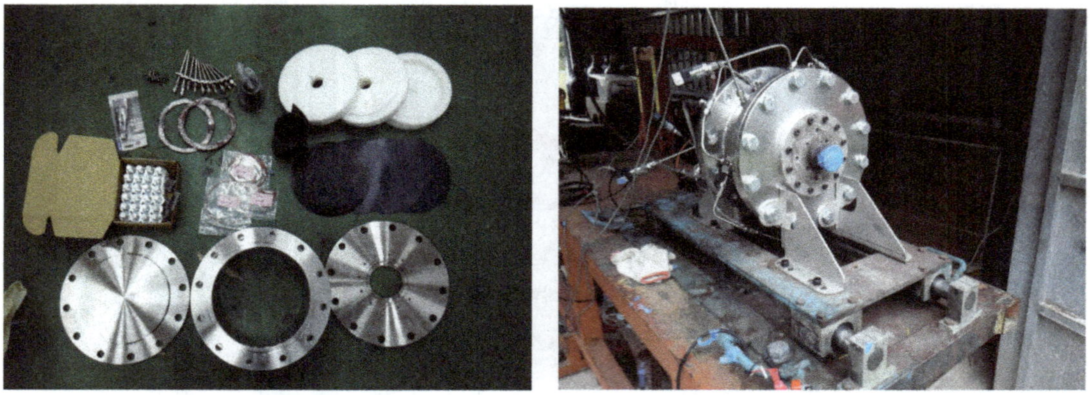

(a) (b)

FIGURE 4: (a) Components of 250 kgf class DVF engine. (b) Assembled 250 kgf class DVF engine.

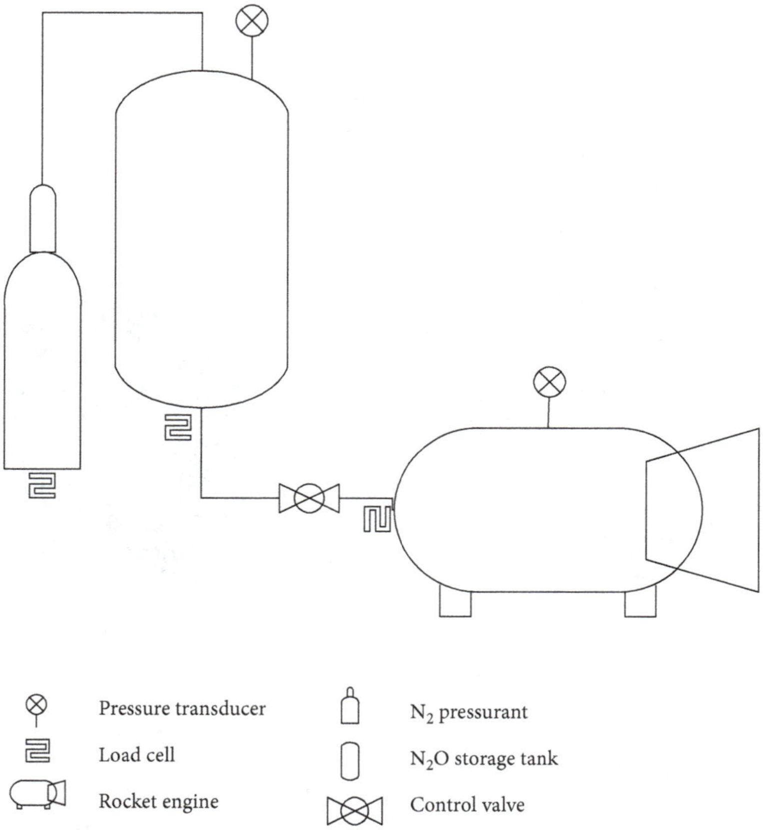


FIGURE 5: Schematic diagram of the test setup of 250 kgf class DVF engine.

which leads to incomplete combustion, the second dominant species is CO. The formation and distribution of the species will be further discussed later in this paper.

Figure 8(a) shows the instantaneous sliced pressure distribution in the engine in the quasi-steady-state after ignition of combustion. The pressure inside the engine ranges from 30 to 38 atm, in which the maximum pressure is located near to the injectors (circumference of the disks) and the lowest pressure is distributed near the nozzle as expected. Noticeably, the pressure in the major central core (between disk 1 and disk 2) is low with a value in the range 32-33 atm due to strong swirling motion. This strong swirling central core region disappears as the counter flows meet somewhere downstream of disk 2. As the flow continues to flow towards the nozzle, the flow accelerates and the pressure drops quickly. In addition, Figure 8(b) shows the same instantaneous sliced distribution of Mach number in the engine. Gas flow inside the engine is subsonic, which is accelerated to sonic speed at the throat of the nozzle and is further accelerated in the diverging part.

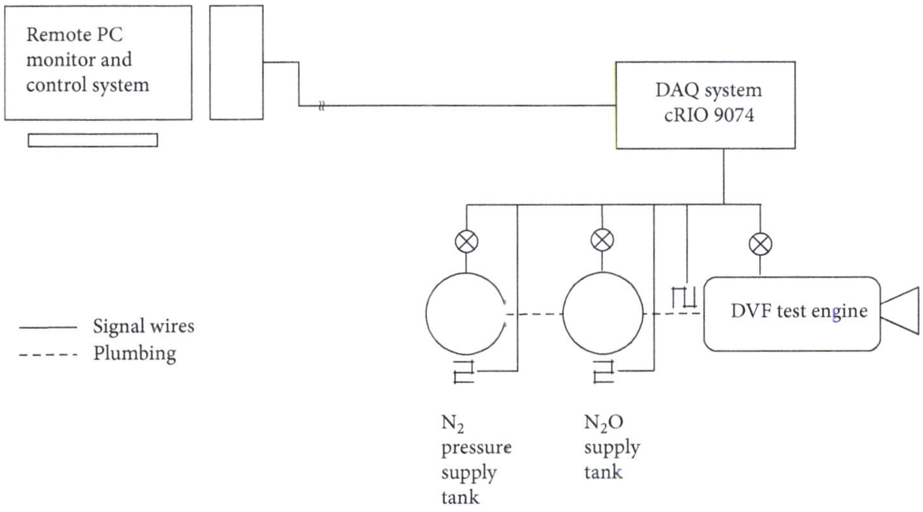


FIGURE 6: Schematic diagram of the EGSE setup.

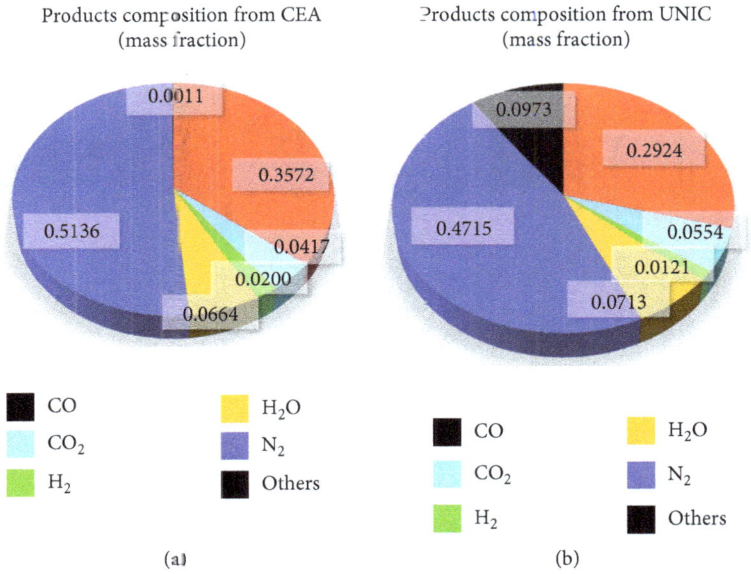


FIGURE 7: Composition of species product at exhaust outlet by (a) NASA CEA and (b) UNIC CFD.

The performance of HRE relies heavily on the combustion efficiency, which temperature of combustion can be considered as a good indicator. Figure 9(a) shows the corresponding instantaneous sliced temperature distribution. Near the injection regions where $N_2O$ are injected, the temperature is at room temperature (300 K), which serves as a natural cooling mechanism for preventing the injectors from melting by high combustion temperature. As the $N_2O$ stream is injected into the engine, the high combustion temperature causes $N_2O$ to decompose directly. This decomposition reaction is an exothermic reaction that helps to sustain the combustion. The main products of $N_2O$ decomposition are $N_2$ and $O_2$ with some related species in radical form. Distributions of some critical species such as $O_2$ and OH radical are shown in Figures 9(b) and 9(c), respectively. Therefore, we can observe that in Figure 9(b), massive $O_2$ are formed and quickly disappear just before the high-temperature region. In hydrocarbon combustion, flame location or highly reacting region generally consists of abundant OH radical generally. This shows that the abundant OH radical distribution corresponds to the high-temperature region very well.

Figure 10 shows the distributions of mass fraction of the five other major species ($N_2$, CO, $H_2O$, $CO_2$, and $H_2$) in the engine. In Figure 10(a), the mass fraction of $N_2$ is about 0.6 in the bulk region of the two major combustion disk chambers due to the direct decomposition of $N_2O$. It decreases rapidly in the very thin region near the fuel grain surface because of the dilution from fuel vapor due to pyrolytic reaction. As the flow goes further downstream, the mass fraction of $N_2$ decreases to about 0.45 before entering the divergent part of the nozzle. Figures 10(b), 10(c), and 10(d) show the distribution of mass fraction of $H_2O$, $CO_2$, and CO,

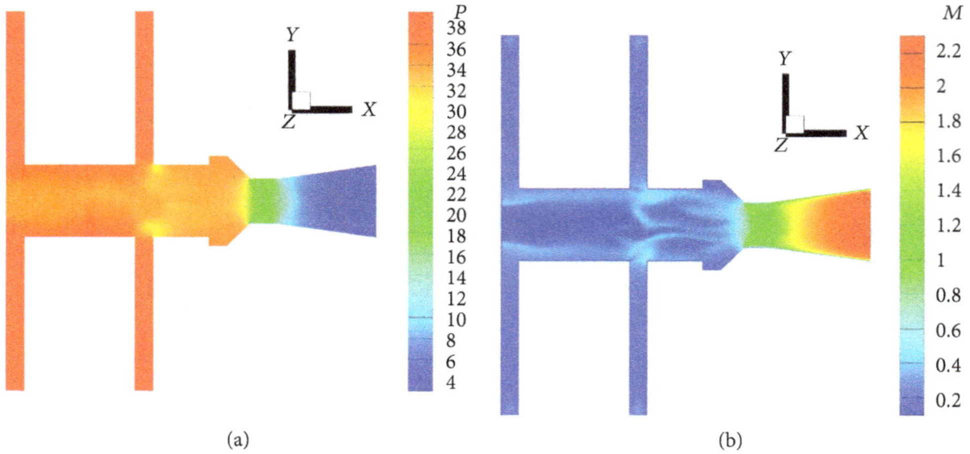


Figure 8: Instantaneous planar distributions on *XY* section: (a) pressure (atm) and (b) Mach number.

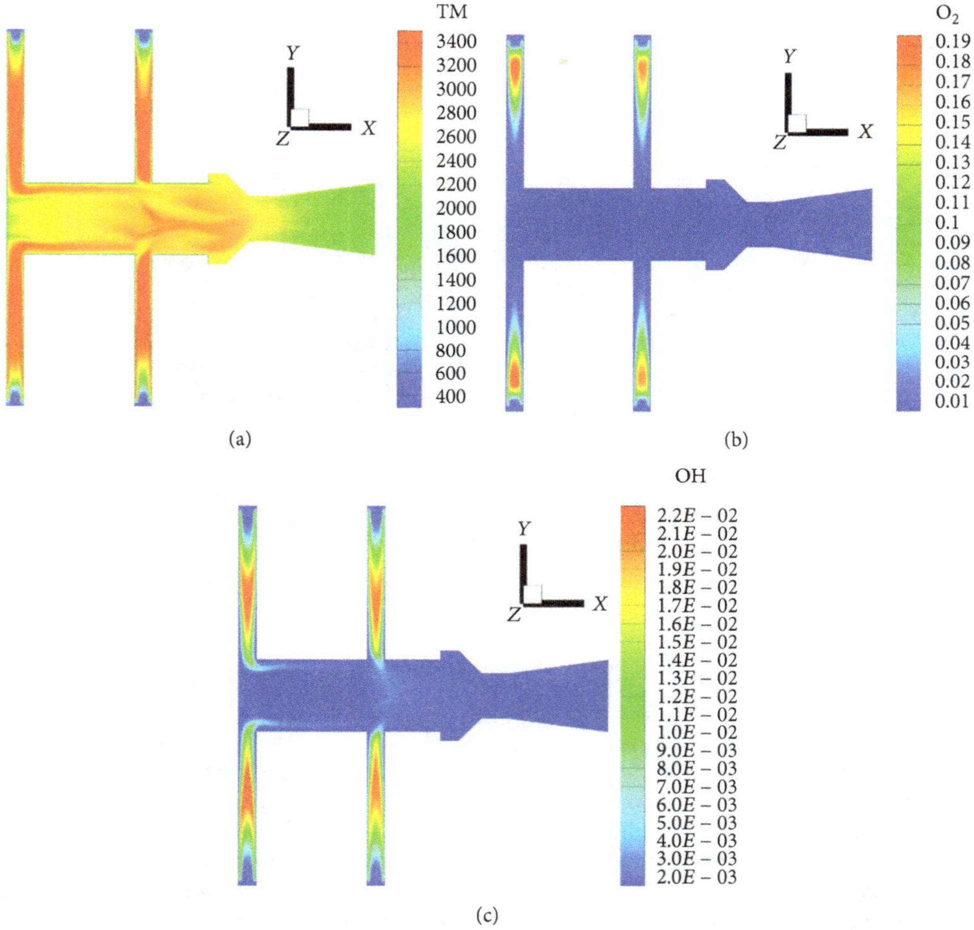


Figure 9: Instantaneous planar distributions on *XY* section: (a) temperature (K), (b) mass fraction of $O_2$, and (c) mass fraction of OH radical.

respectively, which represents the major products of complete combustion of hydrocarbon compound with oxygen. Note that the distribution of $CO_2$ in the disk chambers (Figure 10(c)) is approximately in consistent with that of OH (Figure 9(c)) which corresponds to the flame position where strong chemical reaction occurs. The results show the combustion changes from fuel-lean, stoichiometric, and fuel-rich reactions along the radial direction from injection to central core region. The concentrations of $H_2O$ and $CO_2$ increase along the radial direction inwards because of reactions between decomposed $O_2$ and fuel grain vapor and reach maximal values ($H_2O$: 0.09, $CO_2$: 0.15) somewhere before the

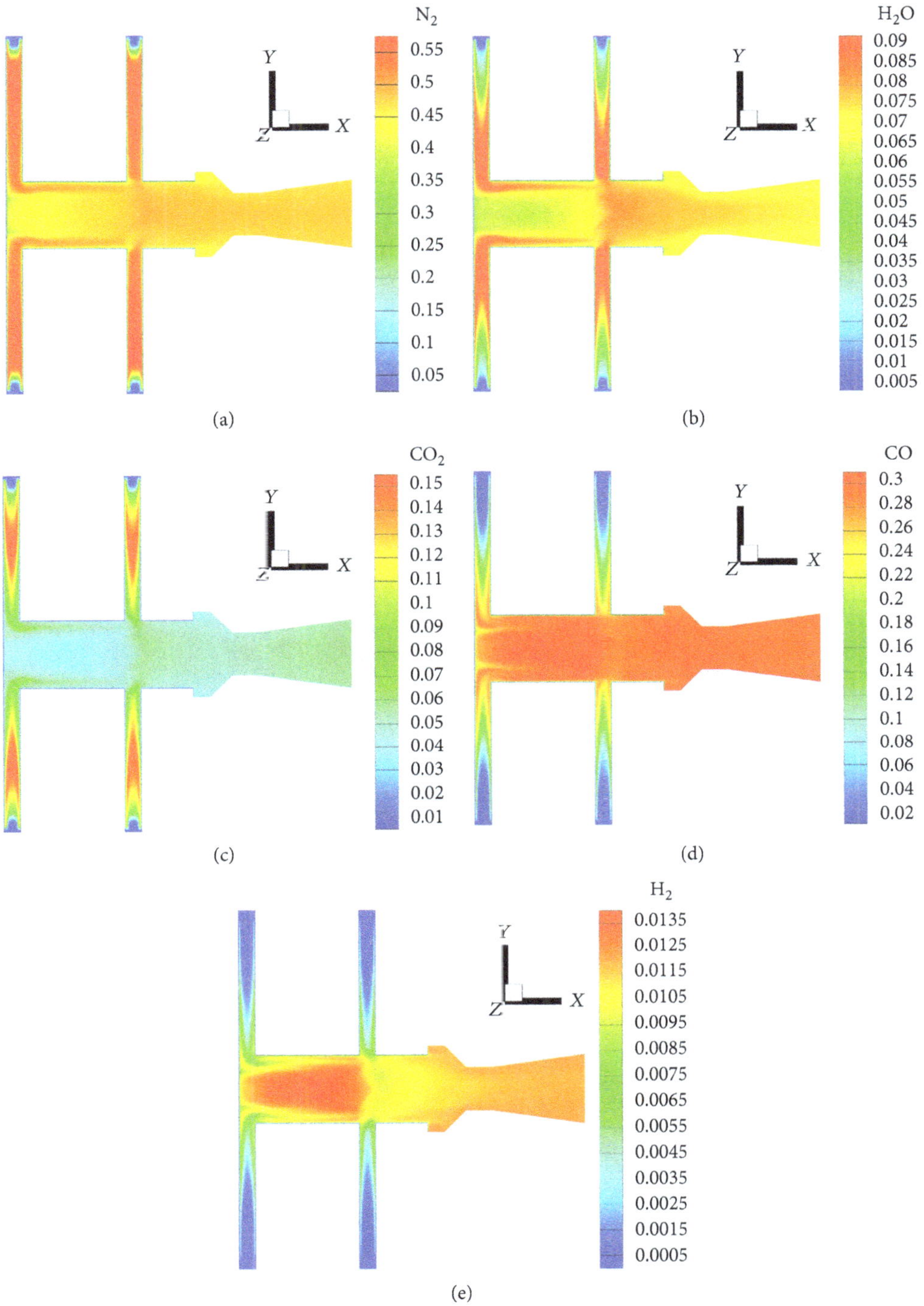


Figure 10: Mass fraction of major products: (a) $N_2$, (b) $H_2O$, (c) $CO_2$, (d) CO, and (e) $H_2$.

flowing into the central core. In addition, they are brought to concentrate near the fuel grain surface of the central core because of the centrifugal force of swirling flow. Due to the lack of "O" species (Figure 9(b)) (more fuel-rich), both the concentrations of $CO_2$ and $H_2O$ decrease after entering the central port. With the same reason, a large amount of syngas (i.e., CO and $H_2$) is formed because of the high temperature before (in the disk chamber) and near the wall in the central port region (Figure 9(a)). These species (CO and $H_2$) can be further reacted in this central port region as the flow moves further downstream, especially the case of $H_2$ (Figure 10(e)).

We have found that the thrust (or pressure, not shown) oscillates as a function of time in Figure 2. A detailed analysis on this oscillating phenomena was performed in this study. Figure 11 illustrates five instantaneous temperature distribution in the middle sliced sections of both disk combustion chambers at five different times of simulation. The "spikes" of the contour rotate clockwise and counterclockwise at disk

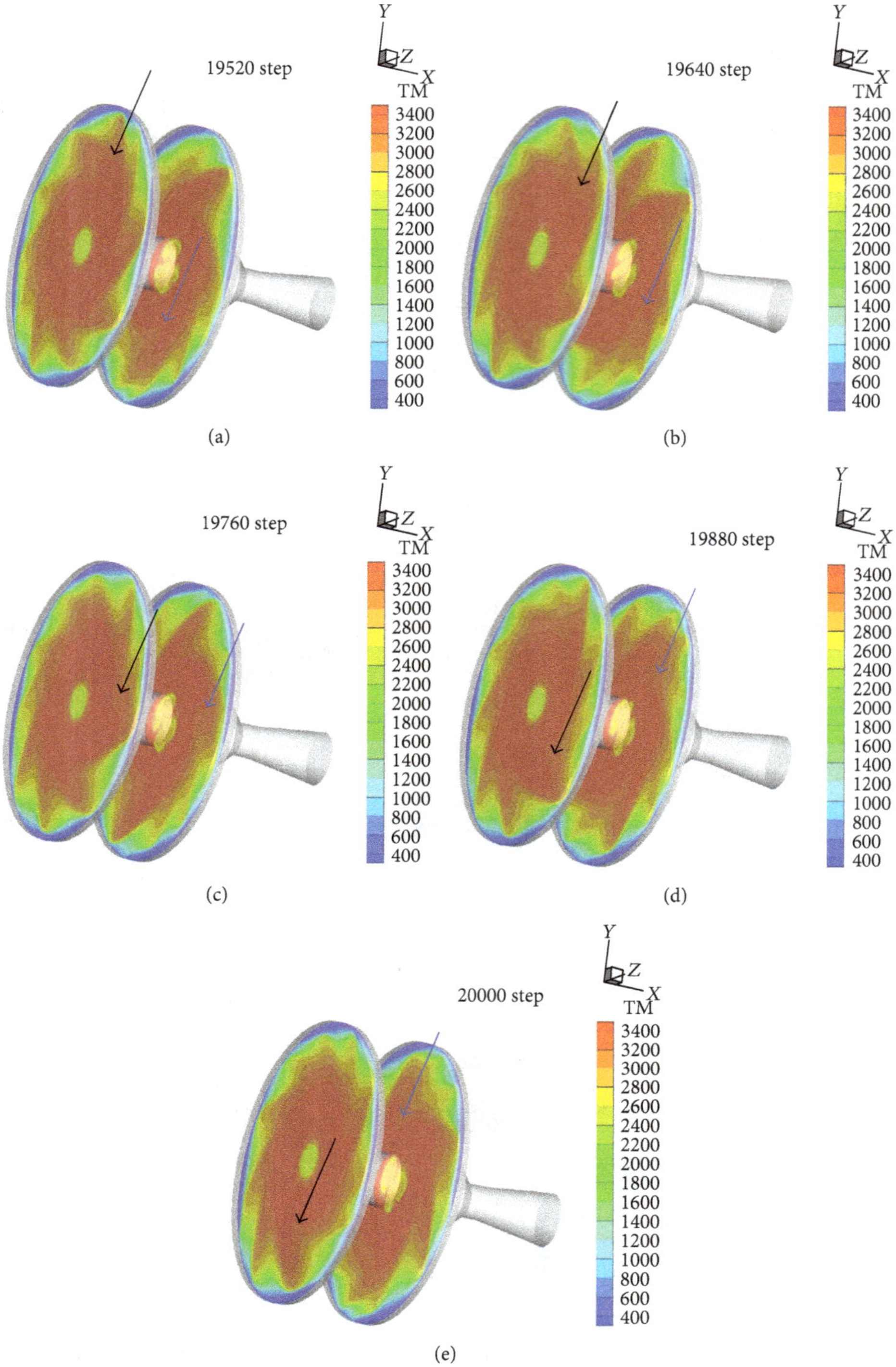


FIGURE 11: Temperature [K] contour of Case $A_2$ at various calculation steps: (a) 19520 step, (b) 16940 step, (c) 19760 step, (d) 19880 step, and (e) 20000 step. Note that the black and blue arrows show the location of the some specific temperature "spike" in disk chamber 1 and disk chamber 2, respectively.

chambers 1 and 2, respectively. The oscillation frequency based on numerical simulations is about 200 Hz which should be due to the lack of flame holding mechanisms in the reacting region near the disks. This oscillation is more distinct as compared to our previous work of DVF HRE with flame holders [10]. The reason without using flame holders as explained earlier is that the implementation of the flame holding mechanism becomes very difficult or even impossible due to the high temperature (exceeds 3000 K) in the reacting region with highly vibrating environment and the fuel shape (internal geometry) changes as it operates. In spite of the lack of flame holders, the performance of the engine is considered to be fairly stable with the maximal oscillation amplitude of 2 kg over 245 kg, which is less than 1%.

Figure 12 shows the stream traces of the flow patterns at the surface of the fuel grain (0.1 mm above the surface) of Case $A_2$ (Figure 12). The injected flow streams revolve in the disk combustion chambers around the central port before entering the central port. This provides a relatively longer flow path for the flow (relative to straight radial injection) which greatly increases the residence time for the combustion reactions to take place. For a specific case, the stream trace marked with the blue arrow in Figure 12 was being observed. We define the stream trace entering the swirl pattern when it reaches the radial position where $R$ is at 95% of $R_{max}$ (about 5 mm from the circumference, indicated by the long red arrow) and exit when $R$ is at 25% of $R_{max}$ (about 5 mm larger than the central port, indicated by the short red arrow). This figure shows that the flow revolved about 180 degrees starting from injection until it entered the central port. The path that the flow takes in this situation is about twice the length of the radius of the disk. As shown in Figure 12, the gap between disk chamber walls is small and the averaged flow speed is larger. This also indicates that the tangential momentum is larger; therefore, the flow revolves almost 180 degrees before reaching the central port. As the gap grows wider, the cross-sectional area increases and the tangential momentum becomes smaller. This will be further compared and discussed with the hot-fire test results.

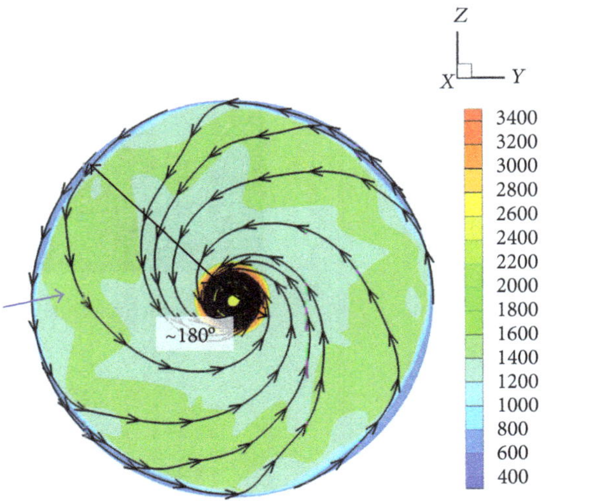


FIGURE 12: Surface stream trace of the flow pattern of Case $A_2$ at $X = -122.6$ mm.

*3.2. Hot-Fire Test Results.* After the rocket engine is set up following Figure 5 with all connections carefully checked, we have followed the following procedure to perform the hot-fire tests. A snapshot of the hot-fire test during combustion is shown in Figure 13, in which the exhausting plume is slightly underexpanded because of clear further expansion leaving the lip of the nozzle. Figure 14 shows the measured thrust and pressures at many locations for a typical run. A pyrograin in the engine is used to ignite the engine at $t = -4$ s. After the pyrograin burns out, main valve opens at $t = 0$ s to allow the $N_2O$ oxidizer to flow into the chamber. A delay of 0.5 s could be observed based on the rise of measured thrust and pressure data, which is caused by the speed of valve opening. The liquid $N_2O$ flow depletes at $t = 9.5$ s, and finally, the valve totally closes at $t = 12$ s to shut down the engine, during which both thrust and pressures decrease rapidly. Note that the $N_2O$ becomes a gaseous state and flows into chamber between $t = 9.5$ and $t = 12$ s. The thrust is relatively stable in the range of 240~245 kgf in the period of 1.25~9.5 s. The thrust decreases almost exponentially from $t = 9.5$ s to the end. The running tank pressure starts at 57 atm and decays slowly to about 50 atm at $t = 9.5$ s due to cooling effect caused by thermal expansion of $N_2O$ flow during operation. Both disk chamber pressures show clear oscillations (even different) probably due to the lack of flame holding mechanism in the engine which coincides with the simulation results. However, the oscillation frequency is only 4 Hz based on the measured pressure data in Figure 14, which is much lower than the experimentally observed 100 Hz of oscillation using 600 fps high-speed camera most probably caused by the limitation of the setup of pressure sensor that may damp out the high-frequency component. In addition, the simulated oscillation frequency is ~200 Hz which is roughly consistent with the measured ~100 Hz. The difference, however, requires further investigation.

The swirl pattern is also observed on the surface of the fuel grain after hot-fire test. Figure 15 shows the photo of cavity wall of chamber disk 1 after hot-fire test. The flow pattern was indicated by the red arrows. Similar to the simulated case, the two red arrows in Figure 15 indicate the entering and exiting of the flow into the chamber disk. The revolved angle of this case is about 80 degrees which is a lot smaller than the simulated Case $A_2$. The main reason was that the gap after firing (~26 mm) is a lot larger than the initial case (10 mm). As the gap increases, the ratio of tangential momentum and radial momentum decreases. Therefore, the angle revolved by the flow is expected to decrease before entering the central port.

The fuel surface contour was measured by a bridge-type 3D Coordinate Measuring Machine (CMM) (Model PIONEER, Hexagon Manufacturing Intelligence). We have scanned three fuel grain surfaces using an automatic mode. Figure 16 shows the averaged regression rates at different radial positions by taking average from 6 to 12 scans per radius along different radial directions, considering minimal measured positions (0, 45, and 48 mm for fuel grain 1, 2, and 3, resp.) to the outer radial position of 100 mm. The averaged regression rate at the disk surface is 0.6 to 0.8 mm/s. The regression rate at the center region of grain 1 (central port) is two times the value of those at the disk surface. This highly regressed region may be attributed to the long light of sight of radiation from the combustion flame to the exhausting plume along the axis. In addition, very high regression rates are also observed at the wall of the central port which is about 1.5 to 1.7 mm/s, which should be caused by the very strong swirling in this region that promotes the pyrolysis of the fuel grain.

FIGURE 13: Snapshot of the 250 class DVF engine during operation.

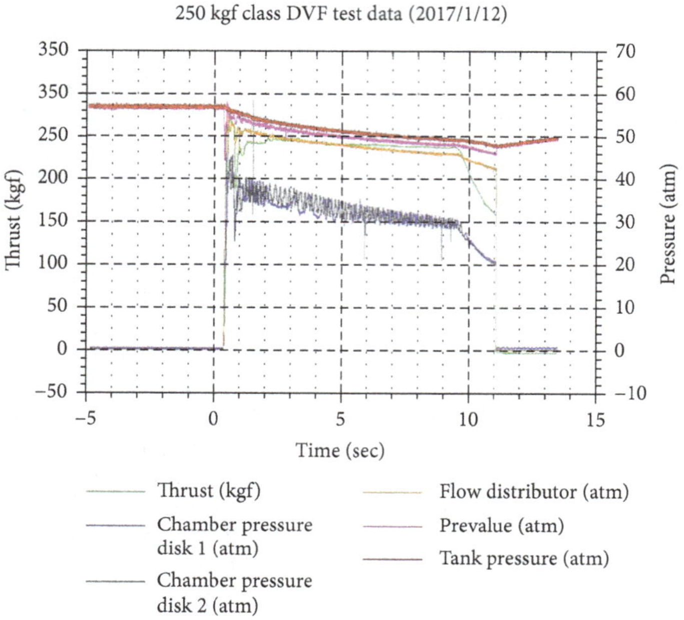


FIGURE 14: Measured data (thrust and pressures) of 250 kgf class DVF engine during the hot-fire test.

FIGURE 15: Swirl pattern on fuel grains of disk chamber 1 after hot-fire test.

*3.3. Comparison of Simulation with Test Data.* Table 3 summarizes the numerically simulated cases with hot-fire test data of the 250 kgf class DVF HRE. For the numerical results, there are two specific cases being discussed. The NASA CEA case assumed 0-D chemical equilibrium condition which we can be considered as the theoretical results, and the UNIC-UNS simulated Case $A_2$ using CFD finite-rate model. For CEA test case, the initial state of the engine is calculated. The resulting *O/F* ratio, ground $I_{sp}$, and $C^*$ are 4.2, 236 s, and 1759 m/s, respectively. The corresponding $C_f$ is calculated to be 1.32 using the nozzle area expansion ratio of 2.56.

For the CFD simulated cases, the simulated Case $A_2$ represents the initial state of this work. The simulated Case $A_2$ has a 98% of $I_{sp}$ efficiency (232 s) and more than 100% of the $C^*$ value of 1814 m/s which exceeds that (1759 m/s) of CEA. For this work, the location $P_c$ used is somewhere near the injector where the pressure sensor of the experimental model could be mounted during tests. Due to the use of this pressure, the $C^*$ value was higher than the one from CEA. As a result, the $C_f$ value of Case $A_2$ is not as high as the one from CEA. This can be easily observed by $I_{sp} \times g_0 = C^* \times C_f$.

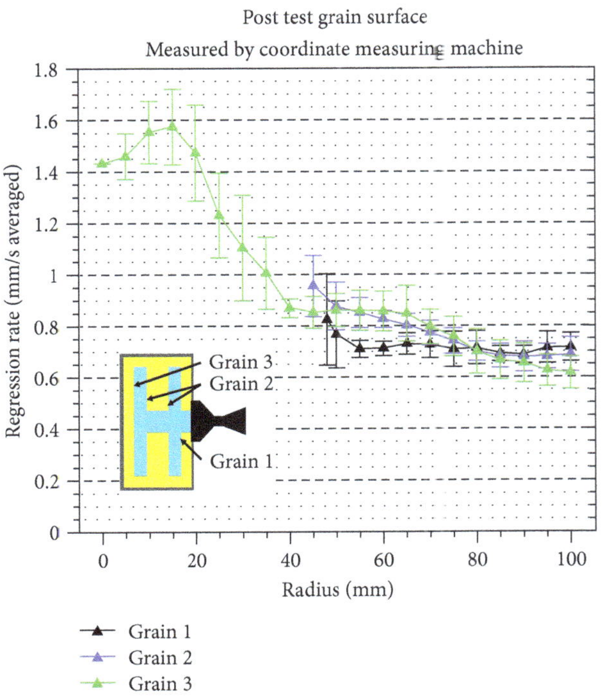


Figure 16: Measured grain regressed thickness for different grain fuel surfaces of the DVF engine after hot-fire test.

Table 3: Comparison of numerical and test data of 250 kgf class DVF engine.

| Parameters | Numerical cases NASA CEA | Case $A_2$ (CFD) | Hot-fire test Initial state | Shutdown state | Time averaged |
|---|---|---|---|---|---|
| $P_c$ (atm) | 38.00 | 38.31 | 37.22 | 29.75 | 32.43 |
| $F$ (kgf) | N/A | 246 | 240 | 240 | 240 |
| $\dot{M}$ (kg/s) | N/A | 1.05 | N/A | N/A | 1.03 |
| $\dot{M}_{oxidizer}$ (kg/s) | N/A | 0.85 | N/A | N/A | 0.83 |
| $\dot{M}_{fuel}$ (kg/s) | N/A | 0.20 | N/A | N/A | 0.20 |
| $O/F$ ratio (–) | 4.20 | 4.20 | N/A | N/A | 4.18 |
| $I_{sp,ground}$ (s) | 236 | 232 | N/A | N/A | 221 |
| $C^*$ (m/s) | 1759 | 1814 | N/A | N/A | 1565 |
| $C_f$ (–) | 1.32 | 1.25 | 1.21 | 1.51 | 1.39 |

For the hot-fire test results, the pressure and thrust are obtained as a function of time but the mass difference (especially the fuel part) could only be measured after the test due to the limitation of the facilities. Therefore, the mass flow rate of the hot-fire test was only available in terms of time averaged values. Due to the reasons above, this work could only obtain the averaged value of $I_{sp}$ and $C^*$ of the hot-fire test. But to obtain the value of $C_f$, the instantaneous value can be calculated properly by $C_f = F_{instant}/(P_{instant} \times A_{throat})$. The averaged $I_{sp}$ was calculated by integrating the measured force and divide it by total mass difference. The $C^*$ value was then calculated with $P_c$ taken as time averaged value. As for the $C_f$, it was calculated using the instantaneous force and chamber pressure. The averaged $I_{sp}$ obtained by the hot-fire test in this work was 221, about 93.6% of the theoretical value. The averaged $C^*$ and $C_f$ are 1565 m/s and 1.39, respectively. Since the value of $C_f$ was available as a function of time, we observed an increase of $C_f$ from 1.21 to 1.51, which is mainly due to the decrease of chamber pressure as the test proceeds. As the chamber pressure decreases, the underexpanded flow shifts towards optimal criteria, and the $C_f$ increases. Though the main objective of this DVF HRE was to provide a constant and stable thrust for a specific flight mission, it was surprising, that to us, that the thrust remains almost the same throughout the test period. This was probably due to $O/F$ ratio shift, mixing efficiency, and a lot more reasons that compensate with each other. However, this definitely requires further investigation in the near future.

## 4. Conclusion

This work proposes a DVF HRE without flame holding mechanism for possible sounding rocket application. The length-to-diameter ratio of the engine is only about 1, which is different from conventional lengthy HREs. This design was simulated considering geometrical configurations at the initial state. The resulting simulated $I_{sp}$ at initial state is 232 s with very high combustion efficiency as compared to that calculated by NASA CEA (236 s). A hot-fire test based on proposed design was performed for 12 s by measuring the thrust and pressures at many locations. The maximal thrust of 245 kgf was measured to be relatively constant with a value of ~240 kgf and an averaged ground $I_{sp}$ of 221 s. Measured regression rates are in the range of 0.6–0.8 mm/s at the walls of the two disk chambers and 1.5–1.7 mm/s at the walls of the central port region due to strong vortex motion of the hot gases. In addition, the central end wall, located furthest from the nozzle, also possesses very high regression rate, probably due to high thermal radiation and highly turbulent flow field. The $O/F$ ratio of this specific test was 4.2 which is relatively low compared with the value for optimal $I_{sp}$ from various references [2]. An oscillation frequency of ~100 Hz was observed in numerical simulation, while ~200 Hz in hot-fire test, which definitely requires further investigation. These observed or measured instabilities may be caused by the fact that no physical flame holders were used in the chamber. Despite with the presence of these instabilities, this DVF engine design still shows fairly stable thrust, which should satisfy the use in sounding rocket application.

## Conflicts of Interest

The authors declare that they have no conflicts of interest.

## Acknowledgments

This work was supported by the Advanced Rocket Research Center (ARRC), National Chiao Tung University of Taiwan, and Ministry of Science and Technology of Taiwan (MOST) through Grant nos. 102-2627-E-009-001, 103-2221-E-009-136-MY2, and 105-2221-E-009-066-MY2 and technical assistance of Taiwan Innovative Space Inc. of Taiwan.

## References

[1] G. P. Sutton and O. Biblarz, *Rocket Propulsion Elements*, JOHN WILEY & SONS, INC., 7th edition, 2001.

[2] M. J. Chiaverini and K. K. Kuo, *Fundamentals of Hybrid Rocket Combustion and Propulsion*, AIAA, 2006.

[3] M. J. Chiaverini, N. Serin, D. K. Johnson, Y. C. Lu, K. K. Kuo, and G. A. Risha, "Regression rate behavior of hybrid rocket solid fuels," *Journal of Propulsion and Power*, vol. 16, no. 1, pp. 125–132, 2000.

[4] N. Gascoin, P. Gillard, A. Mangeot, and A. Navarro-Rodriguez, "Detailed kinetic computations and experiments for the choice of a fuel–oxidiser couple for hybrid propulsion," *Journal of Analytical and Applied Pyrolysis*, vol. 94, pp. 33–40, 2012.

[5] H. Nagata, M. Ito, T. Maeda et al., "Development of CAMUI hybrid rocket to create a market for small rocket experiments," *Acta Astronautica*, vol. 59, no. 1-5, pp. 253–258, 2006.

[6] W. J. Knuth, D. J. Gramer, M. J. Chiaverini, J. A. Sauer, R. H. Whitesands, and R. A. Dill, "Preliminary CFD analysis of the vortex hybrid rocket chamber and nozzle flow field," in *34th AIAA/ASME/SAE/ASEE Joint Propulsion Conference and Exhibit*, Cleveland, OH, USA, July 13–15, 1998.

[7] N. Wall, *Characterisation of Multiple Concentric Vortices in Hybrid Rocket Combustion Chambers, Ph.D. Thesis*, Department of Mechanical Engineering, University of Sheffield, South Yorkshire, England, 2013.

[8] S. Yuasa, O. Shimada, T. Imamura, T. Tamura, and K. Yamamoto, "A technique for improving the performance of hybrid rocket engines," in *35th ALAA/ASME/SAEASEE Joint Propulsion Conference and Exhibit*, Los Angeles, CAL, USA, June 20–24 1999.

[9] G. S. Haag, *Alternative Geometry Hybrid Rockets for Spacecraft Orbit Transfer, Ph.D. Dissertation*, School of Electronic Engineering, Information Technology and Mathematics, University of Surrey, Surrey, United Kingdom, 2001.

[10] A. Lai, Y. C. Lin, S. S. Wei et al., "Numerical investigation of a $N_2O$/HTPB hybrid rocket motor with a dual-vortical-flow (DVF) design," *Journal of Mechanics*, vol. 33, no. 06, pp. 853–862, 2017.

[11] J. Anderson, E. Dick, G. Degrez et al., *Computational Fluid Dynamics an Introduction*, Springer-Verlag, Berlin Heidelberg, 2009.

[12] Y. S. Chen, T. H. Chou, B. R. Gu et al., "Multiphysics simulations of rocket engine combustion," *Computers & Fluids*, vol. 45, no. 1, pp. 29–36, 2011.

[13] G. C. Cheng and R. Farmer, "Real fluid modeling of multiphase flows in liquid rocket engine combustors," *Journal of Propulsion and Power*, vol. 22, no. 6, pp. 1373–1381, 2006.

[14] http://www.jetec.com.tw/.

[15] http://www.sensolink.com/.

[16] V. Zakirov, M. Sweeting, T. Lawrence, and J. Sellers, "Nitrous oxide as a rocket propellant," *Acta Astronautica*, vol. 48, no. 5-12, pp. 353–362, 2001.

[17] https://www.grc.nasa.gov/WWW/CEAWeb/.

[18] S. Gordon and B. J. McBride, *Computer Program for Calculation of Complex Chemical Equilibrium Compositions and Applications, I Analysis*, NASA RP-1311, 1994.

[19] B. J. McBride and S. Gordon, *Computer Program for Calculation of Complex Chemical Equilibrium Compositions and Applications, II Users Manual and Program Description*, NASA RP-1311, 1996.

# Development of a Novel Launch System Microwave Rocket Powered by Millimeter-Wave Discharge

**Kimiya Komurasaki and Kuniyoshi Tabata**

*Department of Aeronautics and Astronautics, The University of Tokyo, 7-3-1, Hongo, Bunkyo, Tokyo 113-8656, Japan*

Correspondence should be addressed to Kuniyoshi Tabata; k.tabata@al.t.u-tokyo.ac.jp

Academic Editor: Angel Velazquez

This paper presents the state of art of *Microwave Rocket* development and related researches on atmospheric discharge in a high-power millimeter-wave beam. Its operational mechanisms, thruster design, history of development, and flight path and cost analyses are introduced along with millimeter-wave discharge observations and numerical simulations. A thruster model of 126 g weight with no on-board propellant was launched to 1.2 m altitude using a 1 MW class gyrotron. A flight analysis that shows 77% cost reduction is possible using *Microwave Rocket* as the first stage of H-IIB heavy. A millimeter-wave discharge with unique plasma structure such as a quarter-wavelength microstructure and a comb-shaped filamentary structure was observed and reproduced by a two-dimensional numerical model.

## 1. Introduction

Many recent studies have been devoted to wireless power transfer systems for flying objects using electromagnetic-wave radiation such as lasers and microwaves [1–3]. The use of electric power presents many benefits because of its efficiency, nontoxicity, and nonexplosive features. Beamed energy propulsion [4–15] used as a space transportation system gains its propulsive energy wirelessly from outside of the vehicle using beamed energy. Various propulsion mechanisms have been proposed as alternatives to conventional chemical rocket propulsion.

Space launch systems are powered by an electromagnetic-wave beam irradiating them from the ground, as depicted in Figure 1. In 2000, Myrabo demonstrated the launch of a thruster model using a 10 kW class pulsed $CO_2$ laser up to an altitude of 71 m in a flight lasting 12.7 s [9]. The vehicle, *Lightcraft*, weighed about 50 g and was driven by a laser-supported detonation (LSD). Parkin and Murakami proposed a microwave thermal thruster; the concept of which is to accelerate on-board propellant heated through a heat exchanger irradiated by a millimeter-wave beam from the ground [10]. The hydrogen propellant reaches double the exhaust velocity of LOX/LH2 rocket propellant.

In detonation-type thrusters, a laser or millimeter-wave induces atmospheric discharge. An ionization front propagates at a supersonic speed toward the energy source accompanying a shock wave. Such waves are designated as LSD or millimeter-wave-supported detonation (MSD), by which electromagnetic wave energy is converted directly and efficiently to thrust.

As described in this paper, a thruster driven by an atmospheric millimeter-wave discharge, *Microwave Rocket*, is introduced. *Microwave Rocket* is characterized by three features. First, it can achieve a high payload ratio because the vehicle need not load fuel and oxidizer on-board. Instead, it uses atmospheric air as a propellant during flight in a dense atmosphere. Second, once an electromagnetic-wave generator facility, such as gyrotrons, is built on the ground, it is reusable in multiple launches. Third, high-pressure gas for thrusting is produced through a millimeter-wave-supported detonation. Therefore, no turbo-pump system is required. For those reasons, *Microwave Rocket* is expected to achieve drastic launch cost reduction.

Figure 1: Artist image of beamed energy propulsion launcher.

A gyrotron is a high-power millimeter-wave oscillator that uses a cyclotron resonance maser phenomenon for energy conversion from electrical energy. The original purpose of the gyrotron development is heating plasma for nuclear fusion using electron cyclotron heating and current drive. In Japan at the National Institute for Quantum and radiological Science and Technology (QST, formerly the Japan Atomic Energy Agency), a gyrotron oscillating at 170 GHz has been developed for the International Thermonuclear Experimental Reactor (ITER) project [16–18]. The improvement of oscillation efficiency using the collection of depressed electrical energy enabled the output power of greater than 1 MW. Detailed characteristics of the QST gyrotron for the ITER project are presented in Table 1. A millimeter-wave beam is inferior to a laser beam in directivity, but millimeter-waves are attractive because the cost of gyrotron manufacture is lower than the laser oscillator cost by 2–4 orders of magnitude.

## 2. Thruster Structure and Engine Cycle

*Microwave Rocket* comprises a cylindrical tube in which an MSD wave propagates and a closed end, called a thrust wall, where high-pressure conditions are sustained. For air-breathing, reed valves are installed on the tube wall. Its engine cycle is identical to a typical pulse detonation engine cycle, as presented in Figure 2. (1) A millimeter-wave beam is line-focused by a conical concentrator. Atmospheric breakdown occurs near the thrust wall. (2) An MSD wave propagates toward the thruster exit, absorbing the incident millimeter-wave power. (3) At the time when the MSD wave is exhausted through the tube exit, the incident millimeter-wave pulse is suspended. At the same time, an expansion wave begins to propagate upstream in the tube from the exit to the thrust wall. (4) When the expansion wave reaches the thrust wall, it is reflected on the wall, generating negative gauge pressure inside the tube. Because of this negative pressure, reed valves open. Fresh air is taken passively. Impulsive thrust is generated intermittently by repeating this cycle.

A chemical detonation is settled down to a steady-state called a Chapman–Jouguet (C–J) detonation, which gives the highest pressure increase for a certain heat input [19]. On the other hand, the propagation velocity of a MSD wave is known to exceed the C–J velocity, as we call over-driven detonation. In order to predict the condition behind a blast wave, it is of much importance to examine the relationship between beam intensity and propagation velocity.

Table 1: Characteristics of QST gyrotron [16].

| | |
|---|---|
| Frequency (wavelength) | 170 GHz (1.77 mm) |
| Output power | >1 MW |
| Output duration | 1 ms to 3600 s |
| Beam transverse mode | Gaussian |
| Electric efficiency | >50% |

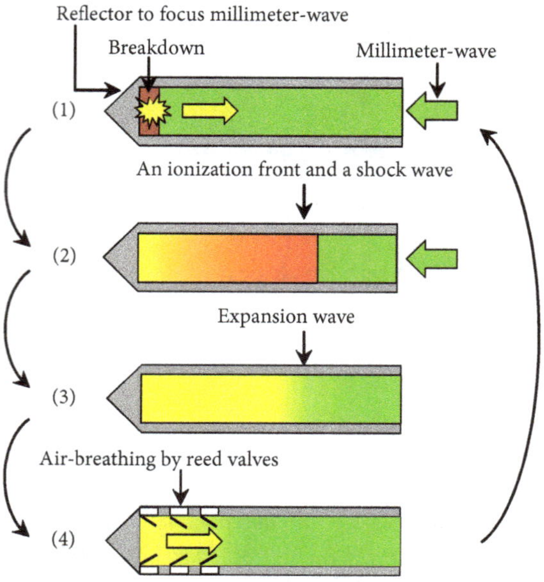


Figure 2: Pulse detonation engine cycle in *Microwave Rocket*.

## 3. Thruster Development and Launch Demonstrations

In 2001, plasma was ignited by focusing a beam launched from the QST gyrotron using a parabolic reflector under atmospheric conditions. In 2003, the first launch experiment used a 930 kW millimeter-wave beam in a single pulse operation. A miniature rocket model weighing 10 g was lifted to 2 m altitude (Figure 3) [20]. In the experiment, the maximum momentum coupling coefficient of 395 N/MW, defined as a ratio of total obtained impulsive thrust impulse to input energy, was achieved. This coefficient was comparable to that of a laser detonation thruster achieved using a solid-state laser with 2.0 J pulse energy in 2013 [21].

After optimizing operational parameters in a repetitive pulse mode and after developing a beam expander, a 126 g thruster model was launched to 1.2 m altitude in 2009 (Figure 4) [22]. Continuous generation of impulsive thrust was confirmed. In the repetitive pulse mode, impulses after the first impulse were found to decrease with residual air density inside the thruster tube, where air density decreases because of heating. However, experiments using a forced air-breathing system with a pressurized air tank proved that the impulse could be recovered with air refreshment inside the tube during the pulse interval [22].

In 2011, thrust was augmented by increases in output power of the gyrotron and in the thrust duty cycle, defined

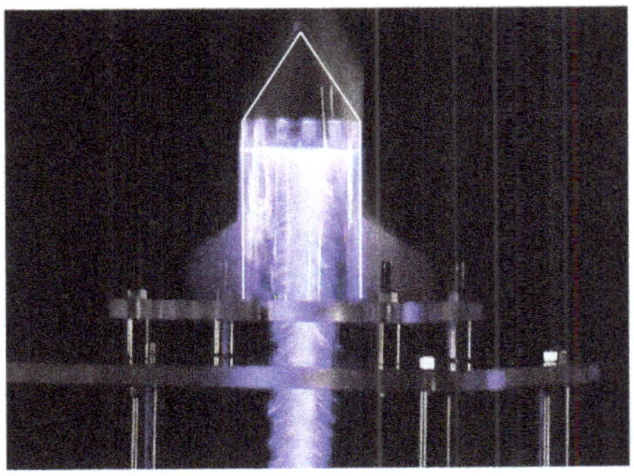

FIGURE 3: Launch demonstration with a 10 g thruster model conducted in 2003 [20]. Millimeter-wave discharge was photographed through a transparent thruster tube.

FIGURE 4: Launch demonstration with a 126 g thruster model in a multipulse operation in 2009 [22].

as a product of millimeter-wave pulse duration and pulse repetition frequency, as presented in Table 2 [23]. This experiment was made possible by the QST gyrotron, which had a high-voltage IGBT switch with switching speed that was higher than earlier versions. Because the achieved time-averaged thrust was as high as 30 N, a MW-class gyrotron can launch a kg-class vehicle. Therefore, a 1 kg thruster model launch is planned as the next step.

Reed valves with an inlet plenum are under development for efficient air-breathing during subsonic and supersonic flights at high altitudes. Reed valves open passively by the pressure difference between those inside and outside of the tube. A tapered-shape reed valve, which is capable of opening widely toward the inside without a stopper, was designed and tested [24]. A stopper designed to prevent plastic deformation of the valve was not used because it interacts with intense electromagnetic waves inside the tube. Air-breathing performance is evaluated by the partial filling rate (PFR) as represented by

$$\mathrm{PFR} = \frac{\text{Refreshed air volume}}{\text{Thruster tube volume}}. \quad (1)$$

TABLE 2: Thrust performance and operation conditions [23].

| Year | 2009 | 2011 |
|---|---|---|
| Beam power @pulse repetition frequency | 270 kW @ 50 Hz | 570 kW @ 200 Hz |
| $C_m$ | 100 N/MW | 360 N/MW |
| Time averaged thrust | 2.3 N | 30 N |

According to a CFD analysis [25], a 10–15-fold thrust augmentation is expected using reed valves on the ground. At an altitude of 10 km and a flight Mach number of 2.0, the inlet plenum stagnates the flow and the thruster breathes the compressed air. The amount of compressed air to be breathed increases with the inlet plenum diameter, improving its air-breathing performance. The optimum inlet plenum diameter is determined by the balance between aerodynamic drag and air-breathing performance [26].

The altitude to which a vehicle continues to accelerate is theoretically proportional to the square of beam diameter: an approximately 40 cm beam diameter is necessary to achieve 10 m altitude; 6 m of beam diameter is necessary to achieve 100 km altitude. The beam intensity inside the tube is adjustable to achieve optimum detonation speed using a beam concentrator [27]. In 2012, a launch experiment with long beam transmission was conducted using a beam expander and concentrator. A $\phi$ 240 mm millimeter-wave beam was led into a thruster as shown in Figure 5, and the momentum coupling coefficient of 204 N/MW was obtained in a single-pulse operation, which was half as high as that in the short beam transmission experiment in 2003 [20]. In a multipulse operation, abnormal air breakdown was observed after the second pulse at the outlet of the thruster, resulting in a decrease of thrust force [28]. Because this phenomenon is considered due to the electrons remaining in the thruster, the pulse repetitive frequency is limited so that the pulse interval is long enough for the remaining electrons to be exhausted out of the thruster.

## 4. Feasibility Studies and Launch Cost Estimation

*4.1. Replacement of H-IIB Launch Vehicle's First Stage by Microwave Rocket.* Feasibility studies of transportation to low earth orbit and launch cost analyses for the *Microwave Rocket* have been conducted through subsonic to supersonic flight analyses. Fukunari et al. showed that the launch cost of the H-IIB heavy, which is capable of orbiting 19-ton payload to LEO, will be reduced by 77% by replacing its first stage including four SRBs by *Microwave Rocket*. The vehicle is significantly accelerated in a dense atmosphere and cut off at an altitude of 20.7 km with a velocity of 2 km/s. Using the conventional second stage, the payload ratio is enhanced to 0.155 which is 4.5 times as high as the one by the H-IIB heavy. The manufacturing cost as the first stage is estimated to be around 3 M$ and is considerably lower than the conventional H-IIB first stage which costs about 85 M$. For this flight, about 94,000 gyrotrons are necessary to generate 188 GW output power. The construction cost of gyrotron

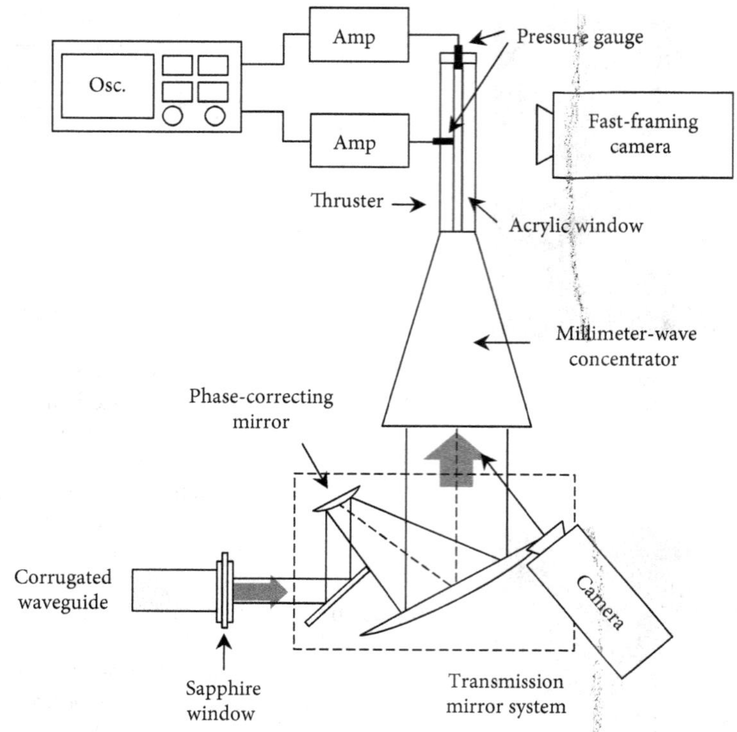


Figure 5: Setup for a launch experiment with long beam transmission using a concentrator [28].

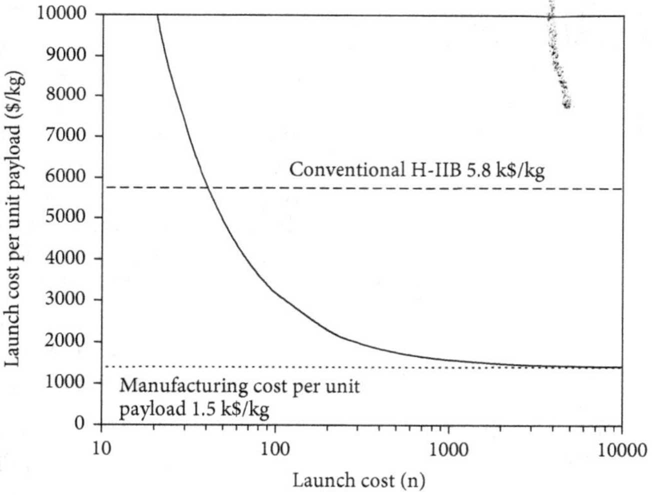


Figure 6: Estimated launch cost to LEO per unit payload as a function of launch counts. Dashed line: cost for H-IIB heavy; dotted line: manufacturing cost for *Microwave Rocket* and H-IIB second stage [14].

accounts for the major part of the new launch system. Although this initial cost could be as much as 3350 M$ including an energy storage facility, it will be amortized with launch counts and the cost per launch is decreased to the same level of conventional launch cost with about 42 launches, and finally 77% cost reduction is expected as shown in Figure 6 [14].

*4.2. Small Payload Launch System Combining Microwave Rocket and NASA Microwave Thermal Thruster.* Another feasibility study introduces a combination of *Microwave Rocket* as the first stage and a microwave thermal rocket studied by NASA as the second stage aiming at small payload launch to low earth orbit. The original NASA concept was to launch a vehicle by an unmanned aerial vehicle (UAV), and then initiate the microwave thermal rocket composed of a heat exchanger and a hydrogen propellant tank [10]. Replacement of the UAV by *Microwave Rocket* enables the second stage to start at high altitude with large initial velocity, resulting in low tank weight and low overall cost. Launch cost

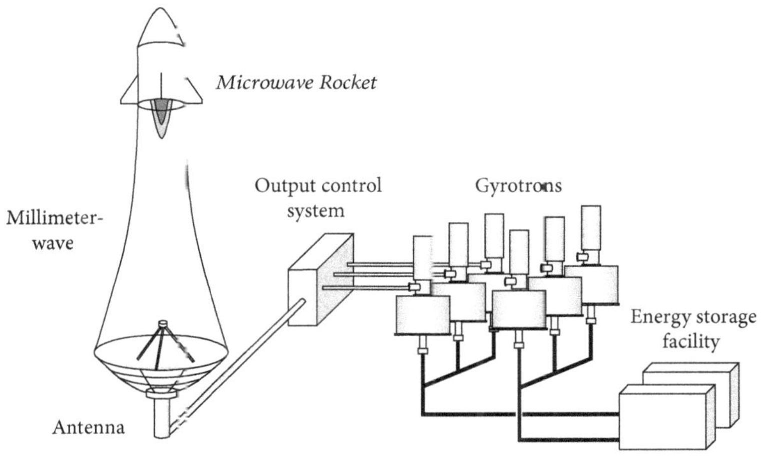


FIGURE 7: System Configuration of a beam facility.

per unit payload mass will be 5.8 k$/kg which is one-fourth of the UAV-assisted launch. The beam facility cost for launching an 8 kg satellite is estimated: the necessary output power is about 100 MW, and the gyrotrons and power supply cost 22 M$, while a transmitter antenna of 180 m in diameter costs 466 M$ [15].

*4.3. Beam Facility Configuration on the Ground and Required Technologies in the Future.* The beam facility consists of gyrotrons and a transmitter antenna. If a phased array antenna system is employed for beamforming to reduce the output power, an output power control system for synthesizing and dividing with phase controlling is required. The beam facility configuration is depicted in Figure 7.

In the ITER project, 24 gyrotrons are combined to have 20 MW heating in total [18]. There will be several restrictions like adjacent distance between gyrotrons, for instance, to avoid the interference of leakage magnetic field from superconducting magnets [29]. The gyrotron operation time of one launch is about 200 seconds. Even if ten launches are carried out every day, total operation time is only 200 hours per year, which is much shorter than expected lifetime of typical gyrotron system 5000 hours [5]. Flywheels utilized in nuclear fusion [30] and more advanced system, superconducting magnetic energy storage (SMES) system, are the candidates for an energy storage facility.

A phased array system which needs to control output power, frequency, and phase will make the required output power even smaller [31]. It will be effective to utilize a phase-locking technology to stabilize gyrotron's frequency and phase by inserting external signals to gyrotrons [32]. When a phased array system is not employed, a huge parabolic antenna will be used as an antenna. An antenna whose diameter is 50 m ~ 100 m can be constructed by hundreds of millions of dollars. The stratospheric platform airship program conducted by JAXA intended to transmit 1 MW electric power to the airship, using a parabolic antenna [33]. An optimal beam frequency has to be chosen by taking into account the resulting size of an antenna and a launch vehicle and the atmospheric attenuation of a beam. When a beam is transmitted by a Gaussian profile, the transmission distance theoretically determines the spread of the beam diameter. Although the directivity of a Gaussian beam is improved with a larger frequency, the atmospheric attenuation rate increases. Applied frequency will be chosen from the transmission windows in the range of 100 GHz calculated by a line by line method [6]. For *Microwave Rocket* launches in the future, it is necessary to amend the current Radio law which restricts the use of electromagnetic wave.

## 5. Propagation of Ionization Front and Filamentary Plasma Structure in Millimeter-Wave Discharge

Experimental studies of atmospheric microwave discharge have been conducted since the 1940s. However, studies using a millimeter-wave band have begun because of the widening use of gyrotrons. According to past studies, the propagation velocity of an ionization front in a millimeter-wave discharge is revealed to have a very different tendency from that in laser discharge, as presented in Figure 8, where the measured propagation velocities in a millimeter-wave discharge using a 170 GHz (wavelength, $\lambda = 1.76$ mm) gyrotron [34, 35] and a 110 GHz ($\lambda = 2.73$ mm) gyrotron [36] are shown along with those in laser discharge obtained using a $CO_2$ laser ($\lambda = 10.6\,\mu$m) with sufficiently large beam spot size [37–39]. The propagation velocity of an ionization front in a millimeter-wave discharge is greater than those in laser discharge by one order of magnitude.

In addition to its high propagation velocity, a millimeter-wave discharge plasma is known to have unique and fine structures as depicted in Figure 9. The plasma structure observed by Oda et al. [34, 35] in a 170 GHz, 1.5 GW/m$^2$ millimeter-wave beam is depicted in Figure 9(a). The branching structure is referred to as a comb-shaped structure. The structure called a fishbone shape is depicted in Figure 9(b). It was observed by Vikharev et al. [40] in a 35 GHz, 0.14 GW/m$^2$ millimeter-wave beam in helium gas at reduced pressure.

The microstructure called a quarter-wavelength structure was observed by Hidaka et al. [41] in a focused 110 GHz,

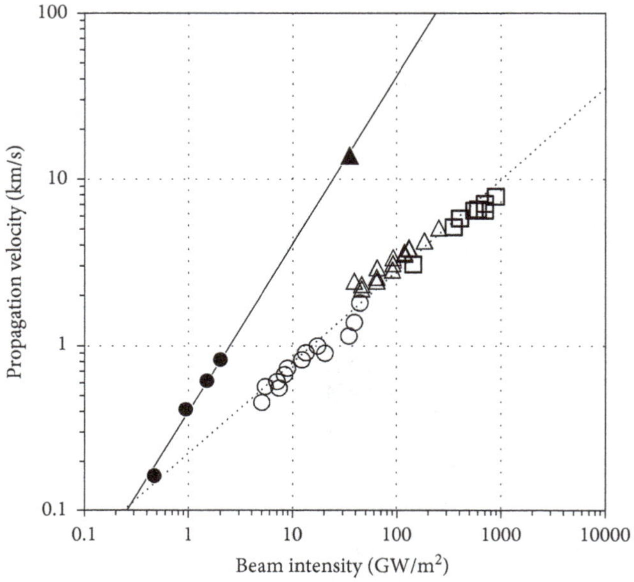


$CO_2$ laser, $\lambda = 10.6\ \mu m$
○ I. A. Bugetov *et al.*
△ K. Matsui *et al.*
□ P. Bournot *et al.*
⋯⋯ Fitting line

Millimeter-wave
▲ $\lambda = 1.77$ mm Y. Oda *et al.*
● $\lambda = 2.73$ mm Y. Hidaka *et al.*
—— Fitting line

FIGURE 8: Ionization-front propagation velocity in an atmospheric millimeter-wave discharge [34, 36] and laser discharge [37–39].

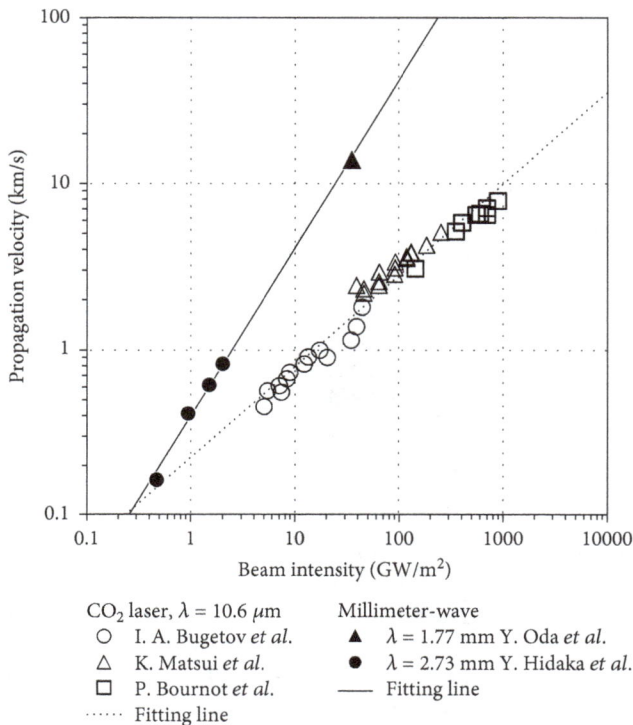

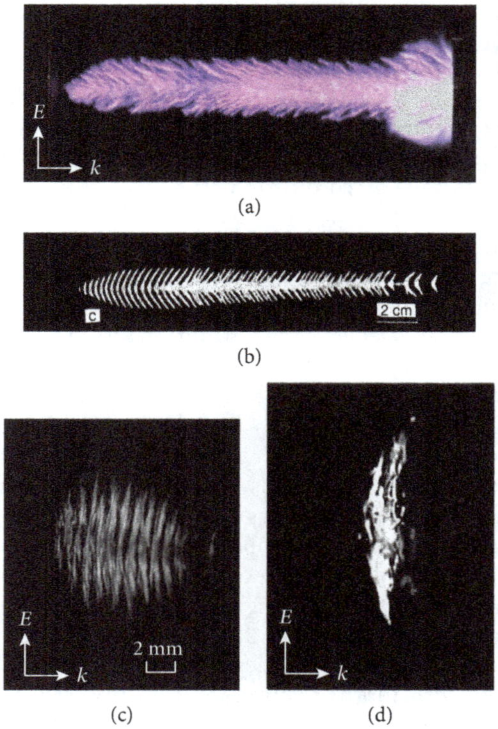


FIGURE 9: Filamentary plasma structures observed in experiments: (a) 170 GHz, 1.5 GW/m$^2$ [34, 35]; (b) 35 GHz, 0.14 GW/m$^2$ [40]; (c) 110 GHz, 35 GW/m$^2$, 0.8 atm [41]; (d) 28 GHz, 0.59 GW/m$^2$, 1 atm. [44]. *k* and *E*, respectively, denote the wave number vector and the electric field vector of an incident beam.

35 GW/m$^2$ beam, as depicted in Figure 9(c). Cook et al. also observed the filamentary structure in a 110 GHz millimeter-wave [42]. In this structure, the interval between neighboring filamentary plasmas extending in the electric field direction was about a quarter-$\lambda$.

In atmospheric discharges in a microwave beam, a streamer-like structure was observed at subcritical conditions and numerically reproduced by Khodataev [43]. It was found that breakdown occurred in an intense electric field near the surface of a thin plasma filament. A similar structure was obtained in a 28 GHz millimeter-wave beam as shown in Figure 9(d) [44].

Atmospheric millimeter-wave discharge physics remains unclear. Especially, the propagation mechanism of a millimeter-wave discharge at a beam intensity far below the breakdown threshold is under investigation.

## 6. Numerical Simulation of Millimeter-Wave Discharge Plasma

Numerical simulations are effective to clarify the formation mechanisms of these distinctive structures and in the end for an optimal thruster designing. Many researchers have been working on simulations. Millimeter-wave discharges are classifiable into two conditions, subcritical or undercritical, in which electric field intensities of the incident beam are, respectively, close to or far below the breakdown threshold.

The quarter-$\lambda$ structure was obtained when the irradiated millimeter-wave is at the subcritical intensity. Boeuf et al. [45] reproduced this quarter-$\lambda$ structure in an *E*-*k* plane. They coupled Maxwell's equations with an electron diffusion equation with effective diffusion coefficients and ionization frequencies computed using Bolsig+. Because of the reflection of incident waves on the plasma, an antinode of a standing wave was generated at quarter-$\lambda$ ahead of the plasma resulting in the quarter-$\lambda$ plasma structure as observed.

The comb-shaped filamentary structure was formed at much lower intensity than the breakdown threshold. Takahashi et al. [46, 47] proposed a numerical model considering compressibility of the ambient gas. Results showed that expansion behind a precursor shock wave enhances the effective field intensity to near the threshold in front of the plasma.

For cases in which shock expansion is not expected, we assumed hypothetical ionization frequencies higher than those obtained using Bolsig+. We reproduced a comb-shape filamentary plasma structure [48]. Figure 10 portrays the numerically computed plasma structure along with the experimentally observed structure, in both of which plasmoid pitch was approximately 0.9 $\lambda$. This structure is created by the wave reflections on the discrete plasmoid. The pitch was invariable for plasma density, millimeter-wave wavelength, and electric field intensity. The quarter-$\lambda$ microstructure is apparent in each plasmoid when the ionization frequency was high and the diffusion coefficient was low.

Millimeter-wave discharge occurs even at the beam intensity far below breakdown threshold [44]. This discharge cannot be explained by field concentration nor by ambient gas expansion behind a blast wave [45, 46] and requires other

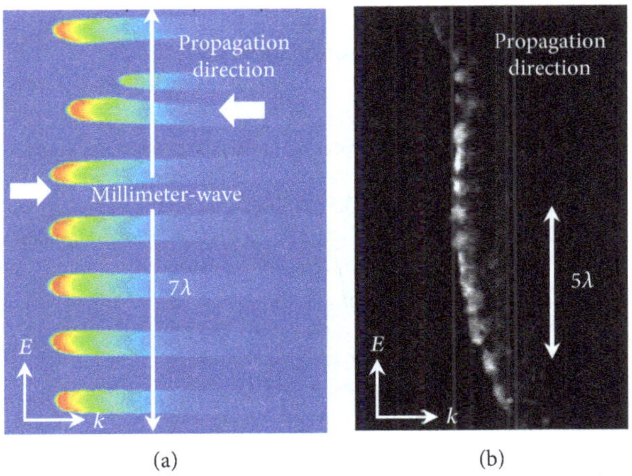


FIGURE 10: Filamentary plasma structures in an atmospheric millimeter-wave discharge at the undercritical intensity: (a) computed density profile and (b) observed luminosity profile. 170 GHz; 1 GW/m$^2$ [48].

discharge sustention mechanisms. To derive the ionization frequencies at the undercritical intensity theoretically, an ionization model considering neutral gas excitation must be examined in the future work.

## 7. Summary

*Microwave Rocket*, one of beamed energy propulsion systems, is under development. Millimeter-wave-supported detonation driven by an atmospheric millimeter-wave discharge generates high pressures in a thruster tube, imparting impulsive thrust on a thrust wall. The repetitively pulsed detonation engine cycle produces continuous thrust. A thruster model of 126 g weight was launched to 1.2 m altitude using a QST gyrotron.

Flight analyses showed that *Microwave Rocket* can replace conventional chemical rockets with a significant improvement in a payload ratio and bring a drastic launch cost reduction. By utilizing *Microwave Rocket* as the first stage of the H-IIB heavy, the total mass of the vehicle is expected to be one-fifth of the conventional one which can directly improve the payload ratio by five times. The manufacturing cost of the gyrotrons is the substantial part of the *Microwave Rocket*, which can be recovered through 2000 launches, and as a result, the launch cost can be 77% lower than the conventional launch cost. Regarding the beam facility on the ground, conventional technologies utilized in the field of nuclear fusion researches can be used.

The physics of an atmospheric millimeter-wave discharge, especially the propagation velocity of the plasma front, is of interest as the thrust performance is directly related to it. The propagation velocity of the ionization front is much higher than those in laser discharge when shown as a function of beam intensity. However, the discharge at a beam intensity far below the breakdown threshold remains unclarified. A new ionization model would be an indispensable tool for it.

## Conflicts of Interest

The authors declare that they have no conflicts of interest.

## Acknowledgments

This work was presented at the 1st Asia-Pacific Conference on Plasma Physics, 18–23 September 2017, Chengdu, China, and was supported by JSPS KAKENHI Grant number JP15H05770

## References

[1] C. Phipps, M. Birkan, W. Bohn et al., "Review: laser-ablation propulsion," *Journal of Propulsion and Power*, vol. 26, no. 4, pp. 609–637, 2010.

[2] G. W. Jull, *Summary Report on SHARP (Stationary High-Altitude Relay Platform) Part A-Technical Feasibility of Microwave-Powered Airplanes*, CRC Report No. 1393, Communications Research Centre, Ottawa, 1985.

[3] K. Shimamura, H. Sawahara, A. Oda et al., "Feasibility study of microwave wireless powered flight for micro air vehicles," *Wireless Power Transfer*, vol. 4, no. 2, pp. 146–159, 2017.

[4] K. Komurasaki and B. Wang, "Laser propulsion," in *Encyclopedia of Aerospace Engineering*, R. Blockley and W. Shyy, Eds., pp. 1351–1360, John Wiley & Sons Ltd., Chichester, UK, 2010.

[5] P. George and R. Beach, "Beamed-energy propulsion (BEP) study," US Patent NASA/TM-2012-217014, 2012.

[6] K. Komurasaki, K. Shimamura, and M. Fukunari, "Microwave rocket "propelled by millimeter wave"," *Journal of Plasma and Fusion Research*, vol. 93, no. 10, pp. 465–490, 2017, (In Japanese).

[7] M. Fukunari, K. Komurasaki, Y. Nakamura, Y. Oda, and K. Sakamoto, "Rocket propulsion powered using a gyrotron," *Journal of Energy and Power Engineering*, vol. 11, no. 6, pp. 363–371, 2017.

[8] A. Kantrowitz, "Propulsion to orbit by ground-based laser," *Astronautics and Aeronautics*, vol. 10, pp. 74–76, 1972.

[9] L. Myrabo, "World record flights of beam-riding rocket lightcraft - demonstration of "disruptive" propulsion technology," in *37th Joint Propulsion Conference and Exhibit, Joint Propulsion Conferences*, Salt Lake City, UT, USA, July 2001.

[10] K. L. Parkin and D. D. Murakami, "An overview of the NASA Ames Millimeter-Wave Thermal Launch System," in *48th AIAA/ASME/SAE/ASEE Joint Propulsion Conference & Exhibit*, Atlanta, GA, USA, July-August 2012.

[11] H. Katsurayama, K. Komurasaki, and Y. Arakawa, "A preliminary study of pulse-laser powered orbital launcher," *Acta Astronautica*, vol. 65, no. 7-8, pp. 1032–1041, 2009.

[12] T. J. Kare, "Modular laser options for HX laser launch," in *Proceeding of 3rd International Symposium on Beamed Energy Propulsion, AIP Conference Proceedings*, pp. 128–139, Troy, NY, USA, October 2005.

[13] E. W. Davis and F. B. Mead Jr., "Review of laser lightcraft propulsion system," in *Proceeding of 3rd International Symposium on Beamed Energy Propulsion, AIP Conference Proceedings*, pp. 283–294, Kailua-Kona, Hawaii, November 2008.

[14] M. Fukunari, A. Arnault, T. Yamaguchi, and K. Komurasaki, "Replacement of chemical rocket launchers by beamed energy propulsion," *Applied Optics*, vol. 53, no. 31, pp. I16–I22, 2014.

[15] K. Kakinuma, M. Fukunari, T. Yamaguchi et al., "Two-stage-to-orbit transporting system combining microwave rocket and microwave thermal rocket for small satellite launch," *Transactions of the Japan Society for Aeronautical and Space Sciences, Aerospace Technology Japan*, vol. 14, no. 30, pp. Pb_99–Pb_103, 2016.

[16] K. Sakamoto, A. Kasugai, K. Takahashi, R. Minami, N. Kobayashi, and K. Kajiwara, "Achievement of robust high-efficiency 1 MW oscillation in the hard-self-excitation region by a 170 GHz continuous-wave gyrotron," *Nature Physics*, vol. 3, no. 6, pp. 411–414, 2007.

[17] T. Kariya, T. Imai, R. Minami et al., "Development of gyrotrons for fusion with power exceeding 1 MW over a wide frequency range," *Nuclear Fusion*, vol. 55, no. 9, article 093009, 2015.

[18] M. A. Henderson and G. Saibene, "Critical interface issues associated with the ITER EC system," *Nuclear Fusion*, vol. 48, no. 5, article 054017, 2008.

[19] T. Endo, J. Kasahara, A. Matsuo, S. Sato, K. Inaba, and T. Fujiwara, "Pressure history at the thrust wall of a simplified pulse detonation engine," *AIAA Journal*, vol. 42, no. 9, pp. 1921–1930, 2004.

[20] T. Nakagawa, Y. Mihara, K. Komurasaki, K. Takahashi, K. Sakamoto, and T. Imai, "Propulsive impulse measurement of a microwave-boosted vehicle in the atmosphere," *Journal of Spacecraft and Rockets*, vol. 41, no. 1, pp. 151–153, 2004.

[21] B. Wang, K. Michigami, K. Komurasaki, and Y. Arakawa, "Thrust measurement for laser-detonation propulsion with a solid-state laser," *Journal of Propulsion and Power*, vol. 29, no. 1, pp. 276–278, 2013.

[22] Y. Oda, T. Shibata, K. Komurasaki, K. Takahashi, A. Kasugai, and K. Sakamoto, "Thrust performance of microwave rocket under repetitive-pulse operation," *Journal of Propulsion and Power*, vol. 25, no. 1, pp. 118–122, 2009.

[23] R. Komatsu, T. Yamaguchi, Y. Oda et al., "Thrust augmentation of microwave rocket with high-power and high-duty-cycle operation," *Journal of the Japan Society for Aeronautical and Space Sciences*, vol. 60, no. 6, pp. 235–237, 2012.

[24] M. Fukunari, T. Yamaguchi, K. Komurasaki et al., "Air-breathing system using reed valve for pulse detonation microwave rocket," *Transactions of the Japan Society for Aeronautical and Space Sciences, Aerospace Technology Japan*, vol. 14, pp. 1–7, 2016.

[25] M. Fukunari, R. Komatsu, A. Arnault, T. Yamaguchi, K. Komurasaki, and Y. Arakawa, "Air-breathing performance of microwave rocket with reed valve system," *Vacuum*, vol. 88, pp. 155–159, 2013.

[26] K. Tabata, F. Nguyen, Y. Harada et al., "Numerical calculation on air-inlet design of microwave rocket," *Journal of the Japan Society for Aeronautical and Space Sciences*, vol. 66, no. 5, pp. 128–134, 2018.

[27] M. Fukunari, N. Wongsuryrat, T. Yamaguchi, Y. Nakamura, K. Komurasaki, and H. Koizumi, "Design of a millimeter-wave concentrator for beam reception in high-power wireless power transfer," *Journal of Infrared, Millimeter, and Terahertz Waves*, vol. 38, no. 2, pp. 176–190, 2017.

[28] M. Fukunari, T. Yamaguchi, Y. Nakamura et al., "Thrust generation experiments on microwave rocket with a beam concentrator for long distance wireless power feeding," *Acta Astronautica*, vol. 145, pp. 263–267, 2018.

[29] K. Kajiwara, Y. Oda, A. Kasugai et al., "Effect of the stray magnetic field on the gyrotrons for ITER," in *2009 34th International Conference on Infrared, Millimeter, and Terahertz Waves*, Busan, South Korea, September 2009.

[30] R. Shimada, "Power conversion and energy storage system for a fusion reactor 3. Performance of large electric power equipment and future view 3.3 kinetic energy storage (flywheel, compressed air energy storage)," *Journal of Plasma and Fusion Research*, vol. 80, no. 7, pp. 572–577, 2004.

[31] K. Kakinuma, F. Nguyen, M. Fukunari, K. Komurasaki, and H. Koizumi, "Application of active phased array antenna for millimeter-wave transfer to beamed energy launch vehicles," *Space Solar Power Systems*, vol. 3, pp. 30–37, 2018.

[32] G. G. Denisov and A. G. Litvak, "New results of development of gyrotrons for plasma fusion installations," in *26th IAEA Fusion Energy Conference*, Kyoto, Japan, 2016.

[33] T. Fujihara and K. Eguchi, "Research and development on regenerative fuel cells for stratospheric platform airship: ground-based testing of 1 KW RFC system models," JAXA Tech. Rep. JAXA-RM-07-014, 2008, (In Japanese).

[34] Y. Oda, K. Komurasaki, and K. Sakamoto, "Dynamics and structure of ignition process in plasmas," *Journal of Plasma and Fusion Research*, vol. 84, no. 6, pp. 343–347, 2008.

[35] Y. Oda, K. Komurasaki, K. Takahashi, A. Kasugai, and K. Sakamoto, "Plasma generation using high-power millimeter-wave beam and its application for thrust generation," *Journal of Applied Physics*, vol. 100, no. 11, article 113307, 2006.

[36] Y. Hidaka, E. M. Choi, I. Mastovsky et al., "Plasma structures observed in gas breakdown using a 1.5 MW, 110 GHz pulsed gyrotron," *Physics of Plasmas*, vol. 16, no. 5, article 055702, 2009.

[37] I. A. Bufetov, A. M. Prokhorov, V. B. Fedorov, and V. K. Fomin, "Optical discharge accompanying a restriction imposed on lateral expansion of gas and a reduction in the threshold of light-induced detonation," *JETP Letters*, vol. 39, pp. 258–261, 1987.

[38] K. Matsui, T. Shimano, J. A. Ofosu, K. Komurasaki, T. Schoenherr, and H. Koizumi, "Accurate propagation velocity measurement of laser supported detonation waves," *Vacuum*, vol. 136, pp. 171–176, 2017.

[39] P. Bournot, P. A. Pincosy, G. Inglesakis, M. Autric, D. Dufresne, and J. P. Caressa, "Propagation of a laser-supported detonation wave," *Acta Astronautica*, vol. 6, no. 3-4, pp. 257–267, 1979.

[40] A. L. Vikharev, V. B. Gil'denburg, S. V. Golubev et al., "Nonlinear dynamics of a freely localized microwave discharge in an electromagnetic wave beam," *Soviet Physics - Journal of Experimental and Theoretical Physics*, vol. 67, no. 4, pp. 724–728, 1988.

[41] Y. Hidaka, E. M. Choi, I. Mastovsky, M. A. Shapiro, J. R. Sirigiri, and R. J. Temkin, "Observation of large arrays of plasma filaments in air breakdown by 1.5-MW 110-GHz gyrotron pulses," *Physical Review Letters*, vol. 100, no. 3, article 035003, 2008.

[42] A. M. Cook, J. S. Hummelt, M. A. Shapiro, and R. J. Temkin, "Observation of plasma array dynamics in 110 GHz millimeter-wave air breakdown," *Physics of Plasmas*, vol. 18, no. 10, article 100704, 2011.

[43] K. V. Khodataev, "Microwave discharges and possible applications in aerospace technologies," *Journal of Propulsion and Power*, vol. 24, no. 5, pp. 962–972, 2008.

[44] Y. Harada, Y. Nakamura, K. Komurasaki et al., "Structural change of plasma at various ambient pressures in 28 GHz millimeter-wave discharges," *Frontier of Applied Plasma Technology*, vol. 10, p. 1, 2017.

[45] J.-P. Boeuf, B. Chaudhury, and G. Q. Zhu, "Theory and modeling of self-organization and propagation of filamentary plasma arrays in microwave breakdown at atmospheric pressure," *Physical Review Letters*, vol. 104, no. 1, article 015002, 2010.

[46] M. Takahashi, Y. Kageyama, and N. Ohnishi, "Joule-heating-supported plasma filamentation and branching during subcritical microwave irradiation," *AIP Advances*, vol. 7, no. 5, article 055206, 2017.

[47] M. Takahashi and K. Komurasaki, "Discharge from a high-intensity millimeter wave beam and its application to propulsion," *Advances in Physics: X*, vol. 3, no. 1, pp. 113–144, 2018.

[48] Y. Nakamura, K. Komurasaki, M. Fukunari, and H. Koizumi, "Numerical analysis of plasma structure observed in atmospheric millimeter-wave discharge at under-critical intensity," *Journal of Applied Physics*, vol. 124, no. 3, article 033303, 2018.

# 10

# Experimental Verification of a Simple Method for Accurate Center of Gravity Determination of Small Satellite Platforms

**Dario Modenini, Giacomo Curzi, and Paolo Tortora**

*Department of Industrial Engineering, University of Bologna, Via Fontanelle 40, 47121 Forlì, Italy*

Correspondence should be addressed to Dario Modenini; dario.modenini@unibo.it

Academic Editor: Franco Bernelli-Zazzera

We propose a simple and relatively inexpensive method for determining the center of gravity (CoG) of a small spacecraft. This method, which can be ascribed to the class of suspension techniques, is based on dual-axis inclinometer readings. By performing two consecutive suspensions from two different points, the CoG is determined, ideally, as the intersection between two lines which are uniquely defined by the respective rotations. We performed an experimental campaign to verify the method and assess its accuracy. Thanks to a quantitative error budget, we obtained an error distribution with simulations, which we verified through experimental tests. The retrieved experimental error distribution agrees well with the results predicted through simulations, which in turn lead to a CoG error norm smaller than 2 mm with 95% confidence level.

## 1. Introduction

The growing interest for the development of light, small, highly capable spacecraft (S/C) platforms for a wide range of missions demands for a boost in performance from the standards established by the multitude of low-cost micro/nanosatellites. Often developed as part of university educational programs, they have been dominating this segment in the last two decades. In this respect, it is known that accurate attitude and orbit control systems rely on the precise knowledge of the spacecraft CoG. However, the development of such a class of S/C is highly cost-driven, whereas methods for measuring the CoG commonly employed for larger platforms [1], being highly accurate, require rather complex and expensive equipment. Thus, cost-effective and easy-to-implement alternatives shall be pursued.

Typically, the methods for measuring CoG of an S/C fall into two broad categories, that is, static methods and dynamic methods [2]. Static methods are often based on the pivoting axis system: the payload under test (PUT) is mounted on an instrument featuring a pivoting axis. In principle, the offset of the CoG from the pivoting axis can be retrieved by measuring the force acting on a point at a certain distance from the axis itself, once the total mass of the payload is known. Complete CoG localization is then obtained by repeating the measurement after rotating the PUT. The most accurate instruments exploiting the static balancing principle consist of rotary platforms featuring a closed loop self-balancing controller, to hold the platform to its neutral position [3]. The torque required for rebalancing is the measured output from which the CoG location can be retrieved, leading to submillimeter accuracies. Another common static measurement method is the one of multipoint weighting, achieved by placing the PUT over a multipoint weight platform equipped with 3 (or 4) high accuracy force transducers. The forces measured by the transducers, whose locations are known, allow to compute the in-plane coordinates of the CoG. This concept is employed at NOVA test facility (Utah University), to measure the mass properties of nanosatellites, with a reported accuracy of 1 mm in localizing the CoG [4].

Dynamic methods are based on the principle of dynamic balancing: the PUT is placed on a spin balance which estimates the CoG location by measuring the centrifugal forces. High sensitivity, however, is achieved at high rotational speeds, which makes such method of limited applicability for space vehicles CoG measurements [1].

Despite various measurement instruments based on all methods listed above are commercially available, these are quite expensive: even when aiming at a relatively low total weight capacity and moderate accuracy, the cost reaches several thousands of Euros. The concept of suspending a body for measuring its CoG, which is pursued in this work, is certainly not new, rather one of the oldest. Suspension was employed for example in NASA X-38 project [5]. In that case, the CoG localization was obtained combining weight distribution (as for a multipoint weight method) with inclination measurements. Recent examples involving the suspension concept are the trifilar torsional pendulum [6], and the photogrammetry technique [7], applied by NASA engineers to locate Orion capsule CoG. The trifilar pendulum is a quite simple mechanism, allowing the joint determination of the CoG and the inertia matrix. The reported accuracy in locating the CoG is 1.5 mm, but this was obtained after a careful calibration of the mechanism and the use of a tricoordinate measuring machine to determine the distance between some predefined points [6]. In [7], the authors suspended a full-scale Orion crew module from an asymmetric bifilar lifting strap and retrieved the CoG position from triangulation of the plumb lines. These, in turn, were determined from a set of images, gathered by a multicamera system, and processed through a set of custom-designed data reduction functions. Authors' indications suggest for an accuracy in the order of few millimeters.

In this paper, we aim at the experimental verification of the method devised by the authors in [8], which relies upon two consecutive monofilar suspensions of the object under test to determine its CoG, using as measured quantities the angle output from a dual-axis inclinometer. To this end, we first generalize the method relaxing some of the constraints outlined in the original formulation. The experimental verification approach is that of applying the method to determine the barycenter of a known mass distribution, that is, a proof mass. To enforce experiment repeatability and smooth systematic errors, we perform measurements from several couple of suspension points. The error of the method is then quantified as the distance between the computed barycenter of the proof mass and the true one.

The main contribution of this work is twofold: (1) to investigate an extremely low-cost method for determining the CoG, with minimum hardware and calibration requirements, with an accuracy suitable for many practical applications, and (2) to provide a comprehensive error analysis which is validated through experiments. To this end, the paper is organized as follows: first, the double suspension method is outlined (Section 2). Then, an error budget is presented, first qualitatively to justify the experimental setup design (Section 3) and later quantitatively by introducing the test facility and the assumed statistical distributions of errors (Section 4). The verification method is then presented in Section 5, which combines Monte Carlo error analysis and experiments. Once the theory is set, results are presented in Section 6, and finally, our conclusions are drawn in Section 7.

## 2. The Double Suspension Method

In recalling and generalizing the method presented in [8], we first define the inclinometer frame of reference. Consider the inclinometer in Figure 1, with top face up; $\hat{\mathbf{z}}_i$ is perpendicular to the top face, with outward positive, $\hat{\mathbf{y}}_i$ is directed in the direction of the cable connection, and $\hat{\mathbf{x}}_i$ completes the right-handed frame.

FIGURE 1: Inclinometer adopted in the experiment.

The dual-axis inclinometer selected for the experiment (Posital Fraba ACS-060) provides as output the direction sines of the gravity vector ($\mathbf{g}$) with respect to $\hat{\mathbf{x}}_i$ (call the angle $X$) and $\hat{\mathbf{y}}_i$ (call the angle $Y$), that is,

$$X = -\sin^{-1}\left(\frac{\mathbf{g}\cdot\hat{\mathbf{x}}_i}{\|\mathbf{g}\|}\right),$$
$$Y = -\sin^{-1}\left(\frac{\mathbf{g}\cdot\hat{\mathbf{y}}_i}{\|\mathbf{g}\|}\right). \quad (1)$$

We define body frame $\hat{\mathbf{x}}_b$, $\hat{\mathbf{y}}_b$, and $\hat{\mathbf{z}}_b$ the frame of reference fixed to the proof mass to be suspended. It results from a simple translation of the inclinometer reference frame. We reserve the definition of the location of its origin later in the manuscript, after the justification of the suspension mass shape. Lastly, we define the laboratory reference frame $\hat{\mathbf{x}}_l$, $\hat{\mathbf{y}}_l$, and $\hat{\mathbf{z}}_l$ as a pseudoinertial frame of reference with: $\hat{\mathbf{z}}_l$ is parallel and opposite to the local gravity vector, $\hat{\mathbf{x}}_l$ points northward, and $\hat{\mathbf{y}}_l$ completes the frame.

The CoG determination method can be summarized as follows: for a given suspension point, the body frame components of the upward local vertical can be computed starting from the inclinometer readings. Then, two suspensions determine two of such unit vectors which identify two lines ideally passing through the CoG of the assembly under test. These lines are not going to intersect exactly due to measurement errors; however, the midpoint of the segment of the closest approach can be taken as the estimated CoG. In what follows, a step-by-step procedure towards the computation of such an estimate is presented.

For solving the problem under discussion, we first need to express the direction of the upward local vertical, $\hat{\mathbf{z}}_l$, in body frame components ($\hat{\mathbf{u}}_b$) as a function of $X$ and $Y$, while the orientation of the body about this direction is not important. From (1), it follows:

$$\hat{\mathbf{u}}_b = \begin{bmatrix} \sin(X) \\ \sin(Y) \\ \pm\sqrt{1 - \sin(X)^2 - \sin(Y)^2} \end{bmatrix}, \quad (2)$$

where the last component of $\hat{\mathbf{u}}_b$ is computed to enforce the unit norm, and its sign depends on the orientation of the inclinometer: it is + when the inclinometer is with top face up, − otherwise.

It shall be noticed that, in the form above, (2) may lead to unphysical results. In fact, it is possible to obtain angular measurements such that $\sin(X)^2 + \sin(Y)^2 > 1$ due to measurement errors. We can handle such occurrence by normalizing the sine of the sensor readings through the factor $\sqrt{\sin(X)^2 + \sin(Y)^2}$ whenever $\sin(X)^2 + \sin(Y)^2 > 1$.

Taking two suspension points, $\mathbf{P}_1$ and $\mathbf{P}_2$, on any of the face of the proof mass, we obtain two body frame representations of the upward vertical vector, $\hat{\mathbf{u}}_1$ and $\hat{\mathbf{u}}_2$. If we call $\mathbf{L}_1$ and $\mathbf{L}_2$ the lines stemming from the suspension points and passing through the barycenter, their parametric equations are

$$\begin{aligned} \mathbf{L}_1 &= \mathbf{P}_1 + \hat{\mathbf{u}}_1 t_1, \\ \mathbf{L}_2 &= \mathbf{P}_2 + \hat{\mathbf{u}}_2 t_2, \end{aligned} \quad (3)$$

with $t_1, t_2 \in ]-\infty, +\infty[$. Note that, in (3), we avoided the indication of the body frame of representation for $\mathbf{P}$ and $\hat{\mathbf{u}}$, which is from now on left implicit for ease of notation. The intersection occurs when $\mathbf{L}_1(t_1) = \mathbf{L}_2(t_2)$ or [9] when

$$\mathbf{P}_1 + \hat{\mathbf{u}}_1 t_1 = \mathbf{P}_2 + \hat{\mathbf{u}}_2 t_2. \quad (4)$$

Subtracting $\mathbf{P}_1$ from both sides of (4) and crossing with $\hat{\mathbf{u}}_2$ yields

$$(\hat{\mathbf{u}}_1 \times \hat{\mathbf{u}}_2) t_1 = (\mathbf{P}_2 - \mathbf{P}_1) \times \hat{\mathbf{u}}_2. \quad (5)$$

Equation (5) can be solved for the parameter $t_1$ by dot-multiplying by $(\hat{\mathbf{u}}_1 \times \hat{\mathbf{u}}_2)$ and dividing by $\|\hat{\mathbf{u}}_1 \times \hat{\mathbf{u}}_2\|^2$ to get

$$t_1^* = \frac{(\hat{\mathbf{u}}_1 \times \hat{\mathbf{u}}_2) \cdot (\mathbf{P}_2 - \mathbf{P}_1) \times \hat{\mathbf{u}}_2}{\|\hat{\mathbf{u}}_1 \times \hat{\mathbf{u}}_2\|^2}. \quad (6)$$

And, operating symmetrically for $t_2$ yields

$$t_2^* = \frac{(\hat{\mathbf{u}}_1 \times \hat{\mathbf{u}}_2) \cdot (\mathbf{P}_2 - \mathbf{P}_1) \times \hat{\mathbf{u}}_1}{\|\hat{\mathbf{u}}_1 \times \hat{\mathbf{u}}_2\|^2}. \quad (7)$$

A useful property of the above solution for the point of intersection is that, if the two lines are skew, as it would certainly happen with actual noisy measurements, $t_2^*$and $t_1^*$ represent the parameters of the points of closest approach, that is, the extremal points of the minimum distance segment. This suggests a definition for the CoG estimation from the double suspension technique as the midpoint of the segment of closest approach between the two suspension lines:

$$\mathbf{C}_{\text{tot}} = \frac{1}{2}(\mathbf{P}_1 + \hat{\mathbf{u}}_1 t_1^* + \mathbf{P}_2 + \hat{\mathbf{u}}_2 t_2^*). \quad (8)$$

It is interesting to note that expressing the solution for the CoG location through (8) is not different from performing a triangulation, that is, a point localization from two angular measurements: this is a well-known concept in the field of angle-only navigation ([10] and references therein, [11]).

## 3. Experiment Requirements

To assess the accuracy of the proposed double suspension method, it is fundamental to identify first the error sources and then to design the experiment such in a way to minimize their detrimental effect on the CoG estimate.

Based on the analysis performed in [8], the main error contributions are expected to be

(1) True barycenter shift due to measurement equipment

(2) Measurement errors (inclinometer error plus analog-to-digital conversion)

(3) Knowledge of the suspension point location

(4) Small geometric errors of the proof mass

Having its own mass distribution and being integral to the proof mass, the measurement equipment (ME) induces a shift of the CoG location for the assembly under test resulting to be $\mathbf{C}_{\text{tot}}$. This differs from the CoG of the proof mass alone, according to

$$\mathbf{C}_{\text{tot}} = \frac{\mathbf{C}m + \mathbf{C}_{\text{me}} m_{\text{me}}}{m_{\text{tot}}}, \quad (9)$$

where $\mathbf{C}_{\text{tot}}$ is the estimated CoG location from the double suspension method, $\mathbf{C}_{\text{me}}$ is the CoG of the measuring equipment alone, $m$ is the proof mass mass, $m_{\text{me}}$ is the mass of the ME, and $m_{\text{tot}} = m + m_{\text{me}}$. Clearly, we can use (9) to compensate for the ME presence; however, the outcome of the experiment will be affected by how much accurate our knowledge of the ME mass and its CoG location ($\mathbf{C}_{\text{me}}$) are. Since we expect an accuracy of the method in the order of 1 mm, we shall design the experiment so that the ME introduces an uncompensated perturbation on the measured $\mathbf{C}_{\text{tot}}$ of one order of magnitude lower, that is, 0.1 mm, or less. Indeed, from (9) we derive the proof mass CoG coordinates as

$$\mathbf{C} = \mathbf{C}_{\text{tot}} \frac{m_{\text{tot}}}{m} - \mathbf{C}_{\text{me}} \frac{m_{\text{me}}}{m}. \quad (10)$$

$\mathbf{C}_{\text{me}}$ can be estimated from a CAD model; however, such an estimate is affected by a modelling error. For our purposes, we assume a conservative error window for $\mathbf{C}_{\text{me}}$ and design the experiment as to make this uncertainty negligible on the CoG computation. Minimizing the sensitivity to the above uncertainties reduces to placing the ME as close as possible the CoG and building the proof mass as heavy as possible, that is, a bulk mass is the best choice.

Error source 2 is intrinsic to the measuring equipment and cannot be reduced by proper design once the ME is chosen. Concerning the error source 3, it is caused by both manufacturing precision and suspension mechanism.

The manufacturing precision contribution can easily be minimized by measuring the effective dimensions after manufacturing. The suspension mechanism shall be designed to minimize the uncertainty in the suspension location and hinge moment. We examined different solutions such as universal joint, uniball, and wire suspension. A thin wire resulted as the best compromise between cost, ease of manufacture, and expected accuracy. Error source 4 is the easiest to control: in fact, one can take a homogenous material of known shape (say a parallelepiped of cast iron) and machine it to strict geometric tolerances. This ensures a negligible CoG shift from the geometric centroid, which can thus be assumed as the true barycenter. Considering all the above points, we finally opted for a bulk parallelepiped suspended by means of a wire as the experimental setup to verify the proposed method. A rough computation with pessimistic assumptions on $\mathbf{C}_{\mathrm{me}}$ error lead to a proof mass of about 10 kg with $80 \times 80$ mm base, needed to bound the uncompensated CoG perturbation due to the ME within 0.1 mm. A detailed quantification of all error sources is provided in the next section.

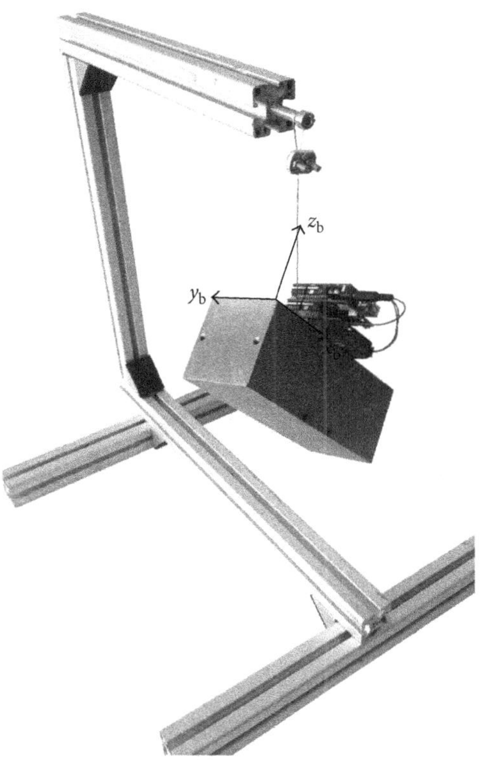


FIGURE 2: Experimental setup; two consecutive suspensions as the one shown are required to locate the CoG.

## 4. Experiment Design and Error Source Models

As outlined in Section 3, for the experimental verification, we employed a steel parallelepiped as proof mass. After the machining, the effective dimensions were $73 \times 75 \times 200$ mm and the mass equal to 9.2 kg. We assumed the body frame origin lying on one of the vertexes with the positive $x$-axis along the longer edge, the positive $y$-axis along the shortest edge, and $z$-axis completing the right-handed frame.

The overall CoG measuring instrumentation can be regarded as the combination of a suspension mechanism plus the measurement equipment (i.e., the inclinometer and the acquisition hardware), see Figure 2. Each of those two parts introduces errors in the measurements, which are discussed, together with the respective implementation details, in the following subsections.

*4.1. Measurement Equipment.* The measurement equipment (ME) consisted of the inclinometer plus some acquisition hardware, namely, an Arduino Uno, a wireless transmitter and power supply. The boards were assembled together with the inclinometer using an Arduino Prototyping Board to form the complete ME. The Arduino reads the analogic inclinometer output, converts it to digital, and sends it through the transmitter to a computer. Wireless transmission was needed to avoid running cables, which would otherwise induce systematic errors in the measurements. The ME was placed onto the $x$ and $y$ plane of the proof mass with the inclinometer's top face up. The exact location, which in principle is free, shall nevertheless be selected accounting for the range of the inclinometer (+/− 60° in our application) and the position of the suspension points, to guarantee that when the body is hanged, the inclination angles lie within the measurement range.

Errors introduced by the ME are of two kinds, namely, the one due to the imperfect knowledge of the location of its own CoG and the inclinometer measurement errors. A 3D CAD model of the ME was used to get an estimate of its CoG. Clearly, the CAD model is never an exact replica of the real ME so that we needed to assign an error to its CoG estimate. The measuring equipment is a stacked structure with nonuniformly distributed mass, as shown in Figure 3.

To justify the assumed error on the $\mathbf{C}_{\mathrm{me}}$, we can think of the ME as built up of 4 volumes: Arduino Uno volume, WiFi transmitter volume, prototyping board plus inclinometer volume, and battery volume. In each of these volumes, we can conservatively assume that the real CoG of the pertinent mass lies wherever inside a cube of 10 mm edge around the CAD's CoG. This is equivalent to say that the error random variable of the CoG in each volume has a uniformly distributed probability density function (PDF) in a 10 mm edge cube. Due to the linearity of the CoG expression, the error in $\mathbf{C}_{\mathrm{me}}$ (i.e., $\Delta\mathbf{C}_{\mathrm{me}}$) is the weighted sum of the 4 random variables:

$$\Delta\mathbf{C}_{\mathrm{me}} = \sum_{i=1}^{4} \frac{m_{v_i}}{m_{\mathrm{me}}} \Delta\mathbf{C}_{v_i}. \tag{11}$$

This means that the PDF of $\Delta\mathbf{C}_{\mathrm{me}}$ is the convolution integral of the 4 weighted random variables $\Delta\mathbf{C}_{v_i}$. Performing a random simulation, we concluded that the total PDF resembles enough a normally distributed PDF with zero mean and standard deviation 1.66 mm (reasonable in force of central limit theorem).

As far as the measurement error is concerned, it is due to both the error affecting the (analog) inclinometer output and the discretization error. The inclinometer accuracy (maximum error) is rated at 0.1°. Since we lack any

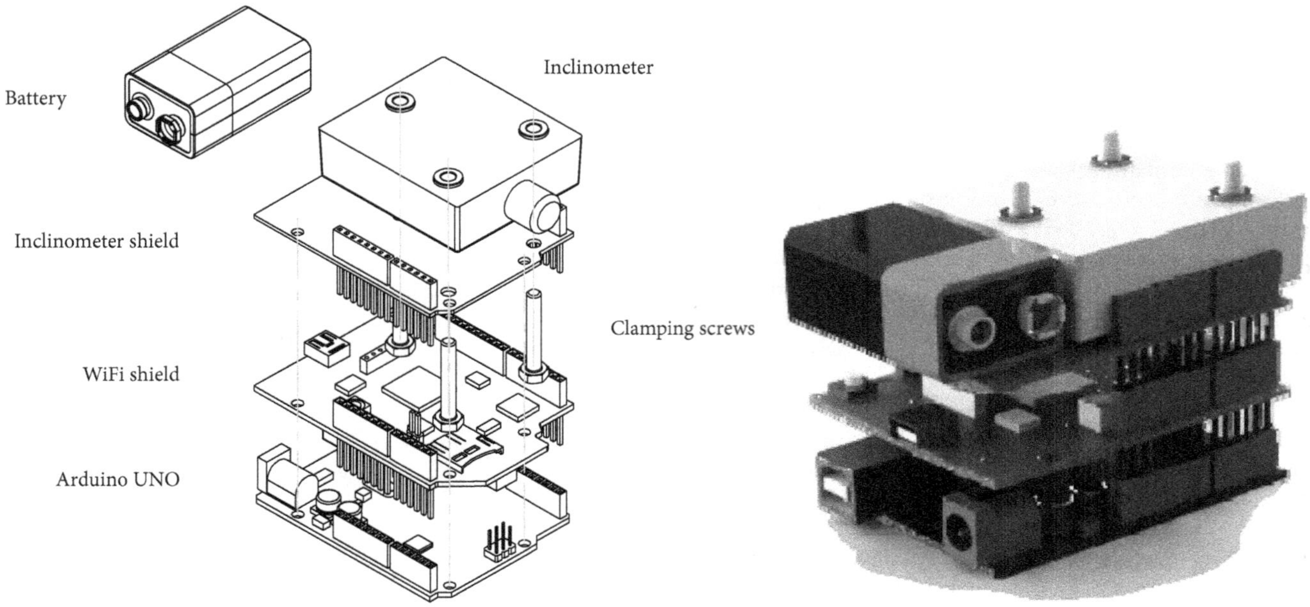


(a) (b)

Figure 3: Measurement equipment: exploded view (a) and CAD rendering (b).

statistical information, we shall assume a uniform distribution for this source in between $-0.1°$ and $+0.1°$. The voltage signal from the inclinometer is then processed by the Arduino UNO 10-bit analog-to-digital converter. The resulting discretization step is $\Sigma_{AD} = 0.16°$, whose effect can be modelled as a uniformly distributed random variable between $\pm 0.08°$, that is, having zero mean and standard deviation equal to $\Sigma_{AD}/\sqrt{12} \approx 0.046°$.

The accuracy and discretization errors can be considered independent and additive so that the global PDF can be computed as the convolution of the single PDF's. The resulting PDF will resemble a triangular distribution with zero mean, since the uniform distributions have comparable widths. To smoothen the effect of random errors, we can average many measurements (say $n = 20$); in force of the central limit theorem, the PDF of the sample mean resembles a normal distribution with standard deviation reduced by a factor $\sqrt{n}$.

*4.2. Suspension Mechanism.* As anticipated in Section 3, a wire suspension was selected. To keep the suspension points as localized as possible, a 0.8 mm multiwire cable and a bolt with a pass-through 1 mm hole in the center were employed, as shown in Figure 4.

Although there is a small play between the wire and the hole (0.2 mm), we can think of the wire as being clamped onto the top of the bolt. This way, we can assume the wire as a clamped beam. The clamping reacts the vertical weight of the proof mass and a moment. The moment is originated from the flexural rigidity of the wire, which prevents the suspension point from aligning (exactly) to the barycenter. Rather, we can more accurately say that the barycenter aligns with the holding point of the wire. The net effect of the bending moment at clamp then would be a shift of the suspension point from the nominal location to a virtual point lying above it by a quantity $\delta$ (see Figure 5). This is the second major contribution to the error in locating the suspension point after the small play around the wire.

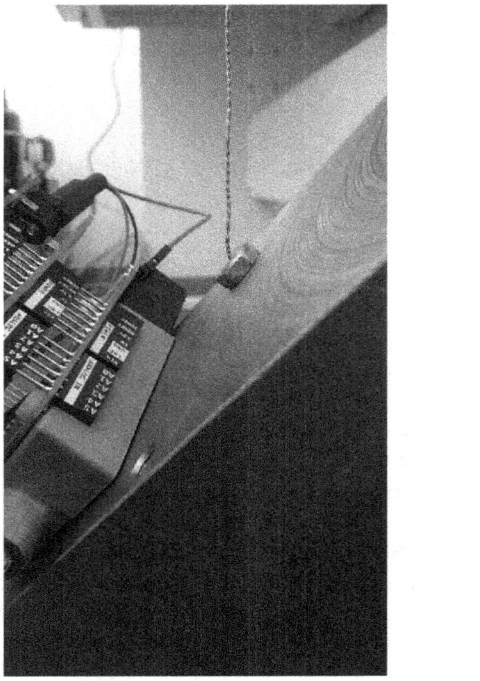

Figure 4: Detailed view of the suspension mechanism.

To have an educated guess of the maximum $\delta$, we express the displacement $\Delta$ as a function of the beam parameters and face inclination with respect to vertical $\alpha_0$ obtaining (see Appendix)

$$\Delta = \frac{\tan(\alpha_0)}{\sqrt{F/EJ}}. \quad (12)$$

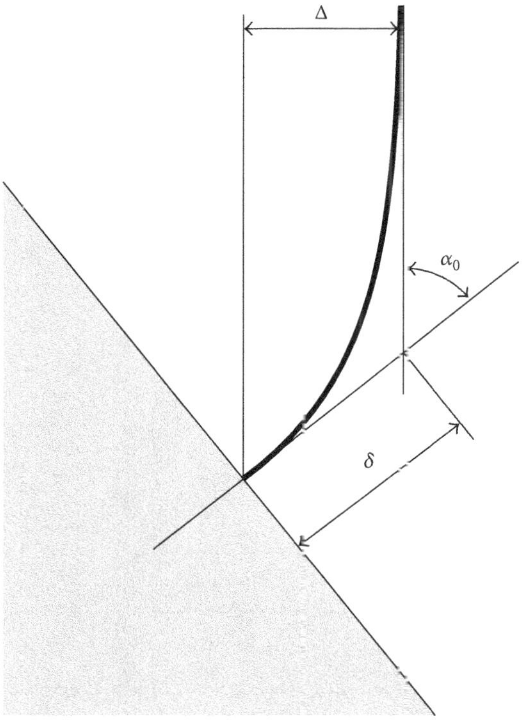


FIGURE 5: Schematic representation of the virtual suspension point concept.

In (12), $F$ is the proof mass weight and the product $EJ$ between the elastic modulus ($E$) and the cross-section moment of inertia ($J$) is the bending stiffness of the beam. In principle, since we use a multiwire cable, $EJ$ is dependent on the load $F$; however, in our experiment, the tension is constant so that we can assume a fixed $\sqrt{F/EJ}$. Thus, the virtual suspension point is offset above the bolt surface by about

$$\delta = \frac{\Delta}{\sin(\alpha_0)} = \frac{\sqrt{EJ/F}}{\cos(\alpha_0)}. \tag{13}$$

Equation (13) shows that $\delta$ gets smaller as $\alpha_0$ decreases (since $EJ$ and $F$ are constant); thus, we can have a worst-case estimate of $\delta$ considering a situation with large $\alpha_0$. Figure 3 depicts an experiment at high $\alpha_0$ from which one can visually estimate $\Delta \cong 0.8$ mm being the cable thickness 0.8 mm. Then, we obtained that $\delta \cong 1$ mm is a conservative upper-bound of the offset. The observations above allow to define an uncertainty volume for the suspension point which can be assumed to have a square base of 1 mm edge on the plane of the bolt head (due to the play), centered in the measured suspension point, with a vertical height of 1 mm. We can assume an error around the theoretic suspension point, belonging to this volume and drawn from a uniform PDF in that volume.

## 5. Verification Process

The verification of the proposed method employs a combination of numerical simulations and experiments according to the following steps:

(1) Take as input the assumed distributions of the error sources outlined in Section 4.

(2) Perform Monte Carlo simulations to estimate the PDF and the cumulative distribution function (CDF) of the CoG distance error $\|\Delta \mathbf{C}\|$.

(3) Verify experimentally the error budget: perform many experiments and check whether the results are compatible (in a sense to be soon specified) with the PDF found in simulation.

Such approach was preferred to a simpler direct error distribution estimation through multiple independent trials. In fact, we aim at experimentally verifying the method rather than characterizing the measuring equipment, for which a more extensive test campaign should be used. Furthermore, characterizing the measuring equipment in absolute would be complicated by (10), as the estimation accuracy depends also on the mass under proof. We considered 6 suspension points, which lead to 15 possible suspension couples, enough for our scope. The number was selected to limit also the perturbation induced by the drilled holes on the true CoG location: we estimated a worst-case shift in the order of 0.01 mm.

A workflow diagram of the entire validation process is given in Figure 6, while the suspension point coordinates are reported in Table 1.

*5.1. Monte Carlo Simulations.* The Monte Carlo simulation scenario was developed in MATLAB® environment, according to the following approach. We can regard the double suspension algorithm as a function taking as input a perturbed vector of parameters, $\mathbf{p} = \mathbf{p}_{\text{true}} + \delta\mathbf{p}$, and whose output is the CoG estimate. We denote by $\mathbf{p}_{\text{true}}$ the vector collecting the true values of such parameters, namely, the suspension point coordinates, the inclinometer readings, and the CoG coordinates of the measuring equipment: $\mathbf{p}_{\text{true}} = [\mathbf{P}_1, \mathbf{P}_2, X_1, Y_1, X_2, Y_2, \mathbf{C}_{\text{me}}]$; by $\delta\mathbf{p}$, we denote the vector containing the error affecting each parameter (masses were not taken as error sources, as we know them accurately enough to safely neglect the impact of their uncertainties on the CoG computation). Within a simulation, the ideal inclinometer output obtained by hanging the proof mass from a given suspension point can be calculated from the reversed application of the algorithm in Section 2. Thus, for a given couple of suspension points, the true parameter vector can be computed and then perturbed with random errors drown from the corresponding PDF (Section 4). The estimated CoG location is retrieved from direct application of the solution algorithm to the perturbed parameters.

When running the Monte Carlo simulations according to the procedure above, we randomly distributed a large number (25) of suspension points on the surface of the proof mass, to avoid as much as possible any dependency of the outcome on the specific geometric configuration. In fact, as pointed out in [8], the accuracy of the double suspension method is also dependent on the mutual configuration of the suspension points, getting worse when the CoG and the suspension points become closely collinear. In such a case,

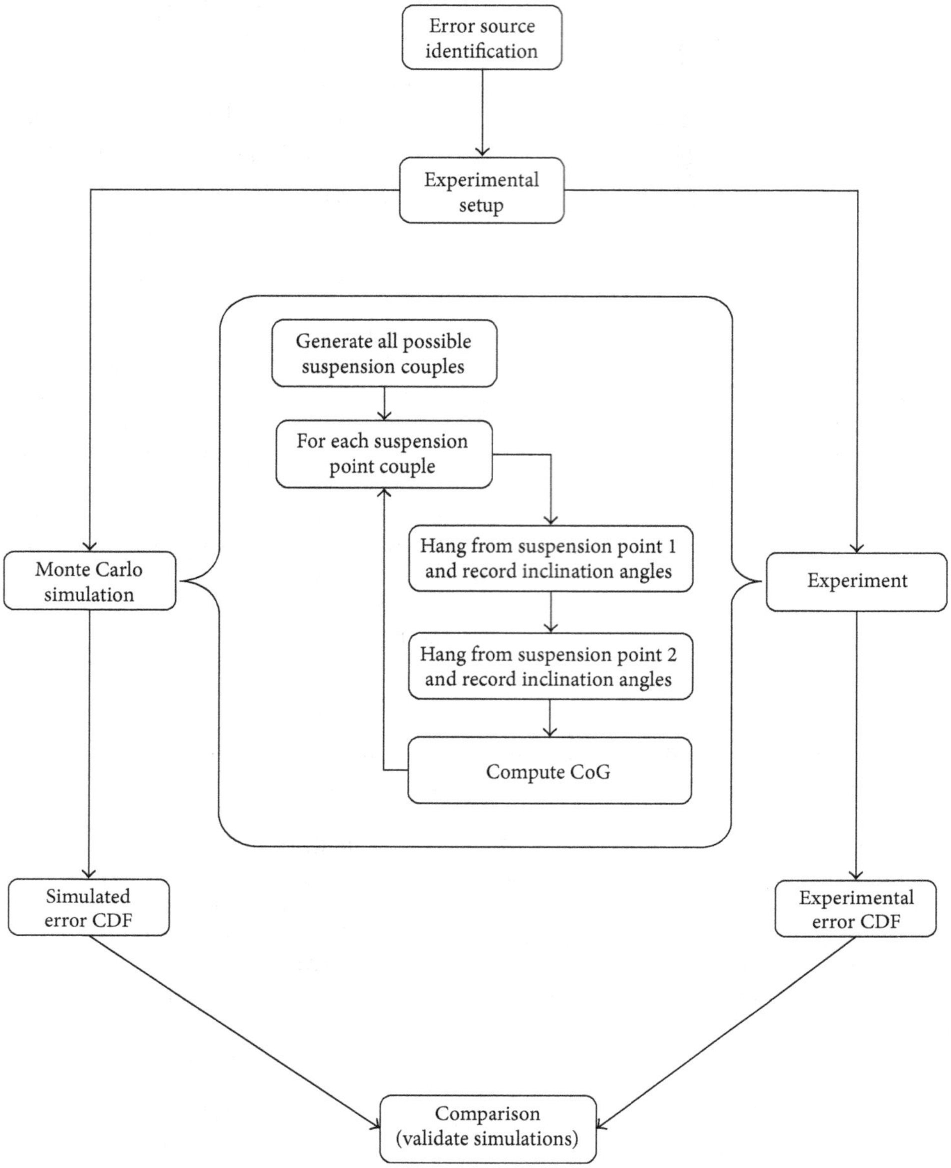


FIGURE 6: Workflow diagram of the validation process for the double suspension method.

TABLE 1: Suspension point coordinates.

| Number | $x$ (body) | $y$ (body) | $z$ (body) |
|---|---|---|---|
| 1 | 49.9 | 17.9 | 4.1 |
| 2 | 139.65 | 47.7 | 4.1 |
| 3 | 81.35 | −4.1 | −73.8 |
| 4 | 121.35 | −4.1 | −73.4 |
| 5 | 67.28 | 68.2 | 4.1 |
| 6 | 153.35 | 37.4 | 4.1 |

we would be attempting to intersect two lines which are almost parallel: the evaluation of the CoG through (6) and (7) would lead to an ill-conditioned operation (when $\hat{\mathbf{u}}_1 \| \hat{\mathbf{u}}_2$, the denominators approach 0).

For each couple of the 25 suspension points, we generated 1000 perturbed input vectors, which were supplied to the CoG estimation algorithm: the resulting estimate was then compared to the geometric center of the proof mass for computing the error $\Delta\mathbf{C}$. Due to the relatively large number of random error contributions, the components of the CoG location error $\Delta\mathbf{C}$ can be regarded as normally distributed, in force of the central limit theorem. As a consequence, the norm of $\Delta\mathbf{C}$ then approximately follows a Rayleigh distribution. Hence, we fitted a Rayleigh distribution to the results, obtaining as output the desired PDF of $\Delta\mathbf{C}$.

*5.2. Experimental Verification Method.* The outcome of the Monte Carlo simulations was checked against a test campaign carried out using the experimental set up in Figure 2. For the subsequent analysis, we can regard the series of suspensions performed as a Bernoulli process. Assuming we performed $n$ trial experiments whose outcome could be either success ($S$) or failure ($F$); in each trial, we had a

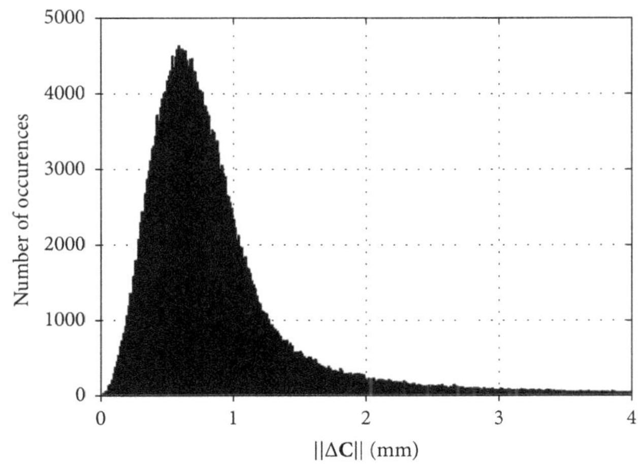


FIGURE 7: Histogram of the occurrences of the CoG estimation error.

probability $p$ to succeed and $q = 1 - p$ to fail. For our experiments, we call success the event in which the error norm is lower than a given threshold $w_e$ and failure otherwise. Then, if we denote by $f_{err}(w_e)$ the PDF of $\Delta\mathbf{C}$ obtained from simulation, and $F_{err}$ the corresponding CDF, taking an error window $w_e$, $F_{err}(w_e)$ provides the probability $p$ that the error belongs to the error window (i.e., success): $p = F_{err}(w_e)$. To validate the CDF obtained through simulations against the experiments, we check that for some $w_e$ is

$$p_{exp}(w_e) \approx p(w_e). \tag{14}$$

In (14), $p_{exp}$ is the experimental probability of success given by the maximum likelihood estimator for the Bernoulli process parameter ($p$) [12], that is,

$$p_{exp}(w_e) = \frac{s(w_e)}{n}. \tag{15}$$

$n$ being the number of trials (15) and $s$ the number of successes given $w_e$.

Note that, the larger the number of experiments $n$, the higher the confidence in the estimator, (15), for $p_{exp}(w_e)$. However, our number of trials is already constrained by the considerations made in the previous section.

## 6. Results

Considering all possible couples of 25 suspension points, the Monte Carlo simulation explored 300 suspensions for a total number of $3 \cdot 10^5$ trials. Figure 7 depicts the histogram of the resulting CoG error.

The best-fit Rayleigh distribution has a shape parameter $B = 0.8$ mm. If we assume the error being isotropic in space, $B$ corresponds to the (common) standard deviation of the three scalar error components.

During the experimental campaign, the CoG was measured through all possible combinations of suspension point couples. The results are depicted in Figure 8 and summarized in Table 2, which compares the CDF obtained from the simulated best-fit Rayleigh distribution ($p$), to the ones from

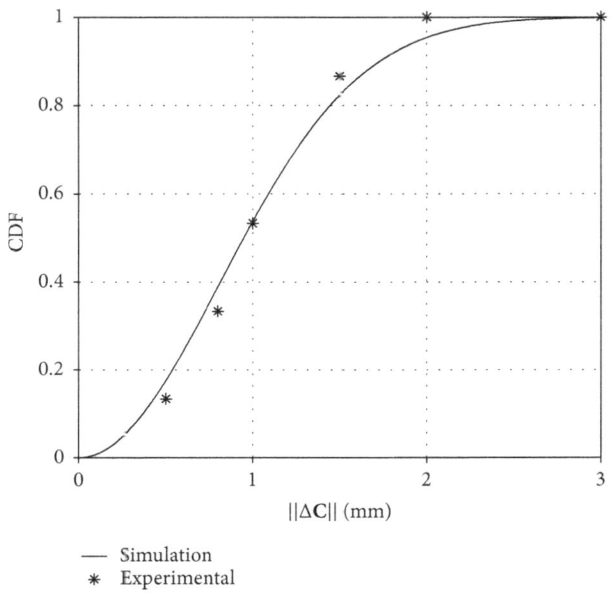


FIGURE 8: Cumulative density function of the CoG measurement error: experimental (markers) versus numerical (full line).

TABLE 2: Bernoulli checks for different error windows.

| $w_e$ (mm) | $p$ (%) | $s$ | $p_{exp}$ (%) |
|---|---|---|---|
| 0.5 | 18 | 2 | 13.3 |
| 0.8 | 39 | 5 | 33.3 |
| 1 | 54 | 8 | 53.3 |
| 1.5 | 82 | 13 | 86.7 |
| 2 | 95 | 15 | 100 |
| 3 | 99.9 | 15 | 100 |

experiments ($p_{exp}$), according to the method outlined in Section 5.2.

The agreement is very good at central $w_e$ and poorer, but still reasonable, at extremal $w_e$. This can be expected, since the "front" and "tail "of the Rayleigh distribution are low probability regions, that is, it is less likely to obtain results in these regions. We can conclude that both experimentally and in simulation the method works as expected, reaching accuracies in the order of 1 mm.

The error analysis performed so far is specific for the assumed ME and on the mass ratio of the proof mass and ME itself; strictly speaking, these conditions are necessary for the estimated error PDF to be valid. It is of interest to briefly assess up to which extent the results obtained can be extrapolated to a generic experiment for estimating the CoG of a small spacecraft. To this end, consider (10) reformulated in terms of error variables:

$$\Delta\mathbf{C} = \Delta\mathbf{C}_{tot}\frac{m_{tot}}{m} - \Delta\mathbf{C}_{me}\frac{m_{me}}{m}, \tag{16}$$

where $\Delta\mathbf{C}_{tot}$ is the error caused by the method when estimating the CoG of the entire assembly (ME + PUT), and $\Delta\mathbf{C}_{me}$ is

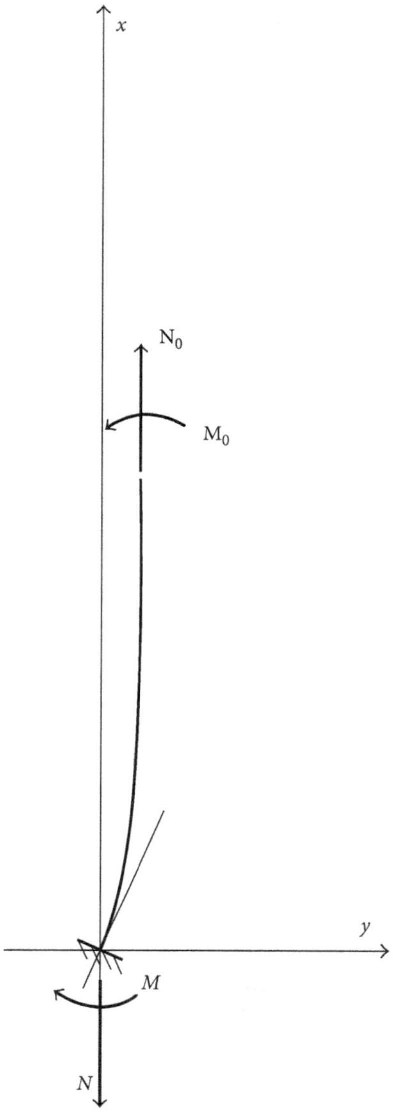


FIGURE 9: Model of the suspension cable as a bending beam.

the error made in locating the measurement equipment CoG, see (11).

From (16), it appears that the uncertainty in $\Delta\mathbf{C}_{me}$ contributes additively to the overall error. We can thus infer that the error PDF estimated in this work can be considered a conservative estimate of the actual PDF for any situation in which either $\Delta\mathbf{C}_{me}$ and $m_{me}/m$ are equal to or smaller than the ones adopted here (having assumed all the rest being the same). Furthermore, despite the error analysis is instrumentation specific, the solution algorithm based on the double suspension method is quite general: it would apply equally if using an inclinometer or suspension mechanism that differ from the ones adopted in this work.

The discussion above suggests that the accuracy of the proposed equipment shall meet the needs of a fairly large class of S/C, entailing, for example, the larger CubeSat form factors (from 6 U and above) as well as microsatellites in the 30–100 kg mass class.

## 7. Conclusions

In this work, we proposed and verified experimentally a simple method to determine the CoG of a small spacecraft, based on two consecutive suspensions. A thorough assessment of the various error sources was performed, and their impact on the estimation error was evaluated through Monte Carlo simulation. This allowed a characterization of the probability density function of the simulated CoG error as a Rayleigh distribution. Afterwards, we verified the agreement with experimental data, by comparing the simulated cumulative density function of the estimation error with the one retrieved from experiments.

We found that the experimental results agree well with the simulated ones, indicating that the error PDF obtained in simulations is a good estimate of the actual one, and showing that the method is capable of accuracies in the order of 1 mm. Extrapolation of the error analysis to different experimental setups was also qualitatively assessed, showing to which extent we can assume the error distribution obtained in this work to be a conservative estimate of the actual one.

The proposed method has the merit of relative simplicity and its accuracy meets the typical needs of microsatellite missions. Nevertheless, it may certainly be improved: future work may involve the development and characterization of a more compact measurement equipment and suspension mechanisms, more suitable for nanosatellites. In fact, the error budget analysis highlights that the mass of the measurement equipment and the quality of the implementation of the suspension mechanisms are crucial for a good estimate.

## Appendix

### A.1. Cog Error due to Bending Stiffness

Consider the beam represented in Figure 9.

We have that $N = N_0 = F$ which is the proof mass weight. Then, every section of the beam must be in equilibrium with the external load, so that it must be

$$\frac{dM}{dy} = F. \tag{A.1}$$

Integrating this equation from the moment $M_0$ and using the differential equation of the deformed beam we have

$$\frac{d^2y}{dx^2} = -\frac{M_0 - Fy}{EJ}. \tag{A.2}$$

We obtain a second-order differential equation of $y(x)$ of the form.

$$\ddot{y} + ay + b = 0, \quad a = -\frac{F}{EJ}, b = \frac{M_0}{EJ} = -a\,\Delta. \tag{A.3}$$

Whose solution is given by

$$y = c_1 e^{\sqrt{-a}x} + c_2 e^{-\sqrt{-a}x} - \frac{b}{a}. \tag{A.4}$$

And the constants $c_1$ and $c_2$ are derived from the boundary conditions:

$$y(0) = 0,$$
$$y(L) = \Delta. \tag{A.5}$$

Notice now that we do not know $\Delta$; however, we have the angle information at clamp that is

$$\frac{dy}{dx} = \tan(\alpha_0). \tag{A.6}$$

Using this equation, it is possible to close the problem and solve analytically for $c_1$, $c_2$, and $\Delta$, obtaining in particular for the latter

$$\Delta = -\frac{1}{2}\frac{\tan(\alpha_0)\sqrt{-a}}{1/2 + \left(e^{2\sqrt{-a}L}/1 - e^{2\sqrt{-a}L}\right)}. \tag{A.7}$$

Which, for $L \to \infty$, simplifies to (12).

## Conflicts of Interest

The authors declare that there are no conflicts of interest regarding the publication of this paper.

## References


[1] K. H. Wiener, "The role of mass properties measurement in the space mission," in *Proceedings of the European Conference on Spacecraft Structures, Materials and Mechanical Testing*, pp. 10–12, Noordwijk, The Netherlands, May 2005.

[2] G. Suresh, J. K. Pandit, K. Ramachandra, and M. R. Thyagaraj, "Estimation of center of gravity (CG) for a spacecraft: a review," *International Journal of Engineering & Science Research*, vol. 2, no. 4, pp. 128–141, 2012.

[3] R. Boynton and K. Wiener, "A new high accuracy instrument for measuring moment of inertia and center of gravity," in *Proceedings of the 23rd Conference of the Society of Allied Weight Engineers*, pp. 23–25, Detroit, Michigan, May 1998.

[4] J. R. Dennison, C. Frazier, E. Stromberg et al., "Small satellite space environment effects test facility with space environment effects ground-testing capabilities," in *Proceedings of the AIAA/USU Conference on Small Satellites, Poster Session IV, N°122*, 2015.

[5] W. L. Peterson, "Mass properties measurement in the X-38 project," in *Proceedings of the 63rd Annual Conference of the Society of Allied Weight Engineers*, pp. 17–19, Newport Beach, CA, USA, May 2004.

[6] L. Tang and W. B. Shangguan, "An improved pendulum method for the determination of the center of gravity and inertia tensor for irregular-shaped bodies," *Measurement*, vol. 44, no. 10, pp. 1849–1859, 2011.

[7] T. W. Jones, T. H. Johnson, D. Shemwell, and C. M. Shreves, "Photogrammetric technique for center of gravity determination," 2012, AIAA-2012-1882.

[8] D. Modenini and P. Tortora, "A simple method for accurate center of gravity determination of small satellite platforms," in *Proceedings of the 23rd Conference of the Italian Association of Aeronautics and Astronautics*, Torino, Italy, 2015.

[9] A. S. Glassner, *Graphics Gems*, Academic Press, Orlando, FL, USA, 1990.

[10] G. H. Kaplan, *A Closed Form Position and Velocity Solution for Angles-Only Navigation*, 2008, http://gkaplan.uscontentnav_by_angles.pdf.

[11] R. Lasagni Manghi, D. Modenini, M. Zannoni, and P. Tortora, "Preliminary orbital analysis for a CubeSat mission to the Didymos binary asteroid system," *Advances in Space Research*, 2017, In press.

[12] P. Sahoo, *Probability and Mathematical Statistics*Department of Mathematics University of Louisville, Louisville, KY, USA.

# Robust Adaptive Output Feedback Control for a Guided Spinning Rocket

**Zhongjiao Shi, [1,2] Liangyu Zhao, [1,2] and Yeqing Zhu [1,2]**

[1] *Beijing Institute of Technology, Beijing 100081, China*
[2] *Key Laboratory of Dynamics and Control of Flight Vehicle, Ministry of Education, Beijing 100081, China*

Correspondence should be addressed to Liangyu Zhao; zhaoly@bit.edu.cn

Academic Editor: Kenneth M. Sobel

An adaptive autopilot is presented for the pitch and yaw channels of a guided spinning rocket. Firstly, the uncertain dynamic model of a guided spinning rocket is established, which is used to evaluate the performance of the proposed adaptive autopilot. Secondly, a robust adaptive output feedback autopilot containing a baseline component and an adaptive component is designed. The main challenge that needs to be addressed is the determination of a corresponding square and strictly positive real transfer function. A simple design procedure based on linear matrix inequality is proposed that allows the realization of such a transfer function, thereby allowing a globally stable adaptive output feedback law to be generated. Finally, numerical simulations are performed to evaluate the robustness and tracking performance of the proposed robust adaptive autopilot. The simulation results showed that the robust adaptive output autopilot can achieve asymptotic command tracking with significant uncertainty in control effectiveness, moment coefficient, and measurement noise.

## 1. Introduction

Traditional artillery ballistic gun-launched munitions cannot satisfy more and more stringent performance requirements about precision-strike capability, dispersion error reduction, and range augmentation required on modern battlefields. Complex guided systems such as missiles can meet these requirements, but they remain expensive due to the integration of high-quality actuators and sensors. The idea is to develop a guided rocket, which permits to reach a compromise between the low cost of ballistic projectiles and the necessity to have high-performance systems with efficient control algorithms. The guided rocket chosen in this work is a canard-controlled spinning configuration, which makes use of an existing multiple launch rocket system (MLRS). This one has the advantage of simplifying the structure of the control system, avoiding asymmetric ablation, relaxing the manufacturing error tolerance, and improving the penetration ability. It is also easy to implement and does not require the design of a new launch system. However, this kind of configuration has the disadvantages of cross coupling due to the spinning airframe. Besides, uncertainties in control effectiveness and moment coefficients are among the practical challenges in the control system design. Finally, the sensors are of limited performance, due to the low cost and small size specification.

These disadvantages can be handled by developing a flight autopilot using modern multivariable control methods, such as robust control [1, 2] and gainscheduling control [3, 4]. For the dual-channel-controlled spinning rocket, various autopilots were designed, such as rate loop autopilot [5], attitude autopilot [6], acceleration autopilot [7], and three-loop autopilot [8]. However, these related works were carried out under the nominal condition without considering uncertainties which may be experienced during the whole flight trajectory. Additionally, it is difficult to design an autopilot for a guided spinning rocket with excellent performance using the traditional separate channel design method.

Adaptive control is known as a proper method to deal with uncertainties and has been used in numerous

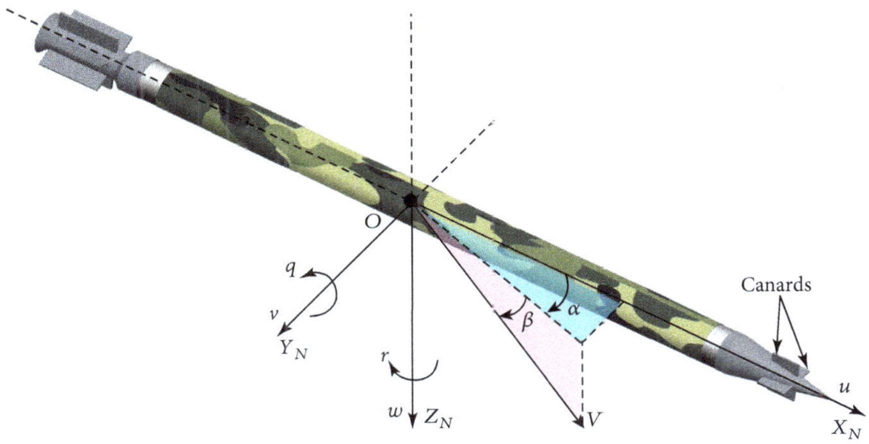


FIGURE 1: Sketch of guided spinning projectile.

applications. Therefore, this is aimed at investigating the potential of robust adaptive control for improving stability and performance of the spinning rocket autopilot. A generic transport aircraft autopilot was designed by employing a modified MRAC scheme to guarantee the transient performance [9]. A Lyapunov-based model reference adaptive PD/PID controller for Satellite Launch Vehicle (SLV) systems was proposed to improve the tracking performance and robustness under wind disturbances [10]. An adaptive integral feedback controller for pitch and yaw channels of an autonomous underwater vehicle (AUV) was designed to handle the actuator saturations [11]. These controllers require that the system state must be measurable [12], which may not always be possible. For this reason, there has been an increasing motivation to develop an adaptive output feedback controller. Existing classical methods of multi-input and multi-output (MIMO) output feedback adaptive control are applicable for square systems; that is, the plant has the same number of inputs and outputs [13, 14]. Recently, an output feedback adaptive spinning rocket autopilot was designed, but only the acceleration information was used to construct the autopilot [15]. In order to make full use of all the measurable information, the output feedback adaptive autopilot for a spinning rocket with four measurable outputs and two control inputs, which is a nonsquare system, needs to be designed. The main challenge is to make the closed-loop transfer functions satisfy the strictly positive real (SPR) property or guarantee the strict passivity of the closed-loop system. These two properties were proved to be equivalent for a linear time-invariant system [16]. For the square system, necessary and sufficient conditions for passifiability of the linear system by output feedback were presented in [17, 18]. An observer-based method was also included in the design of controllers to guarantee the that closed-loop system is SPR [19]. For the nonsquare system, the squaring-down procedure was employed for passification design [20]. And necessary and sufficient conditions for passifiability of nonsquare systems by output feedback were given in [21].

The main contribution of this paper is to combine the observer-based method and squaring-down method to design an adaptive output feedback autopilot for guided spinning rockets. First, a mixing matrix $M$ is designed to make the modified error dynamic model a square system. Then, a Luenberger observer which also serves as the reference model is employed to make the error dynamic model satisfy the SPR property.

The remainder of this paper is structured in the following manner. Section 2 develops the uncertain dynamic model of a dual-channel-controlled spinning rocket. Section 3 presents the robust adaptive output-feedback autopilot design using robust SPR lemma. Section 4 demonstrates the performance of the robust adaptive output-feedback autopilot via numerical simulations. Finally, Section 5 concludes this paper.

## 2. Mathematical Model

The guided spinning rocket considered in this paper is an axially symmetric rolling airframe, as is shown in Figure 1. Two pairs of canard rotating with the airframe are employed as control surfaces twisting and steering the rocket. The maneuver of the airframe requires an autopilot that is able to track different kinds of signals under a different dynamic environment, with slight or even no change of the autopilot architecture.

*2.1. Dynamical Equation.* According to Newton's second law, the translational motion of center of gravity (cg) can be described in the nonrolling body coordinate system as follows [22]:

$$m\frac{d\mathbf{V}_A}{dt} = m\left(\frac{\partial \mathbf{V}_N}{\partial t} + \omega_r \times \mathbf{V}_N\right) = \mathbf{F} + m\mathbf{g}, \tag{1}$$

where $m$ is the mass of the spinning rocket, $\mathbf{V}_A = (x, y, z)^T$ is the velocity of cg represented in inertial coordinate, $\mathbf{V}_N = (u, v, w)^T$ is the velocity of cg in the nonrolling body coordinate, and $\omega_r$ is the angular rate of the nonrolling body coordinate system with respect to the inertial coordinate system. $\mathbf{F}$ is the aerodynamic force, and $\mathbf{g}$ is the acceleration of gravity, which are all represented in the nonrolling body coordinate system. All the variables in (1) can be modeled as

$$\omega_r = (-r \tan\theta, q, r)^T, \tag{2}$$

$$\mathbf{F} = \begin{bmatrix} F_x \\ F_y \\ F_z \end{bmatrix} = QS \begin{bmatrix} -C_x \\ -C_{N\beta}\beta + C_{N\delta}\delta_z \\ -C_{N\alpha}\alpha - C_{N\delta}\delta_y \end{bmatrix}, \tag{3}$$

$$\mathbf{g} = \begin{bmatrix} g_x \\ g_y \\ g_z \end{bmatrix} = \begin{bmatrix} -g\sin\theta \\ -0 \\ g\cos\theta \end{bmatrix}, \tag{4}$$

where $q$ and $r$ are angular rates along $OY_N$ and $OZ_N$ defined in Figure 1; $\theta$ is the pitch angle of the spinning rocket; $Q$ is the dynamic pressure; $S$ is the reference area; $C_x$ is the drag coefficient slope; $C_{N\alpha}$ and $C_{N\beta}$ are the normal force coefficient slopes, due to the character of symmetric $C_{N\alpha} = C_{N\beta}$, so as other coefficients related with $\alpha$ and $\beta$; $C_{N\delta}$ is the control force coefficient slope; $\delta_y$ and $\delta_z$ represent the actuator deflection angles in the nonrolling coordinate system; and $\alpha$ and $\beta$ are the angle of attack and the sideslip angle, respectively. The expressions of $\alpha$ and $\beta$ are presented as

$$\begin{aligned} \alpha &= \arctan\left(\frac{w}{u}\right) \approx \frac{w}{V}, \\ \beta &= \arcsin\left(\frac{v}{u}\right) \approx \frac{v}{V}. \end{aligned} \tag{5}$$

By substituting (2), (3), and (4) into (1), the force equations are obtained:

$$\begin{aligned} \dot{u} + wq - vr &= \frac{F_x}{m} + g_x, \\ \dot{v} + ur - wr\tan\theta &= \frac{F_y}{m} + g_y, \\ \dot{w} - uq - vr\tan\theta &= \frac{F_z}{m} + g_z. \end{aligned} \tag{6}$$

According to the theorem of angular momentum, the rotational motion of the airframe can be described in the nonrolling body coordinate system as

$$\frac{d\mathbf{H}}{dt} = \frac{\partial\mathbf{H}}{\partial t} + \omega_r \times \mathbf{H} = \mathbf{M}, \tag{7}$$

where $\mathbf{H}$ is the angular momentum and $\mathbf{M}$ is the moment acting on the rocket. All the variables in (7) are described in the nonrolling body coordinate system as

$$\mathbf{H} = \left(I_x p, I_y q, I_z r\right)^T, \tag{8}$$

$$\frac{\partial\mathbf{H}}{\partial t} = \left(I_x \dot{p}, I_y \dot{q}, I_z \dot{r}\right)^T, \tag{9}$$

where $I_x$, $I_y$, and $I_z$ are moments of inertia, due to the character of symmetric $I_y = I_z$; $p$ is the spinning rate in the nonrolling coordinate.

For a canard-controlled spinning rocket, positive $\delta_z$ and $\delta_y$ create positive moments in $M_z$ and $M_y$, respectively. Therefore, the aerodynamic moment can be expressed as

$$M = \begin{bmatrix} M_x \\ M_y \\ M_z \end{bmatrix} = QSl \begin{bmatrix} C_{l\delta x}\delta_x - C_{lp}\frac{pl}{V} \\ C_{m\alpha}\alpha - C_{mq}\frac{ql}{V} - C_{mp\alpha}\beta\frac{pl}{V} + C_{m\delta}\delta_y \\ -C_{m\alpha}\beta - C_{mq}\frac{rl}{V} - C_{mp\alpha}\alpha\frac{pl}{V} + C_{m\delta}\delta_z \end{bmatrix}, \tag{10}$$

where $l$ is the reference length; $C_{l\delta x}$ is the rolling moment coefficient due to the canted angle of tails, which is denoted as $\delta_x$; $C_{lp}$ is the rolling damping moment coefficient; $C_{m\alpha}$ is the static moment coefficient; $C_{mq}$ is the damping moment coefficient; $C_{mp\alpha}$ is the Magnus moment coefficient; and $C_{m\delta}$ is the control moment coefficient.

*Remark 1.* The guided spinning rocket considered in this paper is a fin-stabilized airframe. For a fin-stabilized rocket, the classic Magnus force is typically ignored in (3) since its effect is rather small for slowly rolling rockets. However, the Magnus moment with physical mechanisms specific to fin-stabilized rockets cannot be ignored in (10), which can be expressed as a dynamic side moment due to spin rate $p$ and angle of attack $\alpha$ or sideslip angel $\beta$ in the form of $C_{mp\alpha}\alpha(pl/V)$ [23].

By substituting (8), (9), and (10) into (7), the moment equations are obtained:

$$\begin{aligned} I_x\dot{p} &= M_x, \\ I_y\dot{q} + I_x pr + I_y r^2\tan\theta &= M_y, \\ I_y\dot{r} - I_x pq - I_y qr\tan\theta &= M_z. \end{aligned} \tag{11}$$

Equations (6) and (11) are the nonlinear dynamical model of a spinning rocket, which represents the angular motion of the spinning rocket.

*2.2. Linearized Lateral Equation.* For flight control which corresponds to the tracking of the dual-canard-controlled spinning rocket acceleration command, only the lateral dynamic system is necessary. In order to simplify the design procedure of the autopilot, some common and reasonable assumptions are made to linearize the lateral dynamic model of the spinning projectile [7]:

*Assumption 1.* The velocity, roll rate, mass, and aerodynamic coefficients of the rocket remain constant over a short time period.

*Assumption 2.* For simplicity, the gravity and the small canard force are ignored. The lateral velocities $v$ and $w$ are small with respect to the axis velocity $u$, so that $u \approx V$.

Applying the above assumptions to (6) and (11), the linearized angular motion of the spinning rocket can be described as

$$\begin{aligned}\dot{\beta} &= -r - c_{N\alpha}\beta,\\ \dot{\alpha} &= q - c_{N\alpha}\alpha,\\ \dot{q} &= c_{m\alpha}\alpha - c_{mp\alpha}P_p\beta - c_{mq}q - P_p r + c_{m\delta}\delta_y,\\ \dot{r} &= -c_{m\alpha}\beta - c_{mp\alpha}P_p\alpha - c_{mq}r - P_p q + c_{m\delta}\delta_z,\end{aligned} \tag{12}$$

where the dimensionless coefficients are defined as $c_{N\alpha} = QSC_{N\alpha}/mV$, $c_{m\alpha} = QSlC_{m\alpha}/I_y$, $c_{mq} = QSl^2C_{mq}/I_yV$, $c_{mp\alpha} = QSl^2C_{mp\alpha}/I_xV$, $c_{m\delta} = QSlC_{m\delta}/I_y$, and $P_p = p(I_x/I_y)$.

Usually, not all the states of a spinning rocket in (12) can be measured by sensors, especially the angle of attack $\alpha$ and the sideslip angle $\beta$. Low-cost inertial measurement units (IMUs) are the general sensors installed in the spinning rocket, which can provide the accelerations $a_y$, $a_z$, and angular rates $q$ and $r$ of the lateral motion. In this condition, the accelerations measured by IMU along the $OY_N$ and $OZ_N$ axes can be expressed as

$$\begin{bmatrix} a_y \\ a_z \end{bmatrix} = \begin{bmatrix} -c_{N\alpha}V & 0 \\ 0 & -c_{N\alpha}V \end{bmatrix}\begin{bmatrix} \beta \\ \alpha \end{bmatrix}. \tag{13}$$

Hence, the overall dynamic equation of the spinning rocket can be rearranged to the following state-space form,

$$\begin{aligned}\dot{x}_p(t) &= A_p x_p(t) + B_p u(t),\\ y_p(t) &= C_p x_p(t),\\ z(t) &= C_z x_p(t),\end{aligned} \tag{14}$$

where $x_p = [\beta \ \ \alpha \ \ q \ \ r]^T \in \mathbb{R}^{n_p}$ is the state vector, $u = [\delta_y \delta_z]^T \in \mathbb{R}^m$ is the control input vector, $y_p = [a_y \ \ a_z \ \ q \ \ r]^T \in \mathbb{R}^{n_p}$ is the measurement output vector, and $z = [a_y a_z] \in \mathbb{R}^r$ is the regulated output vector that is also measured. $A_p \in \mathbb{R}^{n_p \times n_p}$ is the system matrix, $B_p \in \mathbb{R}^{n_p \times m}$ is the input matrix, $C_p \in \mathbb{R}^{p_p \times n_p}$ is the measurement output matrix, and $C_z \in \mathbb{R}^{r \times n_p}$ is the regulated output matrix, and they are all known matrices.

*2.3. Uncertain Dynamic Equation.* The dynamic model presented in (14) is the ideal case where all the matrices are known. In reality, these matrices are unknown and are obtained through various methods. The system matrix $A_p$ and output matrices $C_p$ and $C_z$ can be determined through wind-tunnel tests fairly accurately. In contrast, the input matrix $B_p$ may not be accurate, as control inputs are subjected to perturbations in the flight period. First, the orientation of the canards can lead to the dynamical parameters different from the trim condition as represented by the nominal model. This effect can be modeled as an additive term $\Psi_p^T x_p$. Second, canard failure caused by electronic circuit or control surface damage is another effect which may cause the inaccuracy of input matrices. This effect can be modeled as constant matched uncertainty weights $\Lambda$. The modified uncertain dynamic model is given as

$$\begin{aligned}\dot{x}_p(t) &= A_p x_p(t) + B_p\left(\Lambda u(t) + \Psi_p^T x_p\right),\\ y_p(t) &= C_p x_p(t),\\ z(t) &= C_z x_p(t),\end{aligned} \tag{15}$$

where $\Lambda \in \mathbb{R}^{m \times m}$ and $\Psi_p \in \mathbb{R}^{r_p \times m}$ are all nonsingular matrices.

## 3. Robust Adaptive Autopilot Design

The fight sequence of guided spinning rockets is decomposed into three flight phases: boost phase, free flight phase, and guided phase. In this section, we mainly consider the guidance phase, during which the autopilot begins to work. In the guided phase, guidance law gives the desired commands which are usually in the form of acceleration, to the autopilot based on the target and rocket's relative motion information. Therefore, designing an acceleration autopilot is more reasonable and effective. The underlying problem is to design an acceleration autopilot that is able to track different kinds of signals under different dynamic environments, with slight or even no change of the autopilot architecture. So, the output feedback adaptive control method is employed to design the acceleration autopilot.

It should be noted that the output feedback adaptive autopilot requires the dynamical system in (15) as the controllable, observable, and minimum phases, which can be satisfied by the canard-controlled spinning rocket. The uncertainty $\Psi_p$ is bound with $\|\Psi_p\| < \Omega < \infty$ and $\operatorname{rank}(C_p B_p) = m$.

*3.1. Addition of Integral Error.* Command tracking and disturbance rejection are the main issues to be handled through the integral action [24, 25]. In this light, consider the regulated output signal tracking error which is defined as follows:

$$\dot{x}_e = z_{cmd} - z, \tag{16}$$

where $z_{cmd}$ is a piecewise continuous command signal. An integral error state is defined as

$$x_e = \int_0^t \dot{x}_e dt = \int_0^t (z_{cmd} - z)dt. \tag{17}$$

Appending (17) to the plant in (15), the augmented open-loop dynamics is given by

$$\begin{bmatrix} \dot{x}_p \\ \dot{x}_e \end{bmatrix} = \underbrace{\begin{bmatrix} A_p & 0_{n_p \times r} \\ -C_z & 0_{r\times r} \end{bmatrix}}_{A} \underbrace{\begin{bmatrix} x_p \\ x_e \end{bmatrix}}_{x} \underbrace{\begin{bmatrix} B_p \\ 0_{r\times m} \end{bmatrix}}_{B} \left(\Lambda u + \Psi_p^T x_p\right) + \underbrace{\begin{bmatrix} 0_{n_p \times r} \\ I_{r\times r} \end{bmatrix}}_{B_{cmd}} z_{cmd},$$
$$\underbrace{\begin{bmatrix} y_p \\ x_e \end{bmatrix}}_{y} = \underbrace{\begin{bmatrix} C_p & 0_{n_p \times r} \\ 0_{r\times n_p} & I_{r\times r} \end{bmatrix}}_{C} \begin{bmatrix} x_p \\ x_e \end{bmatrix}. \tag{18}$$

Therefore, (18) can be written more compactly as follows:

$$\begin{aligned} \dot{x} &= Ax + B\left(\Lambda u(t) + \Psi^T x\right) + B_{cmd} z_{cmd}, \\ y &= Cx, \end{aligned} \tag{19}$$

where $\Psi = \left[\Psi_p^T 0_{m\times r}\right]^T$ is unknown. Define $n = n_p + r$, then $x \in \mathbb{R}^n$.

*3.2. Control Architecture.* Following the design procedure in [25], the controller is divided into two parts: a baseline controller and an adaptive controller. The control input $u$ in (15) is described as

$$u = u_{bl} + u_{ad}, \tag{20}$$

where $u_{ad}$ is the adaptive component and $u_{bl}$ is the baseline component, defined as

$$u_{bl} = K_x^T x_m, \tag{21}$$

where $K_x$ is designed by applying the linear quadratic regulator (LQR) technique on the nominal plant model, that is, when $\Lambda = I$ and $\Psi = 0$. $x_m$ is the state of the observer,

$$\begin{aligned} \dot{x}_m &= A_m x_m + B_{cmd} z_{cmd} + L(y_m - y), \\ y_m &= Cx_m, \end{aligned} \tag{22}$$

where $A_m = A + BK_x^T$, and $K_x^T$ is selected such that $A_m$ is Hurwitz.

*Remark 2.* The observer (22) serves as three purposes:

(1) Luenberger observer. Matrix $L$ is equivalent to an observer gain, for which the purpose is to estimate the state vector $x$ in plant (19). The estimated state vector helps to construct the baseline controller $u_{bl}$ .

(2) Reference model. As our ultimate goal is to establish a model reference adaptive controller, (22) serves as a closed-loop reference model (CRM), which has been proved to result in improving transient properties [26–28].

(3) Robust compensator. The output error feedback term $L(y_m - y)$ can be treated as a robust compensator, making the error dynamics satisfy the SPR property.

The adaptive controller $u_{ad}$ is defined as

$$u_{ad} = \Theta^T(t) x_m, \tag{23}$$

where $\Theta(t)$ is the estimated uncertain parameter, to be updated by a well-designed update law. Substituting the controller (20) into the augmented plant (19),

$$\begin{aligned} \dot{x} &= Ax + B\left(\Lambda(K_x + \Theta(t))^T + \Psi^T x\right) + B_{cmd} z_{cmd}, \\ y &= Cx. \end{aligned} \tag{24}$$

If the proposed control architecture can realize accurate command tracking, an ideal uncertain parameter $\Theta^*$ must exists and satisfies the following matching condition,

$$A_m = A + B\Psi^T + B\Lambda\left(\Theta^{*T} + K_x^T\right) \tag{25}$$

with $\Theta^{*T} = (\Lambda^{-1} - I)K_x^T - \Lambda^{-1}\Psi^T$.

The error dynamic model between the reference model (22) and the closed-loop plant (24) is resorted to accomplish the whole design procedure. The state error $e_x = x - x_m$ and the estimate parameter error $\Theta^{\sim}(t) = \tilde{\Theta}(t) - \Theta^*$ satisfy the dynamics

$$\begin{aligned} \dot{e}_x &= \left(A + LC + B\Psi^T\right)e_x + B\Lambda\tilde{\Theta}^T(t)x_m, \\ e_y &= Ce_x, \end{aligned} \tag{26}$$

where $e_y$ is the measured output error. The problem of finding a stabilizing adaptive controller is equivalent to finding an observer gain $L$ and an adaptive law for $\Theta(t)$ in (26), so that the underlying transfer function matrix is SPR.

The following adaptive law is employed to update the estimated uncertain parameter,

$$\dot{\Theta}(t) = -\Gamma x_m e_y^T M^T \operatorname{sgn}(\Lambda), \tag{27}$$

where $\Gamma$, the adaptive gain, is a positive diagonal free design matrix, $M \in \mathbb{R}^{m\times n}$ is a mixing matrix to "square up" the transfer function, and $\operatorname{sgn}(\Lambda)$ represents the sign matrix of the input uncertainty $\Lambda$.

*3.3. Robust SPR Design.* The SPR property is usually employed to design a stable adaptive law for the error dynamic model (26) of the uncertain plant. However, the definition of SPR is restricted to square transfer functions. As the transfer function of (26) is nonsquare, a suitable mixing matrix $M$ has to be chosen to make the SPR properties applicable to the error dynamic model, yielding

$$G(s) = MC\left(sI - A - LC - B\Psi^T\right)^{-1}B, \tag{28}$$

where $G(s)$ is a square transfer function of the modified error dynamic model,

$$\dot{e}_x = \left(A + LC + B\Psi^T\right)e_x + B\Lambda\tilde{\Theta}^T(t)x_m,$$
$$e_m = MCe_x. \tag{29}$$

Thus, the design of an output feedback adaptive controller is converted to selecting mixing matrix $M \in \mathbb{R}^{m\times n}$ and observer matrix $L \in \mathbb{R}^{n\times n}$, such that the overall closed-loop system is stable.

The choice of mixing matrix $M$ is not unique. $M$ can be computed as

$$M = B^T PC^{-1}, \tag{30}$$

which satisfies $PB = C^T M^T$.

**Lemma 1.** *Given the strictly proper transfer matrix $G(s)$ with stabilizable and detectable realization $(A, B, C, D)$, where $A \in \mathbb{R}^{n\times n}$ is asymptotically stable, $B \in \mathbb{R}^{n\times m}$, $C \in \mathbb{R}^{m\times n}$, and $D \in \mathbb{R}^{m\times m}$; then, $G(s)$ is SPR if, and only if, there exists matrices $P = P^\top > 0$, $P \in \mathbb{R}^{n\times n}$, $Q \in \mathbb{R}^{n\times m}$, and $W \in \mathbb{R}^{m\times m}$ such that*

$$A^\top P + PA = -QQ^T,$$
$$PB - C^T = -QW, \tag{31}$$
$$D + D^T = W^T W.$$

*Proof 1.* The complete proof on the Lemma 1 can be found in [29] Lemma 3.1.

**Lemma 2.** *Let $H$ and $E$ be given matrices of appropriate dimensions, and $F$ satisfies $FF^T < I$; for any $\varepsilon > 0$, there is*

$$HFE + (HFE)^T \le \varepsilon HH^T + \varepsilon^{-1}E^T E. \tag{32}$$

*Proof 2.* The complete proof on the Lemma 2 can be found in [30].

The output error feedback term $L(y_m - y)$ can be treated as a robust compensator $u_m$ for the error dynamic model, making the error dynamics satisfy the SPR property. First, rewrite the error dynamic model (29) into the canonical form of robust control,

$$\dot{e}_x = (A + \Delta A)e_x + B_1 w + B_2 u_m,$$
$$e_m = C_m e_x, \tag{33}$$

where $\Delta A = B\Psi^T$ is defined as the uncertain state matrix; $B_1 = B$, $B_2 = I_n$, $C_m = MC$, $w = \Lambda\Theta^T(t)x_m$, and $u_m = L(y_m - y) = LCe_x$ are defined as the modified virtual control inputs for modified error dynamics.

The modified error dynamics is SPR and stable if and only if Lemma 1 is satisfied. Equation (29) is strictly proper; that is, $D = 0$, so Lemma 1 cannot be used directly. For the case $D = 0$, the above set of equations in Lemma 1 reduces to the first two equations with $W = 0$ [29]. The observer gain $L$ is designed by finding a state-feedback control of (33), and the following inequality should be satisfied:

$$(A + \Delta A + LC)^T P + P(A + \Delta A + LC) < 0, \tag{34}$$

$$PB_1 = C^T M^T. \tag{35}$$

$$\begin{bmatrix} XA^T + AX + B_2W + W^TB_2 & (EX)^T & B_1 \\ * & -\gamma & 0 \\ * & * & -\frac{1}{\gamma} \end{bmatrix} < 0, \tag{36}$$

where $E = \Omega I_n$. The matrix in (36) is symmetric, and $*$ represents the transposition of the corresponding item in the matrix.

**Lemma 3.** *Given an uncertain plant (33), the closed-loop system is stable and SPR, if and only if there exist symmetric positive matrices $X$ and $W$ and a constant $\gamma > 0$ satisfies*

*Proof 3.* The state matrix uncertainty $\Delta A$ can be described as follows:

$$\Delta A = B_1\Psi^T = B_1 FE, \tag{37}$$

where $F = \Psi^T/\Omega$, and the inequality $FF^T < I$ holds.

Define $P = X^{-1}$ and $K = LC = WX^{-1}$, and multiply inequality (36) on both sides by the matrix $\mathrm{diag}(P, I, I)$,

$$\begin{bmatrix} (A + B_2K)^T P + (A + B_2K) & E^T & PB_1 \\ * & -\gamma & 0 \\ * & * & -\frac{1}{\gamma} \end{bmatrix} < 0. \tag{38}$$

By the Schur complement, the LMI defined in (36) is equivalent to the following inequality:

$$(A + B_2K)^T P + P(A + B_2K) + \gamma PB_1B_1^T P + \frac{1}{\gamma}E^T E < 0. \tag{39}$$

With Lemma 2, we have

$$PB_1FE + (PB_1FE)^T < \gamma PB_1B_1^T P + \frac{1}{\gamma}E^T E. \tag{40}$$

Thus, inequality (39) is equivalent to the following inequality,

$$(A + B_2K + B_1FE)^T P + P(A + B_2K + B_1FE) < 0, \tag{41}$$

which means $(A + \Delta A + LC)^T P + P(A + \Delta A + LC) < 0$. Combining $PB = C^T M^T$ (34) holds. So, the modified error dynamics is stable and SPR.

The observer gain $L$ can be obtained by solving the feasible LMI with any widely available numerical LMI solver. It can be expressed as

$$L = W^* X^{*-1} C^{-1}, \tag{42}$$

where $W^*$ and $X^*$ are the feasible solution of (36).

Figure 2 shows the structure of the control loop, where the plant is the dynamical model of the spinning rocket in (14), the reference model is (22), and the adaptive law is (23),

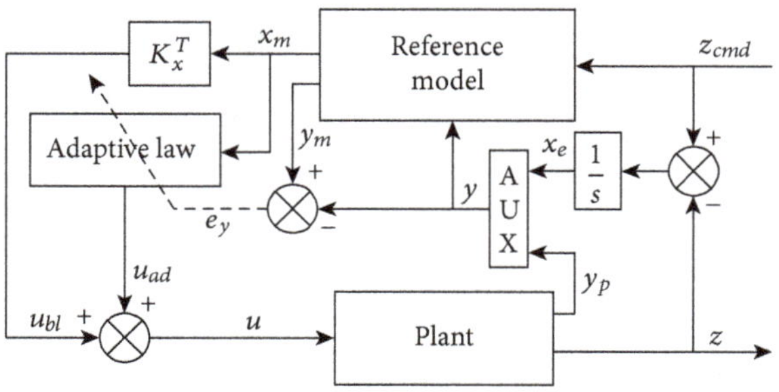


FIGURE 2: Control architecture.

*3.4. Stability Analysis.* Given the uncertain linear system in (19), the reference model in (22) with $L$ as in (42), the control architecture in (20), and the update law in (27) result in global stability, with $\lim_{t\to\infty} e_x = 0$.

Consider the following Lyapunov function candidate,

$$V\left(e_x, \tilde{\Theta}\right) = e_x^T P e_x + tr\left(|\Lambda|\tilde{\Theta}\Gamma^{-1}\tilde{\Theta}\right), \quad (43)$$

where $|\Lambda|$ represents the absolute value of each entry of the matrix element, satisfying $\Lambda = \text{sgn}(\Lambda)|\Lambda|$.

The time derivative of (43) along the system trajectories is given by

$$\dot{V} = -e_x^T Q e_x + 2e_x^T PB\Lambda\tilde{\Theta}^T x_m + 2tr\left(|\Lambda|\tilde{\Theta}^T\Gamma^{-1}\dot{\tilde{\Theta}}\right), \quad (44)$$

where $(A + B\Psi^T + LC)^T P + P(A + B\Psi^T + LC) = -Q < 0$. Substituting the update law given in (27) yields

$$\begin{aligned}\dot{V} &= -e_x^T Q e_x + 2e_x^T PB\Lambda\tilde{\Theta}^T x_m - 2tr\left(|\Lambda|\tilde{\Theta}^T x_m e_y^T M^T \text{ sgn}(\Lambda)\right) \\ &= -e_x^T Q e_x + 2e_x^T PB\Lambda\tilde{\Theta}^T x_m - 2e_y^T M^T \Lambda\tilde{\Theta}^T x_m \\ &= -e_x^T Q e_x + 2e_x^T PB\Lambda\tilde{\Theta}^T x_m - 2e_x^T C^T M^T \Lambda\tilde{\Theta}^T x_m \\ &= -e_x^T Q e_x \le 0,\end{aligned} \quad (45)$$

which implies that $V$ is a Lyapunov function. Since $V > 0$ and $\dot{V} \le 0$, then $V(t) \le V(0) < \infty$. Thus, $V(t) \in L_\infty$, which means $e_x$, $\tilde{\Theta} \in L_\infty$. Since $z_{cmd}$, $e_x \in L_\infty$, and the reference model are stable, $x_m \in L_\infty$, which implies that $x_p \in L_\infty$.

Furthermore, asymptotic stability of the tracking errors is demonstrated by invoking LaSalle's invariance principle, which states that, for a negative semidefinite Lyapunov system in the form of (45), all system trajectories are contained within the domain $\Omega_0 = \{[e_x, \Theta(t)] \mid V([e_x, \Theta(t)], t) \le V([e_x, \Theta(t)], 0)\}$, where the subscript $\{\}_0$ denotes the initial conditions, and the entire state space $[e_x, \Theta(t)]$ ultimately reaches the domain $\Omega_f = \Omega_0 \cap \Omega_z$, where $\Omega_z$ denotes the domain defined by the Lyapunov derivative identical to zero. In other words, the state space ultimately reaches the domain defined by $\dot{V}([e_x, \Theta(t)], t) \equiv 0$ [31, 32]. Because $\dot{V}([e_x, \Theta(t)], t)$ is negative-definite in $e_x$, the system ends with $e_x \equiv 0$. Thus, $x(t) \to x_m(t)$ and the bound reference tracking of $z_{cmd}$ by $z$ follow from the stability of the closed-loop system.

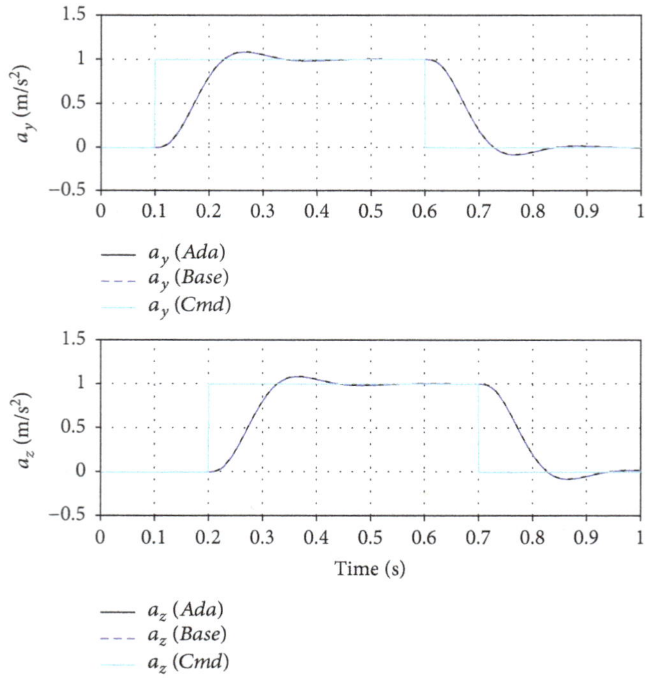


FIGURE 3: The nominal case tracking performance.

*Remark 3.* In this paper, the adaptive parameter $\Theta(t)$ is not guaranteed to converge to its true unknown value $\Theta^*$ nor is it assured to converge to constant value in any way. All that is known is that the unknown parameter remains uniformly bound in time. Sufficient conditions for parameter convergence are known as persistency of excitation.

## 4. Case Study

In this section, numerical examples are performed to evaluate the performance of the output feedback adaptive control scheme applied to the autopilot design for the dual-canard-controlled spinning rocket. All the simulations are performed in MATLAB R2017b with a 64-bit processor, 16 GB memory, and 0.001 s time-step. Robust Control Toolbox 6.4 in MATLAB is employed to obtain a feasible solution of observer (36). And all the simulations are performed with an adaptive controller acting on the nonlinear model.

The nominal model for autopilot design is the linearized lateral dynamics of the spinning rocket at the speed $V$ of 581 m/s and altitude $H$ of 5000 m. The numerical values for the linear system matrices can be found in Appendix A.

All the eigenvalues of the nominal plant system matrix $A_p$ have a negative real part, which means the nominal plant is stable.

We show the response of the closed-loop system with the baseline controller and adaptive controller, respectively. Square-wave commands with amplitude of 1 m/s$^2$ and frequency of 1 Hz are applied to the pitch and yaw channel; there is a 0.1 s time lag between the two channels. It is observed in Figure 3 that both the baseline controller and the adaptive controller are able to guide the spinning rocket following the acceleration command and eventually achieved

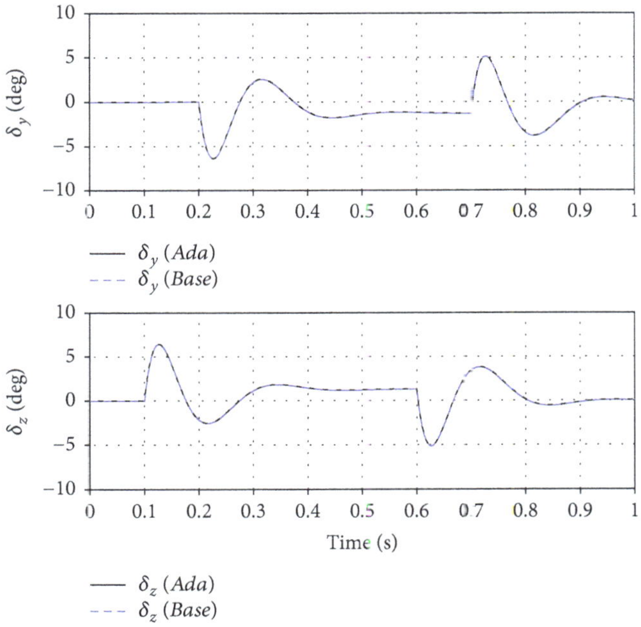


FIGURE 4: The nominal case control signal.

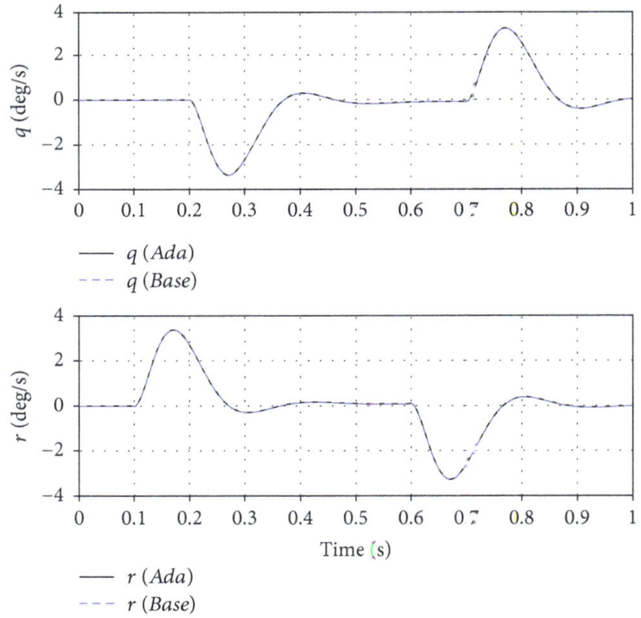


FIGURE 5: The nominal case angular rates.

a zero tracking error. Figure 4 presents the behavior of control signals in the pitch and yaw channels. Only a small amount of canard deflection angles is used to achieve the command tracking. The angular rates of pitch and yaw channels are presented in Figure 5, which shows a good performance in both channels. Considering simulation results for this case, both the proposed adaptive controller and the baseline controller present a proper performance in the presence of the disturbances enforced due to the channel couplings.

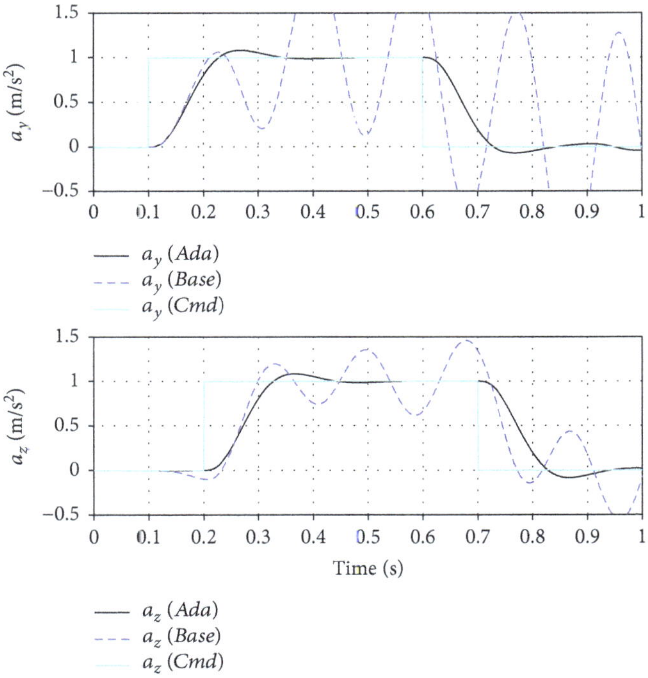


FIGURE 6: The uncertain case tracking performance.

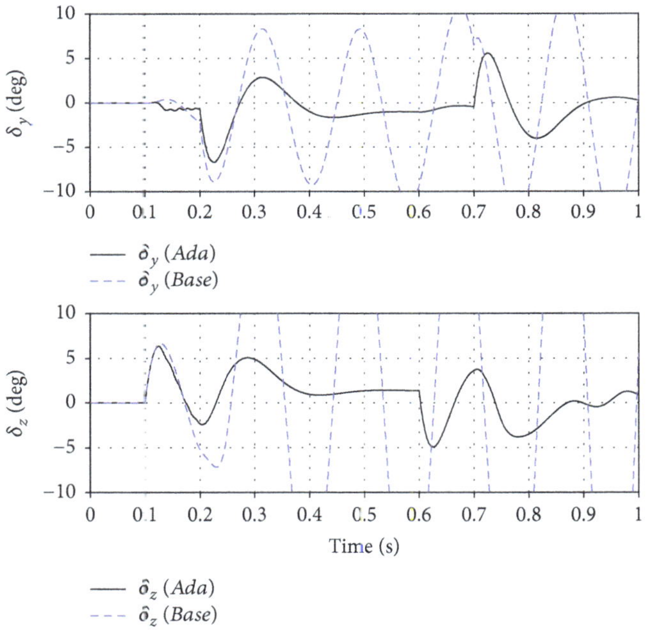


FIGURE 7: The uncertain case control signal.

The same controllers were used in the presence of the following uncertainties:

$$\Lambda = 0.9I, \quad \Psi = \begin{bmatrix} 0 & 0 & 0.12 & 0 & 0 & 0 \\ 0 & 0 & 0 & 0.12 & 0 & 0 \end{bmatrix}. \tag{46}$$

Such uncertainties stem from the fact that the model parameters are expected to vary significantly (by about

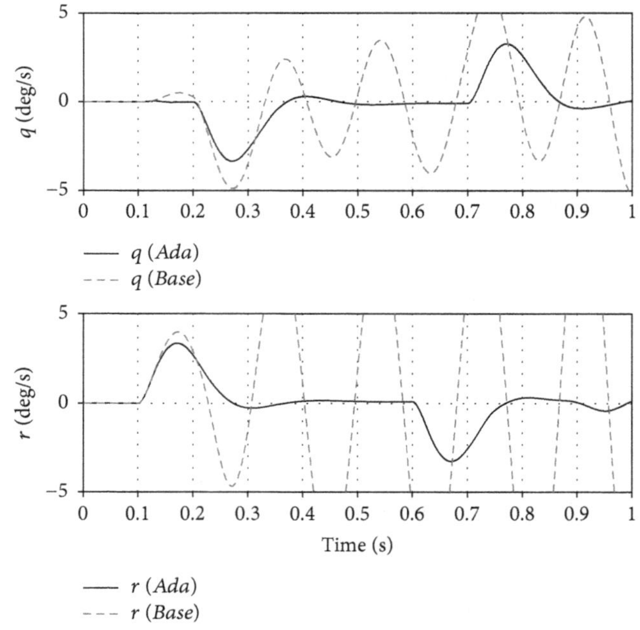


Figure 8: The uncertain case angular rates.

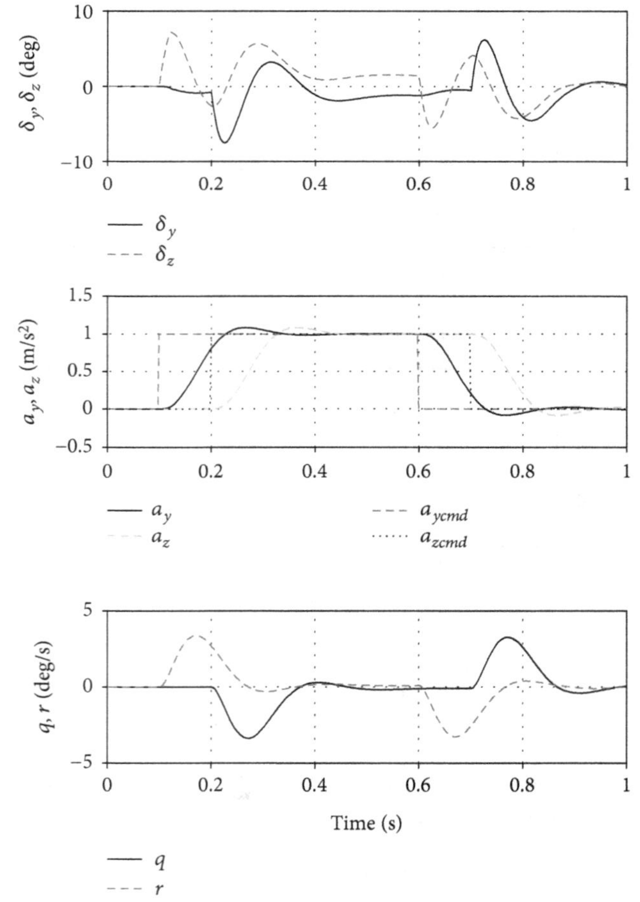


Figure 9: Control performance with $\Lambda = 0.9I$.

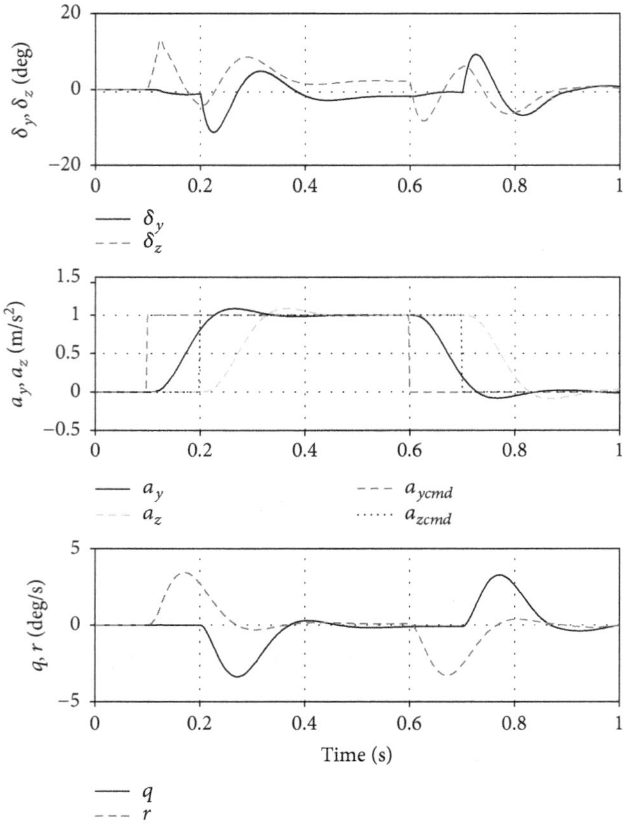


Figure 10: Control performance with $\Lambda = 0.6I$.

30% from the nominal $A$ matrix) compared to those determined from wind channel tests. At this time, the eigenvalues of the system matrix have a positive real part, which means the system is unstable.

Outcomes of the pitch and yaw channel autopilots in the same set-point are presented in Figure 6. In this case, the baseline controller is not able to suppress the unstable pitch and yaw modes, whereas the adaptive controller is able. The behavior of canard angles in the pitch and yaw channels is presented in Figure 7. The amplitude of canard angles for the adaptive controller is slightly bigger than that of the nominal model, due to the uncertainty. The angular rates of pitch and yaw channels are presented in Figure 8, which implies that the adaptive controller shows a better performance.

In order to visualize the destructive effect of the control effectiveness loss, simulations with different control effectiveness are performed. According to Figures 9 and 10, the proposed adaptive controller can autoadjust the canard angles to handle the loss of control effectiveness. Considering simulation results for this scenario, the proposed adaptive controller presents a proper performance in the presence of the disturbances enforced due to the channel couplings and control effectiveness loss.

Measurement noise is also taken into account in order to carry out more investigations in the perturbed conditions. To this end, a worst-case condition is enforced; a white noise with a standard deviation of $0.25(^{\circ}/s)$ is injected into the rate

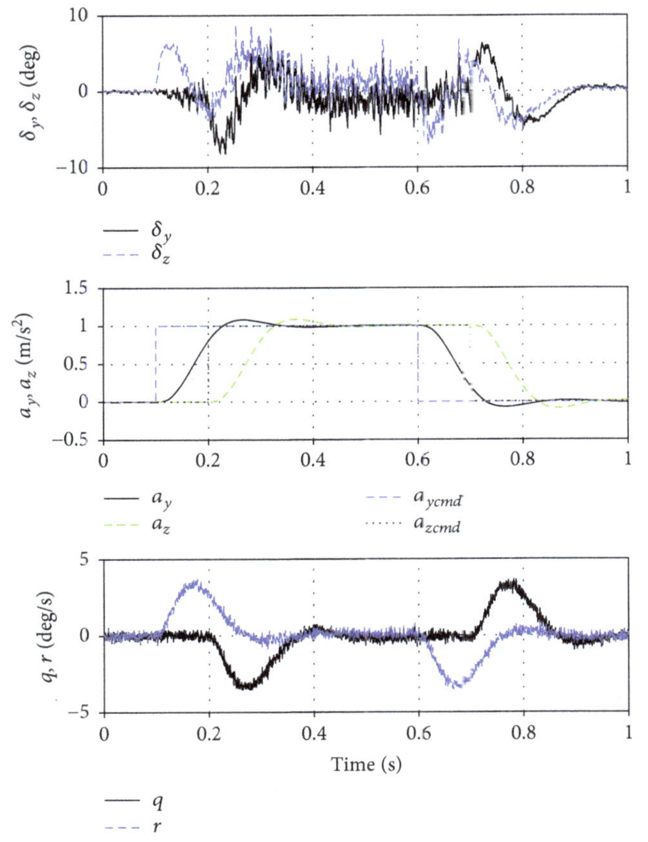


FIGURE 11: Adaptive controller performance with measurement noise.

gyroscopes of each channel, which is corresponding to the noise of low-cost MEMS gyroscopes. Figure 11 shows that the adaptive controller was robust in the presence of measurement noise in the angular rate feedback loop.

We analyze the robustness of the overall closed-loop system with the steady-state gain $K_x^T + \Psi^T$ in what follows. The gain and phase margin for the baseline controller and the closed-loop adaptive controller under the uncertain model are calculated as in Appendix B:

$$\begin{aligned} GM_{b1} &= [-818.1]\text{dB} \quad PM_{b1} = \pm 51.9^{\circ}, \\ GM_{ad} &= [-14.231.6]\text{dB} \quad PM_{ad} = \pm 58.3^{\circ}. \end{aligned} \tag{47}$$

It can be seen that the adaptive controller, in general, is more robust than the baseline controller is.

## 5. Conclusion

This paper presents a robust adaptive output feedback autopilot for a guided spinning rocket to resistant external disturbance and maintain precision command tracking. The autopilot is composed of a baseline controller augmented with an adaptive component to accommodate control effectiveness uncertainty and matched plant uncertainty, and it makes use of the closed-loop reference model to improve the transient properties of the overall adaptive system. Adaptive control is applicable to nonsquare systems through designing mixing matrix $M$ and the observer gain matrix $L$. This procedure only needs to solve a set of linear matrix inequalities, which is given by the robust SPR lemma. The performance of the proposed adaptive output feedback controller is evaluated by numerical simulations when applied to the lateral dynamic model of a guided spinning rocket. The simulation results showed that the robust adaptive output controller can achieve asymptotic command tracking with significant uncertainty in control effectiveness, moment coefficient, and measurement noise.

## Appendix

### A. System Matrices

The nominal dynamical plant matrices for a guided spinning rocket of velocity $V = 581$ m/s, altitude $H = 5000$ m, and spinning rate $p = 25.4$ rad/s are

$$\begin{aligned} A_p &= \begin{bmatrix} -0.35 & 0 & 0 & -1 \\ 0 & -0.35 & 1 & 0 \\ 1.28 & -54.89 & -0.63 & -0.07 \\ 54.89 & 1.28 & 0.07 & -0.63 \end{bmatrix}, \\ B_p &= \begin{bmatrix} 0 & 0 \\ 0 & 0 \\ 12.15 & 0 \\ 0 & 12.15 \end{bmatrix}, \\ C_p &= \begin{bmatrix} -206.59 & 0 & 0 & 0 \\ 0 & -206.59 & 0 & 0 \\ 0 & 0 & 1 & 0 \\ 0 & 0 & 0 & 1 \end{bmatrix}, \\ C_z &= \begin{bmatrix} -206.59 & 0 & 0 & 0 \\ 0 & -206.59 & 0 & 0 \end{bmatrix}. \end{aligned} \tag{A.1}$$

The dynamical system $\{A_p, B_p, C_p, 0\}$ is controllable and observable. There is no transmission zero in the plant. The linear control design parameters are given as $Q_{lqr} = \text{diag}([0.5, 0.5, 0, 0, 0.5, 0.5])$ and $R_{lqr} = \text{diag}([0.005, 0.005])$. Following the design procedure, the baseline controller gain matrix can be computed using the MATLAB command lqr:

$$K_x^T = \begin{bmatrix} 0.12 & -133.98 & -4.64 & 0 & 0.02 & -10 \\ -133.98 & 0.12 & 0 & -4.64 & 10 & 0 \end{bmatrix}. \tag{A.2}$$

Robust Control Toolbox 6.4 is used to solve the LMI in (36), resulting in mixing matrix $M$ and observer $L$

$$M = \begin{bmatrix} 0 & 0 & 0.0467 & 0 & 0 & 0 \\ 0 & 0 & 0 & 0.0467 & 0 & 0 \end{bmatrix},$$

$$L = \begin{bmatrix} 0 & 0 & -0.64 & -26.9 & -103.29 & 0 \\ 0 & 0 & 26.9 & -0.64 & 0 & -103.29 \\ 0 & -0.13 & 0 & 0 & 0 & 0 \\ 0.13 & 0 & 0 & 0 & 0 & 0 \\ 0.5 & 0 & 0 & 0 & -0.5 & 0 \\ 0 & 0.5 & 0 & 0 & 0 & -0.5 \end{bmatrix}. \quad \text{(A.3)}$$

## B. Multivariable Gain and Phase Margins

The multivariable gain and phase margin was calculated along the lines with [25], where the gain and phase margins are defined using the return difference matrix $\underline{\sigma}(I + L_u(s))$ and the stability robustness matrix $\underline{\sigma}(I + L_u^{-1}(s))$.

The operator $\sigma(.)$ corresponds to the minimum singular values; $L_u(s)$ denotes the input loop transfer function.

First, the minimum value of the return difference and the stability robustness transfer function over all frequency $\omega$ are defined as $\alpha_\sigma$ and $\beta_\sigma$, respectively:

$$\begin{aligned} \alpha\sigma &= \min_{\omega} \underline{\sigma}\left(I + L_u(s)\right), \\ \beta\sigma &= \min_{\omega} \underline{\sigma}\left(I + L_u^{-1}(s)\right). \end{aligned} \quad \text{(B.1)}$$

Then, the gain and phase margin of the return difference matrices $\mathrm{GM}_{\alpha\sigma}$ and $\mathrm{PM}_{\alpha\sigma}$ and stability robustness matrices $\mathrm{GM}_{\beta\sigma}$ and $\mathrm{PM}_{\beta\sigma}$ are calculated:

$$\begin{aligned} \mathrm{GM}_{\alpha\sigma} &= \left[(1+\alpha\sigma)^{-1}(1-\alpha\sigma)^{-1}\right], \\ \mathrm{GM}_{\beta\sigma} &= [1-\beta\sigma+1\beta\sigma], \\ \mathrm{PM}_{\alpha\sigma} &= \pm 2\sin^{-1}\left(\frac{\alpha\sigma}{2}\right), \\ \mathrm{PM}_{\beta\sigma} &= \pm 2\sin^{-1}\left(\frac{\beta\sigma}{2}\right). \end{aligned} \quad \text{(B.2)}$$

The union of these gain and phase margins in (B.2) yields the multivariable gain margin GM and phase margin PM, which can be written compactly as

$$\begin{aligned} \mathrm{GM} &= [\mathrm{GM_L}\mathrm{GM_U}], \\ \mathrm{PM} &= [\mathrm{PM_L}\mathrm{PM_U}]. \end{aligned} \quad \text{(B.3)}$$

## Conflicts of Interest

The authors declare that they have no conflicts of interest.

## Acknowledgments

The grant support from the National Natural Science Foundation of China (nos. 11202023 and 11532002) is greatly acknowledged.

## References

[1] J. S. Shamma and J. R. Cloutier, "Gain-scheduled missile autopilot design using linear parameter varying transformations," *Journal of Guidance, Control, and Dynamics*, vol. 16, no. 2, pp. 256–263, 1993.

[2] P. Apkarian, P. Gahinet, and G. Becker, "Self-scheduled $H_\infty$ control of linear parameter-varying systems: a design example," *Automatica*, vol. 31, no. 9, pp. 1251–1261, 1995.

[3] S. Theodoulis and P. Wernert, "Flight control for a class of 155 mm spinstabilized projectile with reciprocating canards," in *AIAA Guidance, Navigation, and Control Conference*, pp. 1–8, Minneapolis, MN, USA, 2012.

[4] S. Theodoulis, F. Seve, and P. Wernert, "Robust gain-scheduled autopilot design for spin-stabilized projectiles with a course-correction fuze," *Aerospace Science and Technology*, vol. 42, pp. 477–489, 2015.

[5] X. Yan, S. Yang, and C. Zhang, "Coning motion of spinning missiles induced by the rate loop," *Journal of Guidance, Control, and Dynamics*, vol. 33, no. 5, pp. 1490–1499, 2010.

[6] X. Yan, S. Yang, and F. Xiong, "Stability limits of spinning missiles with attitude autopilot," *Journal of Guidance, Control, and Dynamics*, vol. 34, no. 1, pp. 278–283, 2011.

[7] K. Li, S. Yang, and L. Zhao, "Stability of spinning missiles with an acceleration autopilot," *Journal of Guidance, Control, and Dynamics*, vol. 35, no. 3, pp. 774–786, 2012.

[8] K. Li, S. Yang, and L. Zhao, "Three-loop autopilot of spinning missiles," *Proceedings of the Institution of Mechanical Engineers, Part G: Journal of Aerospace Engineering*, vol. 228, no. 7, pp. 1195–1201, 2014.

[9] V. Stepanyan and K. Krishnakumar, "Adaptive control with reference model modification," *Journal of Guidance, Control, and Dynamics*, vol. 35, no. 4, pp. 1370–1374, 2012.

[10] A. P. Nair, N. Selvaganesan, and V. R. Lalithambika, "Lyapunov based PD/PID in model reference adaptive control for satellite launch vehicle systems," *Aerospace Science and Technology*, vol. 51, pp. 70–77, 2016.

[11] P. Sarhadi, A. R. Noei, and A. Khosravi, "Adaptive integral feedback controller for pitch and yaw channels of an AUV with actuator saturations," *ISA Transactions*, vol. 65, pp. 284–295, 2016.

[12] Z. Shi and L. Zhao, "Robust model reference adaptive control based on linear matrix inequality," *Aerospace Science and Technology*, vol. 66, pp. 152–159, 2017.

[13] S. Li and G. Tao, "Output feedback MIMO MRAC schemes with sensor uncertainty compensation," in *Proceedings of the American Control Conference*, pp. 3229–3234, Baltimore, MD, USA, 2010.

[14] J. M. Selfridge and G. Tao, "Multivariable output feedback MRAC for a quadrotor UAV," in *Proceedings of the American Control Conference*, pp. 492–499, Boston, MA, USA, 2016.

[15] Z. Shi and L. Zhao, "Adaptive output feedback autopilot design for spinning projectiles," in *Chinese Control Conference*, pp. 3516–3521, Dalian, China, 2017.

[16] D. d. S. Madeira and J. Adamy, "On the equivalence between strict positive realness and strict passivity of linear systems," *IEEE Transactions on Automatic Control*, vol. 61, no. 10, pp. 3091–3095, 2016.

[17] C. H. Huang, P. A. Ioannou, J. Maroulas, and M. G. Safonov, "Design of strictly positive real systems using constant output

feedback," *IEEE Transactions on Automatic Control*, vol. 44, no. 3, pp. 569–573, 1999.

[18] I. Barkana, "Comments on "Design of strictly positive real systems using constant output feedback"," *IEEE Transactions on Automatic Control*, vol. 49, no. 11, pp. 2091–2093, 2004.

[19] R. Johansson and A. Robertsson, "Observer-based strict positive real (SPR) feedback control system design," *Automatica*, vol. 38, no. 9, pp. 1557–1564, 2002.

[20] A. Saberi and P. Sannuti, "Squaring down by static and dynamic compensators," *IEEE Transactions on Automatic Control*, vol. 33, no. 4, pp. 358–365, 1988.

[21] A. Fradkov, "Passification of non-square linear systems and feedback Yakubovich-Kalman-Popov lemma," *European Journal of Control*, vol. 9, no. 6, pp. 577–586, 2003.

[22] C. H. Murphy, *Free Flight Motion of Symmetric Missiles*, Technical Report, U.S.Army Ballistic Research Laboratories, Aberdeen Proving Ground, Maryland, 1963.

[23] M. Pechier, P. Guillen, and R. Cayzac, "Magnus effect over finned projectiles," *Journal of Spacecraft and Rockets*, vol. 38, no. 4, pp. 542–549, 2001.

[24] L. Zhao, Z. Shi, and Y. Zhu, "Acceleration autopilot for a guided spinning rocket via adaptive output feedback," *Aerospace Science and Technology*, vol. 77, pp. 573–584, 2018.

[25] E. Lavretsky and K. A. Wise, "Robust and Adaptive Control with Aerospace Applications," in *Advanced Textbooks in Control and Signal Processing*, Springer, London, UK, 2013.

[26] D. P. Wiese, A. M. Annaswamy, J. A. Muse, M. A. Bolender, and E. Lavretsky, "Adaptive output feedback based on closed-loop reference models for hypersonic vehicles," *Journal of Guidance, Control, and Dynamics*, vol. 38, no. 12, pp. 2429–2440, 2015.

[27] T. E. Gibson, Z. Qu, A. M. Annaswamy, and E. Lavretsky, "Adaptive output feedback based on closed-loop reference models," *IEEE Transactions on Automatic Control*, vol. 60, no. 10, pp. 2728–2733, 2015.

[28] Z. Qu and A. M. Annaswamy, "Adaptive output-feedback control with closed-loop reference models for very flexible aircraft," *Journal of Guidance, Control, and Dynamics*, vol. 39, no. 4, pp. 873–888, 2016.

[29] B. Brogliato, B. Maschke, R. Lozano, and O. Egeland, *Dissipative Systems Analysis and Control, Theory and Applications*, Springer Verlag, London, UK, 2nd edition, 2007.

[30] L. Xie, M. Fu, and C. E. de Souza, "$H_\infty$ infinity control and quadratic stabilization of systems with parameter uncertainty via output feedback," *IEEE Transactions on Automatic Control*, vol. 37, no. 8, pp. 1253–1256, 1992.

[31] I. Barkana, "The new theorem of stability-direct extension of lyapunov theorem," *Mathematics in Engineering, Science & Aerospace (MESA)*, vol. 6, no. 3, pp. 519–550, 2015.

[32] I. Barkana, "Barbalat's lemma and stability-misuse of a correct mathematical result?," *Mathematics in Engineering, Science & Aerospace (MESA)*, vol. 7, no. 1, pp. 197–219, 2016.

# 12

# Coupling Dynamic Behavior Characteristics of a Spacecraft Beam with Composite Laminated Structures and Large-Scale Motions

**Bindi You, Zhihui Gao, Jianmin Wen, Yiming Sun, Peibo Hao, and Dong Liang**

*School of Naval Architecture and Ocean Engineering, Harbin Institute of Technology, Weihai 264209, China*

Correspondence should be addressed to Jianmin Wen; wenjm@hitwh.edu.cn

Academic Editor: Nicolas Avdelidis

A nonlinear dynamic modeling method for a spacecraft body composed of a laminated composite beam undergoing large rotation is proposed in this paper. To study the characteristics of a laminated composite beam attached to a spacecraft body for the dynamic systems, the deformation description of a laminated beam is established with the consideration of laying angles and laying layers, and the displacement-strain relations is acquired based on the global-local higher-order shear deformation theory. Accordingly, a nonlinear dynamic model of the spacecraft body composed of a laminated composite beam is deduced using Hamilton variational principle. And the complete coupling terms for the laminated material properties are considered unlike any other singular or unidirectional materials. Then, the dynamic behavior of the spacecraft system is analyzed by comparison of an orthogonal-symmetric, singular, and unidirectional laminated beam. The results show that the laminated composite structures have significant influences on the dynamics properties of spacecraft compared with conventional equivalent singular or unidirectional materials. Hence, the nonlinear model is well suitable for approaching the problem of coupling relationship between geometric nonlinearity and large rotation motions. These conclusions will have significant theory and engineering practice values for coupling dynamics properties of laminated beams.

## 1. Introduction

Laminated composite materials are formed by combining layers of different materials or the same material by using different laying angles, laying sequence, and laying layers [1–3]. Some of the properties can be improved by forming a composite material. For example, the stiffness and strength of fibrous composites come from fibers which are stiffer and stronger than the same material in bulk form [4]. In addition, some new laminated structures composed of different reinforcements were investigated in [5, 6], such as basalt fiber reinforced polymer and carbon fiber reinforced polymer composites. However, stiffness changes during the service loading of composite laminates can be significantly large, especially as those changes affect deflections, dimensional changes, vibration characteristics, and load or stress distributions [7]. With the development of aerospace exploration technology, spacecraft is becoming so large-scale, complex, flexible, and lightweight that conventional alloy materials cannot meet the practical engineering needs. Because composite materials exhibit not only better strength and stiffness properties but also better fatigue life and vibration resistance. Some laminated structures formed by composite materials have been widely used in aviation, aerospace, and many other engineering fields [8].

Flexible beams undergoing large-scale motions have their own unique theoretical and practical values in many applications. Many slender structures in aerospace engineering, such as space manipulators, solar wings, and satellite antenna, can be idealized to beam structures, and the dynamics properties of those structures are critical to system performance, integrity, and reliability [9–11]. For these reasons, the dynamic behaviors of flexible bodies experiencing a large-scale motion have been studied by many researchers in some academic fields. Accordingly, a lot of research of multibody system dynamics modeling for beam structures has been presented.

Space flexible beams with large-scale motion and nonlinear deformation have been studied in [12]. In particular, the effect of impact-induced damage on the mechanical behavior of laminated composite structures was studied in [13, 14]. Rigid-flexible coupling system dynamic properties with large-scale translational motion were studied in [15]. Next, rigid-flexible coupling system dynamics properties of flexible beams with large-scale deformation were studied in [16]. However, the geometric nonlinearity of laminated structures with composite material has been studied by few researchers. Dynamic properties of rotating composite material plates were studied in [17]. And considering shear deformation, dynamics properties of rigid-flexible coupling composited material beam system were studied in [18]. Moreover, dynamics equations of laminated beams with a large-scale motion were presented in [19], and the influences on dynamics properties for the transverse shear deformation of laminated beams were analyzed. Then, the vibration properties of laminated beams were studied in [20]. However, the geometric nonlinear effect of deformations can significantly influence dynamic behavior of the spacecraft system within the framework of a multibody dynamics, and few researchers have illustrated the specific rigid-flexible coupling characteristics of the laminated composite structures Moreover, most of the above researchers did not consider the influence of laminated composite material beams on its own dynamics properties with large-scale motion. Thus, how to reasonably derive exact models for a rotating laminated beam attached to a spacecraft body within the context of multibody system dynamics has been a subject of several researches.

In order to study dynamic behavior characteristics of a spacecraft beam with laminated composite structures and a large-scale motion, the aim of the present investigation is to develop a dynamic model. Also, the rigid-flexible coupling dynamics modeling are performed with the consideration of laying angles and laying layers of laminated composite beams. Then, the complete expressions of nonlinear terms, coupled deformation terms, and nonlinear elastic forces are developed in this study.

## 2. Formulation for a Laminated Composite Beam with a Large-Scale Motion

*2.1. Deformation Description.* The laminated composite beam can be illustrated in Figure 1. The length, width, and height of the beam are denoted as $a$, $b$, and $h$ ($b \ll a$, $h \ll a$). The laminated beam is composed of $L$ orthotropic layers. $(O_0, e^{(0)})$ and $(O, e)$ are the inertial reference coordinate system and floating coordinate system, respectively. When the laminated composite beam undergoes a large-scale motion with a spacecraft body, the deformation relative to the floating coordinate system will be produced in the laminated beam. An arbitrary point of the laminated composite beam can move from position $P_0$ before deformation to position $P$ after deformation, and the displacement vector can be expressed as $\mathbf{u}$. In Figure 1, $\boldsymbol{\rho}$ and $\boldsymbol{\rho}_0$ denote displacement vectors of points $P$ and $P_0$ relative to floating coordinate origin $O$ of the laminated composite beam, $\mathbf{r}_0$ and $\mathbf{r}$ denote displacement vectors of points $O$ and $P$ relative to inertial reference coordinate origin $O_0$. $z_k$ is the coordinate value of the $k$th layer in the direction of $e_3$, and $h_k$ is the lamina thickness of the $k$th layer by $h_k = z_{k+1} - z_k$. In the inertial reference coordinate system $(O_0, e^{(0)})$, the position vector $\mathbf{r}$ of point $P$ of the laminated composite beam can be expressed as

$$\mathbf{r} = \mathbf{r}_0 + \rho = \mathbf{r}_0 + \mathbf{A}(\rho_0 + \mathbf{u}), \tag{1}$$

where $\mathbf{A}$ is the direction cosine matrix between the floating coordinate system $(O, e)$ and inertial reference coordinate system $(O_0, e^{(0)})$.

Let $\mathbf{u}^{(k)} = [u^{(k)}\ v^{(k)}\ w^{(k)}]$ be the deformation of a point $P_0{}^{(k)}$ on the mid-plane of the layer relative to floating coordinate system, based on the global-local higher-order shear deformation theory, then the deformations $u^{(k)}$, $v^{(k)}$, and $w^{(k)}$ (in Figure 2) of any point on the $k$th layer of the laminated composite beam can be expressed as [21]

$$\begin{aligned} u^{(k)} &= u_0^{(k)} + z\varphi_x^{(k)} - c_1 z^3 \left( \varphi_x^{(k)} + \frac{\partial w_0^{(k)}}{\partial x} \right), \\ v^{(k)} &= v_0^{(k)} + z\varphi_y^{(k)} - c_1 z^3 \left( \varphi_y^{(k)} + \frac{\partial w_0^{(k)}}{\partial y} \right), \\ w^{(k)} &= w_0^{(k)}, \end{aligned} \tag{2}$$

where $c_1 = 4/3h^2$, $\varphi_x = \partial u/\partial z$, $\varphi_y = \partial v/\partial z$, $u_0{}^{(k)}$, $v_0{}^{(k)}$, and $w_0{}^{(k)}$ are the displacements on the mid-plane of the $k$th layer of the laminated composite beam, and $z$ is far away from the mid-plane of laminated composite beam as shown in Figure 2.

The laying angles are denoted as $\theta^{(k)}, k = 1, 2, \ldots, L$, in the laminated composite beam as shown in Figure 3. Then, the total deformation vector of the laminated composite beam can be written as

$$\mathbf{u} = \mathbf{u}^{(1)} + \cdots + \mathbf{u}^{(k)} + \cdots + \mathbf{u}^{(L)} = \sum_{k=1}^{L} \mathbf{B}^{(k)}\mathbf{u}^{(k)}, \quad k = 1, 2, \ldots, L, \tag{3}$$

where $\mathbf{u}^{(k)}$ is the position vector of any point on the mid-plane of the $k$th layer, $\mathbf{B}^{(k)}$ is the direction cosine transformation matrix from floating coordinate system $(O, e)$ of the $k$th layer (shown in Figure 3) to inertial reference coordinate system $(O_0, e^{(0)})$, and

$$\mathbf{B}^{(k)} = \begin{bmatrix} \cos\theta^{(k)} & -\sin\theta^{(k)} & 0 \\ \sin\theta^{(k)} & \cos\theta^{(k)} & 0 \\ 0 & 0 & 1 \end{bmatrix}. \tag{4}$$

The transformation relations between Euler parameters and Euler angles can be written as

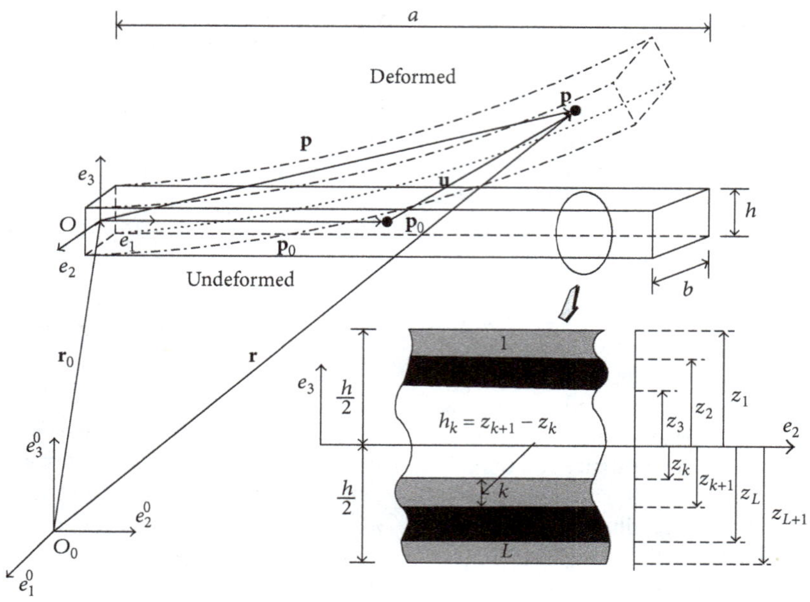


Figure 1: Laminated beam with a large-scale motion.

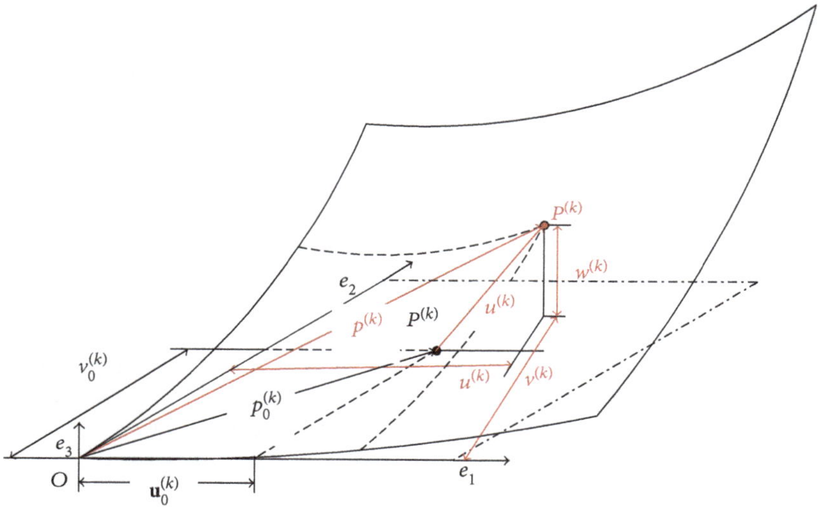


Figure 2: Deformation of the $k$th layer of laminated composite beam.

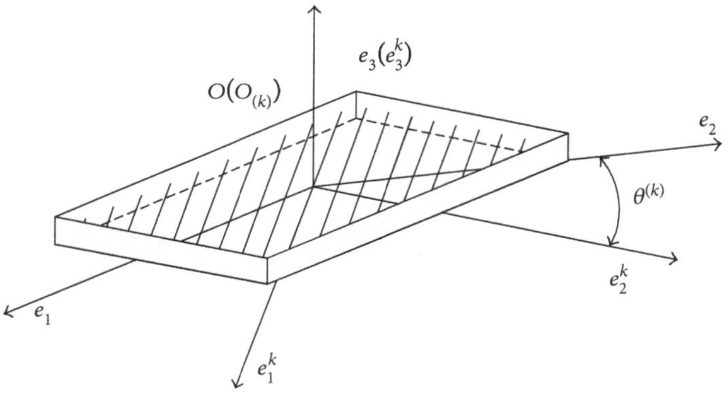


Figure 3: Laying angles of the $k$th layer of a laminated composite beam.

$$\begin{aligned}
\Theta_0 &= \cos\frac{\psi}{2}\cos\left(\frac{\vartheta+\varphi}{2}\right),\\
\Theta_1 &= \sin\frac{\psi}{2}\cos\left(\frac{\vartheta-\varphi}{2}\right),\\
\Theta_2 &= \sin\frac{\psi}{2}\sin\left(\frac{\vartheta-\varphi}{2}\right),\\
\Theta_3 &= \cos\frac{\psi}{2}\sin\left(\frac{\vartheta+\varphi}{2}\right),
\end{aligned}\tag{5}$$

where $\psi$, $\vartheta$, and $\varphi$ are the relative rotating angles of $(O, e)$ around ${e_3}^{(0)}$, ${e_1}^{(0)}$, and ${e_2}^{(0)}$ of $(O_0, e^{(0)})$, respectively.

We introduce a $3 \times 4$ order auxiliary matrix $\widehat{\mathbf{G}}$ defined by

$$\widehat{\mathbf{G}} = \left[-\boldsymbol{\Theta} - \tilde{\boldsymbol{\Theta}} + \Theta_0\mathbf{I}\right] = \begin{bmatrix} -\Theta_1 & \Theta_0 & \Theta_3 & -\Theta_2 \\ -\Theta_2 & -\Theta_3 & \Theta_0 & \Theta_1 \\ -\Theta_3 & -\Theta_2 & -\Theta_1 & \Theta_0 \end{bmatrix}, \tag{6}$$

where $\boldsymbol{\Theta} = [\Theta_1, \Theta_2, \Theta_3]^{\mathrm{T}}$ and antisymmetric matrix

$$\tilde{\boldsymbol{\Theta}} = \begin{bmatrix} 0 & -\Theta_3 & \Theta_2 \\ \Theta_3 & 0 & -\Theta_1 \\ -\Theta_2 & \Theta_1 & 0 \end{bmatrix}. \tag{7}$$

Substituting (2), (3), (5), and (6) into (1), then differentiating with respect to time $t$, we have the velocity vector $\dot{\mathbf{r}}$ of an arbitrary point $P$ in $(O_0, e^{(0)})$:

$$\dot{\mathbf{r}} = [\mathbf{I} \quad -\mathbf{A}\tilde{p} \quad \mathbf{A}]\begin{bmatrix} \dot{\mathbf{r}}_0 \\ \boldsymbol{\omega}' \\ \sum_{k=1}^{L}\mathbf{B}^{(k)}\dot{\mathbf{u}}^{(k)} \end{bmatrix}, \quad k = 1, 2, \ldots, L, \tag{8}$$

where $\tilde{\boldsymbol{\rho}}$ is the antisymmetric matrix,

$$\tilde{\boldsymbol{\rho}} = \begin{bmatrix} 0 & -z & y \\ z & 0 & x \\ -y & x & 0 \end{bmatrix}, \tag{9}$$

and $x$, $y$, and $z$ are the displacement coordinates of point $P_0$ along axis $e_1$, $e_2$, and $e_3$, respectively, in Figures 2 and 3. We introduce a four-dimension vector $\boldsymbol{\Lambda}$ defined by $\boldsymbol{\Lambda} = [\Theta_0\,\Theta_1\,\Theta_2\,\Theta_3]^{\mathrm{T}}$, and $\boldsymbol{\omega}' = 2\widehat{\mathbf{G}}\dot{\boldsymbol{\Lambda}}$ in (8).

Using the finite element discretization method, the laminated composite beam can be divided into $n$ spatial beam elements having two nodes as shown in Figure 4, in which each node has six deformation degrees of freedom.

The node displacement vector $\underline{\mathbf{q}}_k$ of the spatial beam element $i$ in the local reference coordinate system $(O_i, e^{(i)})$ can be written as

$$\underline{\mathbf{q}}_k = [\mathbf{q}_i\mathbf{q}_{i-1}]^{\mathrm{T}}, \tag{10}$$

where

$$\mathbf{q}_k = \begin{bmatrix} u_k & v_k & w_k & \gamma_{xk} & \gamma_{yk} & \gamma_{zk} \end{bmatrix}^{\mathrm{T}}, \quad k = i, i+1. \tag{11}$$

A simple way to obtain the derivatives of the director field is to use interpolation. So, being $N_j$ $(j = 1, 2, \ldots, 6)$, linear Lagrangian shape function coefficients will be used. Then, the deformation vector $\mathbf{u}_i$ of the spatial beam element $i$ in $(O_i, e^{(i)})$ can be written as

$$\mathbf{u}_i = \mathbf{N}\underline{\mathbf{q}}_k, \tag{12}$$

where $\mathbf{N}$ is the shape function of the spatial beam element $i$ with two nodes and six degrees of freedom of two nodes defined by [22]

$$\mathbf{N} = \begin{bmatrix} \mathbf{N}_x & \mathbf{N}_y & \mathbf{N}_z & \mathbf{N}_\gamma \end{bmatrix}^{\mathrm{T}} = \begin{bmatrix} N_1 & 0 & 0 & 0 & 0 & 0 & N_2 & 0 & 0 & 0 & 0 & 0 \\ 0 & N_2 & 0 & 0 & 0 & N_4 & 0 & N_5 & 0 & 0 & 0 & N_6 \\ 0 & 0 & N_3 & 0 & -N_4 & 0 & 0 & 0 & N_5 & 0 & -N_6 & 0 \\ 0 & 0 & 0 & N_1 & 0 & 0 & 0 & 0 & 0 & N_2 & 0 & 0 \end{bmatrix}. \tag{13}$$

The parameters of $\mathbf{N}$ can be written as

$$\begin{aligned}
N_1 &= 1 - \tau,\\
N_2 &= \tau,\\
N_3 &= 1 - 3\tau^2 + 2\tau^3,\\
N_4 &= \tau l_i(1-\tau)^2,\\
N_5 &= 3\tau^2 - 2\tau^3,\\
N_6 &= \tau^2 l_i(\tau - 1),
\end{aligned}\tag{14}$$

where $\tau = \underline{x}/l_i$, and $\underline{x}$ is the total displacement coordinates of point $P$ along axis $e_1$ in $(O_i, e^{(i)})$, $l_i$ is the length of the spatial beam element $i$.

The node deformation displacement vector $\underline{\mathbf{q}}$ in $(O, e)$ can be given by

$$\underline{\mathbf{q}} = \underline{\mathbf{q}}_k + \begin{bmatrix} u_i & v_i & w_i & 0 & 0 & 0 & u_i & v_i & w_i & 0 & 0 & 0 \end{bmatrix}^{\mathrm{T}}. \tag{15}$$

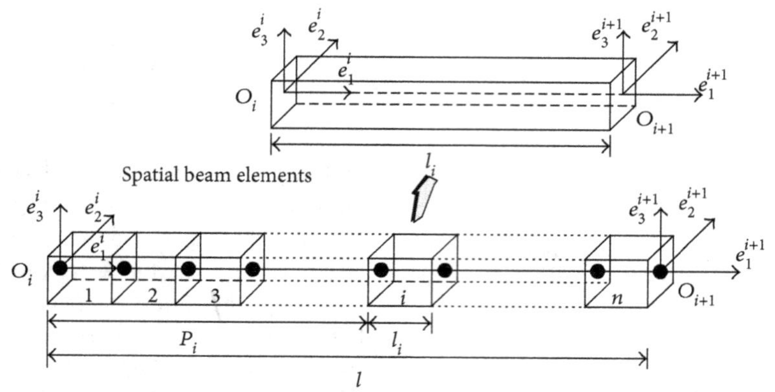


FIGURE 4: Element division of a laminated composite beam.

We introduce a $12 \times 12$ order orientation matrix $\underline{\mathbf{B}}^k$ based on nodes of spatial beam element, and it can be defined by [22]

$$\underline{\mathbf{B}}^k = \begin{bmatrix} 0 & \mathbf{I}_{6\times 6} \\ 0 & 0 \end{bmatrix}_{12\times 12}. \tag{16}$$

Then, the displacement vector of any point in $(O, e)$ can be given by

$$\mathbf{u}_i = \begin{bmatrix} \bar{\mathbf{N}}_1\underline{\mathbf{q}} - 0.5\underline{\mathbf{q}}^{\mathrm{T}}\mathbf{H}\underline{\mathbf{q}} \\ \bar{\mathbf{N}}_2\underline{\mathbf{q}} \\ \bar{\mathbf{N}}_3\underline{\mathbf{q}} \end{bmatrix} = \boldsymbol{\Phi}\underline{\mathbf{q}}, \tag{17}$$

where $\bar{\mathbf{N}}_j (j = 1, 2, 3)$ is a $1 \times 12$ order matrix, $\bar{\mathbf{N}}_1 = \mathbf{N}_x\underline{\mathbf{B}}^k$, $\bar{\mathbf{N}}_2 = \mathbf{N}_y\underline{\mathbf{B}}^k$, $\bar{\mathbf{N}}_3 = \mathbf{N}_z\underline{\mathbf{B}}^k$, and $\mathbf{H}$ is a coupled shape function matrix of order $12 \times 12$.

$$\boldsymbol{\Phi} = \begin{bmatrix} \bar{\mathbf{N}}_1\underline{\mathbf{q}} - 0.5\underline{\mathbf{q}}^{\mathrm{T}}\mathbf{H}\underline{\mathbf{q}} \\ \bar{\mathbf{N}}_2\underline{\mathbf{q}} \\ \bar{\mathbf{N}}_3\underline{\mathbf{q}} \end{bmatrix}. \tag{18}$$

The coupled shape function matrix $\mathbf{H}$ can be written as

$$\mathbf{H} = \int_0^{\underline{x}} \left(\frac{\partial \bar{\mathbf{N}}_1^{\mathrm{T}}}{\partial \underline{x}}\right)\left(\frac{\partial \bar{\mathbf{N}}_1}{\partial \underline{x}}\right) \mathrm{d}\underline{x} + \int_0^{\underline{x}} \left(\frac{\partial \bar{\mathbf{N}}_2^{\mathrm{T}}}{\partial \underline{x}}\right)\left(\frac{\partial \bar{\mathbf{N}}_2}{\partial \underline{x}}\right) \mathrm{d}\underline{x}. \tag{19}$$

*2.2. Displacement-Stress Equations of a Laminated Composite Beam.* According to the deformation displacement of point $P$ in (2) and the displacement-strain relations of elastic mechanics based on the global-local higher order shear deformation theory, the strain of a laminated composite beam can be given by [19]

$$\begin{aligned}
\varepsilon_{xx} &= \frac{\partial u_0}{\partial x} + \frac{1}{2}\left(\frac{\partial w_0}{\partial x}\right)^2 + z\frac{\partial \varphi_x}{\partial x} - c_1 z^3\left(\frac{\partial \varphi_x}{\partial x} + \frac{\partial^2 w_0}{\partial x^2}\right), \\
\varepsilon_{yy} &= \frac{\partial v_0}{\partial y} + \frac{1}{2}\left(\frac{\partial w_0}{\partial y}\right)^2 + z\frac{\partial \varphi_y}{\partial y} - c_1 z^3\left(\frac{\partial \varphi_y}{\partial y} + \frac{\partial^2 w_0}{\partial y^2}\right), \\
\varepsilon_{zz} &= 0, \\
\gamma_{xy} &= \frac{\partial u_0}{\partial y} + \frac{\partial v_0}{\partial x} + \frac{\partial w_0}{\partial x}\frac{\partial w_0}{\partial y} + z\left(\frac{\partial \varphi_x}{\partial y} + \frac{\partial \varphi_y}{\partial x}\right) \\
&\quad - c_1 z^3\left(\frac{\partial \varphi_x}{\partial y} + \frac{\partial \varphi_y}{\partial x} + 2\frac{\partial^2 w_0}{\partial x \partial y}\right), \\
\gamma_{yz} &= \varphi_y + \frac{\partial w_0}{\partial y} - c_2 z^2\left(\varphi_y + \frac{\partial w_0}{\partial y}\right), \\
\gamma_{xz} &= \varphi_x + \frac{\partial w_0}{\partial x} - c_2 z^2\left(\varphi_x + \frac{\partial w_0}{\partial x}\right).
\end{aligned} \tag{20}$$

Using the displacement-strain relations of (20), the displacement-strain equation of the $k$th layer can be given by [19]

$$\begin{Bmatrix} \sigma_{xx} \\ \sigma_{yy} \\ \sigma_{zz} \\ \tau_{xy} \\ \tau_{yz} \\ \tau_{xz} \end{Bmatrix}_k = \begin{bmatrix} \bar{Q}_{11} & \bar{Q}_{12} & 0 & 0 & 0 & 0 \\ \bar{Q}_{12} & \bar{Q}_{22} & 0 & 0 & 0 & 0 \\ 0 & 0 & 0 & 0 & 0 & 0 \\ 0 & 0 & 0 & \bar{Q}_{66} & 0 & 0 \\ 0 & 0 & 0 & 0 & \bar{Q}_{44} & 0 \\ 0 & 0 & 0 & 0 & 0 & \bar{Q}_{55} \end{bmatrix}_k \begin{Bmatrix} \varepsilon_{xx} \\ \varepsilon_{yy} \\ \varepsilon_{zz} \\ \gamma_{xy} \\ \gamma_{yz} \\ \gamma_{xz} \end{Bmatrix}_k, \tag{21}$$

where $Q_{ij}$ $(i, j = 1, 2, \dots, 6)$ is the equivalent stiffness coefficients of two-dimensional layers which can be defined by [21]

$$\begin{aligned}
\bar{Q}_{11} &= Q_{11}n^4 + 2(Q_{12} + 2Q_{66})m^2n^2 + Q_{22}m^4, \\
\bar{Q}_{12} &= (Q_{11} + Q_{22} - 4Q_{66})m^2n^2 + Q_{12}\left(m^4 + n^4\right), \\
\bar{Q}_{22} &= Q_{11}m^4 + 2(Q_{12} + 2Q_{66})m^2n^2 + Q_{22}n^4, \\
\bar{Q}_{44} &= (Q_{11} - Q_{12} - 2Q_{44})mn^3 + (Q_{12} - Q_{22} + Q_{44})m^3n, \\
\bar{Q}_{55} &= (Q_{11} - Q_{12} - 2Q_{55})m^3n + (Q_{12} - Q_{22} + Q_{55})mn^3, \\
\bar{Q}_{66} &= (Q_{11} + Q_{22} - 2Q_{12} - 2Q_{66})m^2n^2 + Q_{66}\left(m^4 + n^4\right),
\end{aligned} \tag{22}$$

where $m = \sin\theta$, $n = \cos\theta$, and $\theta$ is the laying angle, and

$$
\begin{aligned}
Q_{11} &= \frac{E_1}{1-\nu_{12}\nu_{21}},\\
Q_{12} &= \frac{\nu_{12}E_2}{1-\nu_{12}\nu_{21}},\\
Q_{21} &= \frac{\nu_{21}E_2}{1-\nu_{12}\nu_{21}},\\
Q_{22} &= \frac{E_2}{1-\nu_{12}\nu_{21}},\\
Q_{66} &= G_{12},\\
Q_{44} &= G_{23},\\
Q_{55} &= G_{13},
\end{aligned}
\tag{23}
$$

where $\nu_{12}$ is the primary Poisson's ratio of a laminated composite beam, $\nu_{21}$ is the secondary Poisson's ratio, $E_i$ is the elastic modulus, and $G_{ij}$ is the shear modulus of elasticity.

We introduce the stress $\underline{\sigma}$ and the strain $\underline{\varepsilon}$ of point $P$ on the $k$th layer of a laminated composite beam defined by

$$
\begin{aligned}
\underline{\sigma} &= \begin{bmatrix} \sigma_{xx} & \sigma_{yy} & \sigma_{zz} & \tau_{xy} & \tau_{yz} & \tau_{xz} \end{bmatrix}_k^{\mathrm{T}},\\
\underline{\varepsilon} &= \begin{bmatrix} \varepsilon_{xx} & \varepsilon_{yy} & \varepsilon_{zz} & \gamma_{xy} & \gamma_{yz} & \gamma_{xz} \end{bmatrix}_k^{\mathrm{T}},
\end{aligned}
\tag{24}
$$

Then, (21) can be rewritten as the matrix form

$$\underline{\sigma} = \mathbf{Q}\underline{\varepsilon}, \tag{25}$$

where $\mathbf{Q}$ is reduced stiffness matrix of laminated composite beam along the primary fiber direction, and

$$
\mathbf{Q} = \begin{bmatrix}
\bar{Q}_{11} & \bar{Q}_{12} & 0 & 0 & 0 & 0\\
\bar{Q}_{12} & \bar{Q}_{22} & 0 & 0 & 0 & 0\\
0 & 0 & 0 & 0 & 0 & 0\\
0 & 0 & 0 & \bar{Q}_{66} & 0 & 0\\
0 & 0 & 0 & 0 & \bar{Q}_{44} & 0\\
0 & 0 & 0 & 0 & 0 & \bar{Q}_{55}
\end{bmatrix}_k. \tag{26}
$$

*2.3. Rigid-Flexible Coupling Dynamics Modeling.* The coupling dynamics equations of a laminated composite beam undergoing a large-scale motion can be obtained by using the Hamilton variational principle. The Hamilton variational principle can be given by [23]

$$\int_{t_1}^{t_2} (\delta U - \delta T - \delta W)\mathrm{d}t = 0, \tag{27}$$

where $T$ is the kinetic energy of the laminated composite beam, $U$ is the potential energy of the laminated composite beam, and $W$ is the virtual work by external forces of the laminated composite beam.

*2.3.1. Kinetic Energy.* The kinetic energy of the laminated composite beam undergoing a large-scale motion can be written as

$$T = \frac{1}{2}\int_V \hat{\rho}\dot{r}^{\mathrm{T}}\dot{r}\mathrm{d}V, \tag{28}$$

where $\hat{\rho}$ is the mass density of the laminated composite beam, $V$ is the volume of the laminated composite beam, and $\dot{r}$ is the absolute velocity of any point of laminated composite beams.

According to (8) and (17), the velocity vector of point $P$ of the laminated beam in the inertial reference coordinate system $(O_0, e^{(0)})$ can be rewritten as

$$\dot{\mathbf{r}} = \begin{bmatrix} \mathbf{I} & \mathbf{C} & \mathbf{D\Phi} \end{bmatrix}\begin{bmatrix} \dot{\mathbf{r}}_0 \\ \dot{\mathbf{\Lambda}} \\ \dot{\underline{\mathbf{q}}} \end{bmatrix} = \mathbf{L}\dot{\mathbf{q}}, \tag{29}$$

where $\mathbf{C} = -2\mathbf{A}\tilde{\rho}\hat{\mathbf{G}}$ and $\mathbf{D}$ is reduced transformation matrix, $\mathbf{D} = \mathbf{A}\sum_{k=1}^{L}\mathbf{B}^{(k)}$ $(k = 1, 2, \ldots, L)$, $\mathbf{L} = \begin{bmatrix} \mathbf{I} & \mathbf{C} & \mathbf{D\Phi} \end{bmatrix}$, $\mathbf{q}$ is the generalized coordinate of the laminated composite beam, and $\mathbf{q} = \begin{bmatrix} \mathbf{r}_0^{\mathrm{T}} & \mathbf{\Lambda}^{\mathrm{T}} & \underline{\mathbf{q}}^{\mathrm{T}} \end{bmatrix}^{\mathrm{T}}$.

Substituting (29) into (28), the variational form of kinetic energy of the laminated composite beam can be given by

$$\delta T = \frac{1}{2}\delta\dot{\mathbf{q}}^{\mathrm{T}}\mathbf{M}\delta\dot{\mathbf{q}}, \tag{30}$$

where $\mathbf{M}$ is the mass matrix of the laminated composite beam defined by

$$\mathbf{M} = \begin{bmatrix} m_{rr} & m_{r\theta} & m_{rd} \\ m_{\theta r} & m_{\theta\theta} & m_{\theta d} \\ m_{dr} & m_{d\theta} & m_{dd} \end{bmatrix}, \tag{31}$$

in which

$$
\begin{aligned}
m_{rr} &= \int_v \hat{\rho}\mathbf{I}\mathrm{d}V,\\
m_{rd} &= m_{dr}^{\mathrm{T}} = \int_v \hat{\rho}\mathbf{D\Phi}\mathrm{d}V,\\
m_{\theta\theta} &= \int_v \hat{\rho}\mathbf{C}^{\mathrm{T}}\mathbf{C}\mathrm{d}V,\\
m_{\theta d} &= m_{d\theta}^{\mathrm{T}} = \int_v \hat{\rho}\mathbf{C}^{\mathrm{T}}\mathbf{D\Phi}\mathrm{d}V,\\
m_{dd} &= \int_v \hat{\rho}\mathbf{\Phi}^{\mathrm{T}}\mathbf{D}^{\mathrm{T}}\mathbf{D\Phi}\mathrm{d}V,\\
m_{r\theta} &= m_{\theta r}^{\mathrm{T}} = \int_v \hat{\rho}\mathbf{C}\mathrm{d}V.
\end{aligned}
\tag{32}
$$

*2.3.2. Potential Energy.* The potential energy of the laminated composite beam undergoing a large-scale motion can be written as

$$\delta \mathrm{U} = \frac{1}{2}\int_V \left(\underline{\sigma}^{\mathrm{T}}\delta\underline{\varepsilon}\right)\mathrm{d}V. \tag{33}$$

Further, we introduce a $6 \times 6$ order operator matrix $\underline{\widehat{\mathbf{D}}}$ defined by

$$\underline{\widehat{\mathbf{D}}} = \begin{bmatrix} \frac{\partial}{\partial x} & 0 & \frac{1}{2}\left(\frac{\partial}{\partial x}\right)^2 - c_1 z^3 \frac{\partial^2}{\partial x^2} & z\frac{\partial}{\partial x} - c_1 z^3 \frac{\partial}{\partial x} & 0 & 0 \\ 0 & \frac{\partial}{\partial y} & \frac{1}{2}\left(\frac{\partial}{\partial y}\right)^2 - c_1 z^3 \frac{\partial^2}{\partial y^2} & 0 & z\frac{\partial}{\partial y} - c_1 z^3 \frac{\partial}{\partial y} & 0 \\ 0 & 0 & 0 & 0 & 0 & 0 \\ \frac{\partial}{\partial y} & \frac{\partial}{\partial x} & \frac{\partial}{\partial x}\frac{\partial}{\partial y} - c_1 z^3 2\frac{\partial^2}{\partial x \partial y} & \left(z - c_1 z^3\right)\frac{\partial}{\partial y} & \left(z - c_1 z^3\right)\frac{\partial}{\partial x} & 0 \\ 0 & 0 & \left(1 - c_2 z^2\right)\frac{\partial}{\partial y} & 0 & \left(1 - c_2 z^2\right) & 0 \\ 0 & 0 & \left(1 - c_2 z^2\right)\frac{\partial}{\partial x} & \left(1 - c_2 z^2\right) & 0 & 0 \end{bmatrix}. \tag{34}$$

Then, substituting (3), (17), (20), and (25) into (33), the potential energy of the laminated composite beam can be rewritten as

$$\delta U = \frac{1}{2}\mathbf{q}^{\mathrm{T}}\mathbf{K}\delta\mathbf{q}, \tag{35}$$

where $\mathbf{K}$ is the stiffness matrix of the laminated composite beam defined by

$$\mathbf{K} = \begin{bmatrix} 0 & 0 & 0 \\ 0 & 0 & 0 \\ 0 & 0 & \mathbf{k}_{ff} \end{bmatrix}, \tag{36}$$

where $\mathbf{k}_{ff}$ is a coupling stiffness matrix and

$$\mathbf{k}_{ff} = \int_V \left(\underline{\widehat{\mathbf{B}}}^{\mathrm{T}}\underline{\widehat{\mathbf{D}}}^{\mathrm{T}}\mathbf{Q}^{\mathrm{T}}\underline{\widehat{\mathbf{D}}}\underline{\widehat{\mathbf{B}}}\right)\mathrm{d}V, \tag{37}$$

in which $\underline{\widehat{\mathbf{B}}} = \sum_{k=1}^{L}\mathbf{B}^{(k)}\boldsymbol{\Phi}$ $(k = 1, 2, \ldots, L)$. Consequently, $\mathbf{k}_{ff}$ is strongly determined by the constitutive relationships of laminated composite structure and a large rotation motion. In other words, as the beam structure is made of singular or unidirectional laminated materials, $\mathbf{k}_{ff}$ can be condensed statistically to a smaller sized matrix. Therefore, it accurately represents the nonlinear properties unlike other conventional stiffness matrix.

*2.3.3. Virtual Work by External Forces.* The virtual work $\delta W$ by external forces of the laminated composite beam with a large-scale motion can be written as

$$\delta W = \delta W_F + \delta W^*, \tag{38}$$

where $\delta W^*$ is the virtual work by inertial forces and $\delta W_F$ is the virtual work by all the driving and disturbing forces.

Let the external force $\mathbf{F}$ act on the laminated composite beam; the virtual work by external forces can be given by

$$\delta W_F = \mathbf{F}^{\mathrm{T}}\delta\mathbf{r} = \mathbf{F}^{\mathrm{T}}\begin{bmatrix}\mathbf{I} & \mathbf{C} & \mathbf{D}\boldsymbol{\Phi}\end{bmatrix}\begin{bmatrix}\delta\mathbf{r}_0 \\ \delta\boldsymbol{\Lambda} \\ \delta\underline{\mathbf{q}}\end{bmatrix} = \mathbf{F}_q\delta\mathbf{q}, \tag{39}$$

where $\mathbf{F}_q = \begin{bmatrix}\mathbf{F}_r & \mathbf{F}_\theta & \mathbf{F}_f\end{bmatrix}$, $\mathbf{F}_r = \mathbf{F}^{\mathrm{T}}$, $\mathbf{F}_\theta = \mathbf{F}^{\mathrm{T}}\mathbf{C}$, and $\mathbf{F}_f = \mathbf{F}^{\mathrm{T}} \cdot \mathbf{D}\boldsymbol{\Phi}$.

Given the generalized inertial force $\mathbf{F}^*$ acting on the laminated composite beam, the virtual work $\delta W^*$ by the inertial forces can be written as

$$\delta W^* = \mathbf{F}^*\delta\mathbf{q}, \tag{40}$$

where $\mathbf{F}^* = \begin{bmatrix}\mathbf{F}_r^* & \mathbf{F}_\theta^* & \mathbf{F}_f^*\end{bmatrix}$. Accordingly,

$$\begin{aligned} \mathbf{F}_r^* &= -\mathbf{A}\left[4\widehat{\mathbf{G}}\left(\mathbf{I} - \boldsymbol{\Lambda}\boldsymbol{\Lambda}^{\mathrm{T}}\right)\widehat{\mathbf{G}}\mathbf{S}_t + 4\widehat{\mathbf{G}}\dot{\widehat{\mathbf{G}}}^{\mathrm{T}}\mathbf{S}\underline{\mathbf{q}}\right], \\ \mathbf{F}_\theta^* &= -8\dot{\widehat{\mathbf{G}}}^{\mathrm{T}}\boldsymbol{\Gamma}_{\theta\theta}\widehat{\mathbf{G}}\boldsymbol{\Lambda} - 4\dot{\widehat{\mathbf{G}}}^{\mathrm{T}}\boldsymbol{\Gamma}_{\theta f}\underline{\mathbf{q}} - 2\widehat{\mathbf{G}}^{\mathrm{T}}\boldsymbol{\Gamma}_{\theta\theta}\widehat{\mathbf{G}}\boldsymbol{\Lambda}, \\ \mathbf{F}_f^* &= -\int_V \widehat{\boldsymbol{\rho}}\left\{\boldsymbol{\Phi}^{\mathrm{T}}\left[4\widehat{\mathbf{G}}\left(\mathbf{I} - \boldsymbol{\Lambda}\boldsymbol{\Lambda}^{\mathrm{T}}\right)\widehat{\mathbf{G}}\boldsymbol{\rho} + 4\widehat{\mathbf{G}}\dot{\widehat{\mathbf{G}}}^{\mathrm{T}}\underline{\mathbf{q}}\right]\right\}\mathrm{d}V, \end{aligned} \tag{41}$$

where

$$\begin{aligned} \mathbf{S}_t &= \int_V \widehat{\boldsymbol{\rho}}\boldsymbol{\rho}\,\mathrm{d}V, \\ \mathbf{S} &= \int_V \widehat{\boldsymbol{\rho}}\boldsymbol{\Phi}\,\mathrm{d}V, \\ \boldsymbol{\Gamma}_{\theta\theta} &= \int_V \widehat{\boldsymbol{\rho}}\tilde{\boldsymbol{\rho}}^{\mathrm{T}}\tilde{\boldsymbol{\rho}}\,\mathrm{d}V, \\ \boldsymbol{\Gamma}_{\theta f} &= \int_V \widehat{\boldsymbol{\rho}}\tilde{\boldsymbol{\rho}}\boldsymbol{\Phi}\,\mathrm{d}V. \end{aligned} \tag{42}$$

Here, the expressions of $\mathbf{F}^*$ is characterized by more conciseness and formalized for the dynamic model, and it can be easily applied to the derivation of generalized force for arbitrary

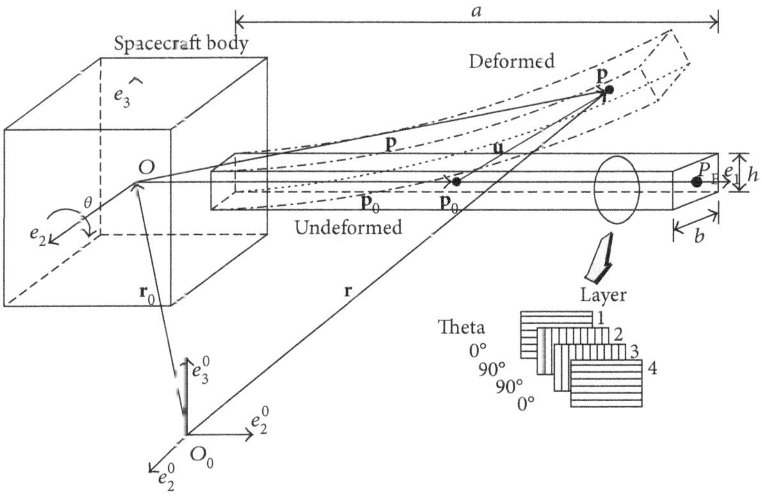


FIGURE 5: A spacecraft body with a laminated beam rotating around a fixed axis.

beam structures. Especially, in dealing with geometric nonlinear problems, the expressions can be used to avoid the derivation of large nonlinear stiffness matrix. Hence, it can significantly simplify the computational procedures.

Substituting (39) and (40) into (38), the virtual work $\delta W$ by external and inertial forces can be given by

$$\delta W = \mathbf{F}_q \delta \mathbf{q} + \mathbf{F}^* \delta \mathbf{q}. \quad (43)$$

*2.3.4. Coupling Dynamic Modeling of the Spacecraft System.* Substituting (30), (35), and (43) into (27), the coupling dynamic equation of the laminated composite beam with a large-scale motion can be written as

$$\begin{bmatrix} \mathbf{M} & \mathbf{K} & \mathbf{\Phi}_q^{\mathrm{T}} \\ \mathbf{\Phi}_q & 0 & 0 \end{bmatrix} \begin{bmatrix} \ddot{\mathbf{q}} \\ \mathbf{q} \\ \lambda \end{bmatrix} = \begin{bmatrix} \mathbf{F}_q + \mathbf{F}^* \\ \underline{\varsigma} \end{bmatrix}, \quad (44)$$

where $\lambda$ is the Lagrange multiplier, $\underline{\varsigma}$ is the right-side term column matrix of the acceleration form constrain equation, and $\mathbf{\Phi}_q$ is the Jacobi matrix of link hinge joints constrain equations of the spacecraft system.

## 3. Numerical Simulations

In this section, numerical simulations of a laminated composite beam rotating around the fixed axis are conducted as shown in Figure 5, including four laying layers and laying style (0°/90°/90°/0°). The size of the beam is $a \times b \times h = 1000.0\,\text{mm} \times 10.0\,\text{mm} \times 10.0\,\text{mm}$, and the thickness of each layer $h_k = 2.5\text{mm}$. Point $P_E$ is on the end of the laminated beam. The influences of gravity, air resistance force, and structural damping force have been neglected in this simulation.

The numerical examples can be governed and verified by a generally large-scale motion. The equation of angular displacement curve can be defined by

TABLE 1: Material characteristic of boron/aluminum composite [24].

| Parameters | Value | Note |
|---|---|---|
| $E_1$ (GPa) | 215.3 | Modulus of elasticity along fiber direction 1 |
| $E_2$ (GPa) | 144.1 | Modulus of elasticity perpendicular to fiber direction 2 |
| $E_3$ (GPa) | 144.1 | Modulus of elasticity perpendicular to fiber direction 3 |
| $G_{12}$ (GPa) | 54.39 | Shear modulus along fiber directions 1 and 2 |
| $G_{23}$ (GPa) | 54.39 | Shear modulus along fiber directions 2 and 3 |
| $G_{13}$ (GPa) | 45.92 | Shear modulus along fiber directions 1 and 3 |
| $\mu_{12}$ | 0.195 | Poisson's ratio along directions 1 and 2 |
| $\mu_{23}$ | 0.255 | Poisson's ratio along directions 2 and 3 |
| $\mu_{13}$ | 0.255 | Poisson's ratio along directions 1 and 3 |

$$\theta = \begin{cases} \dfrac{\Omega}{T}t - \dfrac{\Omega}{2\pi}\sin\left(\dfrac{2\pi}{T}t\right), & 0 \le t \le T, \\ \Omega, & t \ge T, \end{cases} \quad (45)$$

where $\Omega = 180°$ and $T = 20.0$ s.

Boron/aluminum composite materials are used in both the orthogonal-symmetric laminated beam $[0/90]_s$ and unidirectional laminated beam $[0_n]$. The material properties of the boron/aluminum composite are listed as follows in [24]. The density $\rho = 2653.0\,\text{kg/m}^3$; other material characteristic parameters are listed in Table 1. For the single material structure beam with boron aluminum alloy, the modulus of elasticity $E_1 = 215.3\,\text{GPa}$, the density $\rho = 2766.7\,\text{kg/m}^3$, and Poisson's ratio $\mu = 0.3$.

The initial parameters are the following: the displacement and velocity of the spacecraft body is zero, namely, $\mathbf{r}_0 = 0$ and $\dot{\mathbf{r}}|_{t=0} = 0$. The spacecraft body experiences a large rotation motion around a fixed axis $e_2$ in Figure 5, Euler parameter $\vartheta = 0$ and $\varphi = 0$, and the initial motion position $\psi|_{t=0} = 0$. Moreover, the initial velocity of the

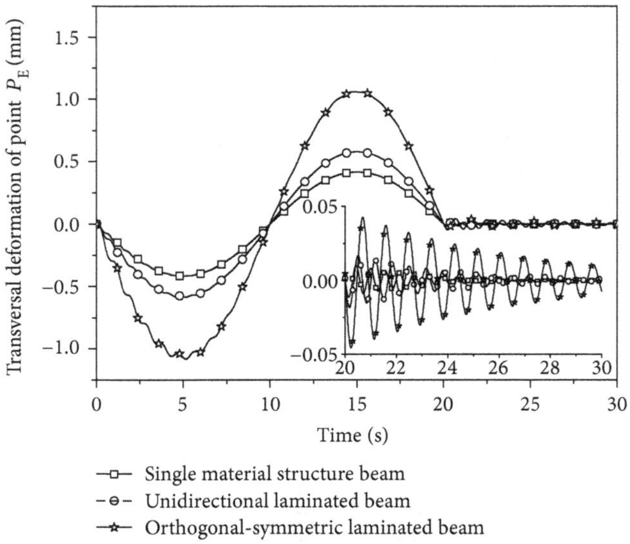


FIGURE 6: Transversal deformation of point $P_E$.

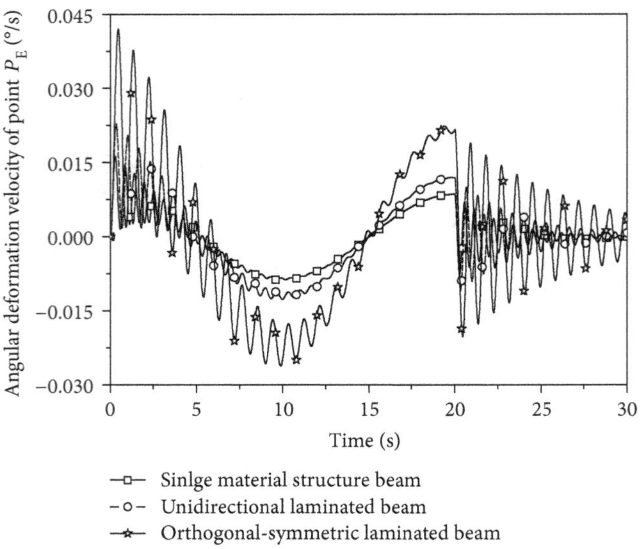


FIGURE 8: Angular deformation velocity of point $P_E$.

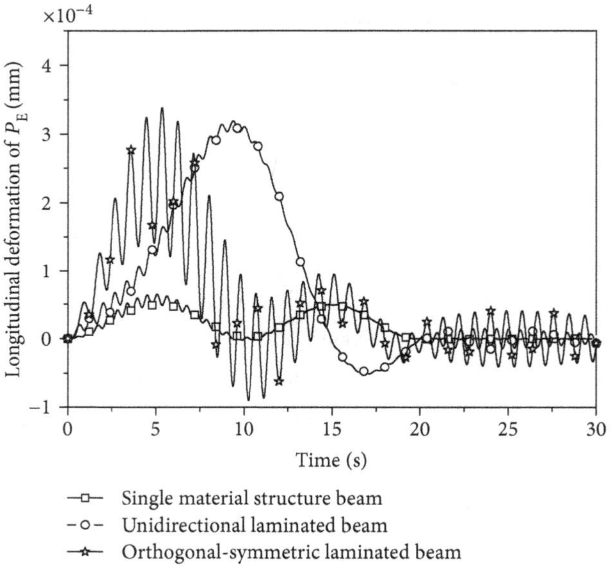


FIGURE 7: Longitudinal deformation of point $P_E$.

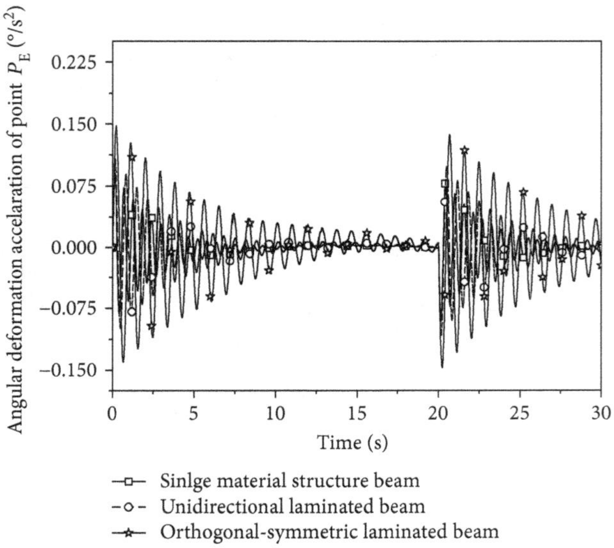


FIGURE 9: Angular deformation acceleration of point $P_E$.

flexible beam is assumed as zero without any external load disturbance during the motion.

To reveal the dynamics characteristics of the spacecraft beam with a large-scale motion as shown in Figure 5, the dynamic behavior of a spacecraft body composed of a rectangle beam with different material properties has been studied in this section. The transversal and longitudinal deformations of the end point $P_E$ in the singular material beam and laminated beam are shown in Figures 6 and 7, respectively. From the two figures, when $0 \le t \le 20.0$ s, the deformation displacement $\mathbf{u}$ of the laminated beam has greater fluctuations for its transversal and longitudinal components. Meanwhile, because the $\mathbf{Q}$ and $\hat{\underline{\mathbf{B}}} = \sum_{k=1}^{L} \mathbf{B}^{(k)} \mathbf{\Phi}$ are introduced in (37) to express the coupling stiffness matrix $\mathbf{k}_{ff}$ of the anisotropy of materials, the deformation of the laminated beam is larger than that of the isotropic beam. Consequently, the geometric nonlinear effect of laminated structures can be described more exactly. Besides, the unidirectional laminated beam and the orthogonal symmetric beam show greater difference because the coupling stiffness matrix $\mathbf{k}_{ff}$ can produce the differences caused by the laying angle and laying layers of the laminated beam. Likewise, when $t > 20.0$ s, the response of the three beams can converge to the same equilibrium position with low-amplitude decay oscillations, but the amplitude of singular material beam is smaller. Thus, it can be seen that modeling without considering the laminated composite structure will lead to greater errors and cannot exactly describe the spacecraft dynamics responses.

The angular deformation velocity and acceleration curves of point $P_E$ of a singular material and laminated beams are illustrated in Figures 8 and 9, respectively. From the two

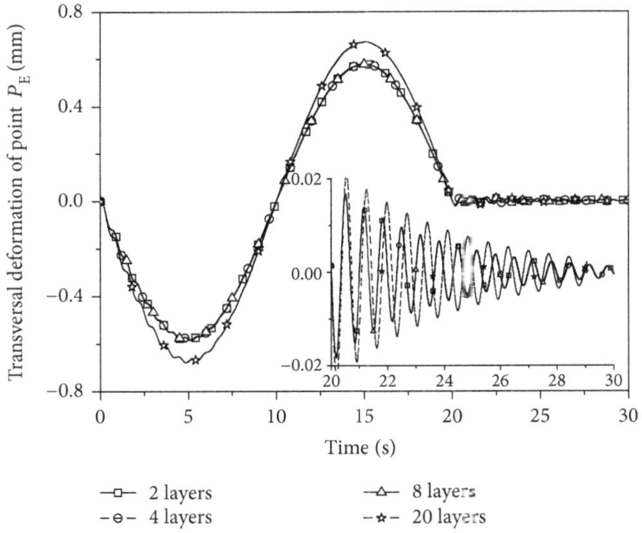


FIGURE 10: Transversal deformation of point $P_E$.

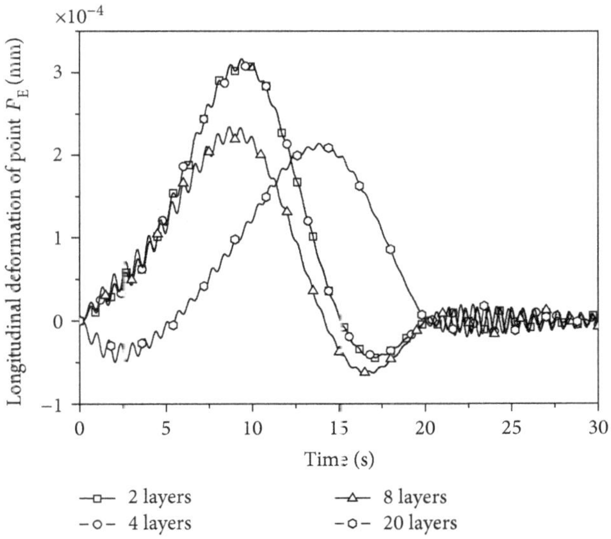


FIGURE 11: Longitudinal deformation of point $P_E$.

figures, the variation trend of those curves is consistent between the total motions of singular material and laminated beams. The velocity of the laminated beam fluctuates obviously, so obvious vibration is induced in the motions. Meanwhile, due to persistent disturbances by the nonlinear elastic force $\mathbf{F}_f^*$ in (41), the value of angular acceleration amplitude of the laminated beam is much larger than that of the singular material beam. Furthermore, the obvious disturbing force will be induced to increase the coupling dynamics effect. Thus, it can be seen that the differences of system dynamics responses between the singular material and laminated beams are much more obvious. For this reason, the geometric nonlinear effect of laminated structures must be considered in the study of the dynamic characteristics of the spacecraft system.

Furthermore, we study the influences of different laying layers of laminated structure on the system dynamics characteristics. The structural dimensions and material parameters of the beam are defined by Figure 5 and Table 1, respectively, and the numbers of layers of the laminated beam $L$ are 2, 4, 8, and 20, and laying angle of the unidirectional laminated beam $\theta$ is 0°.

Figures 10 and 11 show the transversal and longitudinal deformation of the unidirectional laminated beam with different layer thickness. From the two figures, when the number of laying layers $L \leq 8$ during the total motion of the laminated beam, the responses of transversal and longitudinal deformations of point $P_E$ are similar. When $L = 20$, the value of deformation of the laminated beam obviously become greater due to the coupling stiffness matrix $\mathbf{k}_{ff}$ considering the influence of laying layers $L$ in the matrix $\hat{\mathbf{B}}$. Thus, as the thickness of each layer of the laminated beam increases, the differences of deformations tend to become larger.

Figures 12 and 13 show the angular deformation velocity and acceleration of point $P_E$ with different laying layers of the laminated composite beam, respectively. From the two

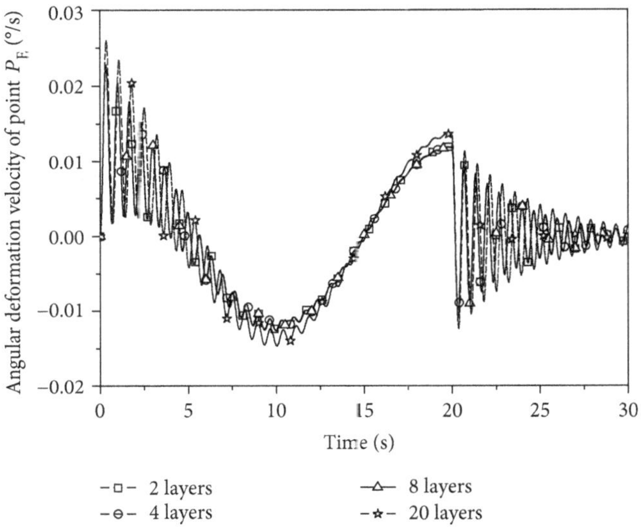


FIGURE 12: Angular deformation velocity of point $P_E$.

figures, in the laminated beam model, the difference between values of angular deformation velocity and acceleration is smaller as the layers $L$ increases. However, as a whole, the vibration amplitudes tend to increase as the thickness of laying layer decreases. For this reason, the influence of thickness of laying layer should be considered carefully when a laminated beam rotates with large rotations.

The transverse and longitudinal deformation of point $P_E$ with different laying angles (0°/45°/90°) are illustrated in Figures 14 and 15, respectively. From the two figures, for $\theta = 0°$, the elastic deformation varies between −2.0 mm and 2.0 mm, and for $\theta = 90°$, the elastic deformation varies between −8.0 mm and 8.0 mm. Accordingly, the elastic deformation of the laminated beam becomes larger as the laying angle increases. In general, the laying angles of laminated beams have significant influences on the dynamic properties of a laminated composite beam with a large-scale motion.

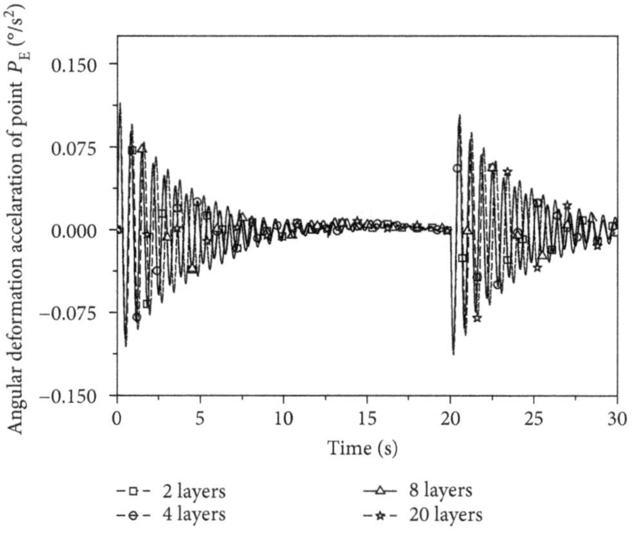


FIGURE 13: Angular deformation acceleration of point $P_E$.

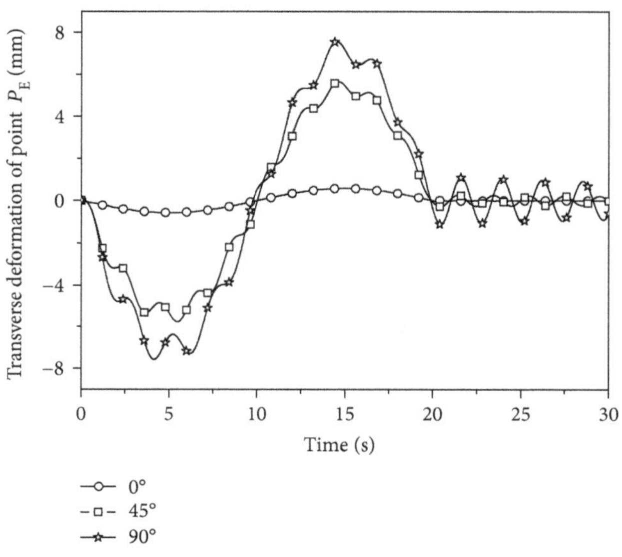


FIGURE 14: Transverse deformation of point $P_E$.

Through the above analysis, the dynamics characteristics of the rectangle beam considering laminated composite material structures and simplification of isotropic material (singular material) from the conventional equivalence have been verified preliminarily by using numerical methods. And, the complete expressions of the coupling stiffness matrix and the nonlinear elastic force are considered in the dynamic modeling of the spacecraft system. The results show that the laminated structure has significant influences on the exact calculation of the dynamic model. This also reveals the importance and correctness of considering laminated composite structures. Meanwhile, the influence of the laminated composite beam with various lamination parameters on the system dynamic behavior is further carefully considered in this section. The results also show that the number of layers and the laying angles have more significant influences on system dynamics properties.

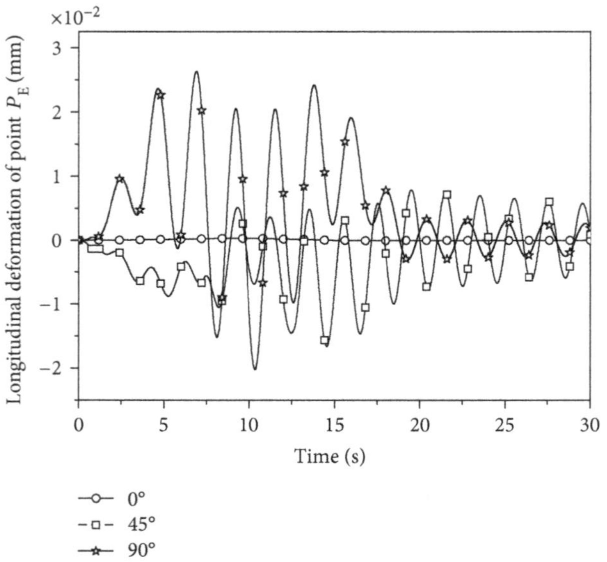


FIGURE 15: Longitudinal deformation of point $P_E$.

## 4. Conclusions

When the spacecraft with laminated beam appendages undergoes large-scale motions, such as attitude adjustment and orbital maneuver, the elastic deformation of laminated composite appendages can be induced, which can influence the dynamics properties of the spacecraft system. From these reasons, the rigid-flexible coupling dynamic model for a spacecraft body with laminated composite beam-shape appendages have been presented in this paper by considering constitutive relationships of anisotropic laminated structures. Accordingly, the important influences of the coupling stiffness matrix and the nonlinear elastic force are considered in this model. Furthermore, the numerical simulations of laminated composite beams with a large rotation motion are conducted by considering the influences of the laminated structures and equivalent laminated beams of a singular material, different laying angles, and layers on the dynamics properties of the spacecraft system. The numerical results indicate that characteristics of material properties, the number of layers, and laying angles cannot be ignored for the dynamic analysis of large-scale rigid-flexible multibody systems with laminated composite structures.

## Conflicts of Interest

The authors declare that they have no conflicts of interest.

## Acknowledgments

This material is based on Projects 51575126 and 51675118 supported by the National Natural Science Foundation of China and Project 2015T80358 supported by the China Postdoctoral Science Foundation.

## References


[1] C. M. C. Roque, D. S. Fidalgo, A. J. M. Ferreira, and J. N. Reddy, "A study of a microstructure-dependent composite laminated Timoshenko beam using a modified couple stress theory and a meshless method," *Composite Structures*, vol. 96, pp. 532–537, 2013.

[2] J. R. Vinson and R. L. Sierakowski, *The Behavior of Structures Composed of Composite Materials*, Springer, Netherlands, 1986.

[3] R. K. Kapania and S. Raciti, "Recent advances in analysis of laminated beams and plates. Part I - shear effects and buckling," *AIAA Journal*, vol. 27, no. 7, pp. 923-935, 1989.

[4] J. N. Reddy, "A general non-linear third-order theory of plates with moderate thickness," *International Journal of Non-Linear Mechanics*, vol. 25, no. 6, pp. 677–686, 1990.

[5] H. Zhang et al., "Optical and mechanical excitation thermography for impact response in basalt-carbon hybrid fiber-reinforced composite laminates," *IEEE Transactions on Industrial Informatics*, vol. PP, no. 99, p. 1, 2017.

[6] H. Zhang, L. Yu, U. Hassler et al., "An experimental and analytical study of micro-laser line thermography on micro-sized flaws in stitched carbon fiber reinforced polymer composites," *Composites Science and Technology*, vol. 126, pp. 17–26, 2016.

[7] A. L. Highsmith and K. L. Reifsnider, "Stiffness-reduction mechanisms in composite laminates," in *Damage in Composite Materials: Basic Mechanisms, Accumulation, Tolerance, and Characterization*, pp. 103–117, ASTM International, West Conshohocken, PA, USA, 1982, STP775.

[8] M. A. Neto, J. A. C. Ambrósio, and R. P. Leal, "Composite materials in flexible multibody systems," *Computer Methods in Applied Mechanics and Engineering*, vol. 195, no. 50-51, pp. 6860–6873, 2006.

[9] J. Li, S. Wang, X. Li, X. Kong, and W. Wu, "Modeling the coupled bending–torsional vibrations of symmetric laminated composite beams," *Archive of Applied Mechanics*, vol. 85, no. 7, pp. 991–1007, 2015.

[10] L. Tian and J. Liu, "An improved dynamic model for hub and laminated composite plate system considering warping effect," *Mechanics Research Communications*, vol. 74, pp. 14–19, 2016.

[11] G. B. Chai and C. W. Yap, "Coupling effects in bending, buckling and free vibration of generally laminated composite beams," *Composites Science and Technology*, vol. 68, no. 7-8, pp. 1664–1670, 2008.

[12] X. S. He, F. Y. Deng, G. Y. Wu, and R. Wang, "Dynamic modeling of a flexible beam with large overall motion and nonlinear deformation using the finite element method," *Acta Physica Sinica*, vol. 59, no. 1, pp. 25–29, 2010.

[13] S. Abrate, "Impact on composite structures," Cambridge University Press, Cambridge, England, 1998.

[14] S. Sfarra, F. López, F. Sarasini et al., "Analysis of damage in hybrid composites subjected to ballistic impacts: an integrated non-destructive approach," in *Handbook of Composites from Renewable Materials*, V. K. Thakur, M. K. Thakur, and M. R. Kessler, Eds., pp. 175–210, John Wiley & Sons, Inc., Hoboken, NJ, USA, 1st edition, 2017, Chapter: 8.

[15] S. B. Wu and D. G. Zhang, "Rigid-flexible coupling dynamic analysis of hub-flexible beam with large overall motion," *Journal of Vibration Engineering*, vol. 24, no. 1, pp. 1–7, 2011.

[16] X. S. Zhang, D. G. Zhang, and J. Z. Hong, "Rigid-flexible coupling dynamic modeling and simulation with the longitudinal deformation induced curvature effect for a rotating flexible beam under large deformation," *Chinese Journal of Theoretical and Applied Mechanics*, vol. 48, no. 3, pp. 692–701, 2016.

[17] G. Y. Wu, X. S. He, and F. Y. Deng, "Dynamic analysis of a rotating composite plate," *Journal of Vibration and Shock*, vol. 27, no. 8, pp. 149–154, 2008.

[18] K. Q. Pan and J. Y. Liu, "Rigid-flexible coupling dynamics of composite beam considering shear deformation," *Journal of Shanghai Jiaotong University*, vol. 43, no. 8, pp. 1293–1297, 2009.

[19] M. Karama, K. S. Afaq, and S. Mistou, "Mechanical behaviour of laminated composite beam by the new multi-layered laminated composite structures model with transverse shear stress continuity," *International Journal of Solids and Structures*, vol. 40, no. 6, pp. 1525–1546, 2003.

[20] H. H. Yoo, S. H. Lee, and S. H. Shin, "Flapwise bending vibration analysis of rotating multi-layered composite beams," *Journal of Sound and Vibration*, vol. 286, no. 4-5, pp. 745–761, 2005.

[21] A. Nosier and J. N. Reddy, "A study of non-linear dynamic equations of higher-order shear deformation plate theories," *International Journal of Non-Linear Mechanics*, vol. 26, no. 2, pp. 233–249, 1991.

[22] M. Xu and G. Q. Xu, "Finite element analysis of a laminated beam with coupled bending-torsional deflections," *Journal of Nanjing University of Aeronautics & Astronautics*, vol. 19, no. 3, pp. 35–44, 1987.

[23] J. N. Reddy, *Mechanics of Laminated Composite Plates and Shells: Theory and Analysis*, CRC Press, Boca Raton, FL, USA, 2006.

[24] C. T. Sun and R. S. Vaidya, "Prediction of composite properties from a representative volume element," *Composites Science and Technology*, vol. 56, no. 2, pp. 171–179, 1996.

13

# Pose and Shape Reconstruction of a Noncooperative Spacecraft Using Camera and Range Measurements

**Renato Volpe,[1] Marco Sabatini,[2] and Giovanni B. Palmerini[3]**

[1] *Department of Mechanical and Aerospace Engineering, University of Rome La Sapienza, Via Eudossiana 18, Rome, Italy*
[2] *Department of Astronautics, Electrical and Energetics Engineering, University of Rome La Sapienza, Via Salaria 851, Rome, Italy*
[3] *Scuola di Ingegneria Aerospaziale, University of Rome La Sapienza, Via Salaria 851, Rome, Italy*

Correspondence should be addressed to Giovanni B. Palmerini; giovanni.palmerini@uniroma1.it

Academic Editor: Enrico C. Lorenzini

Recent interest in on-orbit proximity operations has pushed towards the development of autonomous GNC strategies. In this sense, optical navigation enables a wide variety of possibilities as it can provide information not only about the kinematic state but also about the shape of the observed object. Various mission architectures have been either tested in space or studied on Earth. The present study deals with on-orbit relative pose and shape estimation with the use of a monocular camera and a distance sensor. The goal is to develop a filter which estimates an observed satellite's relative position, velocity, attitude, and angular velocity, along with its shape, with the measurements obtained by a camera and a distance sensor mounted on board a chaser which is on a relative trajectory around the target. The filter's efficiency is proved with a simulation on a virtual target object. The results of the simulation, even though relevant to a simplified scenario, show that the estimation process is successful and can be considered a promising strategy for a correct and safe docking maneuver.

## 1. Introduction

Autonomous rendezvous and docking is a research field which has been extensively studied in the last years. In particular, high interest has been put in visual navigation and pose estimation; various mission architectures have been analyzed, all of which present a different way to estimate the relative pose between a chaser, equipped with some optical device, and a target.

The ATV docking mission to the ISS is an example of a real application involving visual navigation techniques using cameras [1–3]. In particular, the ISS is equipped with optical targets of known shape, which make it easier to estimate the relative attitude between them and the ATV. In other words, the ATV docking mission relies on the cooperativeness of the target spacecraft.

Lately, growing interest in orbit servicing and debris removal switched the focus to autonomous docking to an object that is potentially uncontrolled, free-tumbling, noncooperative, and possibly unknown [4]. An example of a real application in this sense is the PRISMA mission [5, 6], whose focus was to demonstrate the feasibility of autonomous, partially noncooperative in-orbit rendezvous with the use of GPS measurements and a radio frequency sensor, switching to a vision-based control at short range. While the far approach phase can be considered as fully noncooperative, in the short range approach phase, the close range camera of the chaser detects the signal of light emitting diodes located on Tango satellite [7].

Optical navigation has been also used in ESA's Rosetta mission, where the target is a celestial object, the comet 67P/Churyumov-Gerasimenko [8–10]. Navigation cameras in the far approach phase were used to identify significant landmarks on the comet, which were then tracked and used as reference points for the relative position and velocity estimation process. Moreover, images of the comet taken at close range were used to build a 3D reconstruction. However, the landmark identification and estimation process was made by ground operators, thus making the mission not autonomous.

Many studies in open literature have been devoted to prepare and extend these still partial real applications. Longuet-Higgins [11] was one of the first to develop an algorithm for reconstructing a scene by means of stereo vision in 1981. Weng et al. [12, 13] studied the reconstruction of the structure and the motion of a target body by means of the so-called *two-view motion algorithm*, which estimates the rigid rotation between two consecutive images. Johnson and Mathies [14] applied the two-view motion algorithm to a landing maneuver on a small body and made use of an altimeter to compute the scale variation between two images during the descent. Vetrisano et al. [15] studied the use of cameras and intersatellite links between a swarm of spacecraft during the rendezvous to an asteroid, to estimate both the relative state and the relative angular velocity of the body. Oumer and Panin [16] presented a camera-based 3D feature tracking method used to estimate the position and velocities of an observed rigid body without reconstructing its shape.

In such a research frame, this paper aims to present a vision-based feature tracking method to estimate 3D shape, relative rotation, and translation of a free-tumbling, noncooperative, and unknown satellite orbiting Earth. This task will be achieved by using a monocular camera and a distance sensor (such as a LIDAR), with no prior knowledge either of the relative pose or of the target's shape. Such a target is clearly more difficult to analyze with respect to the asteroid dealt in [14, 15], as its dynamics can be far more complex and fast. The addition of the target's shape reconstruction, not tackled in [16], is significant with respect to inspection tasks and to the need to evaluate target's inertia characteristics and possible locations for a possible grasp or dock. From the academic perspective, one of the major contributions brought by this paper is its generality and the possibility of being considered as a valid navigation technique in different scenarios. In fact, no prior information regarding target's shape and attitude is required, since the shape itself is reconstructed after the estimate process. At the same time, no restrictions are required either on the magnitude or on the orientation of the relative attitude motion for the correct convergence of the estimation process. Most importantly, in this paper we propose a technique which makes use of a passive optical sensor (a monocular camera) which is more reliable and less power consuming than time-of-flight sensors [17–20]. The specific contribution of this paper stays also with the detailed description, including the selected dynamics, of a process that, by means of a judicious selection of the reference frames adopted, leads to a quite performing technique. Presented tests confirm the effectiveness and the quality of the estimates of the pose and motion of an unknown target. The results of the research are a first step for an application to autonomous rendezvous and docking.

The paper is organized as follows. The mission scenarios, along with the issues involved with the feature tracking and matching process, are described in Section 2. In Section 3, the Hill, camera, and body-fixed reference frames are defined. The relative and attitude dynamics and kinematics model, along with the definition of every angular velocity, are presented in Section 4. In Section 5, the relation between 2D features and 3D body points is described. The filter's structure is shown in Section 6, while in Section 7 the data used in the simulation are listed. The results of the simulation are discussed in Section 8 and final comments are made in Section 9.

## 2. Mission Scenario

In the following simulations, we assume that two satellites, a chaser and a target, are orbiting Earth. The chaser, a controlled satellite whose attitude with respect to an inertial reference frame is known from its attitude determination system, is equipped with an optical camera and a distance sensor at the scope of performing the relative navigation; the target is a free-tumbling, uncontrolled satellite, whose shape is initially unknown. We assume that the chaser's attitude control system maintains the optical axis of the camera pointing towards the target.

The preliminary steps focus on navigation. In particular, estimating target's shape, relative position, and attitude is of crucial importance to ensure that no collisions between chaser and target are taking place and indeed a correct and safe docking is achieved.

The target's shape and relative state can be estimated by tracking significant features through subsequent frames captured by the camera. A set of features extracted from an image at time $t_k$ by means of a feature extraction algorithm, such as SURF *(Speeded-Up Robust Features)* [21], is compared to a set of features extracted from the following frame at time $t_{k+1}$ in order to find the correspondences. The identified set of matched features is then used as observables for the filter's update process. An example of the feature's extraction and matching process is shown in Figures 1(a)–1(c) where a 3D CAD model of the satellite AQUA [22] has been used as a test. As can be seen, the satellite has accomplished an in-plane rotation between the two frames. Note that in a more generic situation the satellite may rotate out of the image plane, and features may disappear from the camera field of view due to either target's body occlusion or changes in sunlight illumination conditions. Other events that can happen as a result of the matching process are false matches, which occur also in the example presented (Figure 1(c)) for better understanding. These issues and others have to be faced if a robust estimation process must be achieved.

Therefore, given the scenario, the aim is to build a filter which merges the measurements obtained by the camera (i.e., the features detected, tracked, and matched) and by the distance sensor (i.e., the target-to-chaser distance measurements) and combines them with the state prediction given by a dynamics model to estimate in real time the target's relative position, velocity, and attitude and the target's shape in a body-fixed reference frame.

## 3. Reference Frames

*3.1. Hill Frame.* The reference frame in which the relative orbital dynamics is expressed is the *Hill Reference Frame* (HRF), which is centred in the target's centre of mass and is made up by the following axes:

(1) $\hat{\mathbf{x}}^H$ axis is parallel to target's orbital radius, $\mathbf{r}_0(t)$.

(2) $\hat{\mathbf{z}}^H$ axis is parallel to target's orbital angular momentum, $\mathbf{h}_0(t)$.

(3) $\hat{\mathbf{y}}^H$ axis forms an orthonormal, right-handed frame.

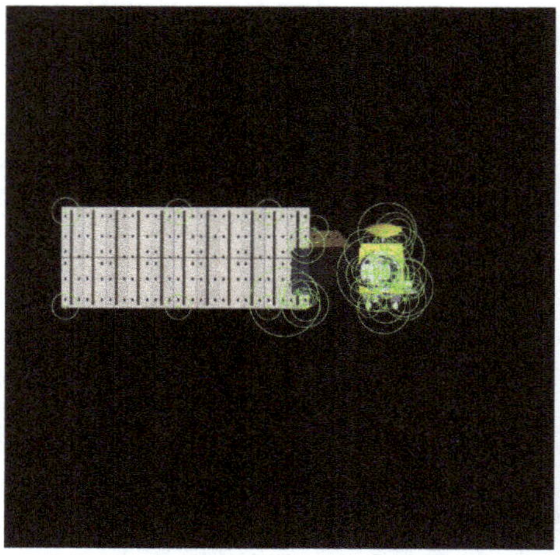

(a) Extracted features at time $t_k$

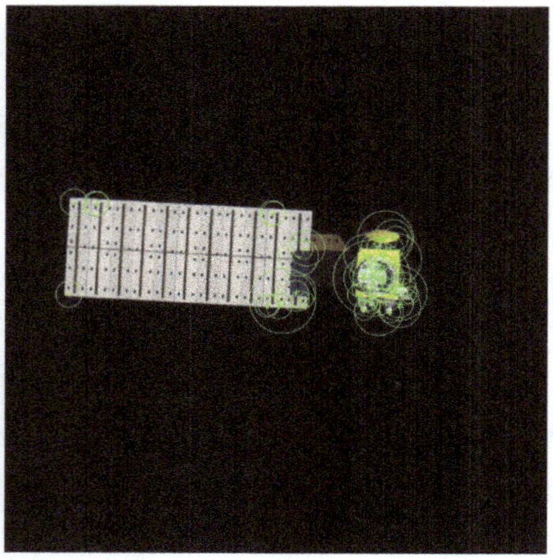

(b) Extracted features at time $t_{k+1}$

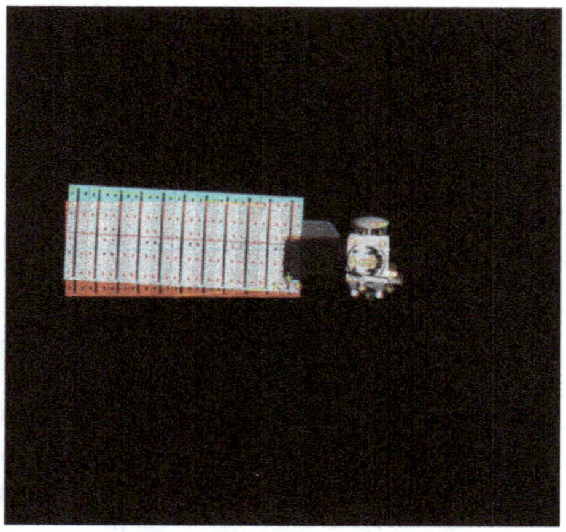

(c) Matched features between features at $t_k$ (red circles) and at $t_{k+1}$ (green crosses)

FIGURE 1: Features' extraction and matching process between two subsequent frames.

The matrix describing the transformation between HRF and ECI *(Earth Centred Inertial)* is a function of target's position and velocity expressed in ECI

$$ {}^{\mathrm{HRF}}\mathscr{R}_{\mathrm{ECI}} = \begin{bmatrix} \hat{\mathbf{r}}_0 \\ \hat{\mathbf{h}}_0 \times \hat{\mathbf{r}}_0 \\ \hat{\mathbf{h}}_0 \end{bmatrix}. \tag{1} $$

*3.2. Camera Reference Frame.* The rotation between chaser and target has to be evaluated with respect to a reference frame in which the measurements are taken. The feature's pixels are 2D projection of 3D points whose coordinates are expressed in the *Camera Reference Frame* (CRF). In the present work, we assume that the origin of CRF, which is the camera's focal point, coincides with the chaser's centre of mass. This hypothesis, which does not affect the validity of the results, is made in order to avoid defining another reference frame parallel to CRF and differing from it only for a rigid translation. The axes are displaced as follows:

(1) $\hat{\mathbf{z}}^C$ is parallel to the *optical axis*. As said in Section 2, we assume to know and control chaser's attitude to our needs. This means that, in order to have continuous and full visibility of the target to acquire significant images, it is needed to make the optical axis point towards the target's centre of mass. Therefore, its direction will be opposite to the relative position vector $\boldsymbol{\rho} = [x, y, z]^T$.

(2) $\hat{\mathbf{x}}^C$ and $\hat{\mathbf{y}}^C$ form a plane perpendicular to $\boldsymbol{\rho}$ but are not bound to any direction since a rotation about $\hat{\mathbf{z}}^C$ is irrelevant to the purpose of framing the target.

The rotation used to switch between HRF and CRF is a 3-1-0, meaning that the first two rotations are needed to bring $\hat{\mathbf{z}}^C$ parallel and opposite to $\boldsymbol{\rho}$, but the third rotation is null ($\theta = 0$), as there is no constraint on $\hat{\mathbf{x}}^C$ and $\hat{\mathbf{y}}^C$ directions. The correspondent rotation matrix is an identity matrix

$$ {}^{\mathrm{CRF}}\mathscr{R}_{\mathrm{HRF}} = R_1(\varphi) R_3(\psi) \tag{2} $$

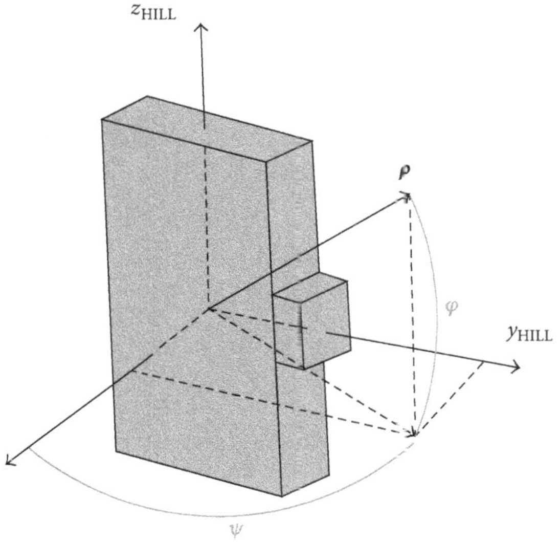


FIGURE 2: Geometrical representation of the azimuth and elevation angles definition. In black: the Hill Reference Frame. In grey: the target's body, centred in the origin of the HRF. In red: the $\boldsymbol{\rho}$ vector, pointing from the target's centre of mass towards the chaser's centre of mass. In green: the azimuth angle $\psi$ and the elevation angle $\phi$.

with

$$R_3 = \begin{bmatrix} \cos(\psi - 90^\circ) & \sin(\psi - 90^\circ) & 0 \\ -\sin(\psi - 90^\circ) & \cos(\psi - 90^\circ) & 0 \\ 0 & 0 & 1 \end{bmatrix} \tag{3a}$$

$$R_1 = \begin{bmatrix} 1 & 0 & 0 \\ 0 & \cos(\varphi + 90^\circ) & \sin(\varphi + 90^\circ) \\ 0 & -\sin(\varphi + 90^\circ) & \cos(\varphi + 90^\circ) \end{bmatrix}. \tag{3b}$$

The angles $\varphi$ and $\psi$, which are, respectively, the azimuth and elevation angles of the vector $\boldsymbol{\rho}$ with respect to HRF, are obtained as follows (see Figure 2 for a geometrical representation of the angles):

$$\varphi = \arcsin\left(\frac{z}{\|\rho\|}\right) \tag{4a}$$

$$\psi = \arctan\left(\frac{y}{x}\right). \tag{4b}$$

*3.3. Target Body-Fixed Reference Frame.* One of the goals of the filter is to evaluate the attitude of the body relatively to a known reference frame. Because the measurements are taken in CRF, which is a chaser-fixed reference frame, it would be worth expressing the attitude of the target with respect to the chaser, that is, relatively to CRF. As the target body reference frame is concerned, a meaningful choice has to be made.

The *Body Principal Reference Frame* (BPRF) is defined as the frame in which the inertia matrix is diagonal. It is centred in the target's centre of mass; the axes are body-fixed and parallel to the target's principal axes. The attitude dynamics equations assume a simple form when written in the BPRF, so it seems the most convenient frame to choose. However, it is not possible to evaluate the attitude of BPRF with respect to CRF, as there is no information about the direction of the target's principal axes at any time.

In order to give an evaluation of the target's attitude with respect to a known frame, it is therefore necessary to introduce a *Body Dummy Reference Frame* (BDRF) that is a reference frame whose origin and axes' directions are arbitrarily chosen by us at the beginning of the estimation process. We fix the origin in the target's centre of mass for convenience. The most clever choice is to assume BDRF being parallel to CRF at $t_0$, that is, the attitude matrix ${}^{\text{BDRF}}\mathscr{C}_{\text{CRF}}$ being an identity matrix at $T_0$

$${}^{\text{BDRF}}\mathscr{C}_{\text{CRF}}(t_0) = I_{3\times 3}. \tag{5}$$

## 4. Dynamics Model

*4.1. Relative Translation.* The relative orbital dynamics is described by the second-order, inertial derivative of the relative position vector $\boldsymbol{\rho}$, whose coordinates are expressed in HRF

$$\left.\frac{d^2\rho^{\text{HRF}}}{dt^2}\right|_{\text{ECI}} = \ddot{\boldsymbol{\rho}} + 2\boldsymbol{\omega}_{H/E} \times \dot{\boldsymbol{\rho}} + \dot{\boldsymbol{\omega}}_{H/E} \times \boldsymbol{\rho} + \boldsymbol{\omega}_{H/E} \times (\boldsymbol{\omega}_{H/E} \times \boldsymbol{\rho}) = \frac{\mu}{\|\mathbf{r}_0\|^2}\widehat{\mathbf{r}}_0 - \frac{\mu}{\|\mathbf{r}_1\|^2}\widehat{\mathbf{r}}_1, \tag{6}$$

where

(i) $\boldsymbol{\omega}_{H/E}$ denotes the angular velocity of HRF with respect to ECI, that is, the target's orbital angular velocity;

(ii) $\mathbf{r}_0$ and $\mathbf{r}_1$ are the target's and chaser's position vectors with respect to Earth's centre in HRF, respectively.

For the sake of simplicity, since the relative distances involved are small compared to the orbital radius, we use the linearisation of (6), which are the well-known *Clohessy-Wiltshire* equations [23], largely adopted in space proximity operations

$$\ddot{x} = 3n^2 x + 2n\dot{y} \tag{7a}$$

$$\ddot{y} = -2n\dot{x} \tag{7b}$$

$$\ddot{z} = -n^2 z, \tag{7c}$$

where $n = \sqrt{\mu/a^3}$ is the target's mean motion.

*4.2. Relative Attitude.* For the purpose of the present research, we want to investigate how the target rotates with respect to the chaser. This means we want to evaluate target's angular velocity between BDRF and CRF, $\boldsymbol{\omega}_{B/C}$.

As said in Sections 2 and 3.2, we assume the chaser's attitude varies in a known way; that is, the CRF-to-HRF angular velocity $\boldsymbol{\omega}_{C/H}$ is known. It can be written as a function of Euler angles' derivatives, which are computed as derivatives of (4a) and (4b) and therefore depend on the relative position and velocity vectors $\boldsymbol{\rho}(t)$ and $\dot{\boldsymbol{\rho}}(t)$

$$\boldsymbol{\omega}_{C/H}^{\mathrm{CRF}} = R_1 R_3 \begin{bmatrix} 0 \\ 0 \\ \dot{\psi} \end{bmatrix} + R_3 \begin{bmatrix} \dot{\varphi} \\ 0 \\ 0 \end{bmatrix} \tag{8}$$

Because the target's orbit is known, we can use its orbital parameters to compute the orbital angular velocity vector, which is the angular velocity between HRF and ECI $\boldsymbol{\omega}_{H/E}$

$$\omega_{H/E} = \frac{\mathbf{h}}{r_0^2} \tag{9}$$

The attitude dynamics equations describe the rotation of BDRF with respect to ECI $\boldsymbol{\omega}_{B/E}$, but the angular velocity is expressed in BDRF so that the inertia matrix $\mathcal{J}$ is constant over time. Besides the target's free dynamics, the effect of the gravity gradient is taken into account

$$\dot{\boldsymbol{\omega}}_{B/E}^{\mathrm{BDRF}} = \mathcal{J}^{-1} \left( -\boldsymbol{\omega}_{B/E} \times \mathcal{J} \boldsymbol{\omega}_{B/E} + 3 \frac{\mu}{\|\mathbf{r}_0\|^3} \widehat{\mathbf{o}} \times \mathcal{J} \widehat{\mathbf{o}} \right) \tag{10}$$

where $\widehat{\mathbf{o}} = -\widehat{\mathbf{r}}_0^{\mathrm{BDRF}}$ and $\mathcal{J}$ is not diagonal because BDRF is not a principal reference frame.

Finally, we can obtain the BDRF-to-CRF angular velocity vector by using the following:

$$\boldsymbol{\omega}_{B/C} = \boldsymbol{\omega}_{B/E} + \boldsymbol{\omega}_{E/H} + \boldsymbol{\omega}_{H/C} = \boldsymbol{\omega}_{B/E} - \boldsymbol{\omega}_{H/E} - \boldsymbol{\omega}_{C/H}. \tag{11}$$

We make use of the quaternions set to describe the attitude of the body frame with respect to the camera frame. The kinematics equations describing the quaternions' variation over time is written as follows:

$$\dot{\mathbf{q}}_{B/C} = \frac{1}{2} \Omega\, \mathbf{q}_{B/C}, \tag{12}$$

where $\Omega$ is a matrix depending on the BDRF-to-CRF angular velocity vector expressed in the body reference frame

$$\Omega = \begin{bmatrix} 0 & -\omega_x & -\omega_y & -\omega_z \\ \omega_x & 0 & \omega_z & -\omega_y \\ \omega_y & -\omega_z & 0 & \omega_x \\ \omega_z & \omega_y & -\omega_x & 0 \end{bmatrix} \tag{13}$$

$$\begin{bmatrix} \omega_x \\ \omega_y \\ \omega_z \end{bmatrix} = \boldsymbol{\omega}_{B/C}^{\mathrm{BDRF}}.$$

## 5. Features Identification

Each feature tracked by the extraction and matching algorithm corresponds to a 3D point of the target's body. We therefore can link every matched 2D feature to a precise 3D point whose coordinates in a certain reference frame have to be estimated in order to accomplish the shape reconstruction problem. The 3D position of the $i$-th point can be identified in BDRF by means of the vector from the target's centre of mass to the point $\mathbf{d}_i^{\mathrm{BDRF}}$, which is constant over time as the body is considered rigid.

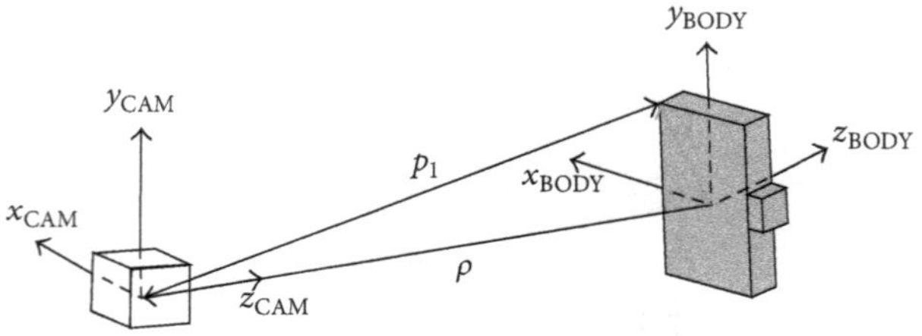


Figure 3: Picture representing the chaser pointing with its optical axes $\widehat{\mathbf{z}}^C$ directed towards the target's centre of mass. The triangle made up by vectors $\boldsymbol{\rho}$ (red), $\mathbf{p}_1$ (orange), and $\mathbf{d}_1$ (green) is shown for the generic point 1 to give a clear explanation of (15).

On the other hand, the same point can be identified in CRF. We define the vector $\mathbf{p}_i^{\mathrm{CRF}}$ which goes from the chaser's centre of mass to the $i$-th point, whose coordinates are expressed in CRF.

These two representations of the $i$-th tracked point can be linked with each other by means of the chaser-to-target relative position vector $\boldsymbol{\rho}$, as can be seen in Figure 3

$$\mathbf{p}^{\mathrm{CRF}} = \mathbf{d}^{\mathrm{CRF}} - \boldsymbol{\rho}^{\mathrm{CRF}}. \tag{14}$$

or equivalently

$$\mathbf{p}_i^{\mathrm{CRF}} = {}^{\mathrm{BDRF}}\mathscr{C}_{\mathrm{CRF}}^{T}\, \mathbf{d}_i^{\mathrm{BDRF}} - {}^{\mathrm{CRF}}\mathscr{R}_{\mathrm{HRF}}\, \boldsymbol{\rho}^{\mathrm{HRF}} \tag{15}$$

The 2D coordinates $(u_i, v_i)$ of the $i$-th feature on the image plane are related to their respective $i$-th point's 3D coordinates by the prospective camera model equations [24]

$$u_i = p_{i,x}^{\mathrm{CRF}} \frac{f}{p_{i,z}^{\mathrm{CRF}}} \tag{16a}$$

$$v_i = p_{i,y}^{\mathrm{CRF}} \frac{f}{p_{i,z}^{\mathrm{CRF}}}, \tag{16b}$$

where $f$ is the focal length of the camera.

## 6. Filter's Structure

The *Unscented Kalman Filter* (UKF) [25] is a recursive estimator built on the unscented transformation and based on the idea that the evolution of a (noisy) process could be known through the associated statistical distribution. The propagation of the statistical distribution is carried out by means of a set of purposely selected points (the so-called sigma points). In this way, the propagation of the state and of the statistical indicators is based on the integration of the dynamics instead of involving the Jacobian of the dynamics and of the measurements equations (as in Extended Kalman Filter), meaning that UKF is especially convenient for complex nonlinear dynamics.

For the purpose of the present study, the state vector has to include parameters regarding target's shape, rotational state, and relative position and velocity. As a consequence, the state vector is defined as

$$\mathbf{x} = \begin{bmatrix} \mathbf{p}_i^{\mathrm{CRF}} \\ \mathbf{d}_i^{\mathrm{BDRF}} \\ \mathbf{q}_{B/C} \\ \boldsymbol{\omega}_{B/E}^{\mathrm{BDRF}} \\ \boldsymbol{\rho}^{\mathrm{HRF}} \\ \dot{\boldsymbol{\rho}}^{\mathrm{HRF}} \\ \mathcal{J} \end{bmatrix} \quad \forall i = 1, \ldots, n_p, \tag{17}$$

where $n_p$ is the number of points tracked; therefore, the state vector's length is $6n_p + 19$.

The dynamics equations are a system of $6n_p + 19$ differential equations

$$\dot{\mathbf{p}}_i^{\mathrm{CRF}} = -\boldsymbol{\omega}_{C/E}^{\mathrm{CRF}} \times \mathbf{p}^{\mathrm{CRF}} + \boldsymbol{\omega}_{B/E}^{\mathrm{CRF}} \times \mathbf{d}_i^{\mathrm{CRF}} - \dot{\boldsymbol{\rho}}^{\mathrm{CRF}} - \boldsymbol{\omega}_{H/E}^{\mathrm{CRF}} \times \boldsymbol{\rho}^{\mathrm{CRF}} \tag{18a}$$

$$\dot{\mathbf{d}}_i^{\mathrm{BDRF}} = \mathbf{0} \tag{18b}$$

$$\dot{\mathbf{q}}_{B/C} = \frac{1}{2}\Omega\, \mathbf{q}_{B/C} \tag{18c}$$

$$\dot{\boldsymbol{\omega}}_{B/E}^{\mathrm{BDRF}} = \mathcal{J}^{-1}\left(-\boldsymbol{\omega}_{B/E} \times \mathcal{J}\boldsymbol{\omega}_{B/E} + 3\frac{u}{\|\mathbf{r}_0\|^3}\hat{\mathbf{o}} \times \mathcal{J}\hat{\mathbf{o}}\right) \tag{18d}$$

$$\ddot{x} = 3n^2 x + 2n\dot{y} \tag{18e}$$

$$\ddot{y} = -2n\dot{x} \tag{18f}$$

$$\ddot{z} = -n^2 z \tag{18g}$$

$$\dot{\mathcal{J}} = 0. \tag{18h}$$

As explained in the previous sections, (18c)–(18g) represent the target's attitude kinematics represented via the quaternions set, the target's attitude dynamics, and the Clohessy-Wiltshire equations written in components, respectively.

Equation (18h) can be written because we express the attitude dynamics equation (18d) in the body-fixed reference frame BDRF, in which the inertia matrix, though not diagonal, is constant over time.

Equation (18b) represents the kinematics of the $i$-th target's body point in BDRF. Because we are considering a rigid target's structure, the vectors $\mathbf{d}_i$ are constant over time if expressed in a body-fixed reference frame.

Equation (18a) is a differential equation which describes the kinematics of the vector $\mathbf{p}_i$, expressed in CRF. It derives from differentiation in time of (14).

Since we want to analyze the variation of the vector $\mathbf{p}$ in the noninertial CRF, we need to derive the quantities in the inertial ECI frame and consider the angular velocities between the frames in which we are deriving

$$\begin{aligned} \left.\frac{d\mathbf{p}}{dt}\right|_{\mathrm{ECI}} &= \left.\frac{d\mathbf{p}}{dt}\right|_{\mathrm{CRF}} + \boldsymbol{\omega}_{C/E} \times \mathbf{p} \\ \left.\frac{d\mathbf{d}}{dt}\right|_{\mathrm{ECI}} &= \left.\frac{d\mathbf{d}}{dt}\right|_{\mathrm{BDRF}} + \boldsymbol{\omega}_{B/E} \times \mathbf{d} \\ \left.\frac{d\boldsymbol{\rho}}{dt}\right|_{\mathrm{ECI}} &= \left.\frac{d\boldsymbol{\rho}}{dt}\right|_{\mathrm{HRF}} + \boldsymbol{\omega}_{H/E} \times \boldsymbol{\rho}, \end{aligned} \tag{19}$$

where the notation $(\cdot)|_R$ means the derivative is computed in the reference frame $R$, while the vectors' components can be expressed in any reference frame.

Differentiating (14) in time leads to

$$\left.\frac{d\mathbf{p}}{dt}\right|_{\mathrm{ECI}} = \left.\frac{d\mathbf{d}}{dt}\right|_{\mathrm{ECI}} - \left.\frac{d\boldsymbol{\rho}}{dt}\right|_{\mathrm{ECI}} \tag{20}$$

Substituting (19) into (20) and simplifying the notation as

$$\begin{aligned} \left.\frac{d\mathbf{p}}{dt}\right|_{\mathrm{CRF}} &= \dot{\mathbf{p}} \\ \left.\frac{d\mathbf{d}}{dt}\right|_{\mathrm{BDRF}} &= \dot{\mathbf{d}} \\ \left.\frac{d\boldsymbol{\rho}}{dt}\right|_{\mathrm{HRF}} &= \dot{\boldsymbol{\rho}} \end{aligned} \tag{21}$$

we get the following expression:

$$\dot{\mathbf{p}} + \boldsymbol{\omega}_{C/E} \times \mathbf{p} = \dot{\mathbf{d}} + \boldsymbol{\omega}_{B/E} \times \mathbf{d} - \dot{\boldsymbol{\rho}} - \boldsymbol{\omega}_{H/E} \times \boldsymbol{\rho}, \tag{22}$$

where all the quantities must be expressed in the same reference frame. We will use CRF for convenience.

We get (18a) from (22) by recalling that the variations of the body vectors over time are null, as previously stated in (18b).

*6.1. Observables.* The distance sensor and camera mounted on board of the chaser provide, respectively, a range measure and the pixels of the features tracked. The observables equations can be written as follows:

$$\begin{aligned} \varrho &= \sqrt{x^2 + y^2 + z^2} + \eta_\varrho \\ u_i &= p_{i,x}^{\mathrm{CRF}} \frac{f}{p_{i,z}^{\mathrm{CRF}}} + \eta_u \\ v_i &= p_{i,y}^{\mathrm{CRF}} \frac{f}{p_{i,z}^{\mathrm{CRF}}} + \eta_v. \end{aligned} \tag{23}$$

The terms $\eta_\varrho$, $\eta_u$, and $\eta_v$ represent the noises affecting the observables, which are considered Gaussian. The variances associated with the measurements are assumed to be $\sigma_\varrho$ = 1 mm, $\sigma_u$ = 0.5 px, and $\sigma_v$ = 0.5 px. Nevertheless, the measurements obtained by the camera are discretized due to the nature of the detector made up by a pixel map.

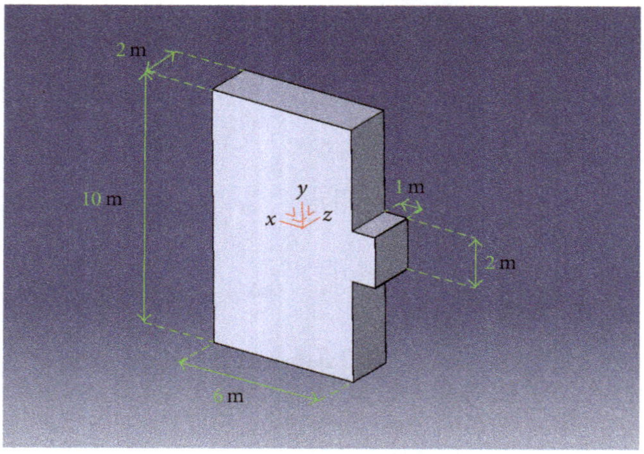


FIGURE 4: CATIA 3D Model of the target. The dimensions of the body, expressed in meters, are given in green, while the Body Dummy Reference Frame is shown in red.

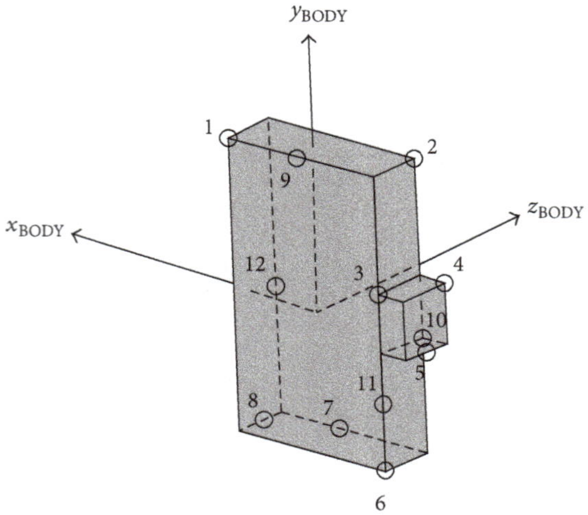


FIGURE 5: Representation of the target and the tracked points at $t_0$. The axes $x_{BODY}$, $y_{BODY}$, and $z_{BODY}$ identify the BDRF, which coincides with BPRF. The 12 tracked points are circled in red.

## 7. A Test Case

In this step of the research we want to focus on target's shape and relative state estimation process. We therefore do not want to pay particular attention at the feature detection and matching process. Thus, no feature extraction and matching algorithm is used here, meaning that no real image is being processed to extract features. It is indeed true that specific problems related to the space environment, such as eclipses, harsh illumination, and occlusions, are not considered in this phase of the research. Nevertheless, we assume that some specific points of the target's body are always visible and perfectly tracked throughout the entire estimation process. This certainly is a simplification to be removed in future developments of the approach, but it allows for a verification of the filter's performance in an ideal case.

The target has a basic shape, which can be seen in Figure 4.

We assume we are tracking 12 points of the target's body ($n_p = 12$), whose positions can be seen in Figure 5. In this figure, the target is represented in its initial attitude.

*7.1. Filter's Initialization.* In order to generate the observables by means of (23), we need to know the true state of the system at each instant $t_k$. Therefore, we need to integrate the system of differential equations (18a)–(18h) with some initial conditions.

For this test case, the target's orbit is assumed to be circular and inclined: $a = 7{,}000$ km, $e = 0$, $i = 40°$, $\Omega = 0°$, $\omega = 0°$, and $\nu_0 = 0°$. Moreover, we decide that the chaser is on a closed and inclined relative orbit around the target. This example is also quite reasonable because it represents the case of a parking orbit, which frequently is an intermediate step before docking.

The initial attitude matrix from BDRF to CRF is an identity matrix, as said in Section 3.3, while the tracked points are chosen as shown in Figure 5. The inertia matrix is computed assuming that the BDRF is initially aligned with BPRF. This assumption is a simplification which makes the inertia matrix diagonal and does not affect the validity of the results because a simple, fixed rigid rotation would be needed to bring BDRF parallel to BPRF if they were not.

The state vector (17) is therefore initialized to its true initial value

$$\begin{aligned}
\mathbf{p}_{i,0}^{CRF} &= {}^{BDRF}\mathscr{C}_{CRF}(t_0)\,\mathbf{d}_{i,0}^{BDRF} - {}^{CRF}\mathscr{R}_{HRF}(t_0)\,\boldsymbol{\rho}_0^{HRF} \\
\mathbf{d}_{i,0}^{BDRF} &= \mathbf{d}_i \\
\mathbf{q}_{0,B/C} &= [1, 0, 0, 0] \\
\boldsymbol{\omega}_{B/E} &= [0.3, 0.2, 0.1] \text{ deg/s} \\
\boldsymbol{\rho}_0^{HRF} &= [40, 0, 5] \text{ m} \\
\dot{\boldsymbol{\rho}}_0^{HRF} &= [0, -2n\rho_{x,0}, 0] \text{ m/s} \\
\mathscr{I} &= \begin{bmatrix} 20708.223 & 0 & 0 \\ 0 & 10508.103 & 0 \\ 0 & 0 & 28321.692 \end{bmatrix} \text{kg m}^2.
\end{aligned} \tag{24}$$

In addition to the true value for the observables generation, we need to give an initial guess to the filter from which it can start the estimation process. We assume to have a certain error on the initial guess of the relative position and we set the initial relative velocity estimate to zero.

$$\hat{\boldsymbol{\rho}}_0^{HRF} = [41, 1, 3] \text{ m} \tag{25}$$

$$\hat{\dot{\boldsymbol{\rho}}}_0^{HRF} = \mathbf{0} \text{ m/s}. \tag{26}$$

As we are arbitrarily setting BDRF to be parallel to CRF at time $t_0$, we have a perfect initial guess on quaternions

$$\hat{\mathbf{q}}_{B/C,0} = \mathbf{q}_{B/C,0}. \tag{27}$$

We assume to have an uncertainty of 1 deg/s on each component of the initial guess of the target's angular velocity

$$\widehat{\boldsymbol{\omega}}_{B/E} = [1.3, 1.2, 1.1] \text{ deg/s} \quad (28)$$

and an uncertainty of 100 kg/m$^2$ on each element of the inertia matrix

$$\widehat{\mathcal{J}} = \begin{bmatrix} 20808.223 & 100 & 100 \\ 100 & 10608.103 & 100 \\ 100 & 100 & 28421.692 \end{bmatrix} \text{kg m}^2. \quad (29)$$

To give an initial estimate of the vectors $\mathbf{p}_i$, we make use of their pixels' coordinates obtained by the measurements at time $t_0$ and go back from the 2D coordinates on the image plane to the 3D coordinates in CRF. However, we need to solve the perspective ambiguity, for which a 3D point corresponding to its 2D representation can be at any depth from the image plane. Therefore, along with the in-plane coordinates obtained through the camera, the depth of each point is needed. As an initial rough estimation, we assume that each point is situated at the same depth and set it to be the norm of the relative position guess (25).

Dropping the $i$ subscript for the sake of simplicity, we can thus write the initial guess for each $\mathbf{p}$ vector as follows:

$$\widehat{p}_{x,0}^{\text{CRF}} = u_0 \frac{\widehat{\rho}_0}{f} \quad (30a)$$

$$\widehat{p}_{y,0}^{\text{CRF}} = v_0 \frac{\widehat{\rho}_0}{f} \quad (30b)$$

$$\widehat{p}_{z,0}^{\text{CRF}} = \widehat{\rho}_0. \quad (30c)$$

From (15) and (30a)–(30c) we can write the initial guess for the $\mathbf{d}$ vector

$$\begin{aligned} \widehat{\mathbf{d}}_0^{\text{BDRF}} &= {}^{\text{BDRF}}\mathscr{C}_{\text{CRF}}(t_0)\,\widehat{\mathbf{p}}_0^{\text{CRF}} \\ &+ {}^{\text{BDRF}}\mathscr{C}_{\text{CRF}}(t_0)\,{}^{\text{CRF}}\mathscr{R}_{\text{HRF}}(t_0)\,\widehat{\boldsymbol{\rho}}_0^{\text{HRF}}. \end{aligned} \quad (31)$$

Along with the initial guess, a certain standard deviation has to be associated with each state component. Because we are tracking 12 points, the covariance matrix $\mathscr{P}_0$ is a $(91 \times 91)$ diagonal matrix. The $\mathscr{P}_{i,i}$ element is the square of the standard deviation associated with the $i$-th component of the state vector

$$\sigma_p = 1 \text{ m} \quad (32a)$$

$$\sigma_d = 1 \text{ m} \quad (32b)$$

$$\sigma_q = 10^{-5} \quad (32c)$$

$$\sigma_\omega = 10^{-1} \text{ deg/s} \quad (32d)$$

$$\sigma_\rho = 1 \text{ m} \quad (32e)$$

$$\sigma_{\dot{\rho}} = 1 \text{ m/s} \quad (32f)$$

$$\sigma_{\mathcal{J}} = 10^3 \text{ kg m}^2. \quad (32g)$$

Because we assume to know the correct dynamics equations, the process noise matrix $\mathcal{Q}$ is set to be a null matrix: in case of significant perturbations, $\mathcal{Q}$ should be tuned accordingly. On the other hand, as the measurements obtained by the camera and the distance sensor are considered to have Gaussian noise, the measures noise matrix $\mathscr{R}$ is a $(25 \times 25)$ diagonal matrix, where the element $\mathscr{R}_{i,i}$ is the square of the variance associated with that measure

$$\mathscr{R}_u = \sigma_u^2 = 0.25 \text{ px}^2 \quad (33a)$$

$$\mathscr{R}_v = \sigma_v^2 = 0.25 \text{ px}^2 \quad (33b)$$

$$\mathscr{R}_\varrho = \sigma_\varrho^2 = 10^{-6} \text{ m}^2. \quad (33c)$$

## 8. Results

The output of the simulation is plotted in Figures 6–10, where the errors between true and estimated values of each component of the state vector are plotted. However, only the first point is analyzed and plotted in Figure 6 for the sake of simplicity. The errors' components at final time are listed in Table 1, while the norms of the errors at the beginning and at the end of the simulation are shown in Table 2. Note that the inertia matrix errors are written as a vector in the following form, which takes into account the fact that the inertia matrix is diagonal:

$$\mathbf{I} = [I_x, I_y, I_z, I_{xy}, I_{xz}, I_{yz}]. \quad (34)$$

The simulation time is set to one orbital period of the target $t_f = T = 5828$ s and the interval between subsequent steps is $\Delta t = 3$ s.

We can see in Figure 6 that the errors of the point number 1 correctly converge. In fact, the simulation improves the estimate of the tracked points of an order of magnitude, as can be noticed in Table 2.

Along with the point number 1, the estimation process is successful for the remaining 11 tracked points, as can be seen in Figure 7, where the blue markers represent the tracked points in their true position (as defined in Figure 5), while the red markers are the points' estimate at the end of the simulation.

The target's relative attitude estimation errors are plotted in Figure 8(a), while the norms of the initial and final errors are, respectively, $\varepsilon_q^0 = 0$ and $\varepsilon_q^f = 8.36 \times 10^{-4}$. In this case, the filter cannot refine the estimate, as the initial attitude is known with 100% accuracy because it has been defined as the difference between a frame whose orientation is known (CRF) and a frame artificially chosen (BDRF).

The errors on the target's angular velocity with respect to ECI reference frame are plotted in Figure 8(b). In this case, the estimation process improves the estimate between the beginning and the end of the simulation of three orders of magnitude, as can be noted in Table 2.

By taking a look at Figure 9, we can see the results of the relative position and linear velocity estimation. The errors are plotted over time in Figures 9(a) and 9(b), while both the

TABLE 1: Errors in components at the end of the simulation $t_f = 5828$ s.

| | Final errors in components |
|---|---|
| $\mathbf{d}_1$ (m) | $(-1.23 \times 10^{-1}, -7.60 \times 10^{-2}, -1.94 \times 10^{-1})$ |
| $\mathbf{p}_1$ (m) | $(-2.76 \times 10^{-2}, -9.27 \times 10^{-2}, -3.93 \times 10^{-2})$ |
| $\mathbf{q}_{B/C}$ | $(8.14 \times 10^{-3}, 1.13 \times 10^{-4}, 2.13 \times 10^{-2}, 1.78 \times 10^{-2})$ |
| $\boldsymbol{\omega}_{B/E}$ (deg/s) | $(6.85 \times 10^{-3}, 1.66 \times 10^{-3}, -5.80 \times 10^{-3})$ |
| $\boldsymbol{\rho}$ (m) | $(8.08 \times 10^{-3}, -2.10 \times 10^{-2}, -1.88 \times 10^{-2})$ |
| $\dot{\boldsymbol{\rho}}$ (m/s) | $(-2.51 \times 10^{-6}, -1.50 \times 10^{-5}, -4.73 \times 10^{-5})$ |
| $\mathbf{I}$ (kg m$^2$) | $(-3.34 \times 10^{3}, -1.60 \times 10^{3}, -4.57 \times 10^{3}, 9.98 \times 10^{1}, 8.16 \times 10^{1}, 2.50 \times 10^{2})$ |

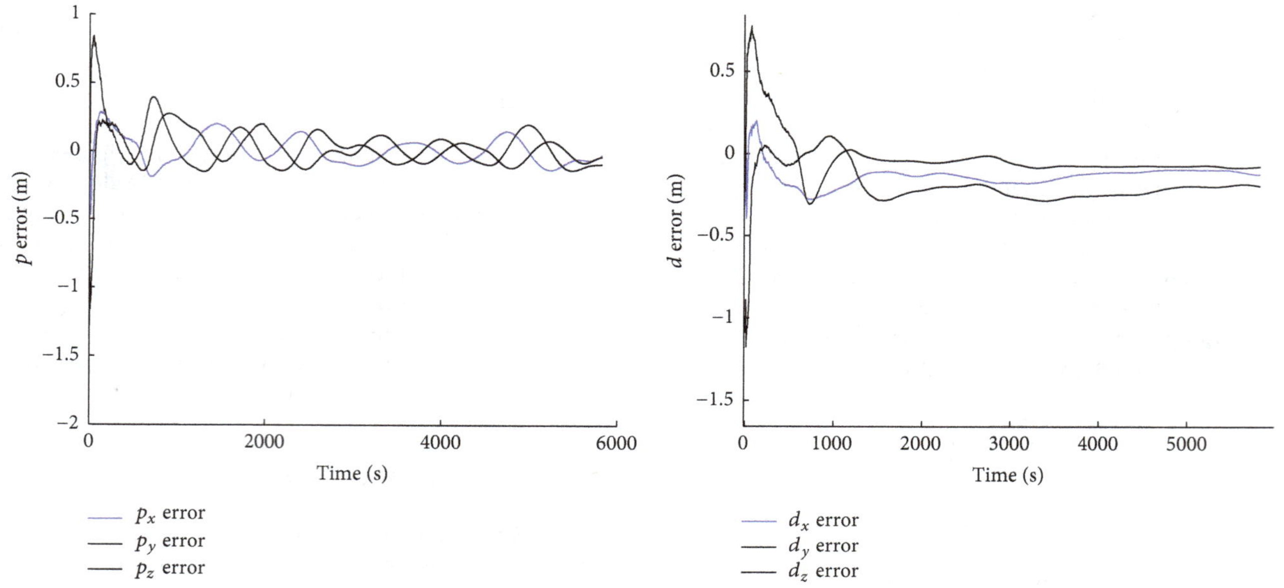


(a) $\mathbf{p}_1$ error

(b) $\mathbf{d}_1$ error

FIGURE 6: (a) $\mathbf{d}_1$ estimation errors in components: $\mathbf{d}_1 - \hat{\mathbf{d}}_1$ (m). (b) $\mathbf{p}_1$ estimation errors in components: $\mathbf{p}_1 - \hat{\mathbf{p}}_1$ (m).

TABLE 2: Comparison between the norms of the errors at the beginning and at the end of the simulation. The 2-norm has been used for the inertia matrix error calculation.

| | Initial errors | Final errors |
|---|---|---|
| $\varepsilon_{d_1}$ (m) | 1.03 | $2.41 \times 10^{-1}$ |
| $\varepsilon_{p_1}$ (m) | 1.83 | $1.04 \times 10^{-1}$ |
| $\varepsilon_q$ | 0 | $8.36 \times 10^{-4}$ |
| $\varepsilon_\omega$ (deg/s) | 1.73 | $9.12 \times 10^{-3}$ |
| $\varepsilon_\rho$ (m) | 2.45 | $2.93 \times 10^{-2}$ |
| $\varepsilon_{\dot{\rho}}$ (m/s) | $8.62 \times 10^{-2}$ | $4.97 \times 10^{-5}$ |
| $\varepsilon_I$ (kg m$^2$) | $3 \times 10^2$ | $4.60 \times 10^3$ |

true and estimated 3D trajectories are shown in Figure 9(c), where it is possible to see the closed and inclined nature of the relative orbit.

In this case the estimation process is satisfactorily achieved in both position and linear velocity, as the improvements of the estimates between the beginning and the end of the simulation are, respectively, of two and three orders of magnitude, as can be seen in Table 2. Moreover, it can be noticed in Figure 9(b) how the convergence time is of few seconds for the linear relative velocity, while, on the other hand, the relative position converges to its final estimate approximately at half orbital period ($t^* \approx 3000$ s).

Conversely to the results shown so far, the filter does not improve the estimate of the target's inertia matrix. This can be clearly seen in Figure 10 and Table 2. While the off-diagonal elements' error roughly remains at the same order of magnitude throughout the estimate process, the error of the diagonal elements increases of an order of magnitude. Moreover, the errors of $\mathcal{I}_x$, $\mathcal{I}_y$, and $\mathcal{I}_z$ seem to converge to a value different from zero (see Figure 10).

The explanation to this result can be found inside (18d). The dependence of the angular acceleration on the inertial matrix is nearly negligible, as it cancels in the right part of (18d), regardless of the inertia's order of magnitude. In fact, the two terms inside the parentheses, which, respectively, represent the free dynamics and the gravity gradient contribution, both depend on the target's inertia, but at the same

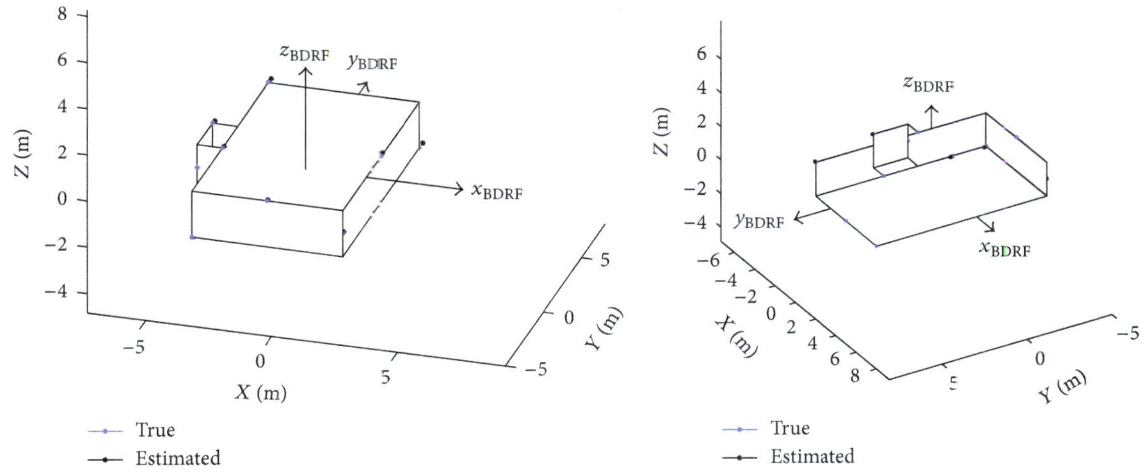


(a) View 1

(b) View 2

FIGURE 7: True versus estimated 3D position of the tracked body points.

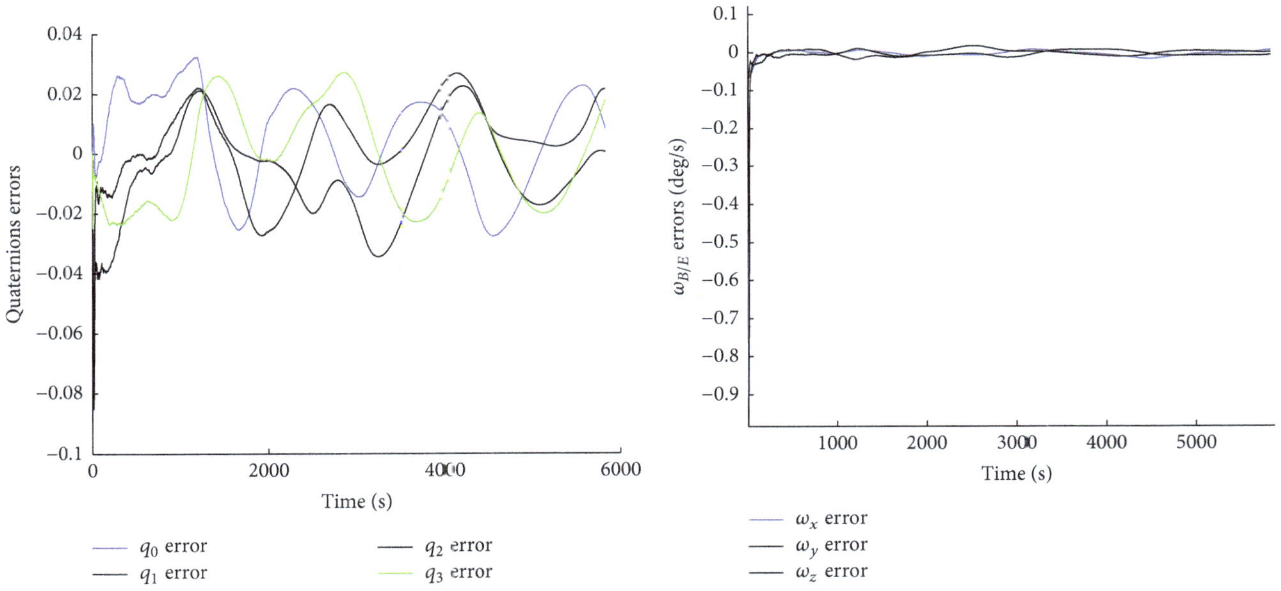


(a) $\mathbf{q}_{B/C}$ error

(b) $\boldsymbol{\omega}_{B/C}$ error

FIGURE 8: (a) Quaternions estimation errors in components: $\mathbf{q}_{B/C} - \widehat{\mathbf{q}}_{B/C}$. (b) Target's angular velocity estimation error in components: $\boldsymbol{\omega}_{B/E}^{BDRF} - \widehat{\boldsymbol{\omega}}_{B/E}^{BDRF}$ (deg/s).

time both are multiplied by the inverse of the inertia matrix to compute the target's angular acceleration. Moreover, the target's inertia is not used in any observable equation (23). This means that the filter has got no information about how a variation on the inertia would affect any state vector's component. In other words, a correct estimation of the inertia is not needed to provide a correct estimation of the other quantities, and, conversely, an accurate estimation of the relative pose cannot either provide or affect the inertia estimate.

## 9. Conclusions

The problem of optical navigation used in a docking maneuver to a noncooperative satellite has been analyzed. In particular, a filter which estimates the relative state (position, velocity, and attitude) together with the shape of the target has been developed and tested in a specific case. It is important to remark that the proposed approach is valid for any attitude and linear motion of the observer spacecraft. Moreover, no previous knowledge of the target's shape is required.

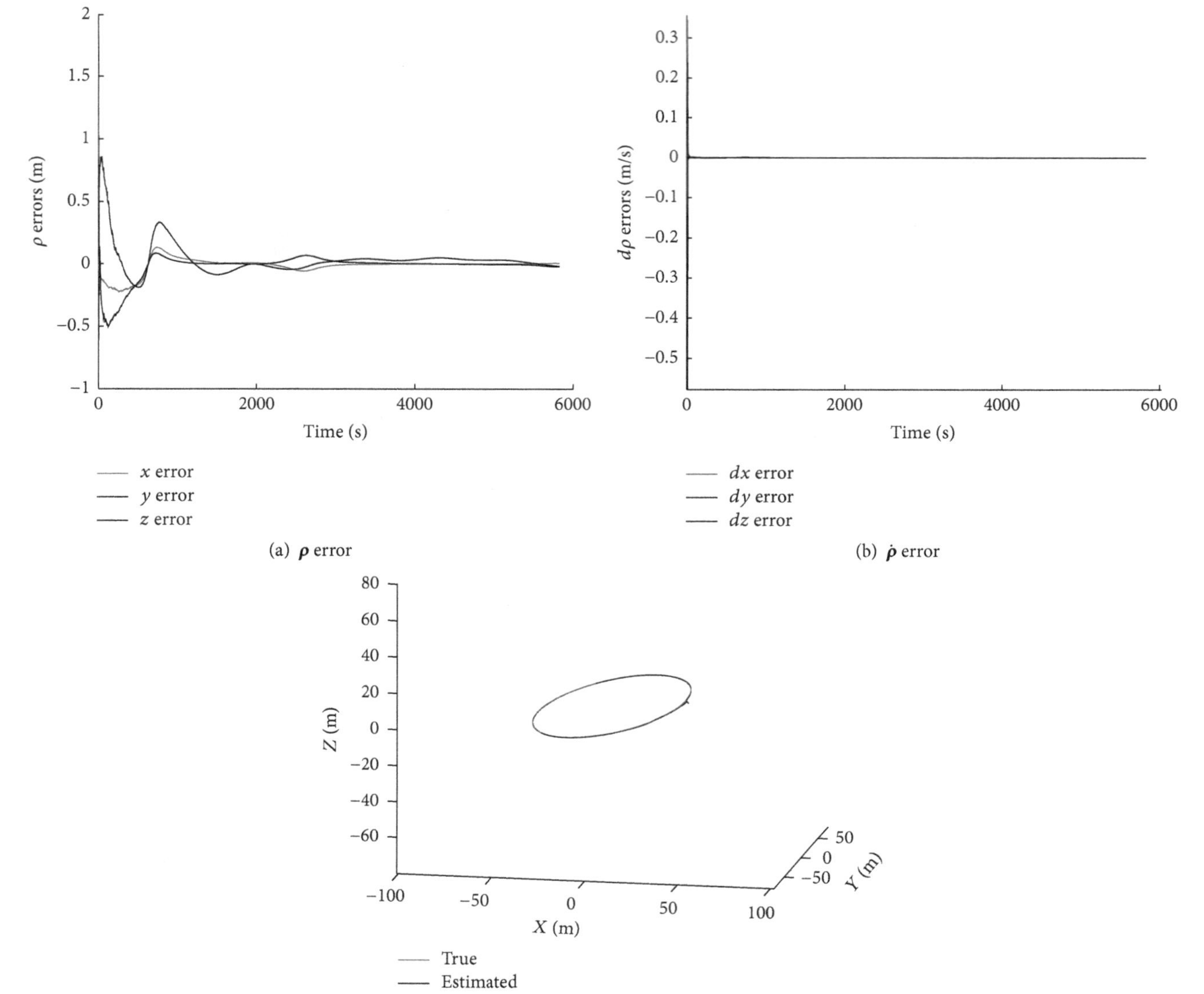


(c) 3D relative trajectory of the chaser with respect to the target

FIGURE 9: (a) Relative position error in components: $\boldsymbol{\rho} - \widehat{\boldsymbol{\rho}}$ (m). (b) Relative position velocity in components: $\dot{\boldsymbol{\rho}} - \dot{\widehat{\boldsymbol{\rho}}}$ (m/s). (c) True versus estimated 3D relative trajectory.

The process of features' extraction and matching between two consecutive frames has been considered as successfully carried out by a prefiltering stage and therefore it has not been implemented in the proposed filter.

The output of the simulation showed how the filter improves five out of seven of the quantities to be estimated. In fact, the relative position, velocity, and angular velocity final estimates have lower errors than the initial one, respectively, of two, three, and three orders of magnitude. Moreover, a 3D shape of target can be reconstructed by using the estimates of the 3D points tracked by the camera, which are evaluated with errors up to tens of centimetres, while starting the simulation with uncertainties of meters. The quaternion vector describing the relative attitude is estimated with errors of the order of $10^{-4}$. The target's inertia is the only parameter whose final estimation is not an improvement of the initial guess, because neither the dynamics nor the observables equations are sensible enough to its variation. Nevertheless, estimating incorrectly the inertia does not affect the other quantities' estimates.

By what has been seen from the simulation's results, the filter can thus solve the problem of real-time pose and shape estimation satisfactorily, even if the considered case is simplified with respect to the real world. Important missing ingredients that will be the object of future studies are relevant to the detection and tracking of features in the case of a real 3D target, where the observables can be noncontinuous (due to occlusion or blinding) and also affected by false matches

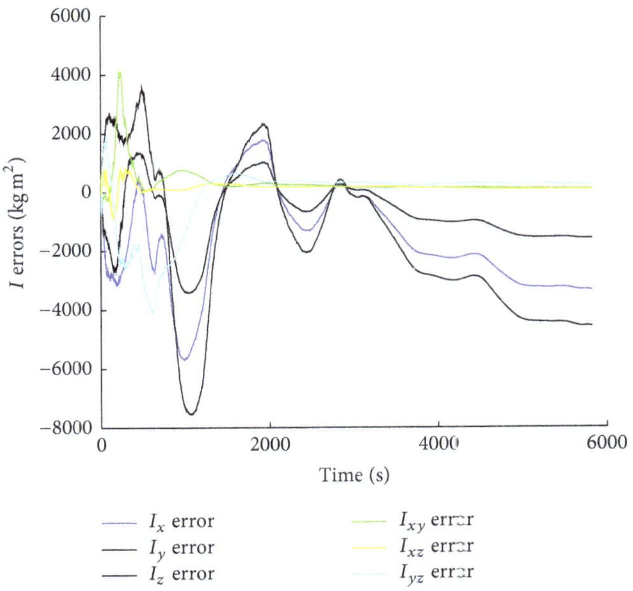


FIGURE 10: Errors in estimating the elements of the inertia matrix.

(due to the statistical nature of the matching process). Preliminary analyses show that the proposed approach should be robust enough with respect to these issues.

Notwithstanding these limitations, the degree of accuracy attained from the present approach can be considered a promising basis for a correct and safe docking maneuver.

## Conflicts of Interest

The authors declare that there are no conflicts of interest regarding the publication of this paper.

## References

[1] M. Ganet, I. Quinquis, J. Bourdon, and P. Delpy, "ATV GNC during rendezvous with ISS," in *Proceedings of the DCSSS Conference*, 2002.

[2] G. Casonato and G. B. Palmerini, "Visual techniques applied to the ATV/ISS rendez-vous monitoring," in *Proceedings of the IEEE Aerospace Conference Proceedings*, vol. 1, pp. 613–624, Big Sky, Mont, USA, March 2004.

[3] D. Pinard, S. Reynaud, P. Delpy, and S. E. Strandmoe, "Accurate and autonomous navigation for the ATV," *Aerospace Science and Technology*, vol. 11, no. 6, pp. 490–498, 2007.

[4] G. B. Palmerini, M. Sabatini, and P. Gasbarri, "Guidelines for active removal of non-functional targets designed to assist rendezvous and capture," in *Proceedings of the IEEE Aerospace Conference (AERO '16)*, pp. 1–13, IEEE, March 2016.

[5] E. Gill, S. D'Amico, and O. Montenbruck, "Autonomous formation flying for the PRISMA mission," *Journal of Spacecraft and Rockets*, vol. 44, no. 3, pp. 671–681, 2007.

[6] P. Bodin, R. Larsson, F. Nilsson, C. Chasset, R. Noteborn, and M. Nylund, "PRISMA: An in-orbit test bed for guidance, navigation, and control experiments," *Journal of Spacecraft and Rockets*, vol. 46, no. 3, pp. 615–623, 2009.

[7] M. Delpech, J.-C. Berges, T. Karlsson, and F. Malbet, "Results of PRISMA/FFIORD extended mission and applicability to future formation flying and active debris removal missions," *International Journal of Space Science and Engineering*, vol. 1, no. 4, pp. 382–409, 2013.

[8] R. P. de Santayana and M. Lauer, "Optical measurements for rosetta navigation around the comet," in *Proceedings of the 25th International Symposium on Space Flight Dynamics (ISSFD '15)*, Munich, Germany, 2015.

[9] P. Mu, F. Budnik, V. Companys et al., "Rosetta navigation during lander delivery phase and reconstruction of Philae descent trajectory and rebound," in *Proceedings of the in 25th International Symposium on Space Flight Dynamics*, 2015.

[10] F. Castellini, D. Antal-Wokes, R. Pardo de Santayana, and K. Vantournhout, "Far approach optical navigation and comet photometry for the rosetta mission," in *Proceedings of the 25th International Symposium on Space Flight Dynamics (ISSFD '15)*, Munich, Germany, 2015.

[11] H. C. Longuet-Higgins, "A computer algorithm for reconstructing a scene from two projections," in *Readings in Computer Vision: Issues, Problems, Principles, and Paradigms*, M. A. Fischler and O. Firschein, Eds., pp. 61-62, 1987.

[12] J. Weng, T. S. Huang, and N. Ahuja, "Motion and Structure from Two Perspective Views: Algorithms, Error Analysis, and Error Estimation," *IEEE Transactions on Pattern Analysis and Machine Intelligence*, vol. 11, no. 5, pp. 451–476, 1989.

[13] J. Weng, N. Ahuja, and T. S. Huang, "Optimal Motion and Structure Estimation," *IEEE Transactions on Pattern Analysis and Machine Intelligence*, vol. 15, no. 9, pp. 864–884, 1993.

[14] E. Johnson and H. Mathies, "Precise image-based motion estimation for autonomous small body exploration," *in Artificial Intelligence, Robotics and Automation in Space*, vol. 440, p. 627, 1999.

[15] M. Vetrisano, C. Colombo, and M. Vasile, "Asteroid rotation and orbit control via laser ablation," *Advances in Space Research*, vol. 57, no. 8, pp. 1762–1782, 2016.

[16] N. W. Oumer and G. Panin, "Tracking and pose estimation of non-cooperative satellite for on-orbit servicing," *i-SAIRAS*, 2012.

[17] J. W. McMahon, S. Gehly, and P. Axelrad, "Enhancing relative attitude and trajectory estimation for autonomous rendezvous using flash LIDAR," in *Proceedings of the AIAA/AAS Astrodynamics Specialist Conference*, San Diego, Calif, USA, August 2014.

[18] J. L. Sell, A. Rhodes, J. O. Woods, J. A. Christian, and T. Evans, "Pose performance of LIDAR-based navigation for satellite servicing," in *Proceedings of the AIAA/AAS Astrodynamics Specialist Conference*, August 2014.

[19] L. Liu, G. Zhao, and Y. Bo, "Point cloud based relative pose estimation of a satellite in close range," *Sensors*, vol. 16, no. 6, article 824, 2016.

[20] J. Ventura, A. Fleischner, and U. Walter, "Pose tracking of a noncooperative spacecraft during docking maneuvers using a time-of-flight sensor," in *Proceedings of the AIAA Guidance, Navigation, and Control Conferen* vol. 875, January 2016.

[21] H. Bay, T. Tuytelaars, and L. van Gool, "SURF: speeded up robust features," in *Proceedings of the European Conference on Computer Vision*, vol. 3951 of *Lecture Notes in Computer Science*, pp. 404–417, Springer, 2006.

[22] https://nasa3d.arc.nasa.gov/models.

[23] W. Clohessy, "Terminal Guidance System for Satellite Rendezvous," *Journal of the Aerospace Sciences*, 2012.

# 14

# Research on Self-Monitoring Method for Anomalies of Satellite Atomic Clock

**Lei Feng[1,2] and Guotong Li[2,3]**

[1]*Shanghai Institute of Microsystem and Information Technology, Chinese Academy of Sciences, Shanghai 200050, China*
[2]*Shanghai Engineering Center for Microsatellite, Shanghai 201203, China*
[3]*Shanghai Tech University, Shanghai 201203, China*

Correspondence should be addressed to Lei Feng; fenglei6036@126.com

Academic Editor: Paolo Tortora

Atomic clock is the core component of navigation satellite payload, playing a decisive role in the realization of positioning function. So the monitoring for anomalies of the satellite atomic clock is very important. In this paper, a complete autonomous monitoring method for the satellite clock is put forward, which is, respectively, based on Phase-Locked Loop (PLL) and statistical principle. Our methods focus on anomalies in satellite clock such as phase and frequency jumping, instantaneous deterioration, stability deterioration, and frequency drift-rate anomaly. Now, method based on PLL has been used successfully in China's newest BeiDou navigation satellite.

## 1. Introduction

The most important function of navigation satellite is to support its users to acquire their position through the satellite signal, during which satellite time is one of the most important factors. Because of the changes in temperature, humidity, radiation, and the aging of the satellite clock, the physical and electric part of clock may both have problems, which will bring anomaly in clock signal, resulting in large error in the prediction of satellite time or even unpredictability, which may lead to disastrous consequence. So the anomaly monitoring of satellite clock is very important.

So far, researchers have proposed schemes to monitor anomalies of clock, such as Interferometric Detection Method [1], Least Square (LS) Detection Method [2, 3], Generalized Likelihood Ratio Test (GLRT) [4–6], Kalman Filtering Method [7–10], and Dynamic Allan Variance (DAVAR) Method [5, 11–13]. Though their schemes have been proved to be effective for some (not all) anomalies, some extra work is still needed to realize Self-Monitoring.

Under normal circumstance, ground station can evaluate the health condition and performance of the clock by continuously tracking satellite signals. But when the satellite flies beyond the ground station's sight, or owing to some reasons, the satellite cannot contact with the ground station in a few hours or even days; the satellite needs to judge the status of the clock all by itself. Self-Monitoring for clock anomaly, which is in the absence of ground station, is that the satellite monitors its clock by itself to make a judgment on the satellite clock running state.

The common anomalies of the satellite clock are signal loss, phase jumping, frequency jumping, instantaneous deterioration, stability, and frequency drift-rate deterioration.

The contributions of this paper can be summarized as follows.

In this paper, a set of Self-Monitoring algorithms is proposed to improve the reliability of satellite. Two methods are put forward to monitor satellite clock anomalies. The first method is based on PLL, and it can detect signal loss and phase and frequency jumping. Based on the measurement data from intercomparison among three clocks, Modified DAVAR is used to detect phase and frequency jumping and instantaneous deterioration; we use windowed overlapping Hadamard variance to evaluate clock stability in real time and the three-state Kalman filter to detect large drift rate.

The method based on PLL has been proved effective and used in newest BeiDou satellite. And the other research on

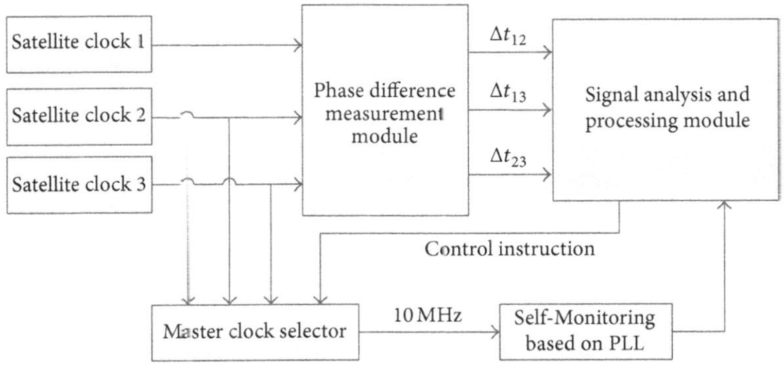


FIGURE 1: Global schematic diagram of Self-Monitoring Method.

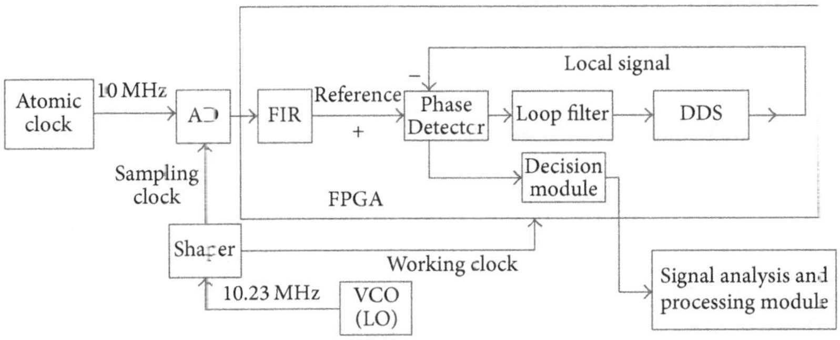


FIGURE 2: Schematic diagram of Self-Monitoring Method based on PLL.

Self-Monitoring Method in this paper can be used in next generation navigation satellites.

## 2. Self-Monitoring Method for Anomaly of Satellite

Generally speaking, there are two methods to evaluate atomic clocks: (1) comparing the clock signal with standard reference whose stability is much better than the evaluated clock and (2) making intercomparison among three or more clocks whose stability is almost the same.

Because there is no standard reference in the satellite and the performance of satellite clocks is similar, we make use of the second method to realize Self-Monitoring for anomalies. The schematic diagram is shown in Figure 1.

Firstly, we define that $\Delta t_1$ is the time error of clock 1, which is the difference between clock time and the standard time. $\Delta t_{12}$ is the time difference between clock 1 and clock 2. As shown in Figure 1, three clocks are all powered up. Their 10 MHz signals act as the input of phase difference measurement module, through which we can get the time difference data $\Delta t_{12}$, $\Delta t_{13}$, and $\Delta t_{23}$ among them. The Signal Processing Module uses the time difference data $\Delta t_{12}$, $\Delta t_{13}$, and $\Delta t_{23}$ to evaluate the health state of three clocks with certain algorithm and then commands the master clock selector to choose suitable clock as the frequency and time source of the entire satellite.

In this paper, $\Delta t_{12}$, $\Delta t_{13}$, and $\Delta t_{23}$ are used in Modified DAVAR to monitor phase and frequency jumping, used to evaluate the stability of three clocks, and used to monitor drift-rate anomaly.

The detailed structure of Self-Monitoring Module based on PLL in Figure 1 can be described in Figure 2.

*2.1. Self-Monitoring Method Based on PLL.* Figure 2 shows the basic schematic diagram of this method. As a phase tracking system, PLL is used to adjust the phase of local signal to trace the reference signal. Voltage Controlled Oscillator (VCO) provides sampling clock and working clock for AD and FPGA, respectively. As the input of the Decision Module, the observed quantity of this method comes from the output of Phase Detector. Output of Decision Module will be sent to Signal Processing Module in Figure 1 to help choose the master clock. At the same time, 10 MHz signal from Figure 1 is sampled by AD in Figure 2. The working frequency of Phase Detector is 1000 Hz.

Once phase or frequency jumping occurs, the output of Phase Detector in PLL will follow. In this section, the response of Phase Detector to these two anomalies will be derived.

According to [14], assuming that the phase of reference signal of PLL is $2\pi f_r t + \varphi_1(t)$ and that of Direct Digital Synthesizer (DDS) output is $2\pi f_r t + \varphi_2(t)$, then we get

$$K\left(\varphi_1(s) - \varphi_2(s)\right) F(s) \frac{1}{s} = \varphi_2(s), \quad (1)$$

where $F(s) = (1 + s\tau_2)/s\tau_1$ is the transfer function of the two-order ideal loop filter, $1/s$ is the normalized transfer function of the DDS, and $K$ is the loop gain. From expression (1), we get

$$\varphi_e(s) = \frac{s}{s + KF(s)} \varphi_1(s), \quad (2)$$

where $\varphi_e(s) = \varphi_1(s) - \varphi_2(s)$ is the phase difference between the reference signal and the local signal, and the error transfer function of the loop can be expressed as

$$H_e(s) = \frac{s}{s + KF(s)} = \frac{s^2}{s^2 + 2\xi\omega_n s + \omega_n^2}, \quad (3)$$

where $\omega_n = \sqrt{K/\tau_1}$ is the undamped oscillation frequency and $\xi = (\tau_2/2)\sqrt{K/\tau_1}$ is the damped coefficient.

In the following, the tracking property of Phase Detector for phase and frequency jumping will be deduced.

*2.1.1. Phase Jumping.* Supposing that the phase jumping can be written as $\varphi_1(t) = \Delta\varphi \cdot \varepsilon(t)$, whose Laplace transform can be expressed as $\varphi_1(s) = \Delta\varphi/s$, then the error response is

$$\begin{aligned} \varphi_e(s) &= \frac{s^2}{s^2 + 2\xi\omega_n s + \omega_n^2} \cdot \frac{\Delta\varphi}{s} \\ &= \frac{s}{s^2 + 2\xi\omega_n s + \omega_n^2} \cdot \Delta\varphi. \end{aligned} \quad (4)$$

Through factorization, (4) is equivalent to

$$\varphi_e(s) = \frac{A}{s - s_1} + \frac{B}{s - s_2}, \quad (5)$$

where

$$\begin{aligned} s_1 &= -\xi\omega_n + i\omega_n\sqrt{1 - \xi^2}, \\ s_2 &= -\xi\omega_n - i\omega_n\sqrt{1 - \xi^2}, \end{aligned} \quad (6)$$

$$\begin{aligned} A &= -\Delta\varphi \frac{\xi - i\sqrt{1 - \xi^2}}{2i\sqrt{1 - \xi^2}}, \\ B &= \Delta\varphi \frac{\xi + i\sqrt{1 - \xi^2}}{2i\sqrt{1 - \xi^2}}. \end{aligned} \quad (7)$$

Considering (6) and (7), the inverse Laplace transform of (5) can be expressed as

$$\begin{aligned} \varphi_e(t) = \Delta\varphi e^{-\xi\omega_n t} \Bigg( &\cos \omega_n t\sqrt{1 - \xi^2} \\ &- \frac{\xi}{\sqrt{1 - \xi^2}} \sin \omega_n t\sqrt{1 - \xi^2} \Bigg). \end{aligned} \quad (8)$$

From (8) we notice that the phase difference at $t = 0$ reaches its peak value which has nothing to do with the loop parameters. Figure 3 is the simulation result, the PLL is locked at the beginning, and phase of reference signal jumps at $t =$ 500 s which leads to obvious jumping in the output of Phase

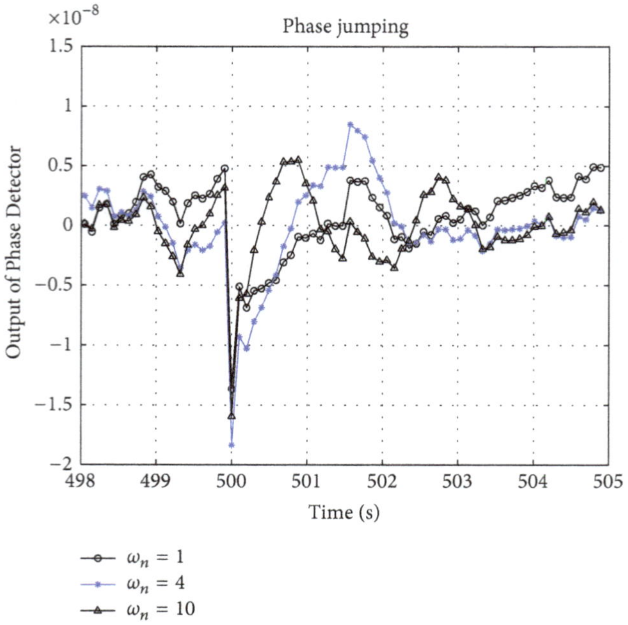


FIGURE 3: Response of Phase Detector to phase jumping during simulation.

Detector. In Figure 3, the loop parameters of three PLLs are $\omega_n = 1, 4, 10$ and $\xi = \sqrt{2}/2$ and the amplitude of phase jumping is equal to $1/10^8$ period of reference signal. With different loop parameters, the relocking process is different. The narrower the loop bandwidth is, the slower the tracking will be.

*2.1.2. Frequency Jumping.* Assuming that frequency jumping is $\varphi_2(t) = \Delta\varphi t \cdot \varepsilon(t)$, whose Laplace transform is $\varphi_2(s) = \Delta\varphi/s^2$, then the error response can be expressed as

$$\varphi_e(s) = \frac{s^2}{s^2 + 2\xi\omega_n s + \omega_n^2} \cdot \frac{\Delta\varphi}{s^2} = \frac{\Delta\varphi}{s^2 + 2\xi\omega_n s + \omega_n^2}. \quad (9)$$

After factorization,

$$\begin{aligned} A &= \frac{\Delta\omega}{2i\omega_n\sqrt{1 - \xi^2}}, \\ B &= \frac{-\Delta\omega}{2i\omega_n\sqrt{1 - \xi^2}}. \end{aligned} \quad (10)$$

$s_1$ and $s_2$ can also be described by expression (6). According to (6) and (10), the inverse Laplace transform of (5) can be expressed as

$$\varphi_e(t) = \frac{\Delta\omega}{\omega_n\sqrt{1 - \xi^2}} e^{-\xi\omega_n t} \sin \omega_n t\sqrt{1 - \xi^2}. \quad (11)$$

As can be seen from (11), the maximum amplitude of the phase difference in tracking is inversely proportional to $\omega_n$. Figure 4 is the simulation result, PLL is locked at the beginning, and frequency of reference signal jumps at $t = 500$ s, which leads to obvious jumping in the output

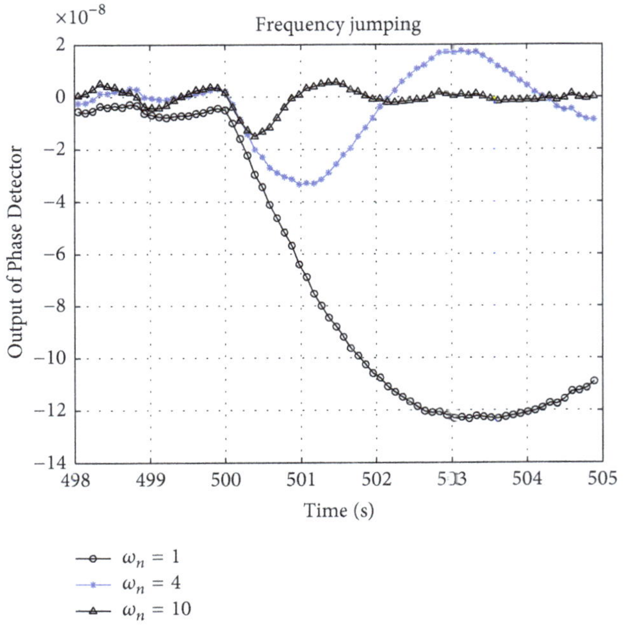


FIGURE 4: Response of Phase Detector to frequency jumping during simulation.

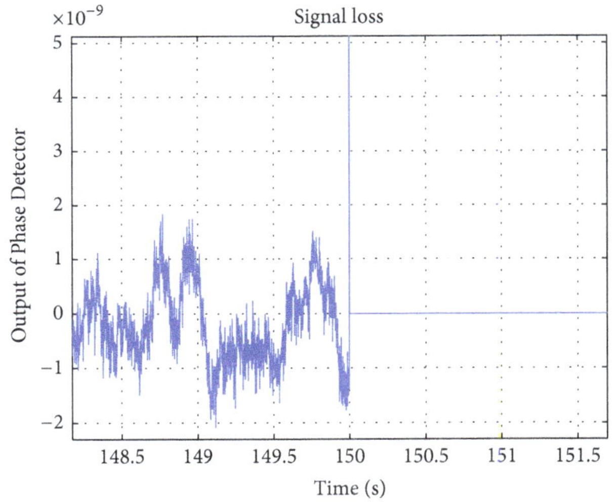


FIGURE 5: Response of Phase Detector to signal loss during simulation.

of Phase Detector. In Figure 4, the loop parameters are the same as in Figure 3 and frequency jumping is equal to $4/10^{11}$ reference signal frequency. With different loop parameters, the relocking process is different. The narrower the loop bandwidth is, the slower the tracking will be; however the jumping is much more obvious.

*2.1.3. Signal Loss.* As shown in Figure 5, assuming that the PLL has been locked and reference signal lost at $t = 150$ s, the jumping amplitude of output of Phase Detector is far larger than threshold in the following section; then it turns to 0 immediately, which is easy to be detected.

TABLE 1: Detection performance for phase jumping.

| Loop parameter | Threshold | PD | PFA | Delay (m) |
|---|---|---|---|---|
| | $8E-9$ | 1.0000 | $6.131e-4$ | 1 |
| | $9E-9$ | 1.0000 | $1.193e-4$ | 1 |
| $\xi = \sqrt{2}/2$ $\omega_n = 4$ | $10E-9$ | 0.9999 | $2.224e-5$ | 1 |
| | $11E-9$ | 0.9978 | $2.412e-6$ | 1 |
| | $12E-9$ | 0.9833 | $8.300e-8$ | 1 |

TABLE 2: Detection performance for frequency jumping.

| Loop parameter | Threshold | PD | PFA | Delay (m) |
|---|---|---|---|---|
| | $8E-9$ | 1.0000 | $6.131e-4$ | 234 |
| | $9E-9$ | 1.0000 | $1.193e-4$ | 296 |
| $\xi = \sqrt{2}/2$ $\omega_n = 4$ | $10E-9$ | 1.0000 | $2.224e-5$ | 337 |
| | $11E-9$ | 0.9933 | $2.412e-6$ | 380 |
| | $12E-9$ | 0.9873 | $8.300e-8$ | 429 |

*2.1.4. Summary.* It can be seen that, from Figures 3, 4, and 5, phase jumping, frequency jumping, and signal loss will all lead to obvious jumping in Phase Detector output, which provides us with chance to monitor anomalies of satellite clock signal.

*2.1.5. Simulations and Detection Performance.* In practice, Probability of False Alarm (PFA) and Detection Probability (PD) are usually used to evaluate the detection method. The basic principle in setting parameters (loop parameters and detection threshold) is to improve PD and minimize PFA at the same time.

The loop parameters and detection threshold are mainly determined by clock noise level and required resolution. We usually use Allan variance (12) to calculate stability to evaluate the size of noise. And resolution is the minimum range of phase and frequency jumping that algorithm can distinguish.

During the following simulations, we simulate 10000 realizations.

*Simulation 1.* During the first simulation with MATLAB, $\xi = \sqrt{2}/2$, $\omega_n = 4$, and we use two-order ideal loop filter. Assuming that the relative frequency deviation $y_i$ (12) of clock signal follows Gauss distribution, whose Allan deviation can be expressed as $3E-12/\sqrt{\tau}$, the detection performance of the method for phase and frequency jumping is shown in Tables 1 and 2. They separately give the PD, PFA, and detection delay in $1/10^8$ period phase jumping and $4/10^{11}$ frequency jumping. Detection delay is defined as $\Delta t = m \cdot T$, where $T$ is the sampling interval and $m$ is the number of sampling points that lasted from the moment anomaly occurred to the moment they are detected by the algorithm. So it is in fact determined by the output frequency of VCO in Figure 2.

*Simulation 2.* During the second simulation, detection threshold and resolution change with the stability of atomic clock. Tables 3 and 4 show it.

TABLE 3: Detection performance of phase jumping for different clock stability.

| Clock stability | Resolution | Threshold | PD | PFA | Delay (m) |
|---|---|---|---|---|---|
| $3E-12/\sqrt{\tau}$ | $1E-8$ | $11E-9$ | 0.9978 | $2.412e-6$ | 1 |
| $3E-11/\sqrt{\tau}$ | $1E-7$ | $11E-8$ | 0.9972 | $6.290e-7$ | 1 |
| $3E-10/\sqrt{\tau}$ | $1E-6$ | $11E-7$ | 0.9955 | $4.370e-7$ | 1 |

TABLE 4: Detection performance of frequency jumping for different clock stability.

| Clock stability | Resolution | Threshold | PD | PFA | Delay (m) |
|---|---|---|---|---|---|
| $3E-12/\sqrt{\tau}$ | $4E-11$ | $11E-9$ | 0.9933 | $2.412e-6$ | 380 |
| $3E-11/\sqrt{\tau}$ | $4E-10$ | $11E-8$ | 0.9947 | $6.290e-7$ | 384 |
| $3E-10/\sqrt{\tau}$ | $4E-9$ | $11E-7$ | 0.9978 | $4.370e-7$ | 381 |

*Analysis.* When $0 < \xi < 1$, PLL is called underdamped system, in which phase and frequency jumping will result in drastic oscillation. If $\xi > 1$, PLL is overdamped and usually more stable and slow to anomaly. In practice, we often set $\xi = 0.707$, which is an acceptable compromise between stability and response speed. From expression (11) and Figure 4, we notice that detection for frequency jumping will become difficult when $\omega_n$ is too large, and suppression for noise will also become weaker. Conversely, if $\omega_n$ is too small, on one hand, the locking process will become difficult, and detection delay becomes longer; on the other hand, the loop will be too sensitive, which leads the Decision Module to regard bottom noise as jumping by mistake frequently, which results in rising in PFA. During simulation, $\omega_n = 4$, which is also a compromise between PD and detection delay and can be adjusted as required.

The noise level of atomic clock directly determines the detection resolution, which we can see from Tables 3 and 4. The relationship between resolution and stability can be described as $\mathrm{Re}\,s(p) \approx (1E4/3)\cdot\sigma(1)$ and $\mathrm{Re}\,s(f) \approx (4E1/3)\cdot\sigma(1)$, while threshold can be set as $\mathrm{Thr} \approx (11E3/3)\cdot\sigma(1)$. From Tables 1 and 2, we can see that PD will be more than 99% and PFA less than 0.001% with appropriate threshold for both phase and frequency jumping. Moreover, it should be noted that PD and PFA listed in the tables are for least phase and frequency jumping that the method can distinguish; the detection performance improves with jumping size increasing. In fact, before the method was used in the satellite, we have tested it in real circuit board for a long time and it works well. Detection delay depends on Phase Detecting frequency, which is 1000 Hz. Delay for phase jumping is 1 ms, and it is less than 0.5 s for frequency jumping.

The method based on PLL can realize Self-Monitoring for phase jumping, frequency jumping, and signal loss. The computation complexity is low and costs little time to detect anomaly. But if we want to enhance its weak anomaly detection performance, we need to lower the working frequency of Phase Detector, which will lead to longer detection delay. In practice, we pay more attention to large frequency jumping in satellite atomic clock, which will obviously affect the positioning accuracy and our PLL method is designed for it.

### *2.2. Self-Monitoring Method Based on Statistics*

*2.2.1. Allan Variance.* We usually use Allan variance [15, 16] to evaluate the stability of atomic clock; it can be expressed as follows:

$$\sigma_y^2(\tau) = \frac{1}{2(M-1)}\sum_{i=1}^{M-1}\left[\bar{y}_{i+1}(m) - \bar{y}_i(m)\right]^2,$$
$$y_i = \frac{x_{i+1} - x_i}{\tau} = \frac{f_o - f_r}{f_r}, \tag{12}$$
$$\bar{y}_i(m) = \frac{1}{m}\sum_{j=1}^{m} y_j,$$

where $\tau = m\tau_0$ is the averaging time and $M$ is the amount of $\bar{y}_i(m)$. What needs to be pointed out is that $y_i$ is the relative frequency deviation. $f_o$ is the instantaneous frequency, $f_r$ is the nominal frequency, and $x_i$ is the clock time error at the $i$th measuring instant.

Power-law spectrum is used to analyze noise property in frequency domain:

$$S_y(f) = h_{-2}f^{-2} + h_{-1}f^{-1} + h_0 + h_1 f + h_2 f^2 = \frac{1}{(2\pi)^2}\sum_{\alpha=-2}^{\alpha=2} h_\alpha f^\alpha, \tag{13}$$

where $S_y(f)$ is spectrum density for relative frequency deviation and $h_\alpha$ is the amplitude corresponding to different noise type. The power-law spectrum model contains five kinds of noise ($\alpha = -2, -1, 0, 1, 2$); they are RW FM, Flicker FM, White FM, Flicker PM, and White PM. $\sigma_y^2(\tau)$ is determined by these five kinds of noise:

$$\sigma_y^2(\tau) = \frac{3f_h h_2}{(2\pi\tau)^2} + \frac{1.038 + 3\ln(2\pi f_h \tau)\,h_1}{(2\pi\tau)^2} + \frac{h_0}{2\tau} + 2\ln(2)\,h_{-1} + \frac{2\pi^2\tau h_{-2}}{3}. \tag{14}$$

The slope of Allan variance gives us a knowledge of noise distribution in different averaging time.

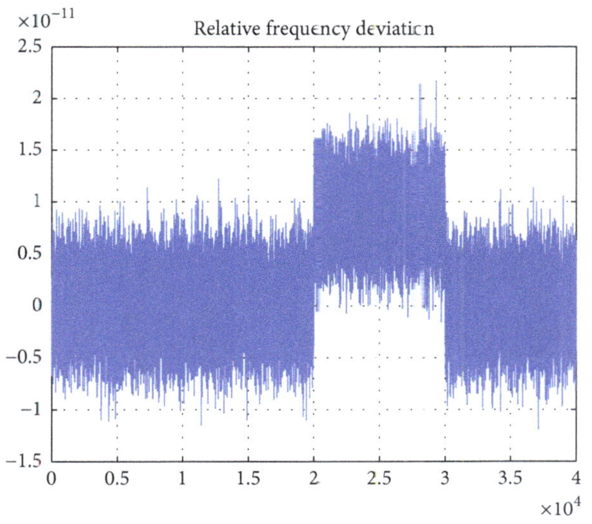


(a) Measured value of relative frequency deviation

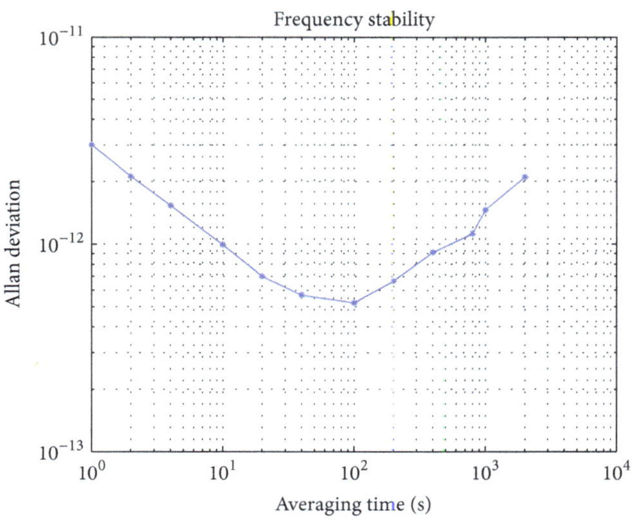


(b) Allan deviation

FIGURE 6: Relative frequency deviation and its Allan deviation when anomaly occurs.

After obtaining a sufficient number of measurement data, Allan variance can be used to calculate the stability of different averaging time. But when anomaly occurs, the results given by Allan variance may lose practical significance. As shown in Figure 6(a), the frequency of clock signal jumped and then returned some time later. We cannot get correct judgment for noise distribution according to the computed result by (12) shown in Figure 6(b). Besides, we do not know the anomaly type and detection delay is also too long.

*2.2.2. Dynamic Allan Variance (DAVAR).* As can be seen from Figure 6(a), the main noise type does not change, but Figure 6(b) gives wrong judgment. Hence the conclusion is not consistent with the actual situation, and we cannot find the sign of frequency jumping either. Therefore, the traditional Allan variance cannot give believable information on such anomaly. In view of this, Galleani and Tavella put forward DAVAR, which can be expressed as (15) and can be used to evaluate the performance of clock in real time:

$$\sigma_y^2(n,k) = \frac{1}{2k^2\tau_0^2}\frac{1}{N/k-1}\sum_{i=0}^{N/k-2}\left[\overline{y}_{i+1,k,n}-\overline{y}_{i,k,n}\right]^2, \qquad \overline{y}_{i,k,n} = \frac{1}{k}\sum_{m=n-N+1+ik}^{m=n-N+1+(i+1)k} y_m. \tag{15}$$

When we calculate DAVAR, a sliding window is used to cut the data. The window length is $N$, and $\sigma_y^2(n,k)$ will be updated when new measurement data $y_n$ comes, so it can tell us health condition of clock in real time. $\tau_0$ is the least measurement period, and $k\tau_0$ is the averaging time.

*2.2.3. Modified DAVAR.* From expression (15), we know that DAVAR can be updated in real time, but in order to guarantee the reliability of long-term stability, $N$ must be large enough, which will greatly reduce the detection probability of instantaneous anomaly. Because when frequency jumping occurs, $\ldots, y_{i-1}-y_{i-2}\approx 0, y_i-y_{i-1}=\delta, y_{i+1}-y_i\approx 0,\ldots$, only one factor is not 0, DAVAR is not sensitive enough to weak frequency jumping. In this paper, we modify Dynamic Allan Variance to improve detection sensitivity for small frequency jumping:

$$\sigma_y^2(n) = \frac{1}{2m}\sum_{i=1}^{m}\left[\overline{y}_{i+n-m}-\overline{y}_{i+n-2m}\right]^2, \tag{16}$$

$$\sigma_y^2(n) = \sigma_y^2(n-1) + \frac{(\overline{y}_n-\overline{y}_{n-m})^2-(\overline{y}_{n-m}-\overline{y}_{n-2m})^2}{2m}. \tag{17}$$

Expression (16) is the Modified DAVAR, and (17) is its iterative calculation method.

*2.2.4. Detection Performance of Modified DAVAR.* In this section, we will firstly analyze and compare the detection performances of DAVAR and Modified DAVAR when facing phase and frequency jumping and then show that Modified DAVAR is also effective in detecting instantaneous stability deterioration.

The monitoring method for phase and frequency jumping is based on statistics. $\Delta t_{12}$, $\Delta t_{13}$, and $\Delta t_{23}$ from Figure 1 will be used here. Assume that only one of the three clocks breaks down. If phase or frequency jumping occurs in clock 1, $\Delta t_{12}$, $\Delta t_{13}$ will be abnormal, while $\Delta t_{23}$ is still normal. Because the anomaly in $\Delta t_{12}$ is the same as that in $\Delta t_{13}$, we only need to analyze $\Delta t_{12}$.

In Figure 7(a), the amplitude of phase jumping is 12 times the standard deviation of the relative frequency deviation data $y_i$. In Figure 7(b), the frequency jumping is 4 times the standard deviation of $y_i$.

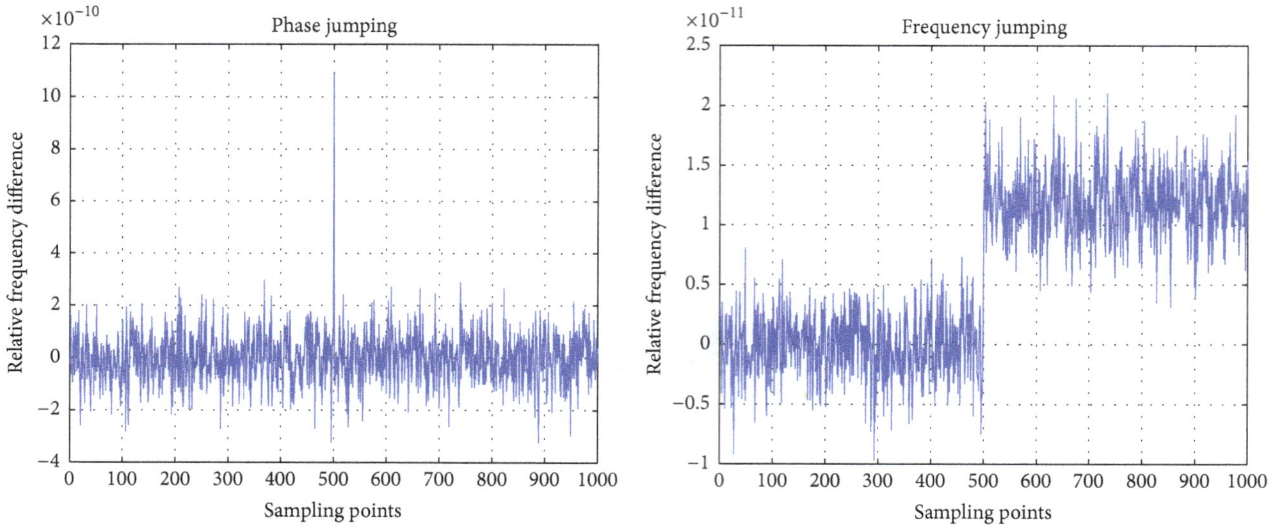


(a) Phase jumping

(b) Frequency jumping

FIGURE 7: Phase and frequency jumping.

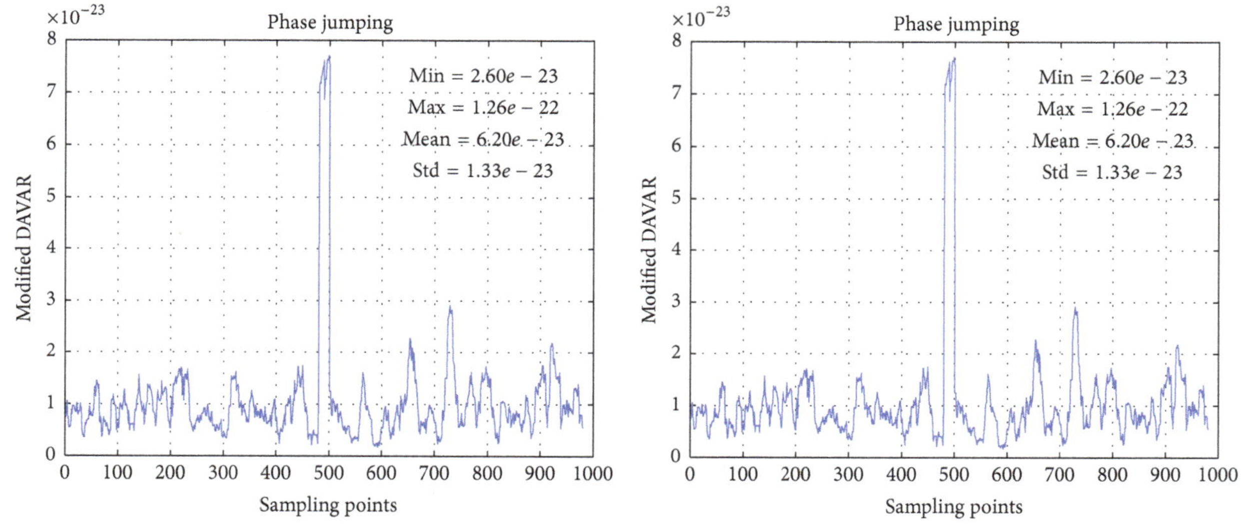


(a) Average response of DAVAR to phase jumping

(b) Average response of Modified DAVAR to phase jumping

FIGURE 8: Comparison of two variances for phase jumping.

We simulated 1000 sampling points and phase and frequency jumping occurred at the 500th point. We simulated one realization and saved the response data of DAVAR and Modified DAVAR to jumping at every time instant; then we repeated 10000 realizations in the same way. Of course the 1000 sampling points' data is different in every realization. Then we got the average response at every time instant that is shown in Figures 8 and 9.

Because the peak value of response of DAVAR and Modified DAVAR to jumping determines if the jumping could be detected, we focus on the peak value in each realization. Assuming that, in one realization, the maximum value of response of DAVAR to frequency jumping is $\sigma_{i,m}^2$, the maximum value of response of Modified DAVAR to frequency jumping is $\upsilon_{i,m}^2$, and then we saved $\sigma_{i,m}^2$ and $\upsilon_{i,m}^2$ $(i = 1, 2, 3, \ldots, 10000)$. We studied the saved data to give the minimum value, maximum value, mean value, and standard deviation of $\sigma_{i,m}^2$ and $\upsilon_{i,m}^2$ $(i = 1, 2, 3, \ldots, 10000)$ in Figures 8 and 9.

To make the statistical result more clear, we list the statistical characteristic of $\sigma_{i,m}^2$ and $\upsilon_{i,m}^2$ $(i = 1, 2, 3, \ldots, 10000)$ in frequency jumping in Table 5. From Table 5 and Figure 9 we know that $\sigma_{i,m}^2$ is not big enough to be distinguished from the base noise, and Modified DAVAR is more sensitive to weak frequency jumping.

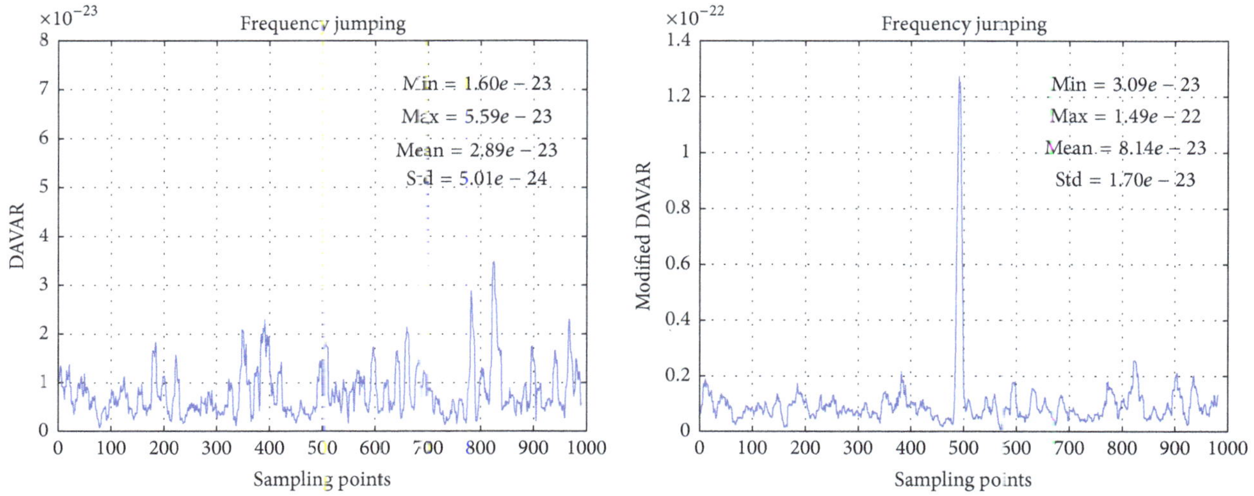


(a) Average response of DAVAR to frequency jumping

(b) Average response of Modified DAVAR to frequency jumping

FIGURE 9: Comparison of two variances for frequency jumping.

TABLE 5: Statistical characteristic of $\sigma^2_{i,m}$ and $v^2_{i,m}$.

| Statistical item | DAVAR | Modified DAVAR |
|---|---|---|
| Minimum value | $1.60e-23$ | $3.09e-23$ |
| Maximum value | $5.59e-23$ | $1.49e-22$ |
| Mean value | $2.89e-23$ | $8.14e-23$ |
| Standard deviation | $5.01e-24$ | $1.70e-23$ |

TABLE 6: Detection performance for phase jumping.

| Detection item | DAVAR | Modified DAVAR |
|---|---|---|
| Resolution | 12 | 12 |
| Threshold | $4.5E-23$ | $4.5E-23$ |
| PD | 1.0000 | 0.9978 |
| PFA | $1.61e-5$ | $3.00e-7$ |
| Detection delay (points) | 1 | 1 |

TABLE 7: Detection performance for phase jumping.

| Detection item | DAVAR | Modified DAVAR |
|---|---|---|
| Resolution | 12 | 4 |
| Threshold | $4.5E-23$ | $4.5E-23$ |
| PD | 0.9874 | 0.9927 |
| PFA | $1.61e-5$ | $3.00e-7$ |
| Detection delay (points) | 1 | 5 |

In Tables 6 and 7, the detection performance of two kinds of variance for phase and frequency jumping is given. Tables 8 and 9 tell us the detection performance for different clock stability.

What should be pointed out is that the unit of resolution in phase and frequency jumping in Tables 6–9 is the standard deviation of the relative frequency deviation data $y_i$, namely, $\sigma_y$. Detection delay is the number of sampling points from anomaly occurring to be detected.

During the simulation, we simulate 10000 realizations and use same threshold for both DAVAR and Modified DAVAR. The window length $N = 10$, and $k\tau_0 = \tau_0 = 1$ s. It should be noted that because $k\tau_0 = \tau_0 = 1$ s, it is precise enough for us to do the approximation to only consider WFM in the simulation. In fact, the Modified DAVAR is still effective in detecting phase and frequency jumping in the presence of other noise types.

Table 6 tells us that PD of Modified DAVAR is almost as good as DAVAR and PFA is lower. From Table 7, we know that Modified DAVAR is more sensitive to weak frequency jumping but detection delay is longer.

From Tables 8 and 9, we notice that the resolution of Modified DAVAR for phase and frequency jumping is the same for different clocks. What we need to do is only to reset the threshold according to expression (18):

$$\mathrm{Thr} = 5 \cdot \sigma^2(1). \tag{18}$$

In addition, the window length $N$ is an important parameter for Modified DAVAR. The longer the window, the weaker the detection performance. However, if the window length is too short, PFA will rise and detection resolution will also deteriorate.

Figure 10 shows that the Modified DAVAR can also monitor instantaneous deterioration effectively.

Modified DAVAR can be considered as a statistical tool; it is effective to detect phase and frequency jumping. Compared with PLL method, we need to measure the time error data firstly and then calculate the statistical characteristics of clock. Modified DAVAR can monitor weaker frequency jumping compared to PLL method, but PLL method is independent of a second standard reference and time-comparison device, which will give us more flexibility. Taking their respective characteristics into account, cooperation between them may be a good choice to improve the Self-Monitoring reliability.

TABLE 8: Detection performance of phase jumping for different clock stability.

| $\sigma(\tau_0)$ | Resolution | Threshold | PD | PFA | Delay (points) |
|---|---|---|---|---|---|
| $3E-12$ | 12 | $4.5E-23$ | 0.9978 | $3.00e-7$ | 1 |
| $3E-11$ | 12 | $4.5E-21$ | 0.9961 | $5.00e-7$ | 1 |
| $3E-10$ | 12 | $4.5E-19$ | 0.9977 | $6.00e-7$ | 1 |

TABLE 9: Detection performance of frequency jumping for different clock stability.

| $\sigma(\tau_0)$ | Resolution | Threshold | PD | PFA | Delay (points) |
|---|---|---|---|---|---|
| $3E-12$ | 4 | $4.5E-23$ | 0.9927 | $3.00e-7$ | 5 |
| $3E-11$ | 4 | $4.5E-21$ | 0.9923 | $5.00e-7$ | 5 |
| $3E-10$ | 4 | $4.5E-19$ | 0.9934 | $6.00e-7$ | 5 |

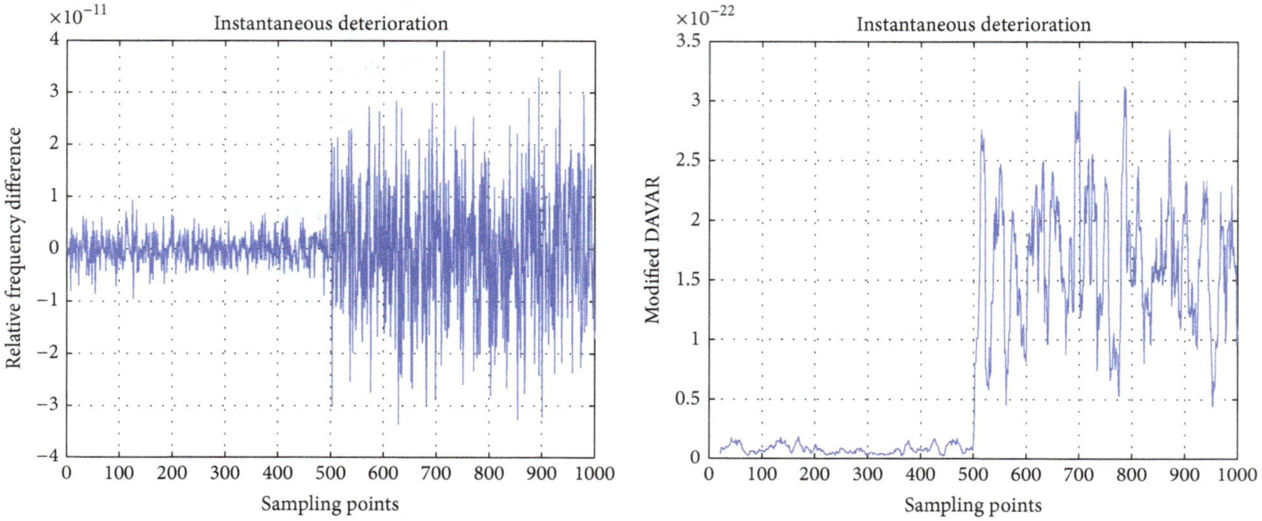


(a) Instantaneous deterioration

(b) Modified DAVAR

FIGURE 10: Instantaneous deterioration of clock.

*2.2.5. LS Method and Kalman Method on Detection for Frequency Jumping.* In this section, we will introduce two existing methods, which are called LS method and Kalman filter method.

*LS Method.* We may make use of LS algorithm to calculate the averaging frequency deviation $\overline{y}$ with $M$ newest saved sampling points and then predict the time error $\Delta\hat{t}_{k+1} = \Delta t_k + \overline{y}\cdot\tau$. After we can compare it with the real measurement $\Delta t_{k+1}$, if the difference $\varepsilon = \Delta\hat{t}_{k+1} - \Delta t_{k+1}$ is beyond the threshold $\gamma$, we think that frequency jumping occurs.

*Kalman Filter Method.* We can make use of Kalman filter to predict the next state $\hat{y}_n$ of clock and then compare it with the real measurement $y_n$. If the difference $\varepsilon = \hat{y}_n - y_n$ is larger than the configurable threshold, we think that anomaly occurs.

*Simulations.* Because the sampling interval $\tau = 1$ s, we only consider WFM noise, whose standard deviation is $\sigma_0 = 3E-12$. During the simulations for LS method, we choose $M = 20$, while, for Kalman filter method, the state transition matrix $\Phi = 1$, observation matrix $H = 1$, system error covariance matrix $Q = (1.0 \times 10^{-13})^2$, and observation error covariance matrix $R = (3.0 \times 10^{-12})^2$.

Figures 11 and 12 show the frequency jumping detection of the two methods.

After we have done numerical simulations, we give Table 10 to show the detection performance of the methods, in which PLL method, DAVAR method, and Modified DAVAR method are included.

*Discussion.* It should be noted firstly that the detection performance will be different with different parameters. We use the same noise level to test different methods to give Table 10, which can be a reference to show different characteristics of different methods.

Different observation quantity is needed for different methods. In our opinion, DAVAR, Modified DAVAR, LS, and Kalman filter are all effective in detecting weak frequency jumping. They need a second standard reference and time-comparison device to get the time error measurement $\Delta t_k$ to

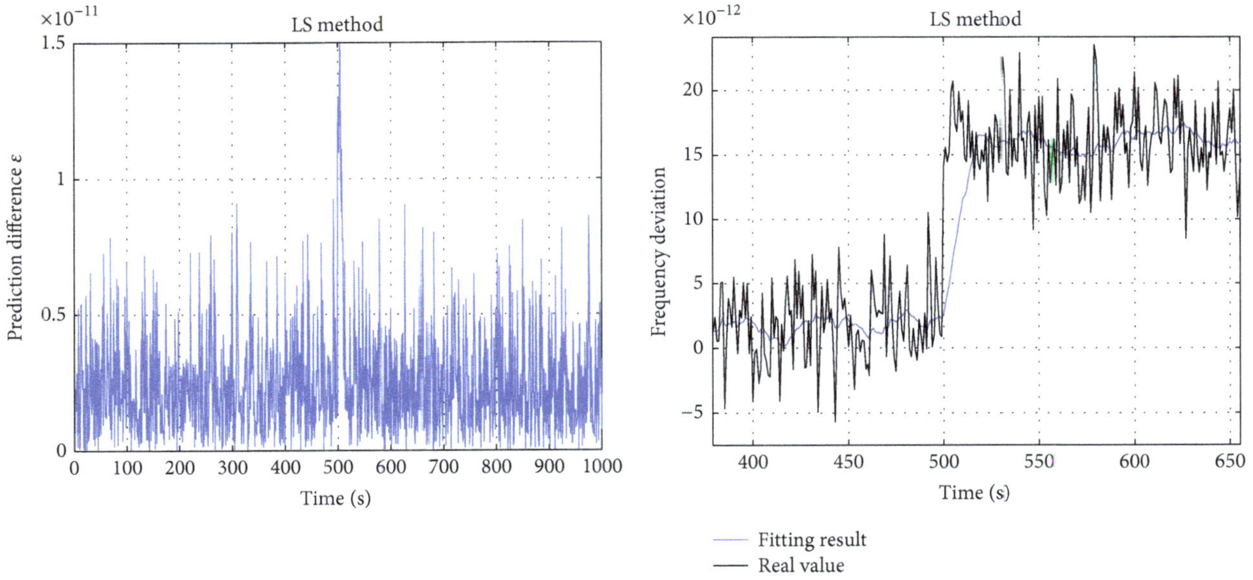


(a) $\varepsilon = \Delta\hat{t}_{k+1} - \Delta t_{k+1}$ (b) $\overline{y}$ and $y$

FIGURE 11: LS method in detecting frequency jumping.

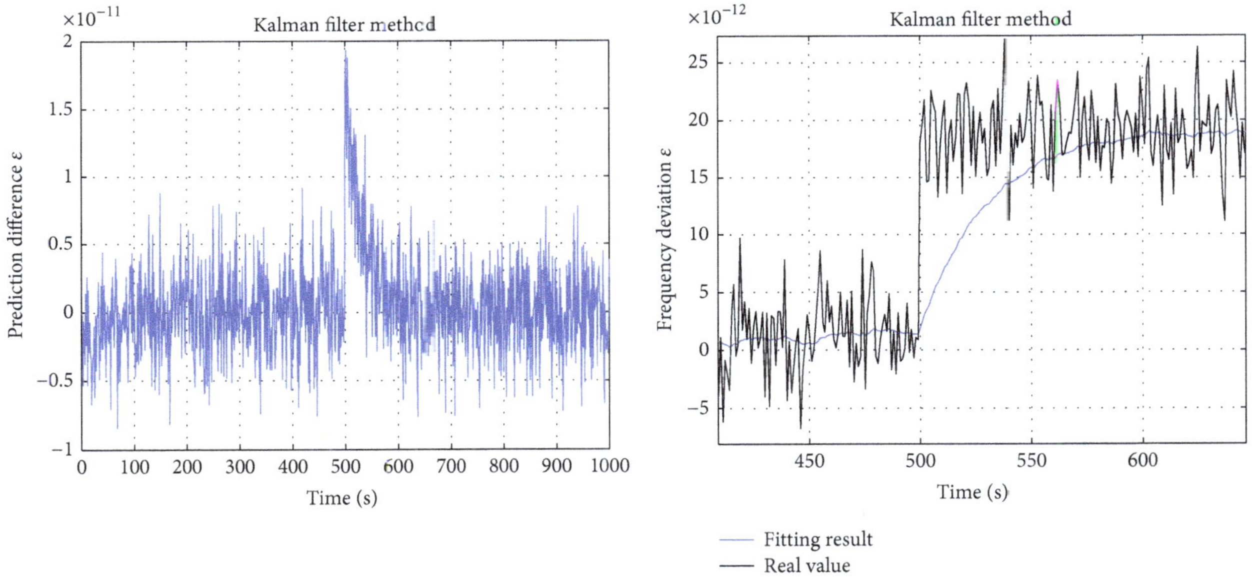


(a) $\varepsilon = \hat{y}_n - y_n$ (b) $\hat{y}_n$ and $y_n$

FIGURE 12: Kalman filter method in detecting frequency jumping.

TABLE 10: Detection performance of frequency jumping for different clock stability.

| Detection method | Resolution | PD | PFA | Delay |
|---|---|---|---|---|
| PLL method | $4.0E-11$ | 0.9933 | $2.41E-6$ | 0.4 s |
| DAVAR method | $3.6E-11$ | 0.9874 | $1.61E-5$ | 1 s |
| Modified DAVAR method | $1.2E-11$ | 0.9927 | $3.00E-7$ | 5 s |
| LS method | $1.5E-11$ | 0.9944 | $2.53E-5$ | 1 s |
| Kalman filter method | $1.8E-11$ | 0.9926 | $2.60E-6$ | 1 s |

TABLE 11: Parameters of generated phase data.

| $\sigma_0$ | $\sigma_{-1}$ | $\sigma_{-2}$ | Drift rate/day | Frequency deviation |
|---|---|---|---|---|
| $1E-12$ | $1E-14$ | $1E-16$ | $1E-12$ | $1E-12$ |

be used as observation quantity to run the algorithm. PLL method can realize Self-Monitoring for frequency jumping without standard source, which can give us more flexibility and has been used on BeiDou satellite. The computation complexity is also different among them. LS method, Kalman filter method, and PLL method should be of less computation quantity. The resolution of Modified DAVAR is the best but at the cost of longer delay, while the PLL method has the shortest delay but at the cost of worst resolution. Sometimes we may have to compromise between resolution and delay.

*2.2.6. Stability Evaluation of Satellite Clock.* In this section, we use the well-known "three-cornered hat" approach [17–19] to evaluate the stability of atomic clocks.

According to the references, we can get $\sigma_1^2(\tau)$, $\sigma_2^2(\tau)$, and $\sigma_3^2(\tau)$ from $\Delta t_{12}$, $\Delta t_{13}$, and $\Delta t_{23}$ in Figure 1. For example, $\sigma_1^2(\tau)$ can be expressed as

$$\sigma_1^2(\tau) = \frac{1}{2}\left[\sigma_{12}^2(\tau) + \sigma_{13}^2(\tau) - \sigma_{23}^2(\tau)\right]. \quad (19)$$

Because Allan variance is convergent for the five kinds of noises at different averaging time, it is often used to evaluate the stability of clock. However Allan variance cannot rule out frequency drift. Especially when the drift is almost equal to Allan variance for certain averaging time, if we use Allan variance to calculate $\sigma_{12}^2(\tau)$, $\sigma_{13}^2(\tau)$, and $\sigma_{23}^2(\tau)$ and then calculate $\sigma_1^2(\tau)$, $\sigma_2^2(\tau)$, and $\sigma_3^2(\tau)$, the result cannot reflect the real situation of atomic clock. In order to avoid the influence of frequency drift, two-order difference for frequency data or three-order difference for phase data is needed, which is just the definition of Hadamard variance.

In order to make full use of the measurement data and also track slow change of satellite clock timely, we use windowed overlapping Hadamard variance, as shown in (20); even though the additional overlapping differences are not all statistically independent, they nevertheless increase the number of degrees of freedom and thus improve the confidence in the estimation. Moreover, by using the latest data, the variance can evaluate the health condition of atomic clock in real time:

$$H\sigma_y^2(n) = \frac{1}{6(N-3k)\tau^2} \cdot \sum_{i=n-N+1}^{n-3k}\left[x_{i+3k} - 3x_{i+2k} + 3x_{i+k} - x_i\right]^2, \quad (20)$$

where $N$ is the length of the window, namely, the amount of data used for each update, and $b\tau_0$ is the time interval between $x_i$ and $x_{i+1}$, which is defined as sampling period. $\tau_0$ is the measurement period, averaging time $\tau = m\tau_0$, and $k = m/b$. We know that, from the characteristics of Hadamard variance, the longer the averaging time, the larger the amount of data needed.

Expression (21) can be used as recursive algorithm to reduce computation complexity in the updating of $H\sigma_y^2(n)$. Moreover, $6(N-3m)\tau^2$ is a constant for each averaging time; division operation can be done only when necessary:

$$H\sigma_y^2(n) = \frac{H\sigma_y^2(n-1) + \Delta_1(n) - \Delta_2(n)}{6(N-3m)\tau^2}, \quad (21)$$

where $\Delta_1(n) = (x_n - 3x_{n-m} + 3x_{n-2m} - x_{n-3m})^2$ and $\Delta_2(n) = (x_{n-N+3m} - 3x_{n-N+2m} + 3x_{n-N+m} - x_{n-N})^2$.

As opposed to the Allan variance, which makes use of a second difference, the Hadamard variance employs a third difference that leads to reduction in the degrees of freedom by one. The Hadamard variance requires more data to produce a single stability calculation, as compared to the Allan variance, given equal averaging time $\tau$. So it will be a better choice to use different statistical tool for different averaging time. When the averaging time is short, it is much more convenient to use an ADEV three-cornered hat method, while HDEV will be a better choice when the linear frequency drift is dominant.

*2.2.7. Detection for Frequency Drift-Rate Anomaly of Clock.* According to [20–24], several drift-rate estimators are discussed and compared. We will firstly compare six different estimators by simulations in the following:

(a) Two points: $\hat{z}_1 = (y(n) - y(1))/(n-1)\tau_0$.

(b) Two groups of points: $\hat{z}_2 = (2/n\tau)[(2/n)\sum_{i=n/2+1}^{n} y(i)z - (2/n)\sum_{i=1}^{n/2} y(i)]$.

(c) LS: $\hat{z}_3 = (6/n(n^2-1)\tau_0)\sum_{i=1}^{n}(2i-n-1)y(i)$.

(d) Three points: $\hat{z}_4 = (x(2n+1) - 2x(1+n) + x(1))/(n\tau_0)^2$.

(e) $\omega_4$: $\hat{z}_5 = (6/N^3\tau_0^2 r_1(1-r_1))[w_N - w_0 - (w_{N-n_1} - w_{n1})/(1-2r_1)]$.

(f) Kalman filter: $X_k = \Phi X_{k-1} + \varepsilon_k$, $Z_k = HX_k + \nu_k$,

where $X_k = [x_k, y_k, z_k]$ is the phase data, frequency deviation and drift rate, and observation matrix $H = [1, 0, 0]$. The state transition matrix is

$$\phi = \begin{bmatrix} 1 & \tau & \dfrac{\tau^2}{2} \\ 0 & 1 & \tau \\ 0 & 0 & 1 \end{bmatrix} \quad (22)$$

and state error covariance matrix $Q$ can be expressed as [25].

To compare these six estimators, we generate simulation data by well-known Stable 32 software. The parameters of phase data are shown in Table 11, and we consider three kinds of noise type which are WFM, FFM, and RWFM.

TABLE 12: Statistical property of the estimation result of six estimators.

| Estimator | $a$ | $b$ | $c$ | $d$ | $e$ | $f$ |
|---|---|---|---|---|---|---|
| $E(z)$ | $1.00E-12$ | $1.00E-12$ | $1.00E-12$ | $1.00E-12$ | $1.00E-12$ | $9.97E-13$ |
| $\sigma(z)$ | $3.48E-13$ | $2.40E-13$ | $2.51E-13$ | $1.74E-13$ | $1.58E-13$ | $3.71E-14$ |

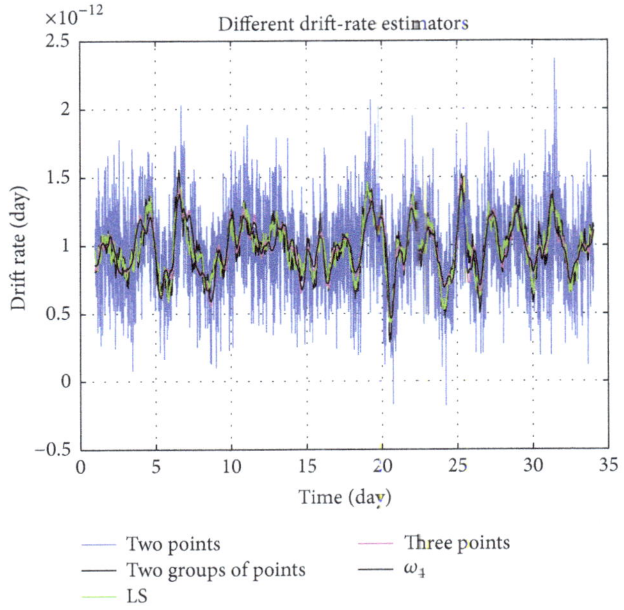


(a) Drift-rate estimation of different estimators

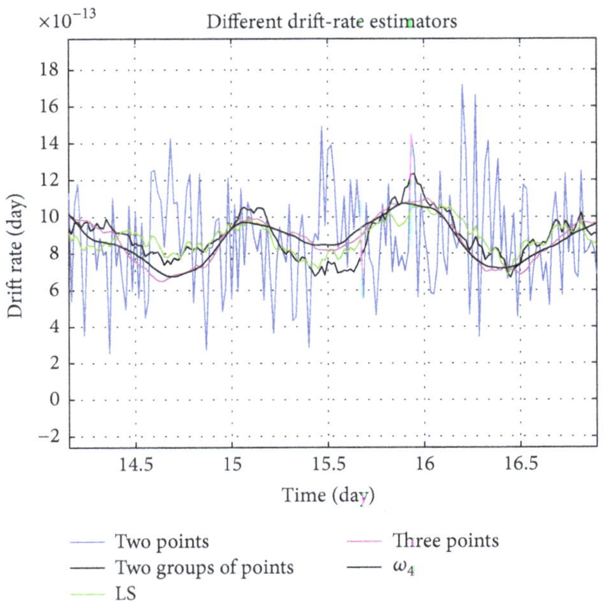


(b) Partial enlarged view of (a)

FIGURE 13: Comparison of different drift-rate estimators.

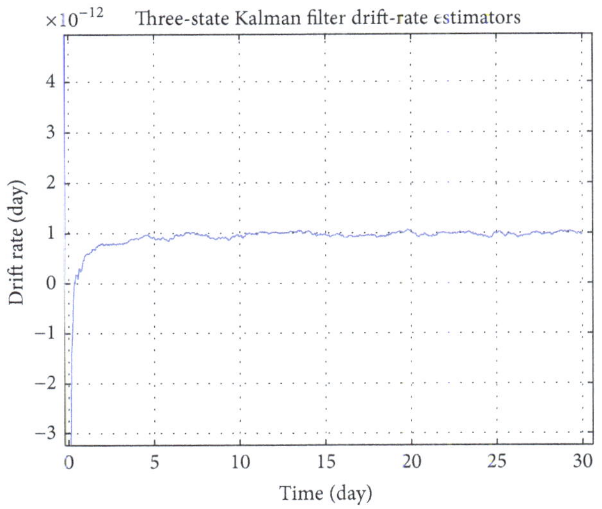


FIGURE 14: The drift-rate estimation of Kalman filter.

From Table 12 and Figures 13 and 14, we know that $\omega_4$ method and Kalman filter should be good choice for drift-rate estimation. To make sure that the simulation result is reliable, we make use of different simulation data to test the performance of the six estimators, and the result is just similar.

In fact, no matter which method we choose to evaluate the drift rate, we must know the time error data $x_k$, which is equivalent to $\Delta t$ in this section.

When the satellite is within the sight of ground station, we can calculate the drift rate with certain estimator by comparing the satellite time with the timescale on the ground. But when the satellite cannot contact with the station, the only available data is the intercomparison data $\Delta t_{12}$, $\Delta t_{13}$, and $\Delta t_{23}$.

In order to evaluate the drift rate, there may be two methods. (1) Three-state Kalman filter can be used to evaluate the drift rate directly with the observation quantity $\Delta t_{12}$, $\Delta t_{13}$, and $\Delta t_{23}$, which is similar to the Kalman timescale algorithm on the ground. (2) Two steps are needed. The first step is to predict $\Delta t_1$, $\Delta t_2$, and $\Delta t_3$ from $\Delta t_{12}$, $\Delta t_{13}$, and $\Delta t_{23}$. And the second step is to evaluate drift rate with $\Delta\hat{t}_1$, $\Delta\hat{t}_2$, and $\Delta\hat{t}_3$ ($\Delta\hat{t}_1$, $\Delta\hat{t}_2$, and $\Delta\hat{t}_3$ are the prediction values of $\Delta t_1$, $\Delta t_2$, and $\Delta t_3$).

In the following, we will compare these two methods. The three-state Kalman filter method is called method 1, and the combination of two-state Kalman filter with $\omega_4$ is called method 2.

The basic Kalman fiter equations are as follows.

System equation is

$$X_k = \Phi X_{k-1} + \varepsilon_k. \tag{23}$$

Observation equation is

$$Z_k = HX_k + \nu_k. \tag{24}$$

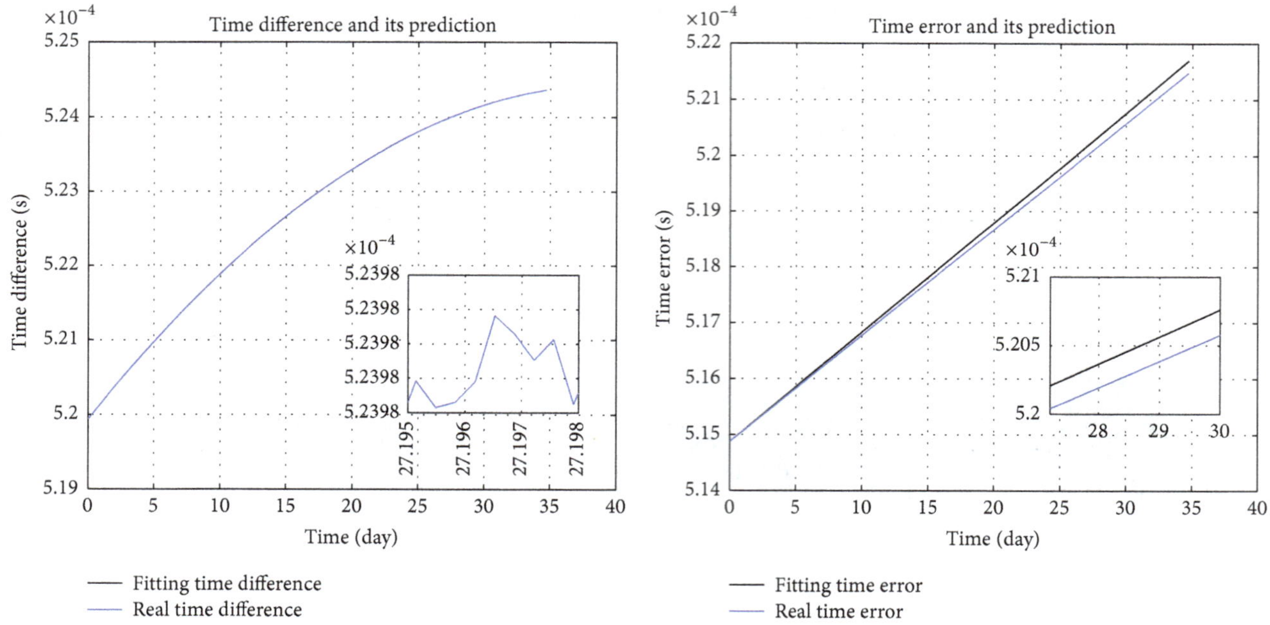


(a) $\Delta t_{12}$ and $\Delta\hat{t}_{12}$

(b) $\Delta t_1$ and $\Delta\hat{t}_1$

FIGURE 15: Prediction performance of Kalman filter for $\Delta t_1$ and $\Delta t_{12}$.

TABLE 13: Parameter setting.

| Parameters | $\sigma_0$ | $\sigma_{-1}$ | $\sigma_{-2}$ | Drift rate/day | Frequency deviation |
|---|---|---|---|---|---|
| Clock 1 | $1e-12$ | $2e-14$ | $3e-16$ | $1e-13$ | $1e-12$ |
| Clock 2 | $2e-12$ | $3e-14$ | $1e-16$ | $1e-12$ | $1e-12$ |
| Clock 3 | $3e-12$ | $1e-14$ | $2e-16$ | $-2e-11$ | $1e-12$ |

For the two-state Kalman filter,

$$X_k = [x_{1,k}, y_{1,k}, x_{2,k}, y_{2,k}, x_{3,k}, y_{3,k}]^T,$$
$$\Phi = \operatorname{diag}(\phi, \phi, \phi),$$
$$\phi = \begin{bmatrix} 1 & \tau \\ 0 & 1 \end{bmatrix}, \tag{25}$$
$$Z_k = [\Delta t_{12,k}, \Delta t_{13,k}]^T,$$
$$H = \begin{bmatrix} 1 & 0 & -1 & 0 & 0 & 0 \\ 1 & 0 & 0 & 0 & -1 & 0 \end{bmatrix}.$$

The system error covariance matrix $Q$ can be expressed as [26–28].

For the three-state Kalman filter,

$$X_k = [x_{1,k}, y_{1,k}, z_{1,k}, x_{2,k}, y_{2,k}, z_{2,k}, x_{3,k}, y_{3,k}, z_{3,k}]^T,$$
$$\Phi = \operatorname{diag}(\phi, \phi, \phi),$$
$$\phi = \begin{bmatrix} 1 & \tau & \frac{\tau^2}{2} \\ 0 & 1 & \tau \\ 0 & 0 & 1 \end{bmatrix},$$
$$Z_k = [\Delta t_{12,k}, \Delta t_{13,k}]^T,$$
$$H = \begin{bmatrix} 1 & 0 & 0 & -1 & 0 & 0 & 0 & 0 & 0 \\ 1 & 0 & 0 & 0 & 0 & 0 & -1 & 0 & 0 \end{bmatrix}, \tag{26}$$

and we can get the system error covariance matrix $Q$ according to [29].

To compare these two methods, we make use of Stable 32 to generate simulation data. The parameters of the phase data are shown in Table 13.

Firstly, we will show the prediction performance of both two-state and three-state Kalman filter for $\Delta t_1$, $\Delta t_2$, and $\Delta t_3$; they are similar, which can be shown in Figure 15.

For method 2, after we have got the prediction of $\Delta t_1$, $\Delta t_2$, and $\Delta t_3$, we will use $\omega_4$ to evaluate the drift rate.

Figure 15(a) is the comparison of the clock difference $\Delta t_{12}$ ($\Delta t_{12} = \Delta t_1 - \Delta t_2$) and its prediction $\Delta\hat{t}_{12}$. Figure 15(b)

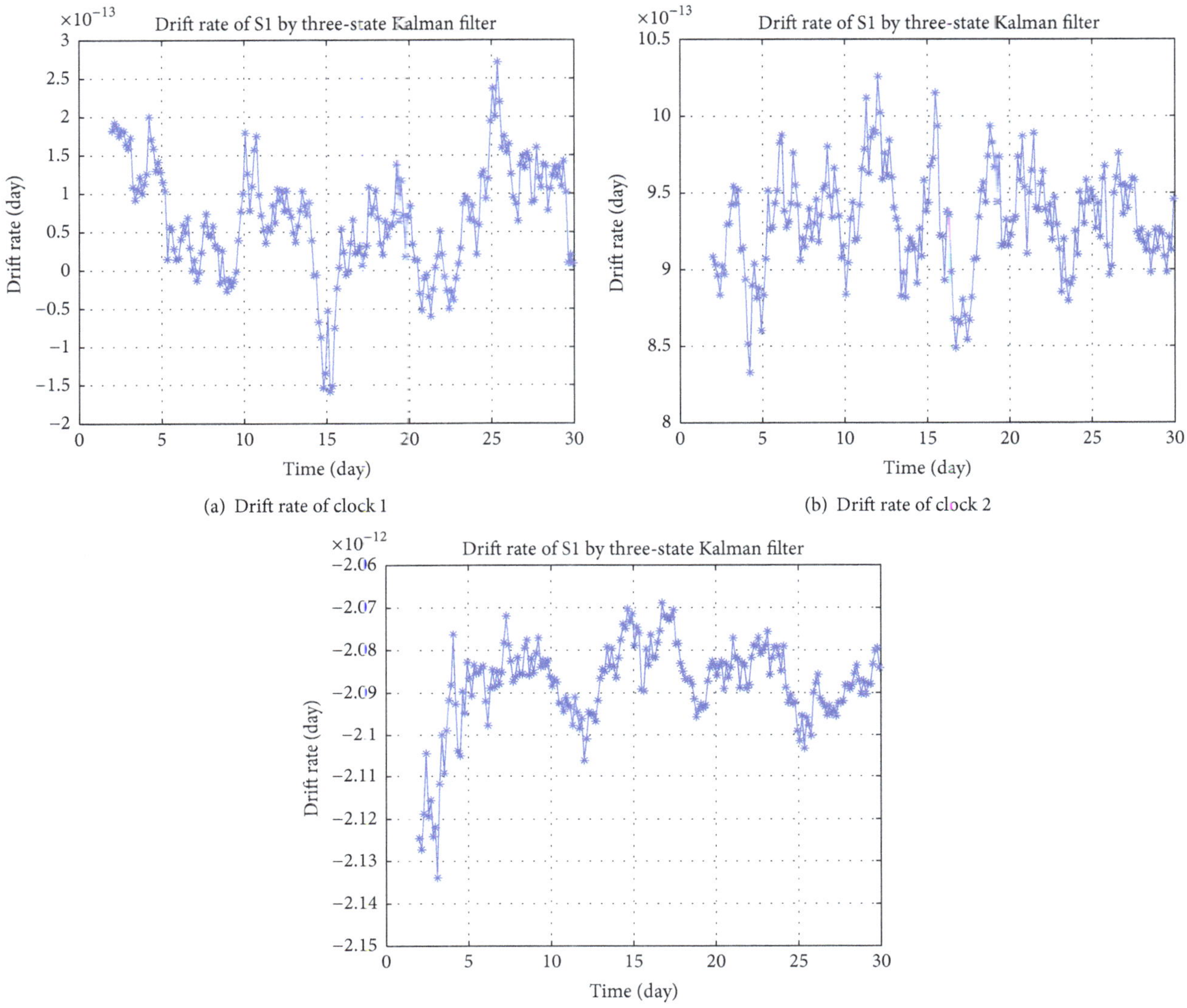


(a) Drift rate of clock 1

(b) Drift rate of clock 2

(c) Drift rate of clock 3

FIGURE 16: Drift-rate estimation of method 1.

compares the real time errors $\Delta t_1$ and $\Delta\hat{t}_1$. From Figure 15, we know that the prediction error $\varepsilon$ ($\varepsilon = \Delta t_1 - \Delta\hat{t}_1$) will increase gradually, while $\Delta\hat{t}_{12}$ is unbiased. In fact, this phenomenon is inevitable owing to the lack of absolute standard reference. No matter which method we choose, the prediction error will increase with time going. That also indicates that we cannot get precise drift rate, but it does not mean we can do nothing. Next we will prove that method 1 can give an alarm when the drift rate of one clock is much larger than the best one. It should be noted that when we make use of $\omega_4$ method to evaluate drift rate for method 2, a sliding window is used to evaluate the drift rate in real time instead of batch processing.

Figures 16 and 17 show the drift-rate estimation of these two methods for three clocks. From the simulation result, we know that if the drift rate of one clock is much larger than the best clock, its drift-rate estimation is very close to its real value, and its computed estimation is much larger than the best clock, which enables method 1 to give an alarm. Besides, method 1 gives more precise and stable computation result compared to method 2.

## 3. Conclusion

This paper puts forward a set of Self-Monitoring Methods for common anomalies. We use PLL to realize Self-Monitoring for signal loss and phase and frequency jumping. Based on the measurement data from intercomparison among three clocks, Modified DAVAR is used to detect phase and frequency jumping and instantaneous deterioration; we use windowed overlapping Hadamard variance to evaluate clock stability in real time and the three-state Kalman filter for large drift-rate anomaly.

The method based on PLL has been proved effective and used in newest BeiDou satellite. And the other research on

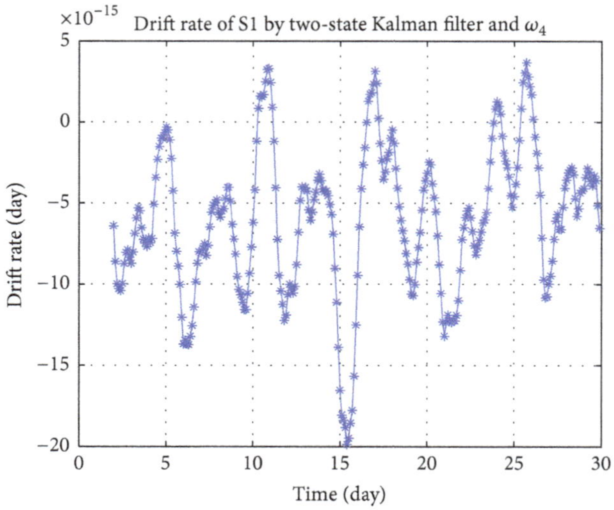


(a) Drift rate of clock 1

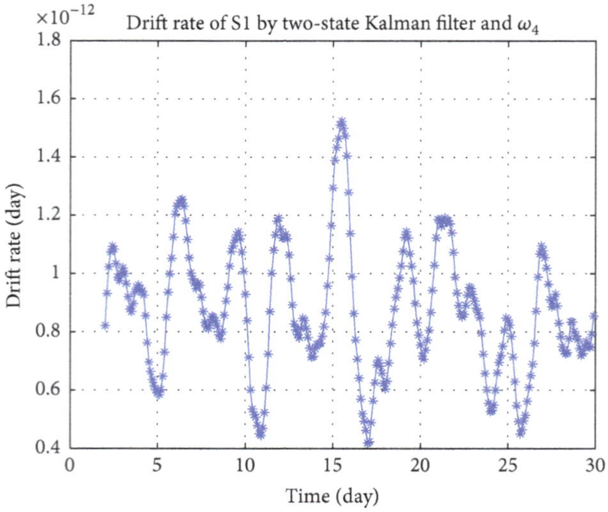


(b) Drift rate of clock 2

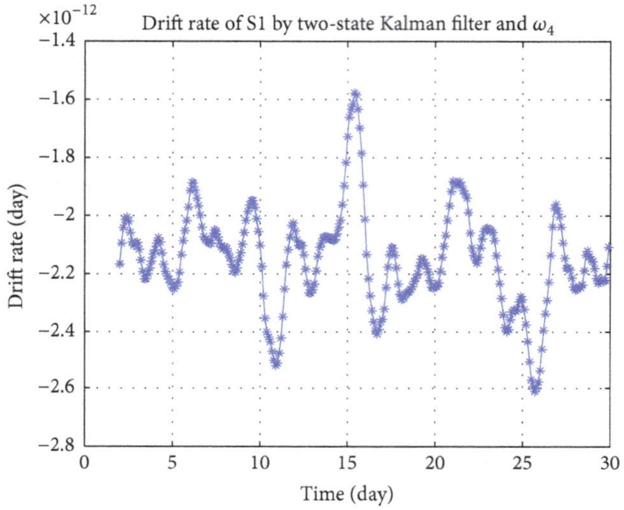


(c) Drift rate of clock 3

FIGURE 17: Drift-rate estimation of method 2.

Self-Monitoring Method in this paper can be used in next generation navigation satellites after year 2019.

## Competing Interests

The authors declare that they have no competing interests.

## References

[1] Y. C. Chan, W. A. Johnson, S. K. Karuza, A. M. Young, and J. C. Camparo, "Self-monitoring and self-assessing atomic clocks," *IEEE Transactions on Instrumentation and Measurement*, vol. 59, no. 2, pp. 330–334, 2010.

[2] Y. Liu and S. Tang, "Studying method of monitoring composite clocks performance," in *Proceedings of the 3th China Satellite Navigation Conference (CSNC '12)*, Guangzhou, China, June 2012.

[3] S. Tang, Y. Liu, and X.-H. Li, "A study on onboard satellite atomic clock autonomous integrity monitoring," *Journal of Astronautics*, vol. 34, no. 1, pp. 39–45, 2013.

[4] E. Nunzi, P. Carbone, and P. Tavella, "Fault detection in atomic clock frequency standards affected by mean and variance changes and by an additive periodic component: the GLRT approach," in *Proceedings of the 2008 IEEE International Instrumentation and Measurement Technology Conference (IMTC '08)*, vol. 34, no 1, pp. 1594–1597, IEEE, Victoria, Canada, May 2008.

[5] E. Nunzi, L. Galleani, P. Tavella, and P. Carbone, "Detection of anomalies in the behavior of atomic clocks," *IEEE Transactions on Instrumentation and Measurement*, vol. 56, no. 2, pp. 523–528, 2007.

[6] U. Bartoccini, G. Barchi, and E. Nunzi, "Methods and tools for frequency jump detection," in *Proceedings of the IEEE International Workshop on Advanced Methods for Uncertainty Estimation in Measurement (AMUEM '09)*, pp. 109–112, Bucharest, Romania, July 2009.

[7] L. Galleani and P. Tavella, "Detection of atomic clock frequency jumps with the Kalman filter," *IEEE Transactions on Ultrasonics, Ferroelectrics, and Frequency Control*, vol. 59, no. 3, pp. 504–509, 2012.

[8] X. Huang, H. Gong, and G. Ou, "Detection of weak frequency jumps for GNSS onboard clocks" *IEEE Transactions on Ultrasonics, Ferroelectrics, and Frequency Control*, vol. 61, no. 5, pp. 747–755, 2014.

[9] S.-W. Lee, J. Kim, and Y. J. Lee, "Protecting signal integrity against atomic clock anomalies on board GNSS satellites," *IEEE Transactions on Instrumentation and Measurement*, vol. 60, no. 7, pp. 2738–2745, 2011.

[10] G. Signorile, "Analysis of a Kalman filter detector for atomic clock anomalies," in *Proceedings of the 31st URSI General Assembly and Scientific Symposium (URSI GASS '14)*, pp. 1–4, Beijing, China, August 2014.

[11] L. Galleani and P. Tavella, "The dynamic allan variance," *IEEE Transactions on Ultrasonics, Ferroelectrics, and Frequency Control*, vol. 56, no. 3, pp. 450–464, 2009.

[12] I. Sesia, L. Galleani, and P. Tavella, "Implementation of the dynamic allan variance for the galileo system test bed V2," in *Proceedings of the IEEE International Frequency Control Symposium Joint with the 21st European Frequency and Time Forum*, pp. 946–949, Geneva, Switzerland, June 2007.

[13] A. Cenigliaro and S. Valloreia, "Analysis on GNSS space clocks performances," in *Proceedings of the Joint European Frequency and Time Forum & International Frequency Control Symposium (EFTF/IFC '13)*, pp. 835–837, Prague, Czech Republic, July 2013.

[14] Z. Ji, *Synchronization Technology and Application in Communication*, Tsinghua University Press, Beijing, China, 2008.

[15] D. W. Allan, "Statistics of Atomic Frequency Standards," *Proceedings of the IEEE*, vol. 54, no. 2, pp. 221–230, 1966.

[16] D. Allan, "Time and frequency (time-domain) characterization, estimation, and prediction of precision clocks and oscillators," *IEEE Transactions on Ultrasonics, Ferroelectrics and Frequency Control*, vol. 34, no. 6, pp. 647–654, 1987.

[17] J. E. Gray and D. W. Allan, "A method for estimating the frequency stability of an individual oscillator," in *Proceedings of the 28th Annual Symposium on Frequency Conrcl*, pp. 243–246, May 1974.

[18] J. Groslambert, D. Fest, M. Olivier, and J. J. Gagnepain, "Characterization of frequency fluctuations by crosscorrelations and by using three or more oscillators," in *Proceedings of the 35th Annual Frequency Control Symposium*, Philadelphia, Pa, USA, May 1981.

[19] A. Premoli and P. Tavella, "A revisited three-cornered hat method for estimating frequency standard instability," *IEEE Transactions on Instrumentation and Measurement*, vol. 42, no. 1, pp. 7–13, 1993.

[20] M. A. Weiss and C. Hackman, "Confidence on the three-point estimator of frequency drift," in *Proceedings of the 24th Annual PTTI Meeing*, pp. 451–460, December 1992.

[21] V. A. Logachev and G. P. Pashev, "Estimation of linear frequency drift coefficient of frequency standards," in *Proceedings of the 50th IEEE International Frequency Control Symposium*, pp. 960–963, Honolulu, Hawaii, USA, June 1996.

[22] Q. Wei, "Estimations of frequency and its drift rate," *IEEE Transactions on Instrumentation and Measurement*, vol. 46, no. 1, pp. 79–82, 1997.

[23] C. A. Greenhall, "Frequency-drift estimator and its removal from modified Allan variance," in *Proceedings of the IEEE International Frequency Control Symposium*, pp. 428–432, May 1997.

[24] W. J. Riley, *Handbook of Frequency Stability Analysis*, U.S. Department of Commerce, 2008.

[25] G. Hairong, "Determination of covariance matrix of kalman filter used for time prediction of atomic clocks of navigation satellites," *Acta Geodaeicaet Cartographica Sinica*, vol. 39, no. 2, pp. 146–150, 2010.

[26] L. Galleani, L. Sacerdote, P. Tavella, and C. Zucca, "A mathematical model for the atomic clock error," *Metrologia*, vol. 40, no. 3, p. S257, 2003.

[27] L. Galleani and P. Tavella, "Time and the Kalman filter," *IEEE Control Systems Magazine*, vol. 30, no. 2, pp. 44–65, 2010.

[28] C. A. Greenhall, "Forming stable timescales from the Jones-Tryon Kalman filter," *Metrologia*, vol. 40, no. 3, pp. S335–S341, 2003.

[29] C. Zucca and P. Tavella, "The clock model and its relationship with the allan and related variances," *IEEE Transactions on Ultrasonics, Ferroelectrics, and Frequency Control*, vol. 52, no. 2, pp. 289–295, 2005.

15

# Model Predictive Control to Autonomously Approach a Failed Spacecraft

**Xun Wang**, [1] **Zhaokui Wang**, [2] **and Yulin Zhang** [1,2]

[1] *College of Aeronautics and Astronautics, National University of Defense Technology, Changsha 410073, China*
[2] *School of Aerospace Engineering, Tsinghua University, Beijing 100084, China*

Correspondence should be addressed to Zhaokui Wang; wangzk@tsinghua.edu.cn

Academic Editor: Seid H. Pourtakdoust

In this study, a model predictive control (MPC) method is developed for a servicer spacecraft autonomously approaching a tumbling failed spacecraft at an ultraclose range. Flight safety and collision avoidance are basic requirements during the approach. Two types of a failed spacecraft with complex configurations are considered, and a double-ellipsoid composite envelope strategy is designed to model their keep-out zones. Given the keep-out zone of the servicer, two expanded ellipsoids are subsequently introduced to determine the collision and sufficient conditions for collision avoidance are derived by using the form of concave constraint. The tumbling motion of the target is considered, and a CW-based translational dynamics and derived attitude dynamics of the target are formulated to predict the motion of the docking point and keep-out zone. The MPC is formulated to drive the servicer tracking the docking point with collision avoidance and handle constraints including control input saturation and relative velocity bound. Convexification of the collision avoidance constraint and sequential convex programming are adopted for the implementation of MPC. Scenarios on the servicer with different initial positions approaching the target with different angular velocities are simulated, and the simulation results indicate that the proposed MPC method is effective.

## 1. Introduction

On-orbit services such as the repair of a failed spacecraft, spacecraft refueling, and spacecraft reorbiting are extremely important. A number of missions were conducted, such as the orbital express (OE) demonstration mission [1], the engineering test satellite VII (ETS-VII) project [2], and the spacecraft for the universal modification of orbits (SUMO) project [3]. Recently, on-orbit services for geosynchronous orbit (GEO) targets have attracted significant attention because of their specific importance. The Phoenix Program [4] and the Robotic Servicing of Geosynchronous Satellites Program (RSGS) [5] have thus been proposed in the last decade. An important stage, namely, "approaching targets at an ultraclose range," is significant for the implementation of on-orbit services. However, it is challenging to ensure an effective and safe approach when the target exhibits a complex configuration and one that especially features uncontrolled tumbling motion.

The study focuses on the control problem for a servicer spacecraft approaching a failed spacecraft with tumbling motion at an ultraclose range. Two types of the on-orbit failed spacecraft with complex configurations are considered. The first is a spacecraft with two large solar panels symmetrically mounted on its body, i.e., most GEO failed spacecraft [6, 7]. The second is a spacecraft with solar panels deployed on only one side where deployment of panels on the other side fails, i.e., SinoSat-II and TV-Sat-I in GEO [8]. The objective of the approach involves tracking a docking point (DP) that is fixed to the body frame of the target and is a few meters from its surface such as the docking mechanism on an apogee motor [5]. The implementation of the approach is conducive to the next step in on-orbit operation, i.e., catching the target with a space manipulator [9]. Flight safety is the major consideration in the approach. The keep-out zone that covers the outer surface of the target is always used for collision avoidance [10, 11]. Most related studies adopted a sphere envelope to model target's keep-out zone [12, 13] although this is unsuitable for the approach problem discussed in the study. Consequently, a double-ellipsoid composite envelope strategy is designed to model the keep-out zone of the above two types of the failed spacecraft, and sufficient conditions for the collision are derived.

Numerous guidance and control methods are used for spacecraft approaching at an ultraclose range [14–16]. However, most of the aforementioned methods do not consider collision avoidance with the target. Guidance methods based on the optimal control principle are widely used. Given the complex state or control constraints, it is difficult to solve analytic solutions to optimal control. In reference [17], the Gauss pseudospectral method was employed to solve the optimal rendezvous trajectory with certain constraints. In references [18–20], Pontryagin's minimum principle was combined with numerical methods to formulate and solve the optimal rendezvous trajectory by considering collision avoidance. Another widely investigated method is the artificial potential function (APF) method. Dong et al. [21] introduced a potential function and two constrained zones to plan a safe rendezvous path. An adaptive control law based on a time-varying sliding manifold was subsequently used to track the desired path. Ge et al. [22] addressed the problem of docking with a tumbling target by using the APF method along with a sliding control. Tumbling motion was classified into three cases, and different safe boundaries were discussed in detail.

Model predictive control is another method that should be considered. Given its working principle, it is extremely suitable for dealing with complex constraints such as multivariable state constraints and control input constraints. Several studies focused on investigating how to conduct close-range rendezvous and docking using the MPC method [23–25]. In reference [24], strategies for handling collision avoidance constraints, control input constraints, velocity constraints, and line-of-sight constraints were discussed in detail. Several studies also investigated the application of MPC to docking with a tumbling target [13, 26–28]. Under the condition, the docking point and the collision avoidance constraints are time variant. Collision avoidance is always modeled as a nonlinear and nonconvex constraint, and thus, it is difficult to ensure computational convergence and efficiency while solving the underlying optimization problem in MPC. Typically, the linearization of the collision avoidance constraint at the desired docking point is adopted [27], which is conducive to solving the optimization problem. However, the derived new collision avoidance zone is subsequently converted into a half-space, which is overly conservative, and the object in safe zones may be misjudged as corresponding to prohibited zones. Other effective methods for handling the collision avoidance constraints such as sequential convex programming and mixed integer programming have been investigated in spacecraft swarm mission and spacecraft rendezvous [29–31]. Based on the aforementioned studies, an MPC is developed for a servicer approaching a tumbling failed spacecraft. A novel collision discrimination method is proposed with the aim of handling the collision avoidance constraint with the modeled double-ellipsoid envelope of the target. Furthermore, a convexification method for the constraints and a sequential convex programming are adopted to implement the MPC.

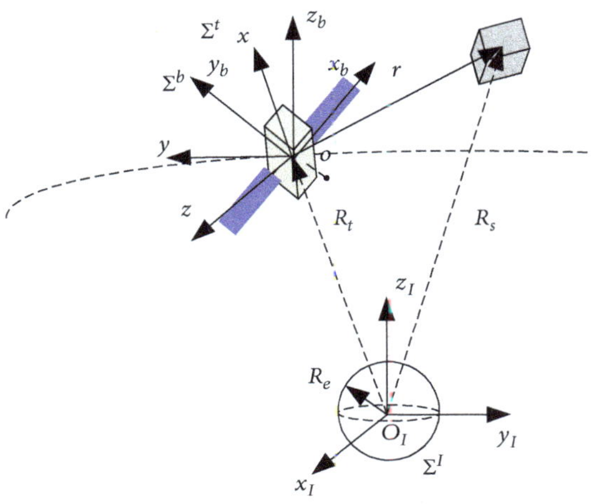


FIGURE 1: Geometrical relationship and coordinate systems.

The remainder of the study is organized as follows: Section 2 presents the mathematical formulation of relative translational dynamics and the attitude dynamics of the target. The control object and constraints on modeling for the approach are stated. Section 3 presents the design of the MPC controller and its solution in detail, and Section 4 provides the results of a numerical simulation to verify the performance of the proposed algorithms. The concluding remarks are discussed in Section 5.

## 2. Mathematical Formulation

In this section, the relative translational dynamics of the servicer and target are set along with the rotational dynamics of the target. Both the aforementioned dynamics form the basis of the state prediction and construction of time-variant constraints used in the design of the MPC.

The relative geometrical relationship and coordinate systems are shown in Figure 1.

As shown in Figure 1, $\mathbf{r} \in \mathbb{R}^3$ denotes the vector directed from the center of mass (c.m.) of the target to the c.m. of the servicer. Additionally, $R_t$ and $R_s$ denote the distances of the target and servicer, respectively, relative to the center of the Earth, and $R_e$ denotes the radius of the Earth.

*2.1. Inertial Coordinate System* ($\sum^I$). The origin of the inertial coordinate system $o_I - x_I y_I z_I$ is centered on Earth, the $z_I$-axis is along the rotational axis, and the $x_I$-axis points toward the vernal equinox. The $y_I$-axis completes the right-handed orthogonal.

*2.2. Target Local Vertical-Local Horizontal (LVLH) Coordinate System* ($\sum^t$). The origin of the LVLH coordinate system $o - x_t y_t z_t$ lies on the c.m. of the target. The $x_t$-axis is in the opposite direction to Earth's center, the $y_t$-axis is along the direction of flight, and the $z_t$-axis completes the right-handed orthogonal.

*2.3. Target Body-Fixed Coordinate System* ($\sum^b$). The origin of the target body-fixed coordinate system $o - x_b y_b z_b$ is in the c.m. of the target. Without loss of generality, it is assumed that $x_b$, $y_b$, and $z_b$ are aligned with its principal axes.

*2.3.1. Relative Translational Dynamics.* It is assumed that the target is on a circular orbit and the distance between the servicer and target is extremely low. Given a short-period approach, the disturbances caused by solar pressure, atmospheric drag, and the effects of nonspherical gravity perturbation are neglected in both orbital and attitude motions. The Clohessy–Wiltshire (CW) equation is subsequently used to describe the relative translational dynamics [13]:

$$\begin{aligned} \ddot{x} - 2n\dot{y} - 3n^2x &= u_x, \\ \ddot{y} + 2n\dot{x} &= u_y, \\ \ddot{z} + n^2\dot{z} &= u_z, \end{aligned} \tag{1}$$

where $[u_x, u_y, u_z]$ denote the components of the control acceleration and are resolved into $\sum^t$, $n = \sqrt{\mu/R_t^3}$ denotes the orbital rate of the target with respect to $\sum^I$, and $\mu$ denotes the gravitational constant.

Eq. (1) is rewritten in the form of a state space model as follows:

$$\dot{X} = AX + BU, \tag{2}$$

where $X \in \mathbb{R}^6$ denotes the state vector, $U \in \mathbb{R}^3$ denotes the control vector, and

$$\begin{aligned} X &= \begin{pmatrix} x \\ y \\ z \\ \dot{x} \\ \dot{y} \\ \dot{z} \end{pmatrix}, \\ A &= \begin{pmatrix} 0 & 0 & 0 & 1 & 0 & 0 \\ 0 & 0 & 0 & 0 & 1 & 0 \\ 0 & 0 & 0 & 0 & 0 & 1 \\ 3n^2 & 0 & 0 & 0 & 2n & 0 \\ 0 & 0 & 0 & -2n & 0 & 0 \\ 0 & 0 & -n^2 & 0 & 0 & 0 \end{pmatrix}, \\ B &= \begin{pmatrix} 0 & 0 & 0 \\ 0 & 0 & 0 \\ 0 & 0 & 0 \\ 1 & 0 & 0 \\ 0 & 1 & 0 \\ 0 & 0 & 1 \end{pmatrix}, \\ U &= \begin{pmatrix} u_x \\ u_y \\ u_z \end{pmatrix}. \end{aligned} \tag{3}$$

With respect to the implementation of the MPC, the discrete-time model of the relative translational dynamics is derived with a sampling period $T_s$ as follows:

$$X(k+1) = A_d X(k) + B_d U(k), \tag{4}$$

where $X(k)$ and $U(k)$ denote the state and control vectors, respectively, at sampling instant $k \in Z^+$. The matrices $A_d \in \mathbb{R}^{6\times6}$ and $B_d \in \mathbb{R}^{6\times3}$ are defined as follows [28]:

$$\begin{aligned} A_d &= e^{AT_s}, \\ B_d &= \left( \int_0^{T_s} e^{A\tau} d\tau \right) B. \end{aligned} \tag{5}$$

*2.3.2. Attitude Dynamics of the Target.* The attitude dynamics of the target are modeled to describe the relationship between $\sum^t$ and $\sum^b$. The attitude of the target is parameterized by using the rotation quaternion $\mathbf{q} = [q_0, \underline{q}^T]^T$, where the first component denotes the scalar part and the other is a three-dimensional vector. The rotation matrix associated with $\mathbf{q}$, denoted by $R_{tb}(\mathbf{q})$, is given as follows [32]:

$$R_{tb}(\mathbf{q}) = \left( q_0^2 - \underline{q}^T \underline{q} \right) \mathbf{I}_3 + 2\underline{q}\underline{q}^T - 2q_0 \left[ \underline{q} \times \right], \tag{6}$$

where $R_{tb}(\mathbf{q})$ transforms a vector from $\sum^t$ into $\sum^b$, $\mathbf{I}_3$ denotes a three-dimensional identity matrix, and $[*\times]$ denotes a skew-symmetric matrix that represents the cross product operator.

The angular velocity $\mathbf{w}$ of $\sum^b$ relative to $\sum^t$ is defined as follows:

$$\mathbf{w} = \mathbf{w}_b - \mathbf{w}_t, \tag{7}$$

where $\mathbf{w}_b$ and $\mathbf{w}_t$ denote the angular velocities of $\sum^b$ and $\sum^t$, respectively, relative to $\sum^I$.

The first time derivative of Eq. (7) in the $\sum^I$ is given as follows:

$$\left.\frac{d\mathbf{w}}{dt}\right|_I = \left.\frac{d\mathbf{w}_b}{dt}\right|_I - \left.\frac{d\mathbf{w}_t}{dt}\right|_I. \tag{8}$$

Based on Coriolis's theorem, the following expression is obtained:

$$\left.\frac{d\mathbf{w}}{dt}\right|_I = \left.\frac{d\mathbf{w}}{dt}\right|_t + \mathbf{w}_t \times \mathbf{w}. \tag{9}$$

Given that $d\mathbf{w}_b/dt\,|_I = d\mathbf{w}_b/dt\,|_b$ and $d\mathbf{w}_t/dt\,|_I = d\mathbf{w}_t/dt\,|_t$, a combination of Eqs. (8) and (9) yields the following expression:

$$\left.\frac{d\mathbf{w}_b}{dt}\right|_b = \left.\frac{d\mathbf{w}}{dt}\right|_t + \left.\frac{d\mathbf{w}_t}{dt}\right|_t + \mathbf{w}_t \times \mathbf{w}. \tag{10}$$

The attitude dynamics of the target are given by the following expression:

$$I_t \left.\frac{d\mathbf{w}_b}{dt}\right|_b + \mathbf{w}_b \times (I_t \mathbf{w}_b) = \mathbf{N}_t, \tag{11}$$

where $I_t$ and $\mathbf{N}_t$ denote the inertial tensors of the target and external torques, respectively. It is assumed that the target is on a circular orbit and environmental perturbations are neglected, and $d\mathbf{w}_t/dt\,|_t = 0$ and $\mathbf{N}_t = 0$ are subsequently obtained, respectively.

Eq. (10) is substituted into Eq. (11) to yield the following expression:

$$I_t(\dot{\mathbf{w}} + \mathbf{w}_t \times \mathbf{w}) + (\mathbf{w}_t + \mathbf{w}) \times (I_t \cdot (\mathbf{w}_t + \mathbf{w})) = 0, \tag{12}$$

where $\mathbf{w}_t$, resolved into $\sum^t$, is given as $\mathbf{w}_t = [0, 0, n]^T$.

Subsequently, the change in target's attitude is given as follows [32]:

$$\dot{\mathbf{q}} = \frac{1}{2} \begin{bmatrix} q_0 \mathbf{I}_3 + [\underline{q} \times] \\ -[\underline{q} \times] \end{bmatrix} \mathbf{w}. \tag{13}$$

According to Eq. (6), the rotation matrix $R_{tb}(k+1)$ at the sampling instant $k+1$ is a function of rotation quaternion $\mathbf{q}(k+1)$. According to Eqs. (12) and (13), $\mathbf{q}(k+1)$ can be obtained through solving the numerical integration with the given $\mathbf{w}(k)$ and $\mathbf{q}(k)$. Thereby, $R_{tb}(k+1)$ can be represented as a nonlinear function expression:

$$R_{tb}(k+1) = f(\mathbf{w}(k), \mathbf{q}(k)). \tag{14}$$

The fourth-order Runge–Kutta algorithm is utilized to solve the numerical integration in Eqs. (12) and (13). It is assumed that there is no attitude maneuvering of the failed target and that there is no external torque applied to the target from the servicer. The offline calculation can be adopted to predict target's attitude, which will be used in MPC to predict the relative translational states and servicer's control input.

*2.3.3. Control Objective.* The study focuses on two types of the on-orbit failed spacecraft. The first includes two large solar panels that are symmetrically mounted on the body as shown in Figure 2. The second includes solar panels on only one side where deployment fails on the other side as shown in Figure 3.

The $x_b$-axis is along the center line of the solar panels, the $y_b$-axis is along the symmetric axis of target's body and is orthogonal to the $x_b$-axis, and the $z_b$-axis completes the dextral triad. In the study, the target is assumed as tumbling freely. It is also assumed that target's DP is located along the $y_b$-axis and the position vector of the DP is $\mathbf{r}_{\mathrm{DP}}^b = [0, -l, 0]^T$ resolved in $\sum^b$. The tracking error is defined as $\delta\mathbf{r}(t) = R_{tb}(t)\mathbf{r}(t) - \mathbf{r}_{\mathrm{DP}}^b$, where $\mathbf{r}(t)$ denotes servicer's trajectories resolved in $\sum^t$. Therefore, the objective of the approach

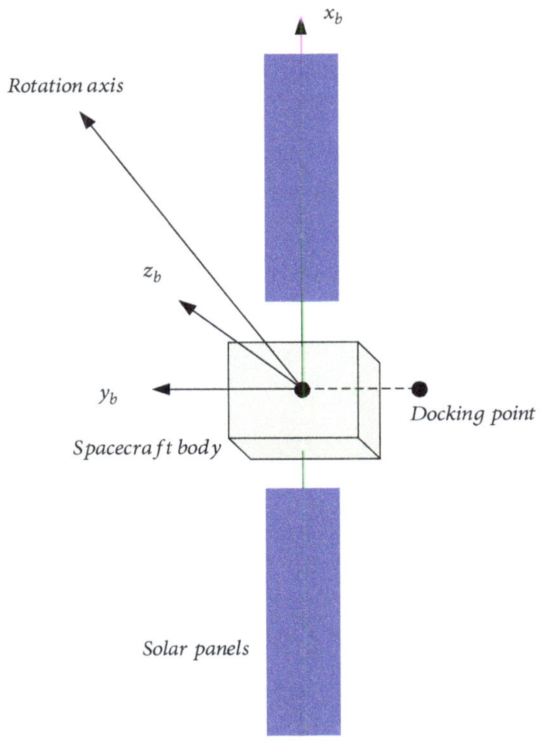


FIGURE 2: Failed spacecraft with two mounted solar panels.

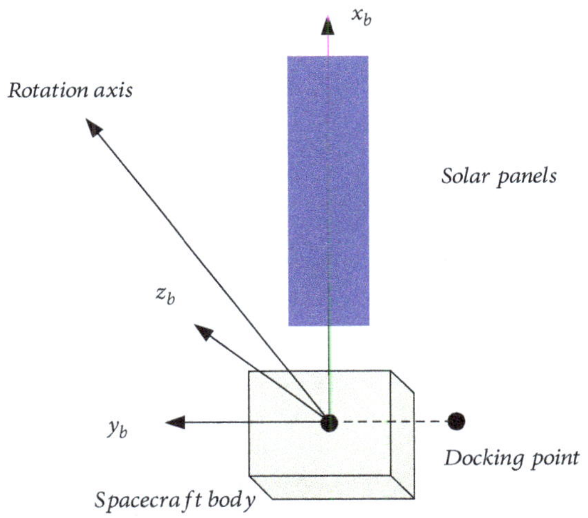


FIGURE 3: Failed spacecraft with solar panels deployed on only one side.

procedure involves driving the servicer while tracking with the target DP, and this is expressed as follows:

$$\lim_{t\to\infty} \|\delta\mathbf{r}(t)\| \le \varepsilon, \tag{15}$$

where $\varepsilon$ is a constant denoting the upper bound of tracking error.

*2.3.4. Constraint Modeling.* With respect to the problem of the ultraclose approach, collision avoidance is the basic requirement. Given the aforementioned two types of the

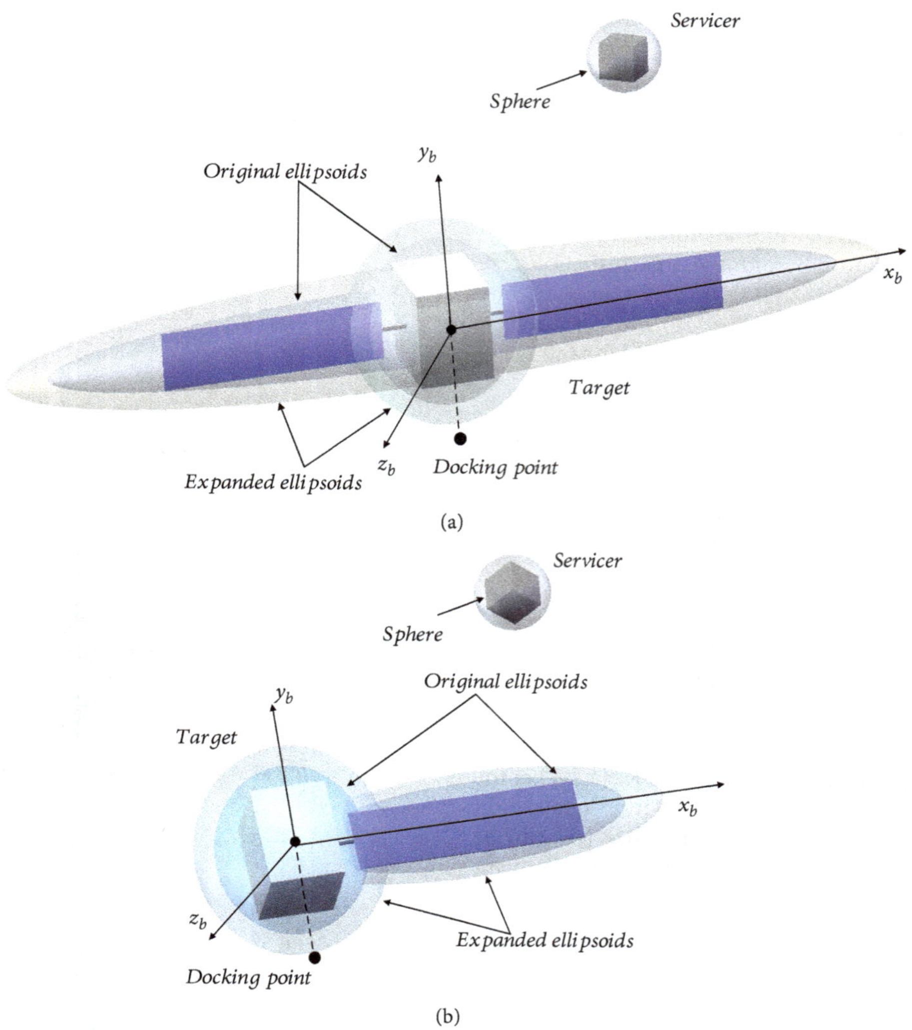


FIGURE 4: The double-ellipsoid composite envelope strategy for (a) the first type of a failed spacecraft and (b) the second type of a failed spacecraft.

failed spacecraft, a double-ellipsoid composite envelope strategy is introduced to model target's keep-out zone in which the spacecraft body and the solar panels are modeled as two different ellipsoids based on their sizes. Furthermore, a sphere envelope with radius $r_e$ is simplified as servicer's keep-out zone.

Intuitively, collision avoidance is attained when the envelopes of the target and the servicer do not intersect. In order to determine collision avoidance, expanded ellipsoids considering the sizes of envelopes of the target and the servicer are introduced, as shown in Figure 4. The original ellipsoids denote target's double-ellipsoid envelope. It is assumed that centers of the original ellipsoids for the first type of the failed spacecraft are located at its mass center. With respect to the original ellipsoid enveloping solar panels of the second type of the failed spacecraft, its center is located at the center of the solar panels.

Furthermore, the scheme for discriminating collision avoidance between the original ellipsoid and the sphere is shown in Figure 5.

Without loss of generality, the original ellipsoid is as follows:

$$\frac{(x-d)^2}{a^2} + \frac{y^2}{b^2} + \frac{z^2}{c^2} = 1, \tag{16}$$

where $a$, $b$, and $c$ are constants designed to envelop the target body or the solar panels based on their sizes and $a > b > c$ is satisfied. If the center of ellipsoid is located at target's mass center, $d = 0$ is satisfied. The parametric equation of the original ellipsoid is given as follows [17]:

$$\begin{aligned} x &= d + a \sin\theta \cos\varphi, \\ y &= b \sin\theta \sin\varphi, \\ z &= c \cos\theta, \end{aligned} \tag{17}$$

where $\theta \in [0, \pi]$ and $\varphi \in [0, 2\pi)$.

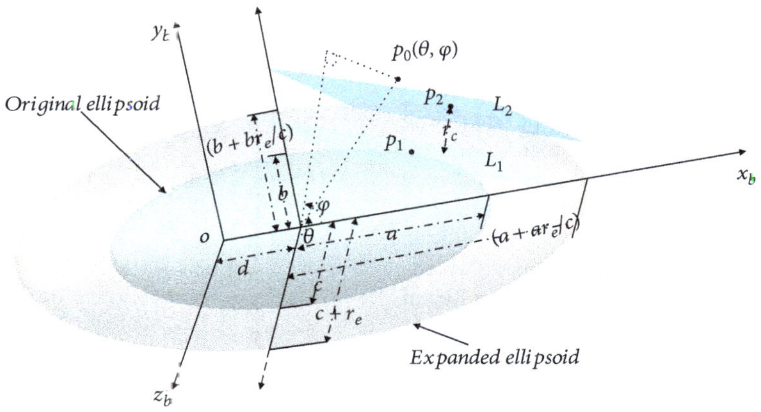


FIGURE 5: Scheme for discriminating collision avoidance between the original ellipsoid and the sphere.

The expanded ellipsoid envelope is as follows:

$$\frac{(x-d)^2}{(a+ar_e/c)^2}+\frac{y^2}{(b+br_e/c)^2}+\frac{z^2}{(c+r_e)^2}=1. \tag{18}$$

It should be noted that the value of the minimum half-axis of the expanded ellipsoid is the summation of the radius of the sphere and the value of the minimum half-axis of the original ellipsoid. The other two half-axes of expanded ellipsoid are augmented based on the product of the radius and the ratio between the corresponding half-axis and minimum half-axis.

The parametric equation of the expanded ellipsoid is given as follows:

$$\begin{aligned} x &= d+\left(a+\frac{ar_e}{c}\right)\sin\theta\cos\varphi,\\ y &= \left(b+\frac{br_e}{c}\right)\sin\theta\sin\varphi,\\ z &= (c+r_e)\cos\theta. \end{aligned} \tag{19}$$

**Proposition 1.** *Given the original ellipsoid in Eq. (16) and the sphere with radius $r_e$, the sufficient condition for collision avoidance between the original ellipsoid and the sphere is defined as follows:*

$$\frac{(x-d)^2}{(a+ar_e/c)^2}+\frac{y^2}{(b+br_e/c)^2}+\frac{z^2}{(c+r_e)^2}-1>0, \tag{20}$$

*where $[x,y,z]$ denote the coordinates of the sphere center. Evidently, Eq. (20) indicates that the center of the sphere is outside the expanded ellipsoid shown in Eq. (18).*

*Proof.* Given a point $p_0$ using spherical coordinates $(\theta,\varphi)$ describing its direction, there are two other corresponding points located at the original ellipsoidal surface and the expanded ellipsoidal surface (points $p_1$ and $p_2$, as shown in Figure 5), and their coordinates are given as follows:

$$\begin{aligned} p_1: x_1 &= d+a\sin\theta\cos\varphi,\\ y_1 &= b\sin\theta\sin\varphi,\\ z_1 &= c\cos\theta,\\ p_2: x_2 &= d+\left(a+\frac{ar_e}{c}\right)\sin\theta\cos\varphi,\\ y_2 &= \left(b+\frac{br_e}{c}\right)\sin\theta\sin\varphi,\\ z_2 &= (c+r_e)\cos\theta. \end{aligned} \tag{21}$$

Subsequently, the tangent planes of the two points (planes $L_1$ and $L_2$, as shown in Figure 5) are given as follows:

$$\begin{aligned} L_1 &: \frac{\sin\theta\cos\varphi}{a}x+\frac{\sin\theta\sin\varphi}{b}y+\frac{\cos\theta}{c}z-\frac{\sin\theta\cos\varphi}{a}d-1=0,\\ L_2 &: \frac{\sin\theta\cos\varphi}{a}x+\frac{\sin\theta\sin\varphi}{b}y+\frac{\cos\theta}{c}z-\frac{\sin\theta\cos\varphi}{a}d-1-\frac{r_e}{c}=0. \end{aligned} \tag{22}$$

According to Eq. (22), it is noted that the two tangent planes are parallel. Subsequently, the minimum distance from the point $p_2$ to the tangent plane $L_1$ is equal to the distance between tangent planes $L_2$ and $L_1$, and this is given as follows:

$$r_c=\frac{r_e}{\sqrt{c^2/a^2\sin^2\theta\cos^2\varphi+c^2/b^2\sin^2\theta\sin^2\varphi+\cos^2\theta}}. \tag{23}$$

Given that $a>b>c$ is defined, $r_c\geq r_e$ is subsequently obtained and it denotes that the minimum distance between the original ellipsoid and the expanded ellipsoid exceeds or is equal to the radius of the sphere. Thus, when the center of the sphere lies outside the expanded ellipsoid, the collision avoidance of the sphere and the original ellipsoid is satisfied. The proof is completed.

*Remark 1.* Given the double-ellipsoid envelope modeling target's keep-out zone, the collision avoidance condition for the target and servicer is the collision avoidance between the sphere and the two original ellipsoids. The modeled ellipsoid envelope is fixed in target's body frame. Therefore, the mathematical description of collision avoidance between the servicer and the target can be conducted in target's body frame. Although this is a sufficient but not a necessary condition to determine collision avoidance, it is derived by using the form of concave constraint, which is conducive to the implementation of MPC as subsequently discussed.

In practice, all thrusters are designed with limited control capability. The infinite norm of vector is applied to describe the thrust saturation constraint, and this is given as follows [24]:

$$\|U\|_\infty \le u_{\max}, \tag{24}$$

where $u_{\max}$ denotes the maximum thrust output.

The relative velocity constraint is also considered to ensure that the servicer exhibits the capability to adjust the thrust output and avoid collision in case of emergency, and this is extremely important in actual on-orbit approaches. The velocity constraint is expressed as follows [27]:

$$\begin{cases} |\dot{x}| \le \dot{x}_{\max}, \\ |\dot{y}| \le \dot{y}_{\max}, \\ |\dot{z}| \le \dot{z}_{\max}, \end{cases} \tag{25}$$

where $\dot{x}_{\max}$, $\dot{y}_{\max}$, and $\dot{z}_{\max}$ denote the components of the maximum approach velocity in each direction.

## 3. Model Predictive Control Formulation

By using the dynamics and constraints discussed in Section 2, the model predictive controller is designed to approach the tumbling failed spacecraft based on the basic principle of finite predictive control. Given the nonconvex constraint of the collision avoidance condition, the convexification method is discussed and the sequential convex programming is subsequently used to implement the MPC.

*3.1. Prediction of the State Variables.* The predicted relative translational state sequence generated by Eq. (4) with state $X(k)$ and control input $U(k)$ is expressed as follows:

$$X_s(k) = A_s X(k) + B_s U_s(k). \tag{26}$$

Subsequently,

$$X_s(k) = \begin{pmatrix} X(k+1|k) \\ X(k+2|k) \\ X(k+3|k) \\ \vdots \\ X(k+N_p|k) \end{pmatrix},$$
$$U_s(k) = \begin{pmatrix} U(k|k) \\ U(k+1|k) \\ U(k+2|k) \\ \vdots \\ U(k+N_c-1|k) \end{pmatrix},$$
$$A_s = \begin{pmatrix} A_d \\ A_d^2 \\ A_d^3 \\ \vdots \\ A_d^{N_p} \end{pmatrix},$$
$$B_s = \begin{pmatrix} B_d & 0 & 0 & \cdots & 0 \\ A_d B_d & B_d & 0 & \cdots & 0 \\ A_d^2 B_d & A_d B_d & B_d & \cdots & 0 \\ \vdots & \vdots & \vdots & \ddots & 0 \\ A_d^{N_p-1} B_d & A_d^{N_p-2} B_d & A_d^{N_p-3} B_d & \cdots & A_d^{N_p-N_c} B_d \end{pmatrix}. \tag{27}$$

where $X(k+i \mid k)$ denotes the predicted state variable at $k+i$ with the given information $X(k)$, $U(k+i \mid k)$ denotes the predicted control input at $k+i$, $U(k \mid k) = U(k)$, $N_p$ denotes the prediction horizon, and $N_c$ denotes the control horizon [12].

As described in Section 2.2, the predicated rotation matrix $R_{tb}(k+i)$ at $k+i$ can be solved offline with the given initial $\mathbf{w}(0)$ and $\mathbf{q}(0)$. Thus, while implementing the MPC, the rotation matrix $R_{tb}(k+i)$ is considered as known.

*3.2. Optimization Index.* The objective of the MPC involves minimizing the tracking error between the predicted states and the desired trajectory. In order to optimize the thruster fuel, the control effort is included in the objective function. Thus, based on Eq. (15), the objective function is defined as follows:

$$\Gamma(k) = \sum_{i=1}^{N_p} \left(R_{tb}(k+i)\mathbf{r}(k+i|k) - \mathbf{r}_{\mathrm{DP}}^b\right)^T Q_i \left(R_{tb}(k+i)\mathbf{r}(k+i|k) - \mathbf{r}_{\mathrm{DP}}^b\right) + \sum_{i=0}^{N_c-1} U^T(k+i|k) R_i U(k+i|k), \tag{28}$$

where $Q_i \in \mathbb{R}^{3\times3}$ denotes a positive-definite state-weighting matrix, $R_i \in \mathbb{R}^{3\times3}$ denotes a positive-definite control-weighting matrix, and $\mathbf{r}(k+i\mid k)$ is available from the predicted state sequence $X(k+i\mid k)$.

$$\mathbf{r}(k+i|k) = C_d X(k+i|k),$$
$$C_d = \begin{pmatrix} 1 & 0 & 0 & 0 & 0 & 0 \\ 0 & 1 & 0 & 0 & 0 & 0 \\ 0 & 0 & 1 & 0 & 0 & 0 \end{pmatrix}. \tag{29}$$

In order to simplify the expression in Eq. (28), matrix $C_s = \mathbf{I}_{Np} \otimes C_d$ is introduced, where $\otimes$ denotes the Kronecker product of two matrices [27], and $\mathbf{I}_{Np}$ denotes a $N_p$-dimensional identity matrix. If we define $\mathbf{r}_d = (\mathbf{r}_{\mathrm{DP}}^{b\ T}, \mathbf{r}_{\mathrm{DP}}^{b\ T}, \dots, \mathbf{r}_{\mathrm{DP}}^{b\ T})^T$, $R_{tb}(k) = \mathrm{diag}\,(R_{tb}(k+1), R_{tb}(k+2), \dots, R_{tb}(k+N_p))$, $Q = \mathrm{diag}\,(Q_1, Q_2, \dots, Q_{N_p})$, and $R = \mathrm{diag}\,(R_1, R_2, \dots, R_{N_c})$, the objective function in Eq. (28) is expressed as follows:

$$\Gamma(k) = U_s(k)^T W(k) U_s(k) + P(k) U_s(k) + M(k)^T Q M(k), \tag{30}$$

where $W(k) = (\tilde{R}_{tb}(k) C_s B_s)^T Q (\tilde{R}_{tb}(k) C_s B_s) + R$, $P(k) = 2M(k)^T Q \tilde{R}_{tb}(k) C_s B_s$, and $M(k) = \tilde{R}_{tb}(k) C_s A_s X_s(k) - \mathbf{r}_d$ are satisfied.

*3.3. Inequality Constraints.* In this section, the constraints discussed in Section 2 are reformulated for the implementation of the MPC. We reconsider the collision avoidance constraint shown in Eq. (20), and the collision avoidance between the servicer and the target is given as follows:

$$g_m(x_b, y_b, z_b) = \frac{(x_b - d_m)^2}{(a_m + a_m r_e/c_m)^2} + \frac{{y_b}^2}{(b_m + b_m r_e/c_m)^2} + \frac{{z_b}^2}{(c_m + r_e)^2} - 1 > 0, \qquad m = 1, 2, \tag{31}$$

where $[x_b, y_b, z_b]$ denote the components of the vector $\mathbf{r}$ resolved in $\sum^b$ and subscript $m$ denotes the two different ellipsoids that model the keep-out zone of the target. Function $g_m$ is defined as the collision threshold, and the collision avoidance is attained when $g_m > 0$. Based on Eq. (31), the collision avoidance constraint for the MPC is expressed as follows:

$$\begin{aligned} g_m^i(U_s(k)) = {} & \left\{R_{tb}(k+i)H(i)C_s[A_s X(k) + B_s U_s(k)] - [d_m, 0, 0]^T\right\}^T \times L_m \times \left\{R_{tb}(k-i)H(i)C_s[A_s X(k) + B_s U_s(k)] - [d_m, 0, 0]^T\right\} - 1 > 0, \\ & i = 1, 2, \dots, N_p, \quad m = 1, 2, \end{aligned} \tag{32}$$

where $L_m = \mathrm{diag}\,((a_m + a_m r_e/c_m)^2, (b_m + b_m r_e/c_m)^2, (c_m + r_e)^2)$ and $H(i)$ is expressed as follows:

$$H(i) = \left[T_1, T_2, \dots, T_j, \dots, T_{N_p}\right] \begin{cases} T_j = \mathbf{I}_3 & \text{if } j = i, \\ T_j = \mathbf{E}_3 & \text{if } j \neq i, \end{cases} \tag{33}$$

where $\mathbf{I}_3$ denotes a three-dimensional identity matrix and $\mathbf{E}_3$ denotes a three-dimensional zero matrix.

After a few manipulations (see Appendix A), Eq. (32) is rewritten as follows:

$$\begin{aligned} g_m^i(U_s(k)) = {} & U_s(k)^T \tilde{F}_m(i) U_s(k) + \tilde{G}_m(i) U_s(k) \\ & + G_m(i)^T L_m G_m(i) - 1 > 0, \quad i = 1, 2, \dots, N_p, \quad m = 1, 2, \end{aligned} \tag{34}$$

where $\tilde{F}_m(i) = F(i)^T L_m F(i)$, $\tilde{G}_m(i) = 2G_m(i)^T L_m F(i)$, $F(i) = R_{tb}(k+i)H(i)C_s B_s$, and $G_m(i) = R_{tb}(k+i)H(i)C_s A_s X(k) - [d_m, 0, 0]^T$ are satisfied.

Based on Eq. (24), the control input constraint is reexpressed as follows:

$$\left(\mathbf{I}_{3N_c}; -\mathbf{I}_{3N_c}\right) U_s \le U_{\max}, \tag{35}$$

where $\mathbf{I}_{3N_c}$ denotes a $3N_c$-dimensional identity matrix and $U_{\max} \in \mathbb{R}^{6N_c}$ is defined as $U_{\max} = (u_{\max}, \cdots, u_{\max})^T$.

The velocity constraint expressed in Eq. (25) is reexpressed as follows (see Appendix B):

$$\left(\mathbf{I}_{3N_p}; -\mathbf{I}_{3N_p}\right) D_s (A_s X(k) + E_s U_s) \le V_{\max}, \tag{36}$$

where the matrices $D_s$ and $V_{\max} \in \mathbb{R}^{6N_p}$ are defined as follows:

$$\begin{aligned} D_d &= \begin{pmatrix} 0 & 0 & 0 & 1 & 0 & 0 \\ 0 & 0 & 0 & 0 & 1 & 0 \\ 0 & 0 & 0 & 0 & 0 & 1 \end{pmatrix}, \\ D_s &= \mathbf{I}_{N_p} \otimes D_d, \\ v_{\max} &= (\dot{x}_{\max} \dot{y}_{\max} \dot{z}_{\max} \dot{x}_{\max} \dot{y}_{\max} \dot{z}_{\max})^T, \\ V_{\max} &= \left(v_{\max}^T, \cdots, v_{\max}^T\right)^T. \end{aligned} \tag{37}$$

*3.4. MPC by Using Sequential Convex Programming.* The design of the MPC for the approach problem is summarized as follows:

*Problem 1.*

$$\begin{aligned} \min_{U_s(k)} \quad & \Gamma(k) = U_s(k)^T W(k) U_s(k) + P(k) U_s(k) \\ & \quad + M(k)^T Q M(k) \\ \text{subject to} \quad & (\mathbf{I}_{3N_c}; -\mathbf{I}_{3N_c}) U_s(k) \le U_{\max} \\ & \left(\mathbf{I}_{3N_p}; -\mathbf{I}_{3N_p}\right) D_s(A_s X(k) + B_s U_s(k)) \le V_{\max} \\ & g_m^i(U_s(k)) = U_s(k)^T \tilde{F}_m(i) U_s(k) + \tilde{G}_m(i) U_s(k) \\ & \quad + G_m(i)^T L_m G_m(i) - 1 > 0, \\ & i = 1, 2, \dots, N_p, \quad m = 1, 2 > 0. \end{aligned} \tag{38}$$

Given that the collision avoidance constraint is a nonconvex constraint, it is difficult to satisfy the convergence and the optimality of the solution to Problem 1. In order to solve Problem 1, it is subsequently converted into a convex optimization problem and sequential convex programming is utilized to implement MPC. Given the collision avoidance constraint in Eq. (38), the linearization technique is adopted to convert it into linear terms as follows:

$$\begin{aligned} g_m^i(U_s(k)) \approx\ & g_m^i\left(U_s^{[n]}(k)\right) + \nabla g_m^i\left(U_s^{[n]}(k)\right)^T \\ & \cdot \left(U_s(k) - U_s^{[n]}(k)\right) > 0, \quad i = 1, 2, \dots, N_p, \quad m = 1, 2, \end{aligned} \tag{39}$$

where $U_s^{[n]}(k)$ denotes the $n$-iterated solution of $U_s(k)$, and $\nabla g_m^i(U_s^{[n]}(k)) = 2\tilde{F}_m(i) U_s^{[n]}(k) + \tilde{G}_m(i)^T$ denotes the gradient of $g_m^i$ at $U_s^{[n]}(k)$.

**Proposition 2.** *Eq. (39) is utilized as the collision avoidance constraint as opposed to Eq. (34), and Problem 1 is subsequently converted to Problem 2 as follows:*

*Problem 2.*

$$\begin{aligned} \min_{U_s(k)} \quad & \Gamma(k) = U_s(k)^T W(k) U_s(k) + P(k) U_s(k) + M(k)^T Q M(k) \\ \text{subject to} \quad & (\mathbf{I}_{3N_c}; -\mathbf{I}_{3N_c}) U_s(k) \le U_{\max} \\ & \left(\mathbf{I}_{3N_p}; -\mathbf{I}_{3N_p}\right) G_s(A_s X(k) + B_s U_s(k)) \le V_{\max} \\ & g_m^i\left(U_s^{[n]}(k)\right) + \nabla g_m^i\left(U_s^{[n]}(k)\right)^T \left(U_s(k) - U_s^{[n]}(k)\right) > 0, \\ & i = 1, 2, \dots, N_p, \quad m = 1, 2. \end{aligned} \tag{40}$$

Subsequently, any $U_s(k)$ that is feasible for Problem 2 is also feasible for Problem 1.

*Proof.* Let $U_s(k)$ denote an arbitrary feasible point for Problem 2, i.e.,

$$g_m^i\left(U_s^{[n]}(k)\right) + \nabla g_m^i\left(U_s^{[n]}(k)\right)^T \left(U_s(k) - U_s^{[n]}(k)\right) > 0. \tag{41}$$

Represent $g_m^i(U_s(k))$ by the second-order Taylor series expansion at $U_s^{[n]}(k)$:

$$\begin{aligned} g_m^i(U_s(k)) =\ & g_m^i\left(U_s^{[n]}(k)\right) + \nabla g_m^i\left(U_s^{[n]}(k)\right)^T \left(U_s(k) - U_s^{[n]}(k)\right) \\ & + \frac{1}{2}\left(U_s(k) - U_s^{[n]}(k)\right)^T \nabla^2 g_m^i\left(U_s^{[n]}(k)\right) \\ & \cdot \left(U_s(k) - U_s^{[n]}(k)\right), \end{aligned} \tag{42}$$

where $\nabla^2 g_m^i$ denotes the Hessian matrix of $g_m^i$. It should be noted that $\nabla^2 g_m^i = 2\tilde{F}_m(i)$ denotes a constant positive-definite matrix where $\tilde{F}_m(i) = F(i)^T L_m F(i)$ is satisfied. We substitute Eq. (41) into Eq. (42), and this results in $g_m^i(U_s(k)) > 0$, and thus, any $U_s(k)$ that satisfies the condition in Eq. (39) also satisfies the condition in Eq. (34). The proof is completed.

It should be noted that Problem 2 is a quadratic programming problem [24]. In order to optimize the control input $U_s(k)$, sequential quadratic programming is utilized and a trust region between $U_s^{[n]}(k)$ and $U_s(k)$ is introduced to ensure the convergence.

$$\left\| U_s(k) - U_s^{[n]}(k) \right\|_\infty \le \rho^n D_0, \tag{43}$$

where $D_0 \in \mathbb{R}^{3N_c}$ denotes the radius of the trust region and $\rho \in (0, 1)$ denotes a parameter that determines the rate of convergence. After a few manipulations (see Appendix C), Problem 2 that considers the trust region is subsequently converted into Problem 3.

*Problem 3.*

$$\begin{aligned} \min_{U_s(k)} \quad & \Gamma(k) = U_s(k)^T W(k) U_s(k) + P(k) U_s(k) + M(k)^T Q M(k) \\ \text{subject to} \quad & (\mathbf{I}_{3N_c}; -\mathbf{I}_{3N_c}) U_s(k) \le U_{\max} \\ & \left(\mathbf{I}_{3N_p}; -\mathbf{I}_{3N_p}\right) G_s B_s U_s(k) \\ & \le V_{\max} - \left(\mathbf{I}_{3N_p}; -\mathbf{I}_{3N_p}\right) G_s A_s X(k) - \nabla g_m^i\left(U_s^{[n]}(k)\right)^T U_s(k) \\ & < -\nabla g_m^i\left(U_s^{[n]}(k)\right)^T U_s^{[n]}(k) + g_m^i\left(U_s^{[n]}(k)\right), \\ & i = 1, 2, \dots, N_p, \quad m = 1, 2 \\ & (\mathbf{I}_{3N_c}; -\mathbf{I}_{3N_c}) U_s(k) \le \left(\rho^n D_0 + U_s^{[n]}(k); \rho^n D_0 - U_s^{[n]}(k)\right). \end{aligned} \tag{44}$$

*Remark 2.* Based on Proposition 2, if there exists a solution feasible for Problem 3, the solution is also feasible for Problem 2. Furthermore, an optimized solution is attained based on sequential quadratic programming for Problem 3. The successive solution process based on Problem 3 requires a first solution $U_s^{[1]}(k)$ to commence. In the study, $U_s^{[1]}(k)$ is set to a zero vector and a maximum iteration number $n$ is designed to end the sequential quadratic programming.

Thus, the constrained optimization problem in the context of the MPC is converted into a sequential quadratic programming problem. The standard quadratic programming algorithm can be employed to solve this problem easily. With the given state vector $X(k)$, rotation quaternion $\mathbf{q}(k)$, and relative angular velocity $\mathbf{w}(k)$ at $k$, the MPC should be solved at each sampling instant $k$ by using sequential quadratic programming to obtain a sequence of the control input where only the first control input is applied.

## 4. Simulation Results and Discussion

In order to test the performance of the proposed MPC method, numerical simulations are described in this section. It is assumed that the target is on GEO, and the simulation parameters relevant to the target are listed in Table 1.

Considering the orbit parameters of the target described in Table 1, $h$ is the orbit height, $e$ is the eccentricity, $\Omega$ is the right ascension of ascending node, $w$ is the argument of perigee, and $f$ is the true anomaly. Given the two failed spacecraft, both their body's keep-out zones are modeled as an ellipsoid envelope with $a_1 = 2$ m, $b_1 = 1.8$ m, $c_1 = 1.5$ m, and $d_1 = 0$m. The parameters of the ellipsoid envelope for the first failed spacecraft's solar panels are set to $a_2 = 6$ m, $b_2 = 1.2$ m, $c_1 = 1.2$ m, and $d_2 = 0$m. The parameters of the ellipsoid envelope for the second failed spacecraft's solar panels are set to $a_2 = 2$ m, $b_2 = 1.2$ m, $c_1 = 1.2$ m, and $d_2 = 4$ m. The radius of the sphere envelope for the servicer spacecraft is $r_e = 0.5$ m.

In our simulations, the sampling period $T_s$ is set to $T_s = 0.1$ s. The weighting matrices are selected as $Q_i = \mathbf{I}_3$ and $R_i = \mathbf{I}_3$. The maximum control input $u_{\max}$ is set to $u_{\max} = 0.1\ \mathrm{m/s^2}$, and the maximum relative velocity $v_{\max}$ is set to $v_{\max} = (0.5, 0.5, 0.5, 0.5, 0.5, 0.5)^T$ m/s. The value of the prediction horizon $N_p$ is set to $N_p = 20$, and the value of control horizon $N_c$ is set to $N_c = 10$. The value of maximum iteration number $n$ is set to $n = 5$, and the trust region related parameters are set to $\rho = 0.9$ and $D_0 = [0.04, 0.04, \cdots, 0.04, \cdots, 0.04]^T$.

*4.1. Approaching the First Type of a Failed Spacecraft.* First, the effectiveness of the proposed MPC is analyzed through the approach simulation. The initial relative translational states $X(0)$ are set to $[25, -25.3, 0]^T$ m and $[0, -0.0027, 0]^T$ m/s. The initial angular velocities of the failed spacecraft are set to $\mathbf{w}_b^b = [2.2, 2.5, 2.3]^T$ deg/s. The simulation time is set to 100 s. The simulation results are shown in Figures 6 and 7.

TABLE 1: Parameters of the failed spacecraft.

| Parameter | Value |
|---|---|
| Orbit parameter | $h = 35786$ km, $e = 0$, $\Omega = 0°$, $i = 0°$, $w = 0°$, $f = 0°$ |
| Inertial moment | $I_t = \mathrm{diag}\,(500, 600, 700)\mathrm{kg \cdot m^2}$ |
| Docking port | $\mathbf{r}_{\mathrm{DP}}^b = [0, -3, 0]^T$ m |
| Initial rotation quaternion | $\mathbf{q}(0) = [1, 0, 0, 0]^T$ |

As shown in Figures 6(a)–6(c), it is evident that the desired trajectory to track the DP is time variant in target's LVLH frame, the approach process is implemented, and the trajectory tracking error in each direction is at $10^{-4}$ m magnitude. As depicted in Figure 6(d), the relative velocity in each direction is lower than 0.5 m/s, and the relative velocity constraint is satisfied. As shown in Figure 6(e), the control input in each direction is lower than $0.1\ \mathrm{m/s^2}$ and the control input constraint is satisfied. As depicted in Figure 6(f), while approaching the target, the collision threshold always exceeds zero, thereby indicating that the center of the servicer lies outside the expanded two ellipsoids and that the collision constraint between the target and servicer is attained. Furthermore, the three-dimensional trajectories of servicer's center relative to the expanded two ellipsoids of the target and servicer's keep-out zone relative to target's keep-out zone are shown in Figure 7. Evidently, the collision constraint is satisfied.

Subsequently, the performances of MPC on target with different initial angular velocities are analyzed. In addition to the initial angular velocities $\mathbf{w}_b^b = [2.2, 2.5, 2.3]^T$ deg/s (case A) as discussed in the previous simulation, initial angular velocities $\mathbf{w}_b^b = [1.2, 1.0, 0.8]^T$ deg/s (case B) and $\mathbf{w}_b^b = [0, 0, 0]^T$ deg/s (case C) are analyzed for comparison purposes and all other simulation parameters are set as the same. Furthermore, the convergence time $T$ and total control input $\Delta V$ are calculated at different initial angular velocities. It should be noted that the convergence time $T$ is recorded when the distance between the servicer and the DP is less than 0.1 m. The analysis results are shown in Figure 8.

The servicer can arrive at the DP while satisfying above control input, relative velocity, and collision avoidance constraints in the aforementioned three cases. As shown in Figure 8, the angular velocity of the target significantly influences servicer's transfer trajectory. When the target exhibits low angular velocity, the total control input of the approaching is lower. Furthermore, the influence of the control horizon and predictive horizon on the MPC is also analyzed. With respect to each control horizon $N_c = \{5, 10, 20\}$, the convergence time and total control input are calculated at different control horizons $N_p = \{20, 30, 40, 50, 60\}$. The initial angular velocities of the target are also set to $\mathbf{w}_b^b = [2.2, 2.5, 2.3]^T$ deg/s. The simulation results are shown in Figure 9.

As shown in Figure 9, with increases in the predictive horizon, the convergence time for approaching increases and the total control input decreases. It is noted that the

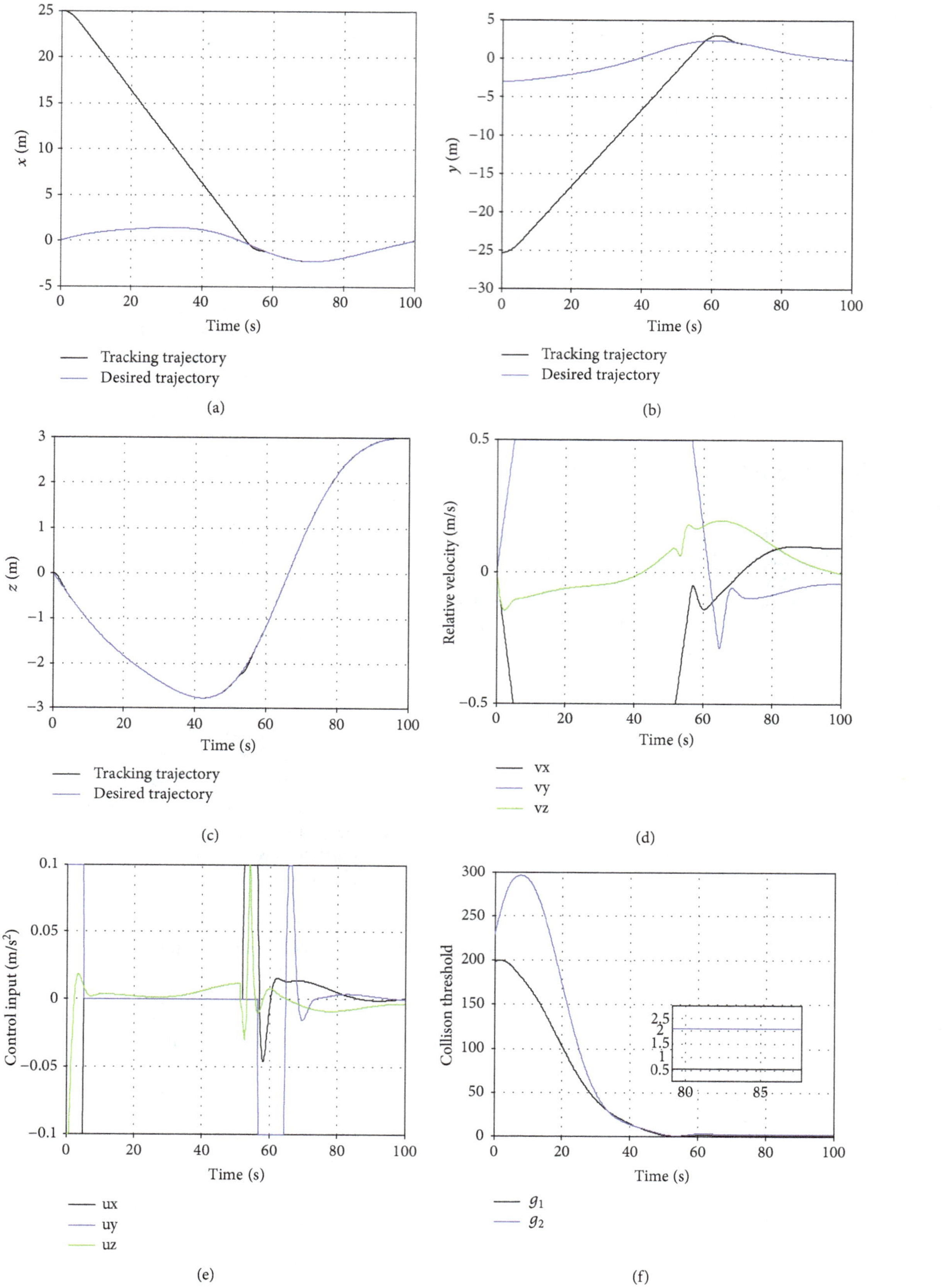


FIGURE 6: Temporal response of the relative position, relative velocity, control input, and collision threshold. (a) Temporal history along the $x$ direction. (b) Temporal history along the $y$ direction. (c) Temporal history along the $z$ direction. (d) Temporal history of relative velocity. (e) Temporal history of the control input. (f) Temporal history of the collision threshold.

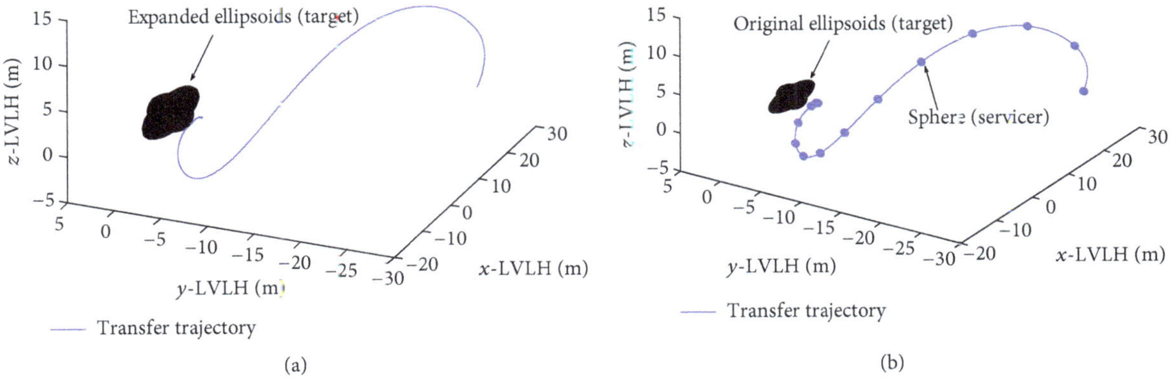


FIGURE 7: Three-dimensional plot of the approaching trajectories resolved in the target body-fixed frame. (a) Servicer's center relative to target's expanded ellipsoids. (b) Servicer's sphere relative to target's original ellipsoids.

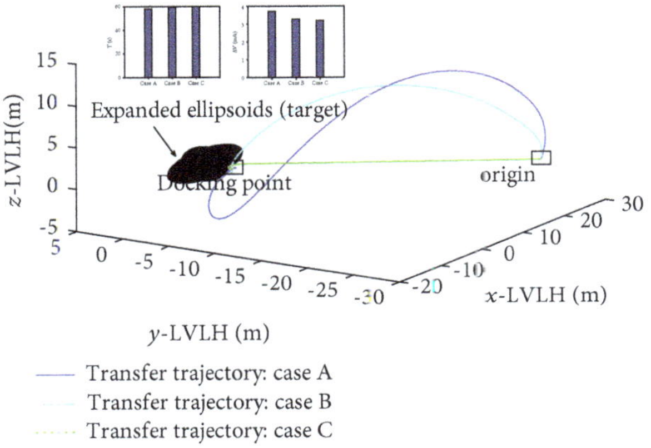


FIGURE 8: Three-dimensional plot of the transfer trajectories at target's different initial angular velocities resolved in the target body-fixed frame.

objective function in Eq. (28) is consist of the tracking error and the control effort. The increase of predictive horizon $N_p$ denotes more predicted states, leading to less state error and less control input to modify the state error. Accordingly, the required total control input is less and the convergence time will be longer. With increases in the control horizon $N_c$, the convergence time for approaching decreases, while the total control input and the accuracy of tracking the DP increase. This is because more predicted control inputs have been used to modify the state error in every sampling instant, leading to a shorter convergence time and higher tracking accuracy. Furthermore, the increases in the predictive horizon and control horizon, denoting more predicted states and control inputs, lead to a higher computation load when solving the underlying optimization problem in MPC. Therefore, while implementing the on-orbit missions, proper MPC parameters should be selected based on mission requirements.

*4.2. Approaching the Second Type of a Failed Spacecraft.* With respect to the simulation that involves approaching the second type of the failed spacecraft, the parameters of the MPC are set as identical to those in the previous simulation. The parameters of the second failed spacecraft's keep-out zone are discussed at the beginning of Section 4. Furthermore, in order to confirm that the proposed method is generally applicable, the designed MPC dealing with the servicer at different initial positions, such as $\mathbf{r}_0 = [25, -25.3, 0]^T$ m (Sat 1), $\mathbf{r}_0 = [25, 25.3, 0]^T$ m (Sat 2), $\mathbf{r}_0 = [-25, 25.3, 0]^T$ m (Sat 3), and $\mathbf{r}_0 = [-25, -25.3, 0]^T$ m (Sat 4), is analyzed. The initial angular velocities of the target are also set to $\mathbf{w}_b^b = [2.2, 2.5, 2.3]^T$ deg/s. Simulation results on the transfer trajectory are shown in Figure 10.

As shown in Figure 10, the proposed MPC exhibits the ability to drive the servicers at different positions to the DP of the target. The convergence times for each servicer are 57.9 s, 59 s, 69.2 s, and 69.2 s. The total control inputs for each servicer are 3.69 m/s, 4.23 m/s, 5.46 m/s, and 4.83 m/s. The temporal response of the collision threshold of each servicer is shown in Figure 11.

As shown in Figure 11, the collision avoidance constraint is evidently satisfied during the approaching process. As shown in Figure 12, required constraints, such as control input and relative velocity, are also satisfied. It should be noted that when compared with linearizing the collision avoidance constraint at the DP in which only half of the plane is the collision avoidance zone, the proposed method for handling collision avoidance constraint exhibits a better performance in which the space outside the expanded ellipsoids is the collision avoidance zone.

In previous simulation analysis, the controller uncertainties are neglected and the measurement of relative states and the control input of the system are assumed to be completely accurate. In order to evaluate the performance of the proposed MPC for real approaching missions, navigation and control noise are considered. Considering unknown but bounded navigation noise, the measurement-relative translational states $\widehat{X}(k)$ can be represented as

$$\begin{cases} \widehat{X}(k) = X(k) + \mathbf{e}(k), \\ \widehat{X}(k+1) = X(k+1) + \mathbf{e}(k+1), \end{cases} \quad (45)$$

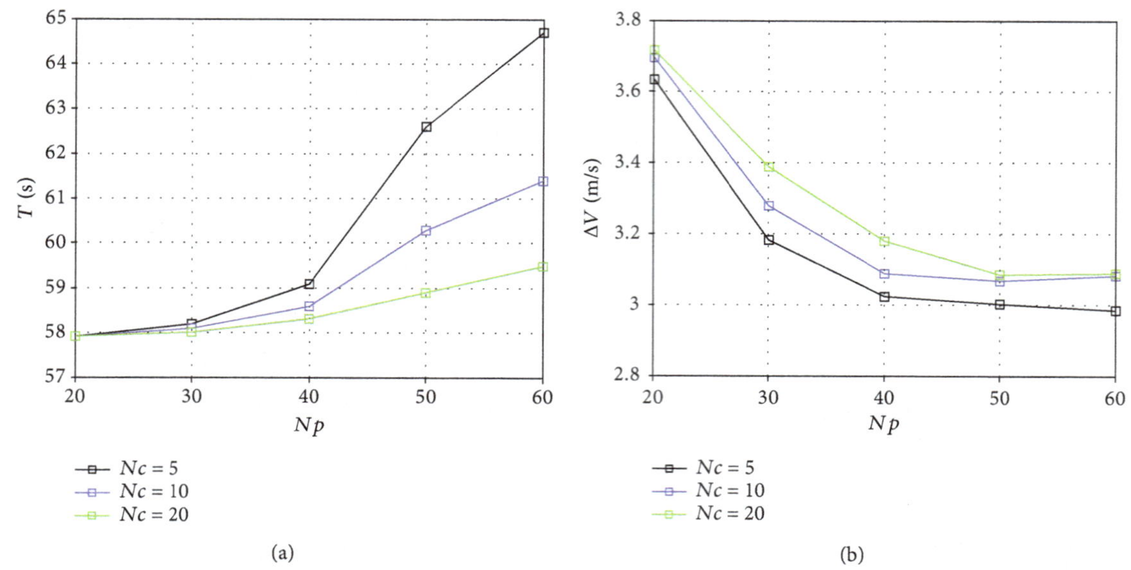


FIGURE 9: Influences of the predictive horizon $N_p$ and control horizon $N_c$ on (a) convergence time and (b) total control input.

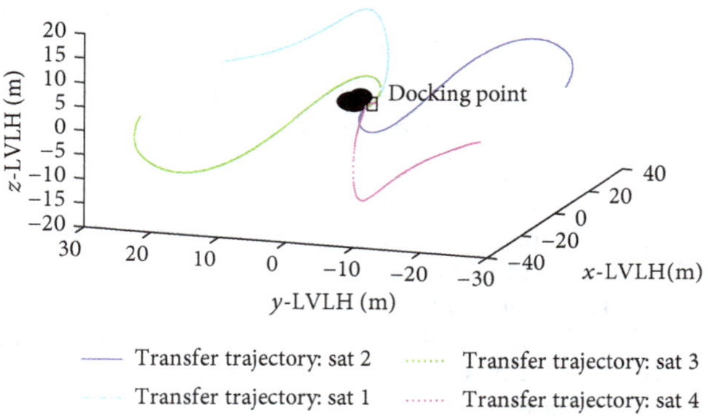


FIGURE 10: Three-dimensional plot of the transfer trajectories with the servicer at different initial positions.

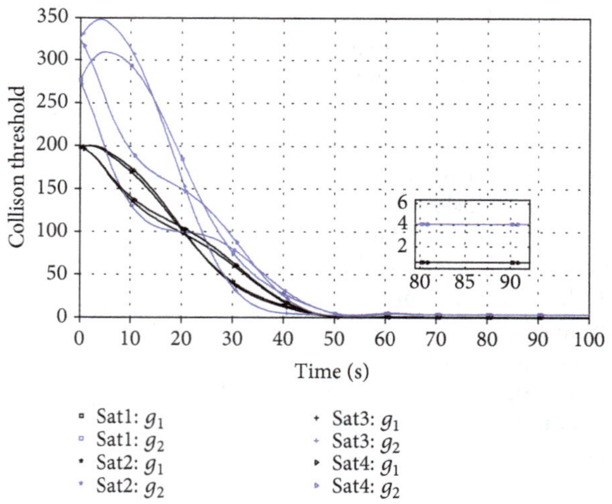


FIGURE 11: Temporal response of the collision threshold of each servicer.

where $\mathbf{e}(k) = [\mathbf{e}_p(k)\,;\mathbf{e}_v(k)]$ and $\mathbf{e}_p(k)$ and $\mathbf{e}_v(k)$ are the measurement errors of position and velocity, respectively. Considering that the control output noise was proportional to the control output [33], the control input $\widehat{U}(k)$ applied to the servicer can be represented as

$$\widehat{U}(k) = (\mathbf{I}_3 + \Lambda(k))U(k), \tag{46}$$

where $\Lambda(k) = \mathrm{diag}\,(\Lambda_x(k), \Lambda_y(k), \Lambda_z(k))$ is a three-dimensional diagonal matrix.

Combining Eqs. (4), (45), and (46), the prediction of relative states should be represented as

$$\widehat{X}(k+1) = A_d\widehat{X}(k) + B_d(I + \Lambda(k))U(k) + \mathbf{e}(k+1) - A_d\mathbf{e}(k). \tag{47}$$

However, navigation and control noise cannot be measured. Then, Eq. (4) with measurement $\widehat{X}(k)$ and control input $U(k)$ was used to predict relative states. The parameters of the MPC are set as identical to those in the previous simulation, and the initial position of the servicer is $\mathbf{r}_0 = [25, -25.3, 0]^T$ m (Sat 1). The initial angular velocities of the target are also set to $\mathbf{w}_b^b = [2.2, 2.5, 2.3]^T$ deg/s.

A Monte-Carlo analysis is performed to illustrate the influence of uncertainties on the MPC performance. It is assumed that elements in $\mathbf{e}_p$, $\mathbf{e}_v$, and $\Lambda$ follow Gaussian distribution, and their accuracies ($3\sigma$) are defined as $E_p$, $E_v$, and $E_u$, respectively. When developing Monte-Carlo analysis, 1000 simulations in each of two conditions such as $[E_p, E_v, E_u] = [0.005\,\text{m}, 10^{-4}\,\text{m/s}, 0.01]$ and $[E_p, E_v, E_u] = [0.02\,\text{m}, 10^{-3}\,\text{m/s}, 0.02]$ are conducted. Reconsidering the tracking error defined in Eq. (15), the average tracking

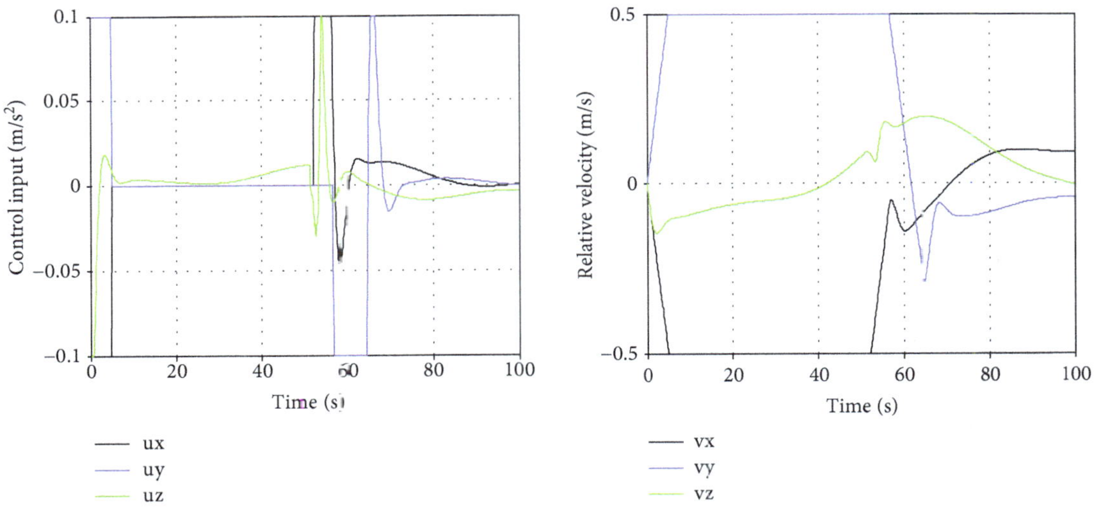


(a)

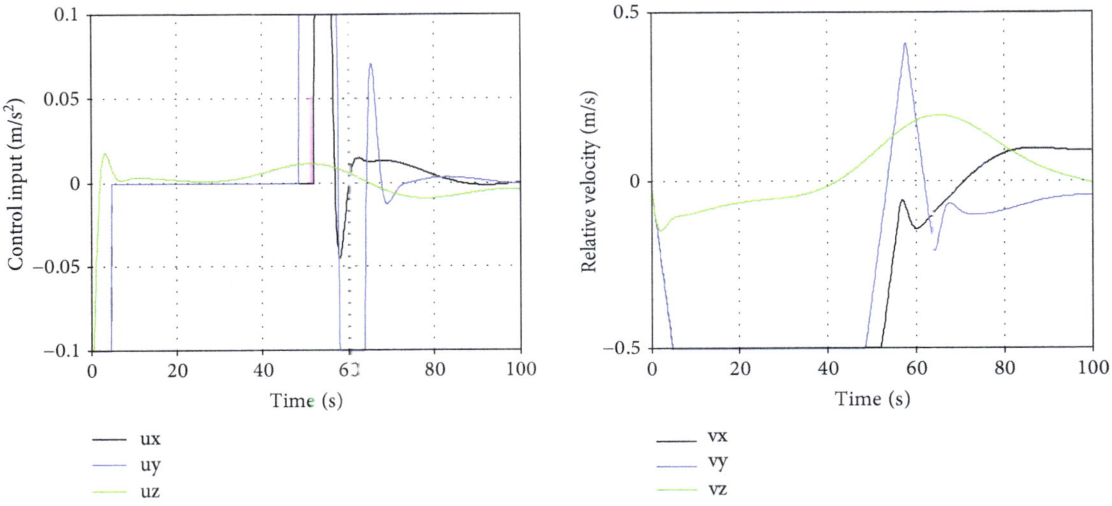


(b)

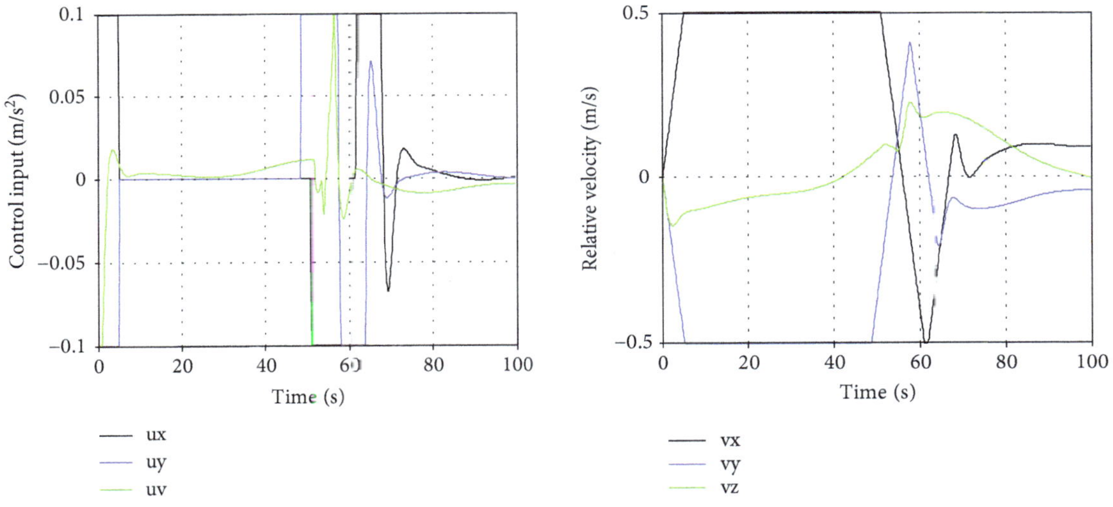


(c)

Figure 12: Continued.

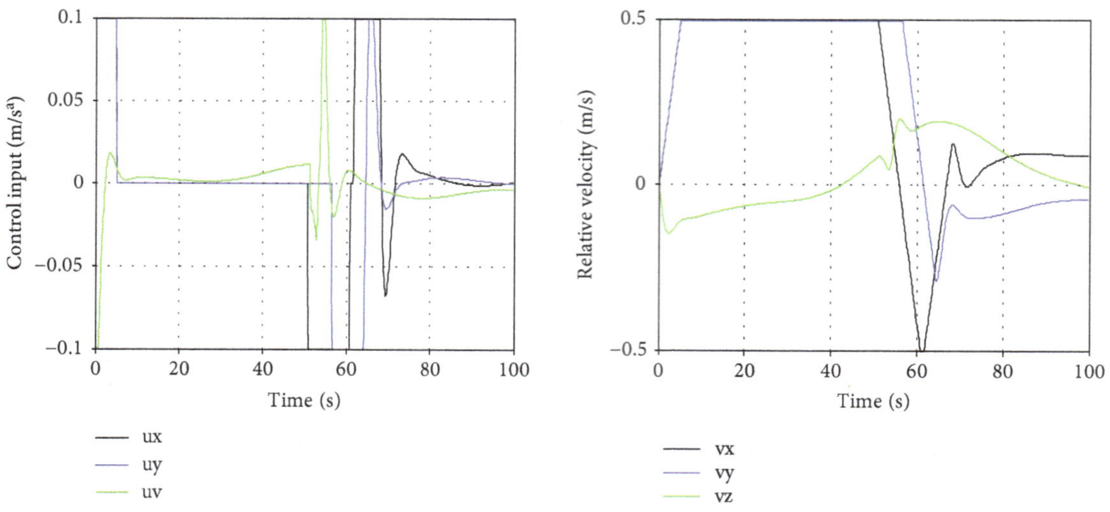


(d)

FIGURE 12: Temporal response of the control input (left) and relative velocities (right) for (a) Sat 1, (b) Sat 2, (c) Sat 3, and (d) Sat 4.

error $\delta\tilde{r}$ is considered to evaluate the influence of uncertainties, which is defined as

$$\delta\tilde{r} = \frac{\int_{t_s}^{t_f} \|\delta\mathbf{r}(t)\| dt}{t_f - t_s}, \tag{48}$$

where $t_f = 100\,\mathrm{s}$ is the end time of simulation and $t_s = 80\,\mathrm{s}$ is designed to start recording the tracking error. The Monte-Carlo simulation results regarding the average tracking error $\delta\tilde{r}$ are shown in Figure 13.

When developing Monte-Carlo analysis, the required constraints such as control input, relative velocity, and collision avoidance in every simulation are satisfied. When there are no navigation and control noise, the average tracking error is approximately 0.5 mm. As shown in Figure 13, when the uncertains are $[E_p, E_v, E_u] = [0.005\,\mathrm{m}, 10^{-4}\,\mathrm{m/s}, 0.01]$ and $[E_p, E_v, E_u] = [0.02\,\mathrm{m}, 10^{-3}\,\mathrm{m/s}, 0.02]$, the average tracking errors are approximately 1.5 mm and 7 mm, respectively. With the increase of uncertains, the average tracking error also becomes larger. The effectiveness of the MPC in handling navigation and control noise is also proved. Based on the above analyses, it is concluded that the proposed MPC strategy can drive the servicer to the DP of a tumbling target while satisfying various constraints and providing capability in handling navigation and control noises.

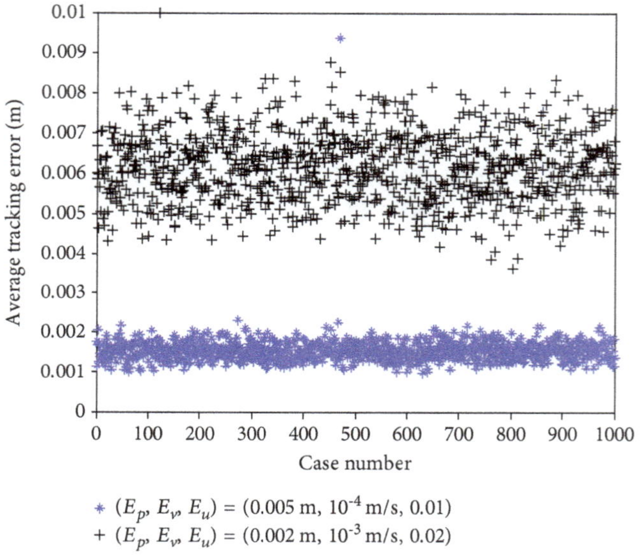


FIGURE 13: Performance comparison in terms of controller uncertainties.

## 5. Conclusion

A model predictive control (MPC) method for a servicer spacecraft autonomously approaching a tumbling failed spacecraft is presented in the study. The objective of the proposed MPC involves driving the servicer tracking the time-variant motion of the docking port of a target by considering collision avoidance, control input saturation, and velocity constraints. The relative translation is predicted by the CW equation, and target's attitude is predicted by a derived attitude dynamics. Sufficient conditions for collision avoidance are derived by using the form of concave constraint. The underlying optimization program for the implementation of MPC is converted into a convex optimization problem, and this is solved by using sequential convex programming. The process of the approach is simulated to evaluate the performance of the MPC strategy, and a few main contributions are obtained as follows:

(1) Given the complex configurations of two types of the failed spacecraft, a double-ellipsoid envelope is designed to model the keep-out zone. Sufficient conditions with simple mathematical expressions for collision avoidance are derived. The proposed modeling method for collision avoidance can be extended to other collision avoidance missions

(2) The convexification method for the collision avoidance constraints and the sequential convex programming for solving the underlying optimization problem in MPC are effective. The proposed MPC exhibits the capacity to handle the problem of tracking the time-varying docking point while handling various constraints

# Appendix

## A. The Derivation of Eq. (34)

Eq. (32) can be rewritten as follows:

$$\begin{aligned} g_m^i(U_s(k)) = &\left\{R_{tb}(k+i)H(i)C_sA_sX(k) - [d_m, 0, 0]^T + R_{tb}(k+i)H(i)C_sB_sU_s(k)\right\}^T \times L_m \\ &\times \left\{R_{tb}(k+i)H(i)C_sA_sX(k) - [d_m, 0, 0]^T + R_{tb}(k+i)H(i)C_sB_sU_s(k)\right\} - 1 > 0, \\ &i = 1, 2, \dots, N_p, \quad m = 1, 2. \end{aligned} \tag{A.1}$$

Then, define $G_m(i)$ as $G_m(i) = R_{tb}(k+i)H(i)C_sA_sX(k) - [d_m, 0, 0]^T$ and $F(i)$ as $F(i) = R_{tb}(k+i)H(i)C_sB_s$, Eq. (A.1) can be rewritten as follows:

$$\begin{aligned} g_m^i(U_s(k)) = &[G_m(i) + F(i)U_s(k)]^T \times L_m \\ &\times [G_m(i) + F(i)U_s(k)] - 1 > 0, \\ &i = 1, 2, \dots, N_p, \quad m = 1, 2. \end{aligned} \tag{A.2}$$

Then, define $\tilde{G}_m(i)$ as $\tilde{G}_m(i) = 2G_m(i)^T L_m F(i)$ and $\tilde{F}_m(i)$ as $\tilde{F}_m(i) = F(i)^T L_m F(i)$, Eq. (A.2) can be expanded as follows:

$$\begin{aligned} g_m^i(U_s(k)) = &U_s(k)^T \tilde{F}_m(i) U_s(k) + \tilde{G}_m(i) U_s(k) \\ &+ G_m(i)^T L_m G_m(i) - 1 > 0, \\ &i = 1, 2, \dots, N_p, \quad m = 1, 2. \end{aligned} \tag{A.3}$$

It is noted that Eq. (A.3) is the same as Eq. (34).

## B. The Derivation of Eq. (36)

Eq. (25) can be rewritten as follows:

$$(\mathbf{I}_3 ; -\mathbf{I}_3) D_d X(k) \le v_{\max}, \tag{B.1}$$

where the matrices $v_{\max}$ and $D_d$ are defined as follows:

$$D_d = \begin{pmatrix} 0 & 0 & 0 & 1 & 0 & 0 \\ 0 & 0 & 0 & 0 & 1 & 0 \\ 0 & 0 & 0 & 0 & 0 & 1 \end{pmatrix}, \tag{B.2}$$
$$v_{\max} = (\dot{x}_{\max} \dot{y}_{\max} \dot{z}_{\max} \dot{x}_{\max} \dot{y}_{\max} \dot{z}_{\max})^T.$$

For the implementation of MPC, the velocity sequence constraints can be represented as follows:

$$\left(\mathbf{I}_{3N_p} ; -\mathbf{I}_{3N_p}\right) \begin{pmatrix} D_d & 0 & 0 & \cdots & 0 \\ 0 & D_d & 0 & \cdots & 0 \\ 0 & 0 & D_d & \cdots & 0 \\ \vdots & \vdots & \vdots & \ddots & 0 \\ 0 & 0 & 0 & \cdots & D_d \end{pmatrix} \begin{pmatrix} X(k+1|k) \\ X(k+2|k) \\ X(k+3|k) \\ \vdots \\ X(k+N_p|k) \end{pmatrix} \le \begin{pmatrix} v_{\max} \\ v_{\max} \\ v_{\max} \\ \vdots \\ v_{\max} \end{pmatrix}, \tag{B.3}$$

where

$$D_s = \begin{pmatrix} D_d & 0 & 0 & \cdots & 0 \\ 0 & D_d & 0 & \cdots & 0 \\ 0 & 0 & D_d & \cdots & 0 \\ \vdots & \vdots & \vdots & \ddots & 0 \\ 0 & 0 & 0 & \cdots & D_d \end{pmatrix} = D_d \otimes \mathbf{I}_{N_p}, \tag{B.4}$$
$$V_{\max} = \left(v_{\max}^T, \cdots, v_{\max}^T\right)^T.$$

Substituting Eq. (B.4) into Eq. (B.3), Eq. (B.3) can be rewritten as follows:

$$\left(\mathbf{I}_{3N_p} ; -\mathbf{I}_{3N_p}\right) D_s (A_s X(k) + B_s U_s) \le V_{\max}. \tag{B.5}$$

It is noted that Eq. (B.5) is the same as Eq. (36).

## C. The Derivation of Eq. (44)

Eq. (40) can be rewritten as follows:

*Problem 2.*

$$\begin{aligned} \min_{U_s(k)} \quad & \Gamma(k) = U_s(k)^T W(k) U_s(k) + P(k) U_s(k) + M(k)^T Q M(k) \\ \text{subject to} \quad & (I_{3N_c} ; -I_{3N_c}) U_s(k) \le U_{\max} \\ & \left(I_{3N_p} ; -I_{3N_p}\right) G_s B_s U_s(k) \\ & \le V_{\max} - \left(I_{3N_p} ; -I_{3N_p}\right) G_s A_s X(k) - \nabla g_m^i\left(U_s^{[n]}(k)\right)^T U_s(k) \\ & < -\nabla g_m^i\left(U_s^{[n]}(k)\right)^T U_s^{[n]}(k) + g_m^i\left(U_s^{[n]}(k)\right), \\ & i = 1, 2, \dots, N_p, \quad m = 1, 2. \end{aligned} \tag{C.1}$$

Considering the inequality constraint about the trust region, Eq. (43) can be rewritten as follows:

$$I_{3N_c}\left(U_s(k) - U_s^{[n]}(k)\right) \le \rho^n D_0 - I_{3N_c}\left(U_s(k) - U_s^{[n]}(k)\right) \le \rho^n D_0. \tag{C.2}$$

Then, substituting Eq. (C.2) into Eq. (C.1), Problem 2 that considers the trust region is subsequently converted into Problem 3:

*Problem 3.*

$$\begin{aligned}
\min_{U_s(k)} \quad & \Gamma(k) = U_s(k)^T W(k) U_s(k) + P(k) U_s(k) + M(k)^T Q M(k) \\
\text{subject to} \quad & \left(I_{3N_c}; -I_{3N_c}\right) U_s(k) \le U_{\max} \\
& \left(I_{3N_p}; -I_{3N_p}\right) G_s B_s U_s(k) \\
& \le V_{\max} - \left(I_{3N_p}; -I_{3N_p}\right) G_s A_s X(k) - \nabla g_m^i \left(U_s^{[n]}(k)\right)^T U_s(k) \\
& < -\nabla g_m^i \left(U_s^{[n]}(k)\right)^T U_s^{[n]}(k) + g_m^i \left(U_s^{[n]}(k)\right), \\
& \qquad i = 1, 2, \ldots, N_p, \quad m = 1, 2 \\
& \left(I_{3N_c}; -I_{3N_c}\right) U_s(k) \le \left(\rho^n D_0 + U_s^{[n]}(k); \rho^n D_0 - U_s^{[n]}(k)\right).
\end{aligned} \tag{C.3}$$

It is noted that Eq. (C.3) is the same as Eq. (44).

## Conflicts of Interest

The authors declare that they have no conflicts of interest.

## Acknowledgments

This paper was sponsored by the National Natural Science Foundation of China (11572168 and 11872034).

## References

[1] A. Ogilvie, J. Allport, M. Hannah, and J. Lymer, "Autonomous satellite serving using the orbital express demonstration manipulator system," in *Proceedings of the Ninth International Symposium on Artificial Intelligence, Robotics and Automation in Space*, pp. 25–29, Los Angeles, CA, USA, 2008.

[2] M. Mokuno, I. Kawano, and T. Suzuki, "In-orbit demonstration of rendezvous laser radar for unmanned autonomous rendezvous docking," *IEEE Transactions on Aerospace and Electronic Systems*, vol. 40, no. 2, pp. 617–626, 2004.

[3] A. B. Bosse, W. J. Barnds, M. A. Brown et al., "SUMO: spacecraft for the universal modification of orbits," in *Proceedings SPIE 5419, Spacecraft Platforms and Infrastructure*, pp. 36–46, Bellingham, USA, 2004.

[4] C. G. Henshaw, "The DAPRA Phoenix spacecraft serving program: overview and plans for risk reduction," in *Proceedings of 12th International Symposium on Artificial Intelligence, Robotics and Automation in Space*, Montreal, Canada, 2014.

[5] Defense Advanced Research Projects Agency, *Program Solicitation for Robotic Serving of Geosynchronous Satellites*, Defense Advanced Research Projects Agency (DARPA), 2016.

[6] T. Flohrer, R. Choc, and R. Jehn, *Classification of Geosynchronous Objects, Issue 13*, European Space Agency, European space operations center, space debris Office, 2011.

[7] Union of Concerned Scientists (UCS), "The UCS satellite database," December 2017, http://www.ucsusa.org/satellite-database.

[8] X. Gao, *Study on Navigation and Guidance of Formation Space Robots for Rendezvous with Non-Cooperative Target of GEO*, Harbin Institute of Technology, 2015.

[9] F. Zhang, Y. Fu, S. Zhu, H. Liu, B. Guo, and S. Wang, "Safe path planning for free-floating space robot to approach noncooperative spacecraft," *Proceedings of the Institution of Mechanical Engineers, Part G: Journal of Aerospace Engineering*, vol. 232, no. 7, pp. 1258–1271, 2017.

[10] D. Zhang, S. Song, and R. Pei, "Safe guidance for autonomous rendezvous and docking with a non-cooperative target," in *AIAA Guidance, Navigation, and Control Conference*, pp. 943–961, Toronto, ON, Canada, 2010.

[11] L. Palacios, M. Ceriotti, and G. Radice, "Close proximity formation flying via linear quadratic tracking controller and artificial potential function," *Advances in Space Research*, vol. 56, no. 10, pp. 2167–2176, 2015.

[12] L. Ravikumar, N. K. Philip, R. Padhi, and M. S. Bhat, "Autonomous terminal maneuver of spacecrafts for rendezvous using model predictive control," in *2016 Indian Control Conference*, pp. 72–78, Hyderabad, India, 2016.

[13] S. Di Cairano, H. Park, and I. Kolmanovsky, "Model predictive control approach for guidance of spacecraft rendezvous and proximity maneuvering," *International Journal of Robust and Nonlinear Control*, vol. 22, no. 12, pp. 1398–1427, 2012.

[14] L. Sun, W. Huo, and Z. Jiao, "Adaptive backstepping control of spacecraft rendezvous and proximity operations with input saturation and full-state constraint," *IEEE Transactions on Industrial Electronics*, vol. 64, no. 1, pp. 480–492, 2017.

[15] M. Navabi and M. R. Akhloumadi, "Nonlinear optimal control of relative rotational and translational motion of spacecraft rendezvous," *Journal of Aerospace Engineering*, vol. 30, no. 5, article 04017038, 2017.

[16] L. Cao, D. Qiao, and J. Xu, "Suboptimal artificial potential function sliding mode control for spacecraft rendezvous with obstacle avoidance," *Acta Astronautica*, vol. 143, pp. 133–146, 2018.

[17] X. Chu, J. Zhang, S. Lu, Y. Zhang, and Y. Sun, "Optimised collision avoidance for an ultra-close rendezvous with a failed satellite based on the Gauss pseudospectral method," *Acta Astronautica*, vol. 128, pp. 363–376, 2016.

[18] G. Boyarko, O. Yakimenko, and M. Romano, "Optimal rendezvous trajectories of a controlled spacecraft and a tumbling object," *Journal of Guidance, Control, and Dynamics*, vol. 34, no. 4, pp. 1239–1252, 2011.

[19] J. Michael, K. Chudej, M. Gerdts, and J. Pannek, "Optimal rendezvous path planning to an uncontrolled tumbling target," in *19th IFAC Symposium on Automatic Control in Aerospace*, pp. 347–352, Wurzburg, Germany, 2013.

[20] J. Ventura, M. Ciarcia, M. Romano, and U. Walter, "Fast and near-optimal guidance for docking to uncontrolled spacecraft," *Journal of Guidance, Control, and Dynamics*, vol. 40, no. 12, pp. 3138–3154, 2017.

[21] H. Dong, Q. Hu, and M. R. Akella, "Safety control for spacecraft autonomous rendezvous and docking under motion constraints," *Journal of Guidance, Control, and Dynamics*, vol. 40, no. 7, pp. 1680–1692, 2017.

[22] J. Ge, J. Zhao, and J. Yuan, "A novel guidance strategy for autonomously approaching a tumbling target," *Proceedings of the Institution of Mechanical Engineers, Part G: Journal of Aerospace Engineering*, vol. 232, no. 5, pp. 861–871, 2017.

[23] H. Park, S. Di Cairano, and I. Kolmanovsky, "Model predictive control of spacecraft docking with a non-rotating platform," in *IFAC 18th World Congress*, pp. 8485–8490, Milano, Italy, 2011.

[24] A. Weiss, M. Baldwin, R. S. Erwin, and I. Kolmanovsky, "Model predictive control for spacecraft rendezvous and docking: strategies for handling constraints and case studies," *IEEE Transactions on Control Systems Technology*, vol. 23, no. 4, pp. 1638–1647, 2015.

[25] U. Eren, A. Prach, B. B. Koçer, S. V. Raković, E. Kayacan, and B. Açıkmeşe, "Model predictive control in aerospace systems: current state and opportunities," *Journal of Guidance, Control, and Dynamics*, vol. 40, no. 7, pp. 1541–1566, 2017.

[26] H. Park, R. Zappulla, C. Zagarisz, J. V. Llop, and M. Romano, "Nonlinear model predictive control for spacecraft rendezvous and docking with a rotating target," in *Proceedings of the 27th AAS/AIAA Spaceflight Mechanics Meeting*, San Antonio, TX, USA, 2017.

[27] Q. Li, J. Yuan, B. Zhang, and C. Gao, "Model predictive control for autonomous rendezvous and docking with a tumbling target," *Aerospace Science and Technology*, vol. 69, pp. 700–711, 2017.

[28] P. Li and Z. H. Zhu, "Model predictive control for spacecraft rendezvous in elliptical orbit," *Acta Astronautica*, vol. 146, pp. 339–348, 2018.

[29] D. Morgan, S. J. Chung, and F. Y. Hadaegh, "Model predictive control of swarms of spacecraft using sequential convex programming," *Journal of Guidance, Control, and Dynamics*, vol. 37, no. 6, pp. 1725–1740, 2014.

[30] X. Liu and P. Lu, "Solving nonconvex optimal control problems by convex optimization," *Journal of Guidance, Control, and Dynamics*, vol. 37, no. 3, pp. 750–765, 2014.

[31] A. Richards, T. Schouwenaars, J. P. How, and E. Feron, "Spacecraft trajectory planning with avoidance constraints using mixed-integer linear programming," *Journal of Guidance, Control, and Dynamics*, vol. 25, no. 4, pp. 755–764, 2002.

[32] X. Wang, Z. Wang, and Y. Zhang, "Stereovision-based relative states and inertia parameter estimation of noncooperative spacecraft," *Proceedings of the Institution of Mechanical Engineers, Part G: Journal of Aerospace Engineering*, pp. 1–14, 2018.

[33] H.-C. Lim and H. Bang, "Adaptive control for satellite formation flying under thrust misalignment," *Acta Astronautica*, vol. 65, no. 1-2, pp. 112–122, 2009.

# 16

# Effective Mechanical Property Estimation of Composite Solid Propellants Based on VCFEM

**Liu-Lei Shen, Zhi-Bin Shen, Yan Xie, and Hai-Yang Li**

*College of Aeronautics and Astronautics, National University of Defense Technology, Changsha, Hunan, China*

Correspondence should be addressed to Zhi-Bin Shen; zb_shen@yeah.net

Academic Editor: Filippo Berto

A solid rocket motor is one of the critical components of solid missiles, and its life and reliability mostly depend on the mechanical behavior of a composite solid propellant (CSP). Effective mechanical properties are critical material constants to analyze the structural integrity of propellant grain. They are estimated by a numerical method that combines the Voronoi cell finite element method (VCFEM) and the homogenization method in the present paper. The correctness of this combined method has been validated by comparing with a standard finite element method and conventional theoretical models. The effective modulus and the effective Poisson's ratio of a CSP varying with volume fraction and component material properties are estimated. The result indicates that the variations of the volume fraction of inclusions and the properties of the matrix have obvious influences on the effective mechanical properties of a CSP. The microscopic numerical analysis method proposed in this paper can also be used to provide references for the design and the analysis of other large volume fraction composite materials.

## 1. Introduction

A composite solid propellant (CSP), a highly packed particulate composite, is a prime structural material of a solid rocket motor in addition to an energetic material. A CSP consists of polymeric binder matrix (e.g., HTPB) and particle inclusions. The particle materials are usually ammonium perchlorate (AP) and aluminum (Al) [1].

Parametric variations of effective material constants have been studied by numerical and analytical methods in the past [2, 3] and present [4–6]. There are a few studied on the microstructural morphology of the CSP. The effective mechanical properties of CSP are critical to study the deformation and fracture characteristics of propellant grain. Various experimental and numerical studies demonstrate that the mechanical properties of the CSP can be highly sensitive to the microstructural morphology such as the dimension, shape, distribution, and properties of the inclusion. It is a kind of natural choice to study the mechanical properties of the CSP from the aspects of microscopic mechanical. In the early stage, some classical theoretical models of conventional composites have been tailored to estimate the effective properties of the CSP. These theoretical investigations are always limited to simple geometries but are often incapable of reflecting the realistic microstructure of the CSP. Instead, the numerical method becomes increasingly popular because its analysis is based on the realistic geometric simulation model. Zhang et al. [7] employed a homogenization theory and the displacement finite element method (FEM) to compute the mean temperature and heat flux of the CSP's representative volume element (RVE). The effect of orientation and shape of oxidizer particles on the burning rate was examined by a direct numerical simulation approach developed by Plaud et al. [8]. Another group of models published devoted to estimate mechanical properties of propellants in recent years. Yang and Liu [9] use coarse triangle meshes and ANSYS to predict the elastic modulus of the composite solid propellant. Zhi et al. [10] use ABAQUS to study the effects of the critical contact stress, initial contact stiffness, and contact failure distance on the damaged interface model. Matous and Geubelle [11] and Matous et al. [12] develop a damage analysis tool at multiple scales from particle packing to failure use of a numerical framework.

However, the conventional finite element method requires complex grid element and huge computational costs, which limits replication and application in microstructure analysis; furthermore, very small elements may occur owing to the fact that the space among the inclusion is too narrow to create a perfect mesh. For example, they have to shrink particles in contact with other particles or reduce the volume fraction to create high-quality meshes [7, 12]. To overcome some of the limitations discussed above, Ghosh and his coworkers proposed a new numerical method known as the Voronoi cell finite element model (VCFEM) to analyze heterogeneous materials. In 1990, they [13] proposed a two-dimensional automatic mesh generation technique to discrete the composite domain to yield an aggregate of convex Voronoi polygons. An assumed stress hybrid formulation has been implemented to utilize the resulting Voronoi polygons as elements in a finite element model in 1993 [14]. In the following years, the developments of VCFEM were presented for linear elastic problems [15] and elastic-plasticity problems [16], as well as failure analysis [17]. Over the last few years, the VCFEM was further developed to address some engineering problems [18, 19]. In addition, some contributions had been devoted by other researchers to develop this method to analyze the thermomechanical [20] and the effective properties [21] of heterogeneous materials. However, very few works have been attempted to tailor the VCFEM to estimate effective mechanical properties of the CSP. Shen et al. [1] introduced a noninclusion VCFEM to analyze material viscoelastic constants.

The feasibility study on the application of VCFEM in the estimation of effective mechanical properties of the CSP is carried out in this paper. We focus on establishing a numerical model for the analysis of composites with high particle volume fraction. A displacement comparison between the results of VCFEM and those of commercial FEM software is carried out to indicate the correctness of the program code. And the validity of the proposed combined method is obtained by employing two classical theory methods from the literature. In addition, a simple case is analyzed to understand the influence of microstructural morphology on the effective modulus and effective Poisson ratio of the CSP.

## 2. Computational Procedure

*2.1. Microstructure Model of CSP.* The CSP is one of the highly particulate composite materials, always with particle volume fraction between 70% and 80%, or even higher [7]. Figure 1 is a SEM image of the HTPB propellant. Spherical particles are bonded together tightly by the polymer matrix. It is necessary to model a RVE to reflect their microstructure features before obtaining effective material constants of the CSP numerically. Voronoi tessellation is a simple but effective geometric representation for charactering the microstructures of the composite. The tessellation can subdivide a plane into many Voronoi cells determined by a set of centers. Each cell may be perceived as the intersection of open half-planes bounded by the perpendicular bisectors of lines joining one inclusion center with each of its neighbors. However, the inclusion will be dissected into multiple segments by

FIGURE 1: SEM image of the HTPB propellant.

element boundaries of the mesh, when the radius of a circular inclusion in a Voronoi cell is greater than the minimum distance from its center to the center of other cells. The centers generated randomly may lead the model to be inhomogeneous usually. If an element center is closed to one adjacent center than any other adjacent centers, the room between the two centers cannot be filled with inclusion. It will result in the failure to get a large volume fraction of circular inclusion.

In fact, the coordinates of circular inclusions are almost determined when the volume fraction of circular inclusions is large and their sizes are almost equal. A new center-generated project is proposed here. In this project, the locations of centers are generated on the basis of the corresponding existing optimal equal circle packing schemes when the center's number is determined. Those circles will be homogeneous in a RVE model. And the distances between any center and other adjacent centers are almost equal.

The procedure to build the microstructure model with a large volume fraction of inclusions is described below.

(1) The size of RVE ($L_{RVE}$) is determined based on the inclusion size and material property. RVE sizes are defined as the minimum size of a microstructural model that meets the requirement of statistical homogeneity [22].

(2) The number of inclusion is generated randomly from the range of $n_{min}$ to $n_{max}$. $n_{min}$ and $n_{max}$ can be obtained using the equation:

$$n_{min} = \frac{V_f L_{RVE}^2}{\pi r_{max}^2}, \quad n_{max} = \frac{V_f L_{RVE}^2}{\pi r_{min}^2}, \tag{1}$$

where $r_{max}$ and $r_{min}$ are the maximum value and the minimum value of mean values of inclusion radius, respectively, and $V_f$ is the volume fraction of inclusion.

(3) The radius and center positions of the circle inclusion are read from the corresponding existing optimal equal circle packing schemes. The RVE is meshed

with the convex Voronoi polygons according to the centers of the system.

(4) Transform the circles into ellipses. The random number generator is used to generate the lengths of major axes, ratio of major to minor axis length, and orientation (angle) of the major axis of each inclusion. The nonoverlapping constraint has to be imposed. To establish this, each inclusion is contained inside a circle having a diameter equal to the length of the major axis.

Figure 2 describes an example. The length and width are 6 mm. The mean values of the radius of inclusions are ranging from 300 $\mu$m to 310 $\mu$m. Since the volume fraction is about 83%, the $n_{\min}$ and $n_{\max}$ are 95 and 101, respectively. We set the number of inclusion to 100. And the RVE model is meshed following the second step and third step. The radius of inclusions is 308 $\mu$m, and the volume fraction is 80%. Further, the inclusions are transformed into ellipses randomly following the fourth step. We can get a model with a volume fraction of 61%.

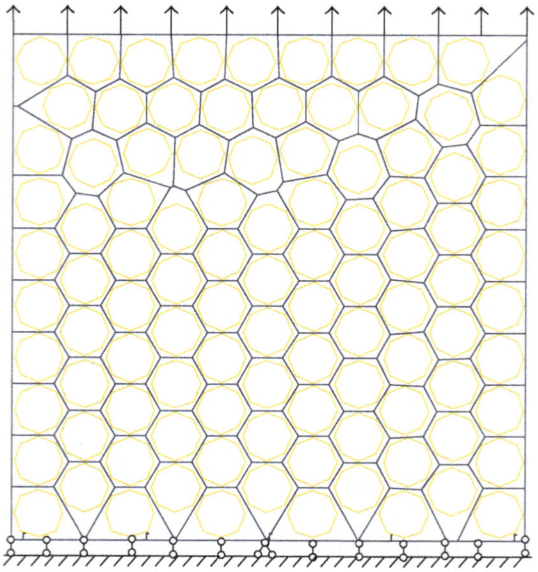


FIGURE 2: A RVE with 100 inclusions.

*2.2. Hybrid Element Formulation.* In this section, the finite element formulation with Voronoi polygons will be reviewed briefly. Figure 3 shows an example of a hybrid element, used in the VCFEM method, with an embedded inclusion. Each node of inclusion may be perceived as the intersection of the circular inclusion and the line joining the inclusion center with each matrix node. The element formulation, as reported in [14], is based upon the stationary of total complementary energy principle. The total complementary energy of the element with an embedded heterogeneity is the addition of the energy of the matrix and inclusion:

$$\begin{aligned}\Pi_{mc} = & \int_{A_e^{\mathrm{M}}} \frac{1}{2} \left(\boldsymbol{\sigma}_e^{\mathrm{M}}\right)^{\mathbf{T}} \mathbf{S}^{\mathrm{M}} \boldsymbol{\sigma}_e^{\mathrm{M}} dA_e^{\mathrm{M}} \\ & + \int_{A_e^{\mathrm{I}}} \frac{1}{2} \left(\boldsymbol{\sigma}_e^{\mathrm{I}}\right)^{\mathrm{T}} \mathbf{S}^{\mathrm{I}} \boldsymbol{\sigma}_e^{\mathrm{I}} dA_e^{\mathrm{I}} \\ & - \oint_{l_{\mathrm{E}}} \left(\mathbf{n}^{\mathrm{E}} \boldsymbol{\sigma}_e^{\mathrm{M}}\right)^{\mathrm{T}} \tilde{\mathbf{u}}_e^{\mathrm{E}} dl \\ & + \oint_{l_{\mathrm{MI}}} \left(\mathbf{n}^{\mathrm{MI}} \left(\boldsymbol{\sigma}_e^{\mathrm{M}} - \boldsymbol{\sigma}_e^{\mathrm{I}}\right)\right)^{\mathrm{T}} \tilde{\mathbf{u}}_e^{\mathrm{I}} dl \\ & + \oint_{l_{\mathrm{T}}} \bar{\mathbf{T}}^{\mathrm{T}} \tilde{\mathbf{u}}_e^{\mathrm{E}} dl,\end{aligned} \tag{2}$$

where the variables with the superscripts M and I correspond to the interior of the matrix and inclusion phases, respectively, while those with superscripts E and MI refer to the variables on the element boundary and internal matrix-inclusion interface, respectively. $\boldsymbol{\sigma}$ is the stress field within the matrix or the inclusion. **S** is the elastic compliance matrix; $A$ represents the area of the matrix or the inclusion. $l_{\mathrm{E}}$, $l_{\mathrm{MI}}$, and $l_{\mathrm{T}}$ represent the displacement boundary of the element, the interface of matrix-inclusion, and traction boundary, respectively. $\tilde{\mathbf{u}}$ represents the compatible displacement fields on the boundary of element or inclusion. $\bar{\mathbf{T}}$ represents the prescribed tractions on the boundary $l_{\mathrm{T}}$. **n** is the outward normal unit vector of the element boundary or matrix-inclusion interface.

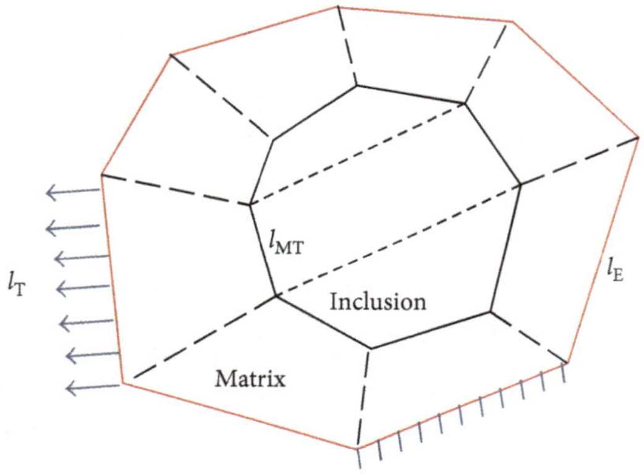


FIGURE 3: A typical Voronoi cell element with an octagonal inclusion.

The displacement field **u** is interpolated by using the nodal displacements **q** and the boundary displacement interpolation functions $\mathbf{L}(\mathbf{u} = \mathbf{L}\mathbf{q})$, while the stress components within the element are assumed to be compatible with the prescribed boundary tractions and satisfy the equilibrium conditions into the region neglecting the body forces. The stress field $\boldsymbol{\sigma}$ is expressed as the polynomial functions of the $x$ and $y$ coordinates, by using complete forms of the stress airy functions. This results in the product of an interpolation matrix **P**, which contains polynomial terms in the $x$ and $y$ coordinates variables as reported in [14], and unknown vectors of coefficients $\boldsymbol{\beta}(\boldsymbol{\sigma}^{\mathrm{M/I}} = \mathbf{P}^{\mathrm{M/I}}\boldsymbol{\beta}^{\mathrm{M/I}})$. The $\Pi_{mc}$ can be simplified as

$$\Pi_{mc}(\boldsymbol{\beta}, \mathbf{q}) = \frac{1}{2}\boldsymbol{\beta}^{\mathrm{T}}\mathbf{H}\boldsymbol{\beta} - \boldsymbol{\beta}^{\mathrm{T}}\mathbf{G}\mathbf{q}. \tag{3}$$

The matrices **H** and **G** are defined as

$$\mathbf{H} = \begin{bmatrix} \mathbf{H}^{\mathrm{M}} & \mathbf{0} \\ \mathbf{0} & \mathbf{H}^{\mathrm{I}} \end{bmatrix}, \quad \mathbf{G} = \begin{bmatrix} \mathbf{G}^{\mathrm{M}} & -\mathbf{G}^{\mathrm{MI}} \\ \mathbf{0} & \mathbf{G}^{\mathrm{I}} \end{bmatrix}, \tag{4}$$

where

$$\mathbf{H}^{\mathrm{M/I}} = \int_{A_{\mathrm{M/I}}} \left[\mathbf{P}^{\mathrm{M/I}}\right]^{\mathrm{T}} \mathbf{S}^{\mathrm{M/I}} \mathbf{P}^{\mathrm{M/I}} dA, \tag{5}$$

$$\mathbf{G}^{\mathrm{E/MI/II}} = \int_{l_{\mathrm{E/MI/I}}} \left[\mathbf{P}^{\mathrm{M/M/I}}\right]^{\mathrm{T}} \left[\mathbf{n}^{\mathrm{E/MI/I}}\right]^{\mathrm{T}} \mathbf{L}^{\mathrm{E/MI/I}} dS. \tag{6}$$

Considering the stationary condition of the total complementary energy, we can get $\partial\Pi_{mc}/\partial\boldsymbol{\beta} = 0$. Consequently, the vector $\boldsymbol{\beta}$ is expressed as

$$\boldsymbol{\beta} = \mathbf{H}^{-1}\mathbf{G}\mathbf{q}. \tag{7}$$

Substituting (7) into (3), with respect to the displacement **q**, the expression of the stiffness matrix of the element is obtained as

$$\mathbf{K}_e = \mathbf{G}^{\mathrm{T}}\mathbf{H}^{-1}\mathbf{G}. \tag{8}$$

Assembling stiffness matrices of each element,

$$\mathbf{K} = \sum_{i=1}^{n} (\mathbf{K}_e)_i. \tag{9}$$

The nodal displacements are the solutions to the following:

$$\mathbf{K}\mathbf{q} = \mathbf{Q}. \tag{10}$$

The Lagrange multiplier method is used to impose additional constraints to avoid rigid body displacement at the interface $l_{\mathrm{MT}}$. The inner node displacements $\mathbf{q}^{\mathrm{I}}$ can be represented by the element boundary nodal displacements $\mathbf{q}^{\mathrm{E}}$, while the inner node displacements are not affected by other elements directly. Therefore, the stiffness matrix of elements and corresponding mechanical load vectors can be eliminated to reduce the computing scale as follows:

$$\mathbf{K}_{\mathrm{E}}^{*} = \mathbf{K}_{11} - \mathbf{K}_{12}\mathbf{K}_{22}^{-1}\mathbf{K}_{12}^{\mathrm{T}}, \quad \mathbf{Q}^{*} = \mathbf{F}_{\mathrm{out}} - \mathbf{K}_{12}\mathbf{K}_{22}^{-1}\mathbf{F}_{\mathrm{in}}, \tag{11}$$

where

$$\mathbf{K}_{11} = \left[(\mathbf{G}^{\mathrm{E}})^{\mathrm{T}}(\mathbf{H}^{\mathrm{M}})^{-1}\mathbf{G}^{\mathrm{E}}\right],$$
$$\mathbf{K}_{12} = \left[-(\mathbf{G}^{\mathrm{E}})^{\mathrm{T}}(\mathbf{H}^{\mathrm{M}})^{-1}\mathbf{G}^{\mathrm{MI}}\left\{\left\{(\boldsymbol{\varphi}^{\mathrm{E}})^{\mathrm{T}}\boldsymbol{\varphi}^{\mathrm{E}}\right\}^{-1}(\boldsymbol{\varphi}^{\mathrm{E}})^{\mathrm{T}}\right\}\right],$$
$$\mathbf{K}_{22} = \left[(\mathbf{G}^{\mathrm{MI}})^{\mathrm{T}}(\mathbf{H}^{\mathrm{M}})^{-1}\mathbf{G}^{\mathrm{MI}} + (\mathbf{G}^{\mathrm{II}})^{\mathrm{T}}(\mathbf{H}^{\mathrm{I}})^{-1}\mathbf{G}^{\mathrm{II}} - \left\{(\boldsymbol{\varphi}^{\mathrm{I}})^{\mathrm{T}}\boldsymbol{\varphi}^{\mathrm{I}}\right\}^{-1}(\boldsymbol{\varphi}^{\mathrm{I}})^{\mathrm{T}}\right], \tag{12}$$

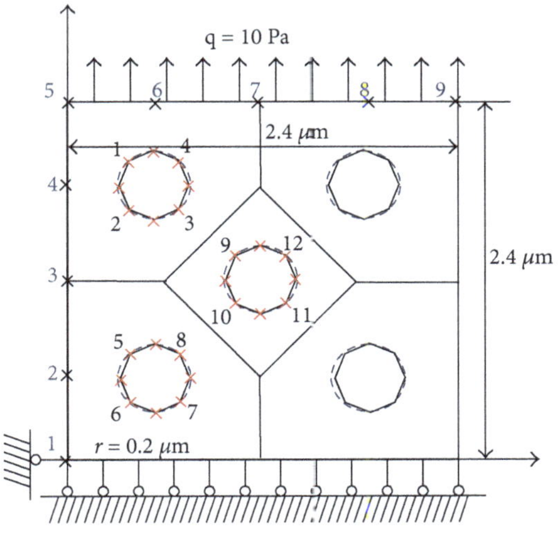

FIGURE 4: RVE model with five inclusions.

where $\boldsymbol{\varphi}$ is the matrix to restrain the rigid body displacements and can be expressed as

$$\boldsymbol{\varphi} = \begin{bmatrix} 1 & 0 & 1 & 0 & \cdots & 1 & 0 \\ 0 & 1 & 0 & 1 & \cdots & 0 & 1 \\ -y_1 & x_1 & -y_2 & x_2 & \cdots & -y_n & x_n \end{bmatrix}^{\mathrm{T}}. \tag{13}$$

*2.3. Homogenization Methods.* As shown in Figure 3, the inclusion and matrix domains of a 2D Voronoi cell element can be divided into triangular and quadrilateral integration regions, respectively. The mean stress and the mean strain are related to the values of every integration cell and can be obtained from the homogenization theory [7]:

$$\bar{\sigma}_{ij} = \sum_{m=1}^{N_{\mathrm{tri}}} \bar{\sigma}_m^{\mathrm{tri}} \frac{S_m^{\mathrm{tri}}}{S_{\mathrm{RVE}}} + \sum_{n=1}^{N_{\mathrm{quad}}} \bar{\sigma}_n^{\mathrm{quad}} \frac{S_n^{\mathrm{quad}}}{S_{\mathrm{RVE}}},$$
$$\bar{\varepsilon}_{ij} = \sum_{m=1}^{N_{\mathrm{tri}}} \bar{\varepsilon}_m^{\mathrm{tri}} \frac{S_m^{\mathrm{tri}}}{S_{\mathrm{RVE}}} + \sum_{n=1}^{N_{\mathrm{quad}}} \bar{\varepsilon}_n^{\mathrm{quad}} \frac{S_n^{\mathrm{quad}}}{S_{\mathrm{RVE}}}, \tag{14}$$

where $N_{\mathrm{tri}}$ ($N_{\mathrm{quad}}$) is the quantity of triangular (quadrangular) elements, $\bar{\sigma}_m^{\mathrm{tri}}/\bar{\varepsilon}_m^{\mathrm{tri}}$ ($\bar{\sigma}_n^{\mathrm{quad}}/\bar{\varepsilon}_n^{\mathrm{quad}}$) is the mean stress/strain of each single triangular (quadrangular) element, $S_m^{\mathrm{tri}}$ ($S_n^{\mathrm{quad}}$) is the area of the $m$th ($n$th) triangular (quadrilateral) integration cell, and $S_{\mathrm{RVE}}$ is the total area of the RVE.

*2.4. Numerical Procedures*

*Remark 1.* Octagons are used to simulate the circular inclusions. To construct the matrix **H**, an integration subdivision scheme is needed to achieve the numerical area integration in (5). An integration subdivision scheme is proposed to reduce the number of integration regions and enhance the precision of numerical integration. As can be seen in Figure 3, the inclusion phase and the matrix phase is subdivided into 3 and 8 quadrilateral regions, respectively.

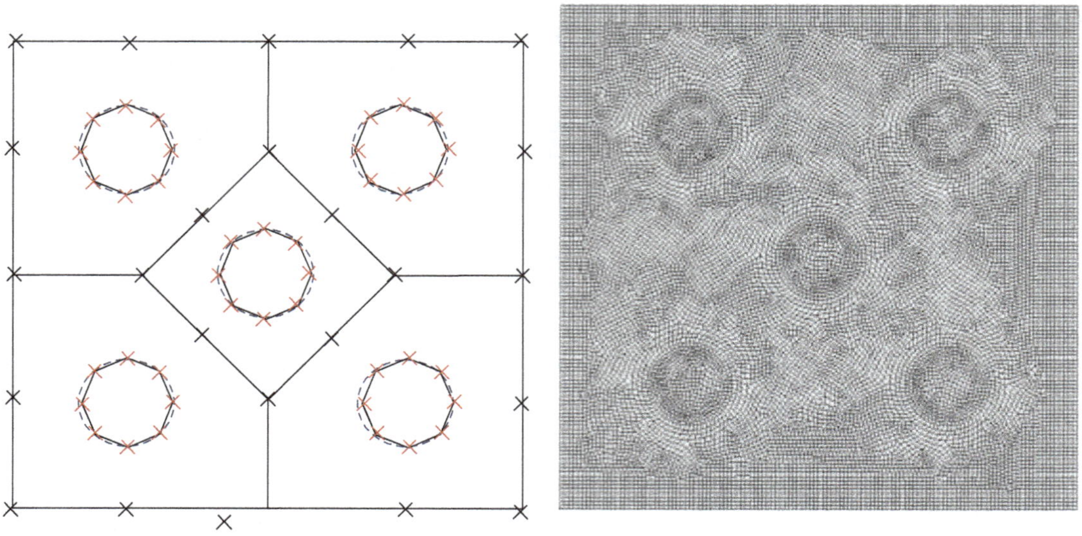

(a) Mesh of VCFEM

(b) Mesh of FEM

FIGURE 5: Mesh of two numerical methods.

*Remark 2.* The rank of the stiffness matrix is determined by the rank of matrix **H** (refer to (8)). And the rank of matrix **H** is equal to the number of the columns of matrix $\mathbf{P}n_\mathbf{P}$ (refer to (5)). For elements with more nodes, it is necessary to increase $n_\mathbf{P}$ to obtain enough rank. The value of $n_\mathbf{P}$ is adopted 25 to analyze octagon element. When $n_\mathbf{P}$ is equal to 25, the highest order of the power of matrix **P** is equal to 4. As shown in (5), the Gauss integral with eight points is used in the calculation of matrix **H** to conquer the ill-conditioned problem of the stiffness matrix [23].

## 3. Numerical Examples

### *3.1. Result Validation*

*3.1.1. Displacement Analysis.* In order to verify the applicability of the Voronoi cell element method developed, the displacement of a heterogeneous material with five inclusions subjected to uniformly tensile load is considered in this example (Figure 4). It is computed using VCFEM, and the results are compared with those obtained from the displacement-based FEM software Nastran. The matrix material is the following: Young's modulus $E = 1000$ Pa and Poisson's ratio $\upsilon = 0.2$. The inclusion material is the following: Young's modulus $E = 3000$ Pa and Poisson's ratio $\upsilon = 0.2$. A uniform tensile stress $q = 10$ Pa at the top of the model is considered. Displacement constraints are imposed on the bottom side of the plate in the vertical direction. The center coordinates of five inclusions are the following: (0.6 $\mu$m,0.6 $\mu$m), (1.8 $\mu$m,0.6 $\mu$m), (1.2 $\mu$m,1.2 $\mu$m), (0.6 $\mu$m,1.8 $\mu$m), and (1.8 $\mu$m,1.8 $\mu$m).

The RVE model is subdivided into 5 cells used in Voronoi tessellation as shown in Figure 5(a). The RVE model is meshed by FEM in the commerce procedure Patran as well. There are 15,062 quadrilateral elements as shown in Figure 5(b).

TABLE 1: Comparison of two methods' displacement results in load direction.

| Node | VCFEM ($\mu$m) | FEM ($\mu$m) | Error* |
|---|---|---|---|
| Matrix node | | | |
| 1 | 0 | 0 | — |
| 2 | 0.005820 | 0.005598 | 3.97% |
| 3 | 0.011190 | 0.010841 | 3.22% |
| 4 | 0.016790 | 0.016027 | 4.76% |
| 5 | 0.022840 | 0.021831 | 4.62% |
| 6 | 0.021620 | 0.020608 | 4.91% |
| 7 | 0.021960 | 0.020928 | 4.93% |
| 8 | 0.021620 | 0.020608 | 4.91% |
| 9 | 0.022840 | 0.021831 | 4.62% |
| Matrix-inclusion interface node | | | |
| 1 | 0.01584 | 0.01568 | 1.03% |
| 2 | 0.01514 | 0.01472 | 2.89% |
| 3 | 0.01594 | 0.01539 | 3.57% |
| 4 | 0.01662 | 0.01634 | 1.72% |
| 5 | 0.00543 | 0.00519 | 4.56% |
| 6 | 0.00442 | 0.00426 | 3.88% |
| 7 | 0.00493 | 0.00504 | 2.17% |
| 8 | 0.00592 | 0.00591 | 0.14% |
| 9 | 0.01058 | 0.01017 | 4.00% |
| 10 | 0.00980 | 0.00937 | 4.54% |
| 11 | 0.01063 | 0.01017 | 4.49% |
| 12 | 0.01130 | 0.01100 | 2.73% |

*Error = |VCFEM − FEM|/FEM × 100%.

Table 1 shows the comparisons of the displacements of nodes on matrix boundary and matrix-inclusion interface in the loading direction. The results computed by the VCFEM and Nastran show a good agreement with each

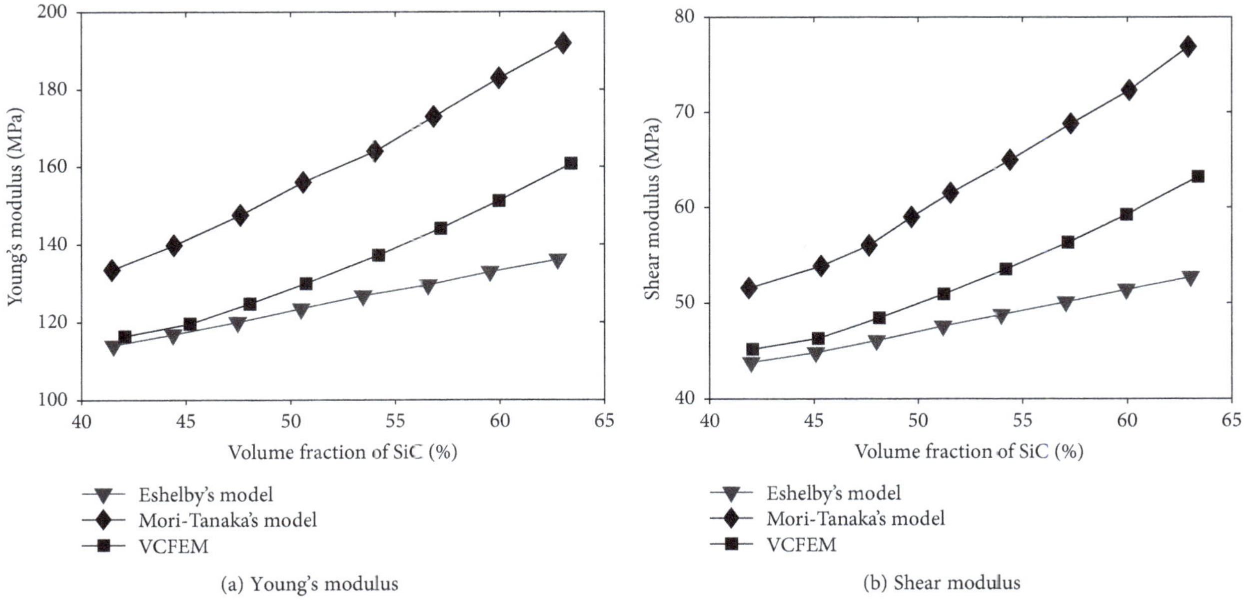


FIGURE 6: Effective modulus of aluminum–SiC composites versus the $V_f$ of SiC.

other. The maximum relative error is less than 5.0%. In consideration, the stiffness matrix obtained by the displacement finite element method based on the minimum principle of potential energy is greater than the real stiffness matrix. So its displacement results are smaller than real displacement results. VCFEM is based on the minimum principle of residual energy. The stiffness matrix obtained by VCFEM is smaller than the real stiffness matrix. So its displacement results are greater than real displacement results. The error between the two displacement results is acceptable.

*3.1.2. Modulus Analysis.* Furthermore, another example is adopted to verify the correctness of the combined method. The effective Young's modulus and shear modulus are analyzed for several volume fractions of the aluminum–SiC composites using two theoretical models and the proposed method [24]. The VCFEM models were generated by using the program introduced in Section 2 and consisted of 100 hybrid elements containing polygonal inclusions as shown in Figure 2. A uniform tensile stress $q = 10$ MPa at the top of the model is considered. Appropriate displacement constraints are imposed on the bottom side of the plate in the vertical direction. The material of the matrix is aluminum: $E = 70.576$ GPa and $\upsilon = 0.33$. The material of the inclusion is SiC: $E = 450.0$ GPa and $\upsilon = 0.17$.

It is clear from Figure 6 that those curves, respectively, obtained by these three methods have the same change trend. Eshelby's model and Mori-Tanaka's model are based on simplifying assumptions. They can only determine a region of real solution. And the result obtained by VCFEM is between the results of two models. It indicates that the method can obtain an effective result and shows its potential in the prediction of mechanical properties of solid propellants.

*3.2. Effective Mechanical Properties of CSP.* In the following case, the effective mechanical properties of the CSP are analyzed for different volume fractions of inclusion and component material properties with the RVE model of Section 3.1.2. The matrix will be assumed as an elastic material: Young's modulus $E_M = 10$ MPa and Poisson's ratio $\nu_M = 0.495$. The AP particle material: Young's modulus, $E_{AP} = 32.4$ GPa, Poisson's ratio $\nu_{AP} = 0.14$. The Al material: Young's modulus $E_{Al} = 68.3$ GPa, Poisson's ratio $\nu_{Al} = 0.33$. The volume fraction of AP and Al particles, the modulus and Poisson's ratio of the matrix and inclusions, will be changed below to understand their relationship with effective mechanical properties. As shown in Figure 7, the effective moduli of different RVEs are calculated here to verify that the RVE size used is large enough. Keep the effective radius of particles (100 $\mu$m) the same; increase the number

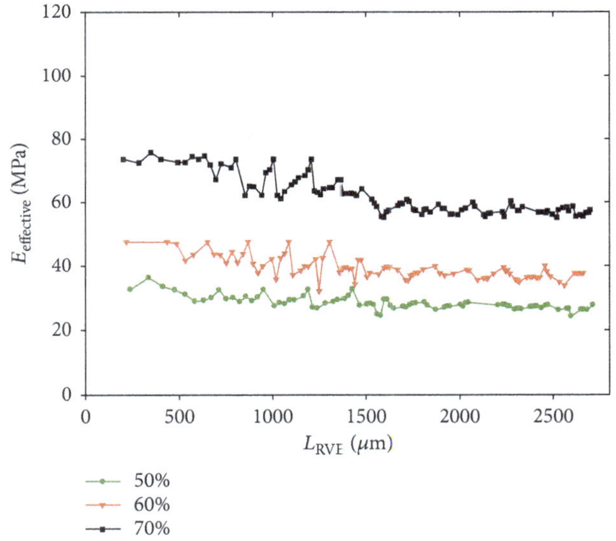


FIGURE 7: Effect of RVE size on effective modulus.

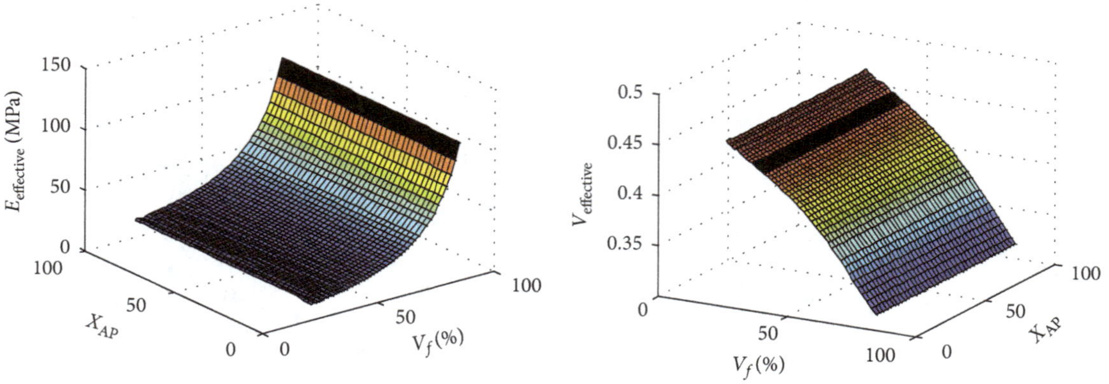


(a) Effective Young's modulus versus $V^f$ and $V^{AP}/V^{Al}$

(b) Effective Poisson's ratio versus $V^f$ and $V^{AP}/V^{Al}$

FIGURE 8: Effective properties of solid propellants versus $V_f$ and $V_{AP}/V_{Al}$.

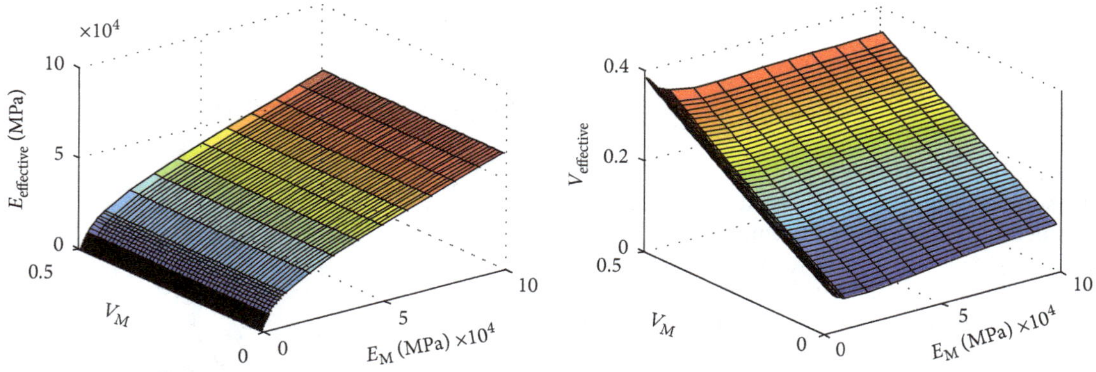


(a) Effective Young's modulus versus $E_M$ and $\nu_M$

(b) Effective Poisson's ratio versus $E_M$ and $\nu_M$

FIGURE 9: Effective properties of solid propellants versus $E_M$ and $\nu_M$.

of particles to obtain RVEs with different sizes. The results showed that when the RVE size is greater than 1500 $\mu$m (contains 48 inclusion particles when the volume fraction is 60%), its effective modulus remained stable. The model used below contains 100 inclusions (as shown in Figure 2), whose size is large enough to ignore the size effect of RVE here. A uniform tensile stress $q = 10$ MPa at the top of the model is considered. Appropriate displacement constraints are imposed on the bottom side of the plate in the vertical direction.

*3.2.1. Effect of Inclusions' Volume Fraction.* The mechanical properties of composites with different volume fractions of AP and Al particles are calculated to study the effect of particle volume fraction. The changing of volume fraction depends on the changing of the size of RVE. The value of $V_{AP}/V_{Al}$ varies with the quantity of AP cell elements $x_{AP}$. The 3D graphs in Figure 7 display the relationship between the effective properties and the volume fraction of inclusions $V_f$ and $V_{AP}/V_{Al}$.

Figure 8(a) illustrates an upward trend with the increase of volume fraction of inclusion $V_f$. Upon further analysis, we can note that the effective modulus is closed to the modulus of the matrix when the $V_f$ is less than 60% and, with the increase of $V_f$, the effective modulus increases more rapidly. Figure 8(b) illustrates a downward trend with the increase of $V_f$. The ratio of $V_{AP}/V_{Al}$ describes little influence on the effective properties.

With the increase of the volume fraction of inclusion, the mechanical properties of composites, such as a composite solid propellant, is closing to properties of the inclusion materials. When the differences between the properties of two inclusion materials are smaller than the differences between theirs and the matrix's, the variation of the proportion of different inclusions is not obvious.

*3.2.2. Effect of Matrix's Material Properties.* The mechanical properties of composites with different modulus and Poisson's ratio of the matrix are calculated to study the effect of the matrix material. The particle volume fraction remains at 65% invariability in this case. The 3D graphs in Figure 8 display the relationship between the effective properties and the matrix modulus $E_M$ and matrix Poisson's ratio $\nu_M$.

It can be seen in Figure 9(a) that the effective modulus has a positive relation with $E_M$, while the influence of $\nu_M$ can be ignored. Figure 9(b) shows that the effective Poisson's ratio has a positive linear relation with $E_M$. Furthermore, the effective Poisson's ratio decreases slightly with the increase of $E_M$. But with the increase of the matrix modulus, the decreasing trend is more and more inconspicuous.

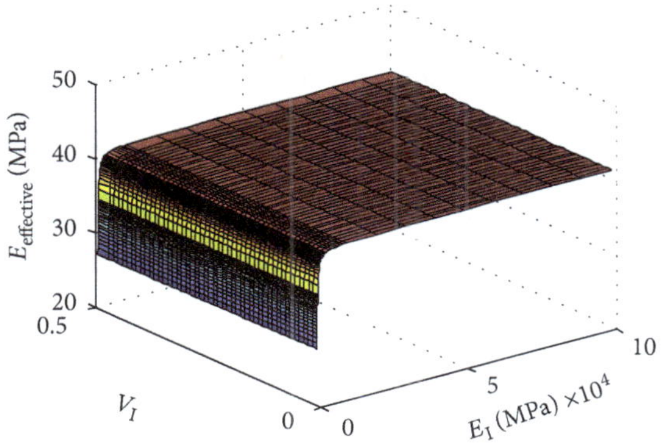


(a) Effective Young's modulus versus $E_I$ and $\nu_I$

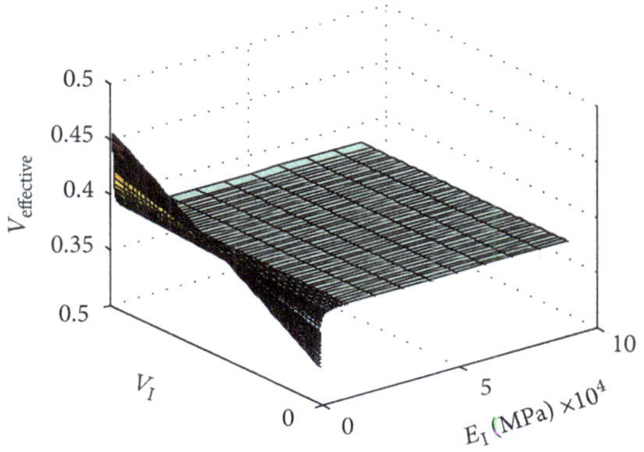


(b) Effective Poisson's ratio versus $E_I$ and $\nu_I$

FIGURE 10: Effective properties of solid propellants versus $E_I$ and $\nu_I$.

The variation of the modulus of the matrix has an effect on both the effective modulus and the effective Poisson's ratio of the CSP, when the variation of the Poisson's ratio of the matrix has a linear effect on the effective Poisson's ratio of the CSP mainly.

*3.2.3. Effect of Inclusion's Material Properties.* The mechanical properties of composites with different modulus and Poisson's ratio of inclusion are calculated to study the effect of the inclusion material. The particle volume fraction remains at 65% invariability in this case. The 3D graphs in Figure 10 display the relationship between the effective properties and the inclusion modulus $E_I$ and the inclusion Poisson's ratio $\nu_I$.

From Figure 10(a), the effective modulus increases sharply with the increase of $E_I$ when the ratio of $E_I/E_M$ is less than 10. As $E_I$ continues to increase, the effective modulus keeps constant steadily. The variation of $\nu_I$ has no significant effect on effective modulus. From Figure 10(b), we can see clearly that a dramatical change has taken place in the growth process of $E_I$. The effective Poisson's ratio increases linearly with $\nu_I$. But the slope of the effective Poisson's ratio and $\nu_I$ is declining with the increase of $E_I$.

Only when the modulus of inclusion is matched by the modulus of the matrix, the effective modulus and effective Poisson's ratio are affected by the modulus of inclusion significantly. However, the modulus of inclusion of the CSP is larger than the matrix's. Hence, the variation of modulus of inclusion does not have as obvious influence as expected.

## 4. Conclusion

The effective modulus and Poisson's ratio of the CSP, which are closely related to the volume fraction and material property of each component, are the critical material parameters to analyze the structural integrity of propellant grains. A strategy for constructing RVE models of highly packed particulate composites is presented here, which is adopted by VCFEM appropriately. A numerical programming method combined with the VCFEM and homogenization method is proposed to investigate the relationship between the microstructural morphology and the effective properties of the CSP. Based on the examples mentioned in the above sections, the following conclusions can be drawn:

(1) The mechanical properties of the CSP are significantly affected by the volume fraction of inclusions, with the increase of the volume fraction of inclusion; the mechanical properties of composites, such as a composite solid propellant, are closing to the properties of the inclusion materials. However, the variation of the proportion of different inclusions has a minor influence. Since the properties different between the inclusion and the matrix are very large, a small change of inclusion's properties does not have a significant effect on the overall effective properties.

(2) Except that the modulus and Poisson's ratio of the matrix directly influence the effective modulus and Poisson's ratio of the CSP, respectively, the variation of the matrix modulus has modest influences on the effective Poisson's ratio of the CSP.

(3) The effective properties are affected by the modulus of inclusion significantly only when the moduli of the inclusion and matrix are close. However, as for the CSP, when the modulus of the inclusion is much larger than that of the matrix, the effects of inclusion's material properties are not obvious as expected.

## Conflicts of Interest

The authors declare that they have no conflicts of interest.


## Acknowledgments

This study was supported by the Science Project of the National University of Defense Technology.


## References


[1] L.-L. Shen, Z.-B. Shen, H.-Y. Li, and Z.-Y. Zhang, "A Voronoi cell finite element method for estimating effective mechanical properties of composite solid propellants," *Journal of Mechanical Science and Technology*, vol. 31, no. 11, pp. 5377–5385, 2017.

[2] Z. Hashin and S. Shtrikman, “A variational approach to the theory of the elastic behaviour of multiphase materials,” *Journal of the Mechanics and Physics of Solids*, vol. 11, no. 2, pp. 127–140, 1963.

[3] T. Steinkopff and M. Sautter, “Simulating the elasto-plastic behavior of multiphase materials by advanced finite element techniques part II: simulation of the deformation behavior of Ag-Ni composites,” *Computational Materials Science*, vol. 4, no. 1, pp. 15–22, 1995.

[4] A. Malyarenko and M. Ostoja-Starzewski, “A random field formulation of Hooke’s law in all elasticity classes,” *Journal of Elasticity*, vol. 127, no. 2, pp. 269–302, 2017.

[5] A. Malyarenko and M. Ostoja-Starzewski, “Spectral expansions of homogeneous and isotropic tensor-valued random fields,” *Zeitschrift für angewandte Mathematik und Physik*, vol. 67, no. 3, 2016.

[6] J. Szafran and M. Kamiński, “Bridges for pedestrians with random parameters using the stochastic finite elements analysis,” *International Journal of Applied Mechanics and Engineering*, vol. 22, no. 1, pp. 175–197, 2017.

[7] J. W. Zhang, S. J. Zhi, and B. Sun, “Estimation of thermophysical properties of solid propellants based on particle packing model,” *Science China Technological Sciences*, vol. 56, no. 12, pp. 3055–3069, 2013.

[8] M. Plaud, S. Gallier, and M. Morel, “Simulations of heterogeneous propellant combustion : effect of particle orientation and shape,” *Proceedings of the Combustion Institute*, vol. 35, no. 2, pp. 2447–2454, 2015.

[9] J. Yang and C. Liu, “A numerical approach for the simulation of composite solid propellant,” in *2009 Second International Conference on Information and Computing Science*, pp. 50–52, Manchester, UK, 2009.

[10] S. J. Zhi, B. Sun, and J. W. Zhang, “Multiscale modeling of heterogeneous propellants from particle packing to grain failure using a surface-based cohesive approach,” *Acta Mechanica Sinica*, vol. 28, no. 3, pp. 746–759, 2012.

[11] K. Matous and P. H. Geubelle, “Multiscale modelling of particle debonding in reinforced elastomers subjected to finite deformations,” *International Journal for Numerical Methods in Engineering*, vol. 65, no. 2, pp. 190–223, 2006.

[12] K. Matous, H. Inglis, X. Gu, D. Rypl, T. Jackson, and P. Geubelle, “Multiscale modeling of solid propellants: from particle packing to failure,” *Composites Science and Technology*, vol. 67, no. 7-8, pp. 1694–1708, 2007.

[13] S. Ghosh and S. N. Mukhopadhyay, “A two-dimensional automatic mesh generator for finite element analysis for random composites,” *Computers & Structures*, vol. 41, no. 2, pp. 245–256, 1991.

[14] S. Ghosh and S. N. Mukhopadhyay, “A material based finite element analysis of heterogeneous media involving Dirichlet tessellations,” *Computer Methods in Applied Mechanics and Engineering*, vol. 104, no. 2, pp. 211–247, 1993.

[15] S. Ghosh, K. Lee, and S. Moorthy, “Multiple scale analysis of heterogeneous elastic structures using homogenization theory and Voronoi cell finite element method,” *International Journal of Solids and Structures*, vol. 32, no. 1, pp. 27–62, 1995.

[16] S. Ghosh and S. Moorthy, “Elastic-plastic analysis of arbitrary heterogeneous materials with the Voronoi cell finite element method,” *Computer Methods in Applied Mechanics and Engineering*, vol. 121, no. 1–4, pp. 373–409, 1995.

[17] S. Ghosh and S. Moorthy, “Particle fracture simulation in non-uniform microstructures of metal–matrix composites,” *Acta Materialia*, vol. 46, no. 3, pp. 965–982, 1998.

[18] S. Ghosh, J. Bai, and D. Paquet, “Homogenization-based continuum plasticity-damage model for ductile failure of materials containing heterogeneities,” *Journal of the Mechanics and Physics of Solids*, vol. 57, no. 7, pp. 1017–1044, 2009.

[19] S. Ghosh, “Adaptive hierarchical-concurrent multiscale modeling of ductile failure in heterogeneous metallic materials,” *JOM*, vol. 67, no. 1, pp. 129–142, 2015.

[20] L. Bruno, F. M. Furgiuele, and C. Maletta, “A hybrid method for the thermo-mechanical analysis of elastic cracks in two-dimensional heterogeneous materials,” *Finite Elements in Analysis and Design*, vol. 43, no. 6-7, pp. 444–452, 2007.

[21] M. Grujicic and Y. Zhang, “Determination of effective elastic properties of functionally graded materials using Voronoi cell finite element method,” *Materials Science and Engineering: A*, vol. 251, no. 1-2, pp. 64–76, 1998.

[22] M. Jiang, K. Alzebdeh, I. Jasiuk, and M. Ostoja-Starzewski, “Scale and boundary conditions effects in elastic properties of random composites,” *Acta Mechanica*, vol. 148, no. 1-4, pp. 63–78, 2001.

[23] S. Moorthy and S. Ghosh, “A model for analysis of arbitrary composite and porous microstructures with Voronoi cell finite elements,” *International Journal for Numerical Methods in Engineering*, vol. 39, no. 14, pp. 2363–2398, 1996.

[24] X. Liu and J. Zhang, “The study of finite element method to predict effective elastic modulus of composites,” *Journal of Shenyang Institute of Engineering (Natural Science)*, vol. 5, no. 2, pp. 175–179, 2009.

# Study of Influential Factors of the Vibration Modal of the Rocket Equipment Bay and Structural Improvement Based on Finite Element Analysis

**Chenghu Li, Guanlin Han, and Chunxu Duan**

*School of Aircraft Engineering, Nanchang Hangkong University, Nanchang 330063, China*

Correspondence should be addressed to Chenghu Li; lichenghu111@126.com

Academic Editor: Paolo Gasbarri

In this paper, the study focuses on influential factors of the vibration modal of the equipment bay of a carrier rocket and the structural improvement of the equipment bay by using finite element modal analysis. The finite element analysis focuses on the influences of the mass of the inertial bracket, the thickness of parts of the inertial bracket, and the stringer thickness on the first modal frequency of the equipment bay. As the analytical results show, the vibration displacement of mounting panels can be greatly reduced when the equipment bay is added with mass, and the vibration of the equipment bay with mass mainly occurs on the inertial bracket (without mass, the maximum vibration displacement occurs on the mounting panel); the top surface thickness of the inertial bracket has the maximum influence on the first modal frequency of the equipment bay, the stringer thickness and the side thickness of the inertial bracket have relatively high influence on the first modal frequency, the rib thickness of the inertial bracket has relatively low influence on the first modal frequency, and the beam thickness of the inertial bracket has the minimum influence on the first modal frequency.

## 1. Introduction

The equipment bay is an extremely important bay section of a carrier rocket, in which most of the rocket's instruments are mounted. The equipment bay has strict requirements for vibratory frequency. It requires a relatively high frequency of the first modal for avoiding sympathetic vibration and a relatively small vibration amplitude of its mounting panels. In the design of the equipment bay, finite element analysis is used to ensure that the structure of the equipment bay satisfies the requirements. Some applications of finite element analysis in simulation analyses of rockets have been presented. Choi et al. [1] evaluated the structural integrity and weight of the rocket motor with a hemispherical dome by analyzing the improved 2-D axisymmetric finite element model. Engberg and Korde [2] verified the possibility of 3-D full-scale finite element modeling by analyzing the two-dimensional finite element acoustic model of the fairing of a launch vehicle. Mahyari et al. [3] established a finite element model of the launch vehicle tank to analyze the effectiveness of the rotor crusher in reducing the critical height of propellant. Schwane and Xia [4] established a finite element model of the rocket engine nozzle to calculate the lateral load generated by the nozzle. Mense et al. [5] studied the thermal response of unprotected structural steel under solid rocket propellant flame by finite element thermal analysis. Howard et al. [6] verified the effect of the absorber in the rocket structure by analyzing the finite element model of the rectangular plate. Salvador and Xu [7] established a two-dimensional finite element model of the rocket burner with electrodes to analyze the influence of an electric field on flame combustion stability. Sim et al. [8] established finite element modeling techniques for computational modal analyses by considering the liquid propellant and flange joints of launch vehicles. Baldesi [9] introduced some studies that the dynamics analyses of various types of launch vehicles were executed by using finite element analysis. Hutchinson and Olds [10] estimated the structural weight of the propellant tank of a rocket by establishing a simplified beam finite element model. Marimuthu and Rao [11]carried out dynamic

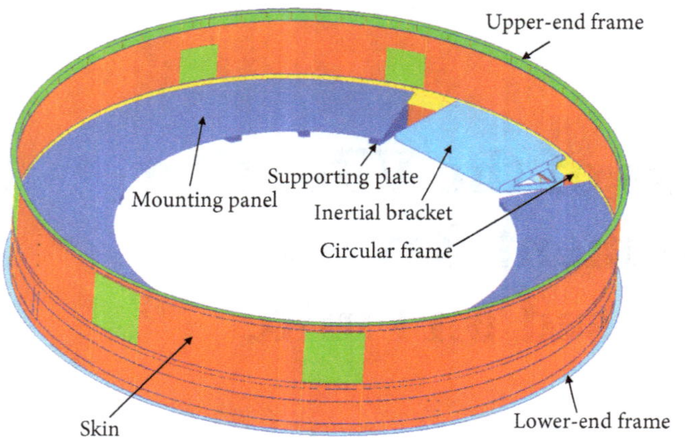


FIGURE 1: Geometry of the equipment bay.

analysis of the open cylindrical storage tank of a launch vehicle by means of finite element dynamic analysis. Roh and Kim [12] solved the trajectory optimization problem of a multistage launch vehicle by using an indirect time finite element method. Li and Zhang [13] researched the connection strength and slip characteristics of the rocket bolt by using finite element analysis. Elhefny and Liang [14] built a 2-D axisymmetric rocket gas turbine disk model to analyze the stress of the gas turbine disk. Zhang and Jiang [15] simulated the flight response of a rocket sled by using finite element coupling mechanical analysis. Chandana et al. [16] analyzed six mode shapes of the launch vehicle payload fairing structure by finite element analysis.

However, these studies focus on the application of finite element analysis in the simple rocket part structure, and the exhibit of finite element analysis in the complex rocket structures is less. This paper will present the vibration modal analysis of an equipment bay that has 7 parts and 572 thousand elements. The materials of the equipment bay include aluminum alloy, cast iron, carbon fiber composite laminates, and honeycomb sandwich. The parts in the finite element model of the equipment bay are linked by adhesive contacts, including in total 13 contact bodies that mutually adhere with each other. And the influence of some structural parameters of the equipment bay on its first modal and the structural improvement of the equipment bay will be investigated.

## 2. Finite Element Model of the Equipment Bay

The equipment bay includes 7 major parts such as the skin, upper-end frame, lower-end frame, supporting plate, mounting panel, circular frame, and inertial bracket, as shown in Figure 1. The skin is a reinforced cylindrical shell structure, with an outer diameter of 3000 mm and an overall height of 650 mm. It includes a cylindrical shell, stringer, window, and window cover. The skin is made of aluminum alloy. The function of stringers is to enhance the bearing capacity of the cylindrical shell, and the role of windows is to facilitate the installation and maintenance of equipment. The window includes a big window and a small window. The big window is a 300 mm × 300 mm square frame, and the small window is

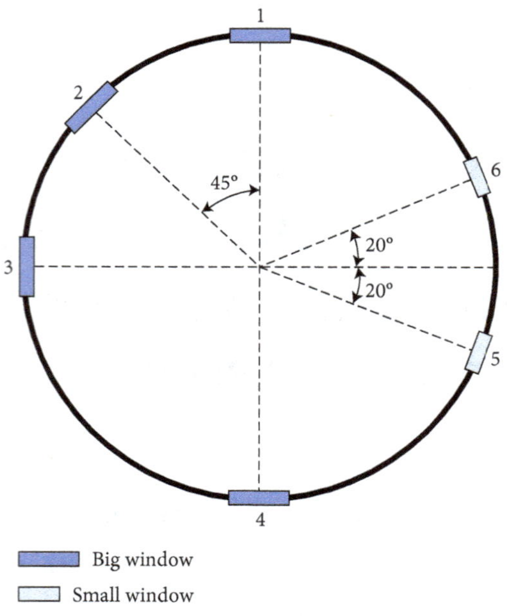


FIGURE 2: Location of windows.

a 200 mm × 200 mm square frame. The lower-end of all the windows is 260 mm away from the lower-end of the skin. The location of the window on the skin is shown in Figure 2. Windows 1, 2, 3, and 4 are big windows and Windows 5 and 6 are small windows. The symmetry axis of Windows 1 and 4 coincides with the vertical symmetry axis. The symmetry axis of Window 3 coincides with the horizontal symmetry axis. The angle between the symmetry axis of Window 2 and the vertical symmetry axis is 45° and the angle between Windows 5 and 6 and the horizontal symmetry axis is 20°. The stringers have two types of cross sections, as shown in Figure 3. The parameters of the I-type stringer are $H_1 = 25$ mm, $W_1 = 20$ mm, $t_1 = 1.5$ mm, and $t_2 = 1.5$ mm. The parameters of the II-type stringer are $H_2 = 25$ mm, $W_2 = 20$ mm, $W_3 = 20$ mm, and $t_3 = 1.5$ mm. The number of I-type stringers and II-type stringers is 12 and 76, respectively. The cylindrical shell is simulated with a shell element of which the thickness is 1 mm. The stringer is simulated with a beam element. The skin finite element model is shown in Figure 4. In the figure, the 3-D beam is the result of the 3-D display of the 1-D beam element. The left and right sides of the window are reinforced by two I-type beams. The cylindrical shell of the skin is linked with stringers and a window cover by using merging nodes.

The inertial bracket is used to control the overall dynamic characteristics, which is made of cast iron as a whole. The finite element model of the inertial bracket is shown in Figure 5. In the figure, all the elements are plate shell elements. The top surface and rib have a thickness of 3.5 mm, the side surface and beam have a thickness of 2 mm, and the top surface and rib, top surface and beam, and rib and beam have the conodes on their intersection. The rib, beam, top surface, and side surface in the inertial bracket are linked with merging nodes, while the inertial bracket is linked with the skin and circular frame by adhesive contact. For the

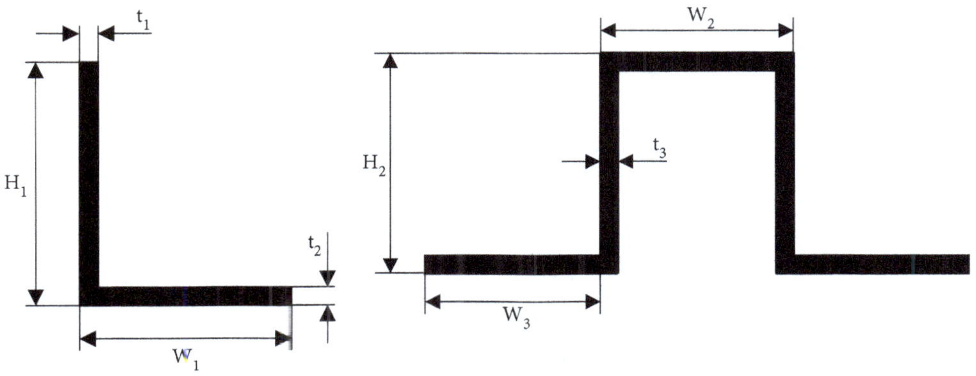


(a) Cross section of I-type stringer

(b) Cross section of II-type stringer

FIGURE 3: Cross sections of I-type stringer and II-type stringer.

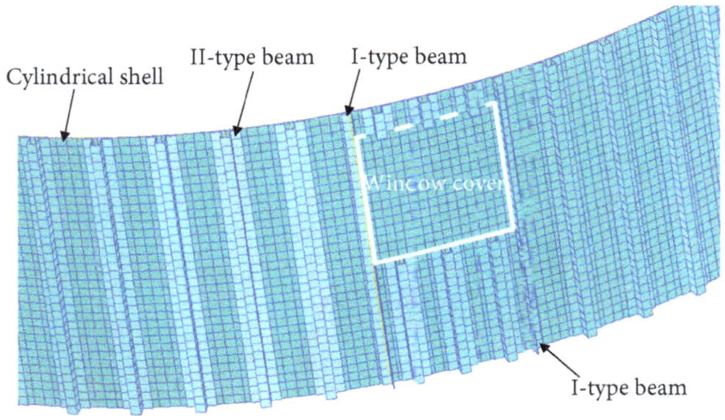


FIGURE 4: Finite element model of the skin.

successful setting of the adhesive contact, the distance between shell elements must be considered or the offset of the shell element is set or the distance between surfaces should be calculated when establishing the surfaces.

The supporting plate is used to support the mounting panel, and it is made of carbon fiber composite laminates. The stacking sequence of the laminates is $[45/0/-45/90]_{3S}$. Each layer has a thickness of 0.125 mm. The shell element is used for simulating the supporting plate. The finite element model of the supporting plate is shown in Figure 6. The blue surface of the supporting plate is in contact with the mounting panel, and its red surface and yellow surface are in contact with the skin and the circular frame, respectively.

The circular frame is used to enhance the stiffness and strength in the circumference direction of the skin. The finite element model of the circular frame is shown in Figure 7. The red surface of the circular frame is in contact with the skin, and its blue surface and yellow surface are in contact with the mounting panel and the supporting plate, respectively. The mounting panel is of a honeycomb sandwich. The upper and lower faces are carbon fiber laminates. The stacking sequence is $[45/0/-45/90]_S$. Each layer has a thickness of 0.125 mm. The middle sandwich is a 10 mm aluminum honeycomb. One 17-layer laminate is used to simulate the honeycomb sandwich. The mounting panel is in contact with the supporting plate and circular frame. The upper-end frame and lower-end frame are used to enhance the stiffness and strength of the upper and lower ends of the skin and are used to link the equipment bay with other bay sections, and their cross sections are both the "L" shape. The upper-end frame and the lower-end frame are in contact with the upper and lower ends of the skin, respectively. The overall finite element model of the equipment bay is shown in Figure 8. The model includes 4 materials: aluminum alloy, cast iron, CFRP (carbon fiber-reinforced polymer) tape, and aluminum honeycomb. Aluminum alloy and cast iron are isotropic materials, and the CFRP tape and aluminum honeycomb are 2-D orthotropic materials; their characteristics are shown in Table 1.

The parts of the equipment bay are linked by adhesive contact. In the finite element model, it is necessary to define the contact body and contact table. In the model, 13 contact bodies are defined in total: the portion of mounting panel that contacts with the circular frame is defined as Contact body 1, the portion of the mounting panel that contacts with the supporting plate is defined as Contact body 2, the portion of the inertial bracket that contacts with the skin is defined as Contact body 3, the portion of the inertial bracket that contacts with the circular frame is defined as Contact body 4, the portion of the skin that contacts with the upper-end frame is defined as Contact body 5, the portion of the skin that contacts with the lower-end frame is defined as Contact body 6 (the portion contacts also with the inertial bracket and supporting plate), the portion of the skin that contacts with the circular frame is defined as Contact body 7, the portion of the upper-end frame that contacts with the skin is defined as Contact body 8, the portion of the lower-end frame that contacts with the skin is defined as Contact body 9, the circular frame is defined as Contact body 10, the portion of the supporting plate that contacts with the mounting panel is defined as Contact body 11, the portion of the supporting plate that contacts with the skin is defined as Contact body 12, and the portion of the supporting plate that contacts with the circular frame is defined as Contact body 13. The two contact bodies in mutual contact are classified as the master and the slave. One contact body can be taken as the salve once only but as the master for $n$ times. The contact table of the 13 contact bodies in mutual contact is shown in Table 2, where S represents the slave, M represents the

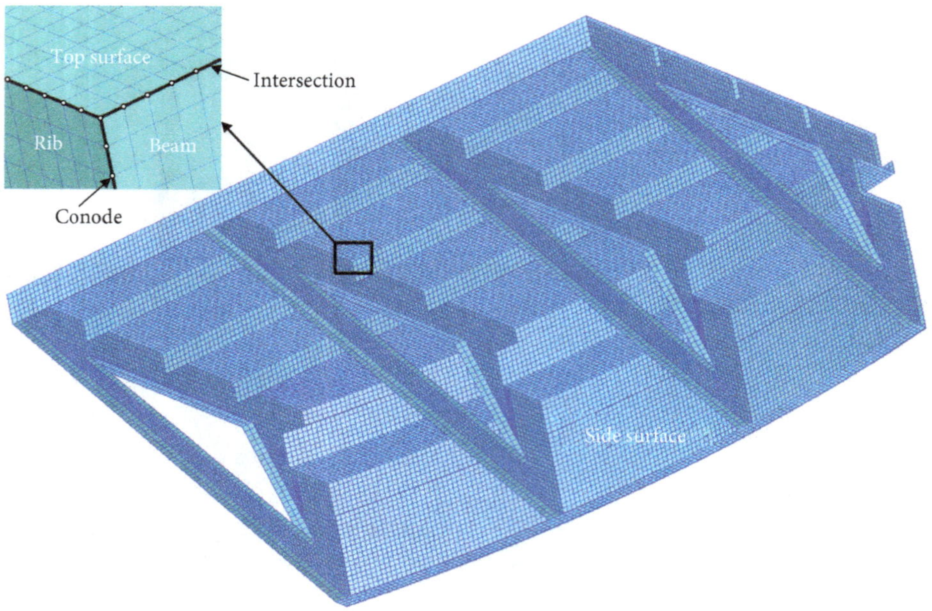


Figure 5: Finite element model of the inertial bracket.

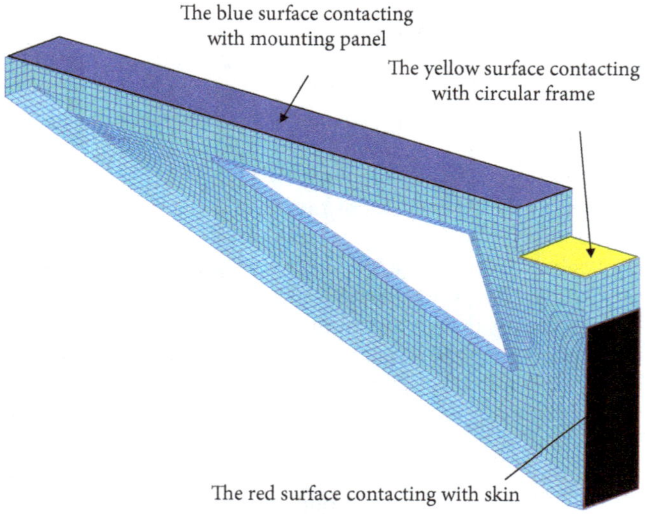


Figure 6: Finite element model of the supporting plate.

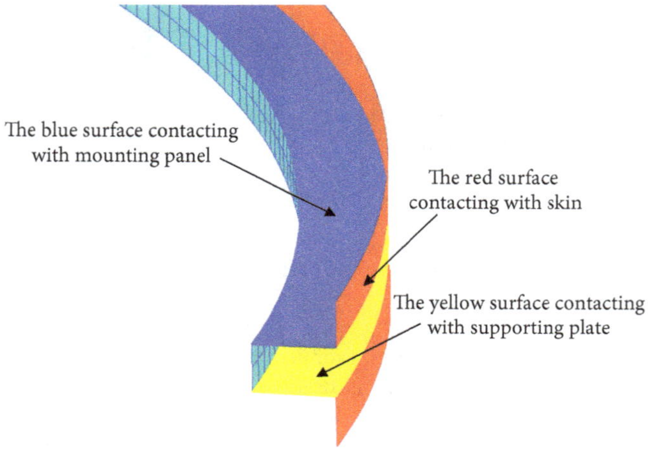


Figure 7: Finite element model of the circular frame.

master, and G represents the adhesive contact between contact bodies. Each row can have one G only, while each column can have more than one G. Column 6 has 3 Gs, indicating that Contact body 6 as the master is in adhesive contact with slave 3, slave 9, and slave 12. If two contact bodies are in contact with each other only, it is unnecessary to identify the master and the slave, for instance, the contact between Contact body 2 and Contact body 11.

## 3. Vibration Modal Analysis of Equipment Bay

The boundary conditions of the modal analysis of equipment bay are as follows: all freedoms of the nodes on the horizontal surface of the upper-end frame are restricted and all freedoms of the nodes on the horizontal surface of the lower-end frame are restricted. The first modal frequency of the equipment bay obtained by the finite element modal analysis is 52.487 Hz. The vibration shape is shown in Figure 9. The location of maximum vibration displacement is on the corner of the mounting panel, with the maximum value of 60.2 mm. Because the equipment is installed on the mounting panel and its excessive vibration displacement may cause damage to the equipment, it is necessary to reduce its vibration displacement. On the top surface of the inertial bracket, add three 20 kg masses, of which each represents the instrument with corresponding quality. The mass is defined with a 0D quality element, which is linked to the relevant nodes on the top surface by RBE3 (Rigid Body Element, Form3), as shown in Figure 10. With the masses added, the first modal frequency of the equipment bay is 28.816 Hz, and the vibration shape is shown in Figure 11. The area with relatively high vibration displacement is mainly on the inertial bracket,

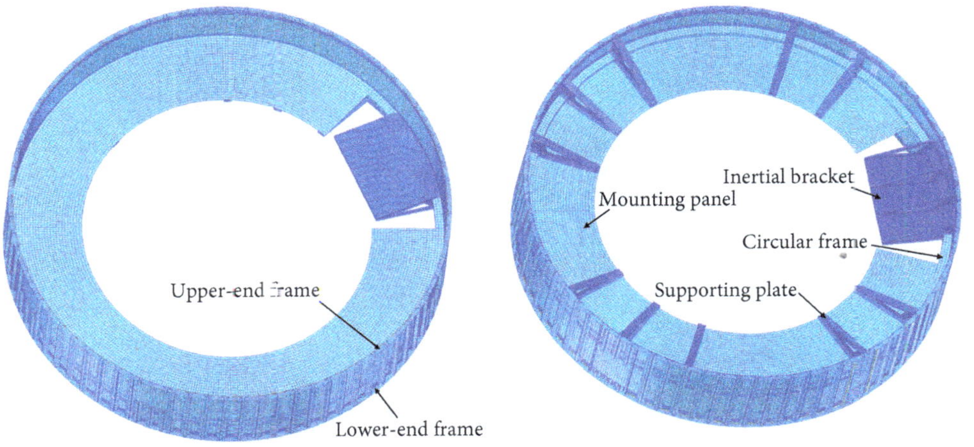


(a) Overall finite element model of the equipment bay (vertical view)

(b) Overall finite element model of the equipment bay (upward view)

FIGURE 8: Overall finite element model of the equipment bay.

TABLE 1: Material characteristics.

| Aluminum alloy | Cast iron | CFRP tape | Aluminum honeycomb |
|---|---|---|---|
| $E = 79$ GPa<br>$\mu = 0.3$<br>$\rho = 2600\ \text{kg/m}^3$ | $E = 160$ GPa<br>$\mu = 0.26$<br>$\rho = 7400\ \text{kg/m}^3$ | $E_{11} = 117$ GPa<br>$E_{22} = 3$ GPa<br>$G_{12} = 4$ GPa<br>$\mu_{12} = 0.3$<br>$\rho = 1520\ \text{kg/m}^3$ | $E_{11} = 120$ MPa<br>$E_{22} = 120$ MPa<br>$G_{12} = 11$ MPa<br>$\mu_{12} = 0.3$<br>$\rho = 102\ \text{kg/m}^3$ |

TABLE 2: Contact table of 13 contact bodies.

| M<br>S | 1 | 2 | 3 | 4 | 5 | 6 | 7 | 8 | 9 | 10 | 11 | 12 | 13 |
|---|---|---|---|---|---|---|---|---|---|---|---|---|---|
| 1 | | | | | | | | | | G | | | |
| 2 | | | | | | | | | | | G | | |
| 3 | | | | | | G | | | | | | | |
| 4 | | | | | | | | | | G | | | |
| 5 | | | | | | | | | | | | | |
| 6 | | | | | | | | | | | | | |
| 7 | | | | | | | | | | | | | |
| 8 | | | | | G | | | | | | | | |
| 9 | | | | | | G | | | | | | | |
| 10 | | | | | | | G | | | | | | |
| 11 | | G | | | | | | | | | | | |
| 12 | | | | | | G | | | | | | | |
| 13 | | | | | | | | | | G | | | |

with the maximum value as 5.32 mm (being 8.8% of the maximum value before the mass is added). The vibration displacement of the two corners of the mounting panel is also relatively high. In order to reduce further the vibration displacement of the mounting panel, one supporting plate is added at the two sides of the mounting panel, so that the first modal frequency of the equipment bay becomes 29.615 Hz, with the vibration shape as shown in Figure 12. The area with

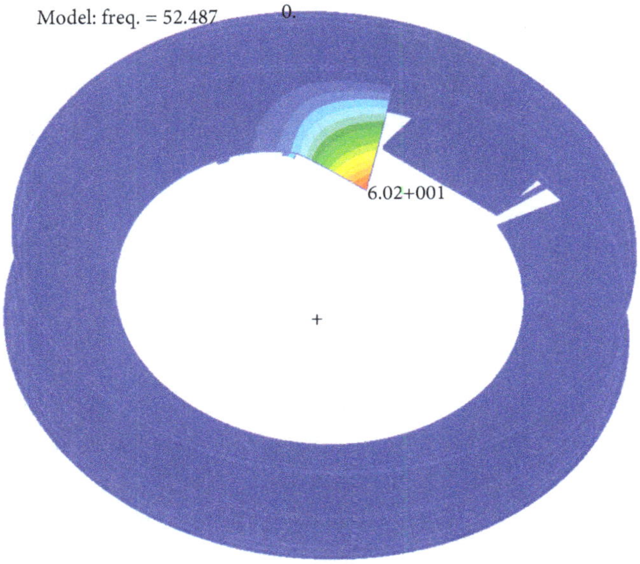


FIGURE 9: First modal shape of the equipment bay.

relatively high vibration displacement is mainly on the inertial bracket, with the maximum value as 5.19 mm.

In order to simulate more truly the vibration of the equipment bay in the rocket, extend the upper end and lower end of the skin by 200 mm (to reduce the influence of the boundary effect on the result). The first modal frequency of the extended equipment bay is 26.855 Hz; the vibration shape is shown in Figure 13. The area with relatively high vibration displacement is mainly on the inertial bracket, with the maximum value as 5.38 mm.

From the foregoing finite element analysis, it is known that the vibration of the equipment bay with the mass mainly occurs on the inertial bracket and its surrounding area, and thus, the influence of the following parameters on the first modal frequency of the equipment bay is further studied: the bracket top surface (as shown in Figure 14(a)) thickness $T_1$, the bracket rib (as shown in Figure 14(b)) thickness $T_2$,

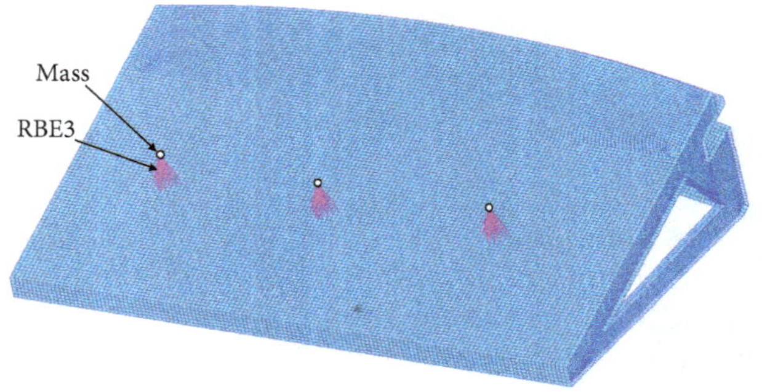


FIGURE 10: Mass on the inertial bracket.

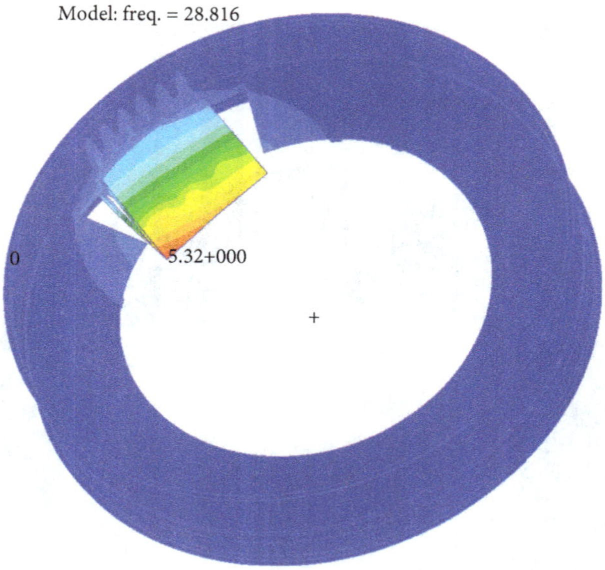


FIGURE 11: First modal shape of the equipment bay (with mass).

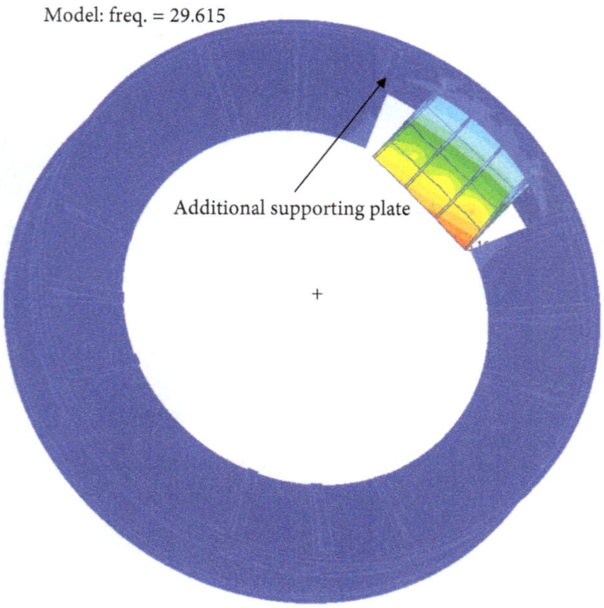


FIGURE 12: First modal shape of the equipment bay (with additional supporting plates).

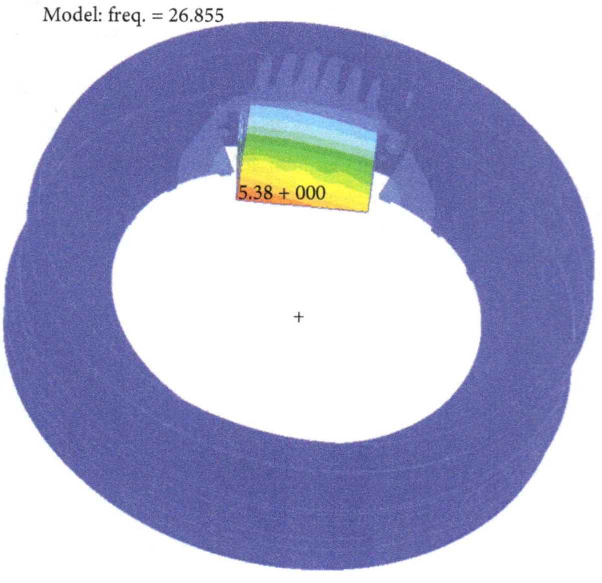


FIGURE 13: First modal shape of the equipment bay (extended).

the bracket side surface (as shown in Figure 14(c)) thickness $T_3$, the bracket beam (as shown in Figure 14(d)) thickness $T_4$, and the stringer (the relevant 6 stringers on the bracket side) thickness $T_5$. The initial value of these thicknesses is 3.5 mm for $T_1$, 3.5 mm for $T_2$, 2 mm for $T_3$, 2 mm for $T_4$, and 1.5 mm for $T_5$.

In order to study the influence of each parameter on the first modal of the equipment bay, keep 4 parameters unchanged and change one of them. Change $T_1$ (with the changing range of 2–4.5 mm) and keep the other 4 parameters unchanged. The influence of $T_1$ on the first modal frequency of the equipment bay is shown in Figure 15. The first modal frequency of the equipment bay increases with the increasing $T_1$. When $T_1$ is 2 mm, the frequency is 22.307 Hz. When $T_1$ is 4.5 mm, the frequency is 27.45 Hz. Change $T_2$ (with the changing range of 2–4.5 mm) and keep the other 4 parameters unchanged. The influence of $T_2$ on the first modal frequency of the equipment bay is shown in Figure 16. The first modal frequency of the equipment bay increases with the increasing $T_2$. When $T_2$ is 2 mm, the frequency is 26.518 Hz. When $T_2$ is 4.5 mm, the frequency is 26.853 Hz. Change $T_3$ (with the changing range of 1–4 mm) and keep the other 4 parameters unchanged. The influence of $T_3$ on the first modal frequency of equipment bay is shown in Figure 17. The first modal frequency of equipment bay increases with the increasing $T_3$. When $T_3$ is 1 mm, the frequency is 26.308 Hz. When $T_3$ is 4 mm, the frequency is 27.614 Hz. Change $T_4$ (with the changing range of 1–4 mm) and keep the other 4 parameters unchanged. The influence of $T_4$ on the first modal frequency of the equipment bay is shown in Figure 18. When $T_4$ is less than 2.5 mm, the first modal frequency of the equipment bay increases with increasing $T_4$. When $T_4$ is more than 2.5 mm, the first modal

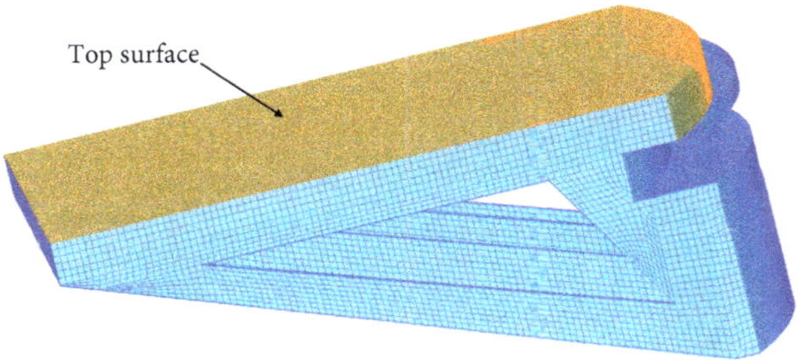


(a) Top surface of the inertial bracket

(b) 4 ribs of the inertial bracket

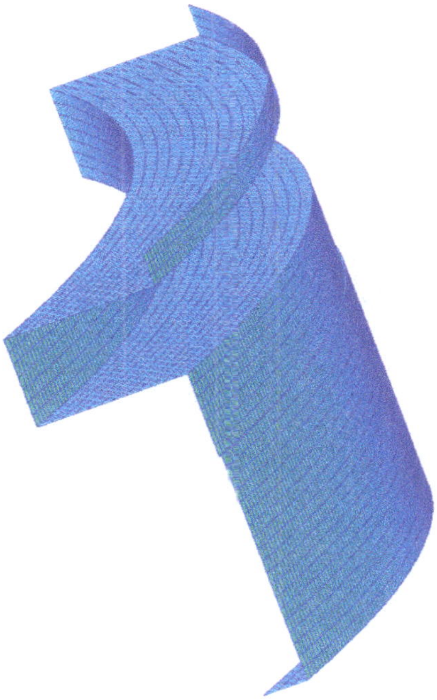

(c) Side surface of the inertial bracket

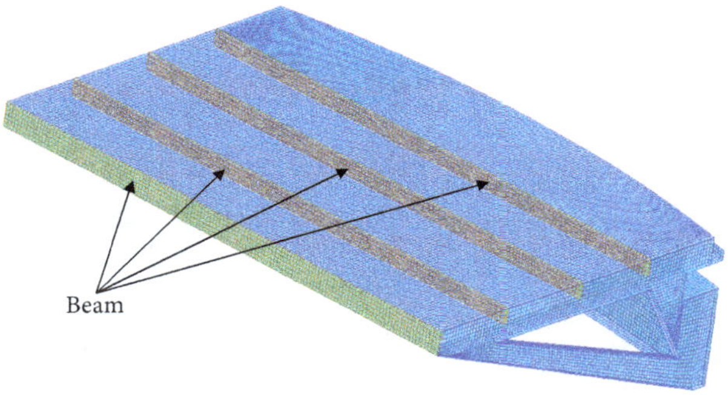


(d) 4 beams of the inertial bracket

FIGURE 14: 4 portions of the inertial bracket.

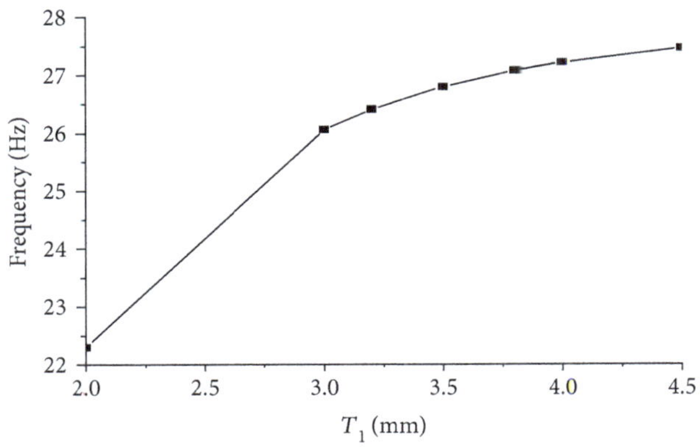


FIGURE 15: Influence of the inertial bracket top surface thickness on the first modal frequency of the equipment bay.

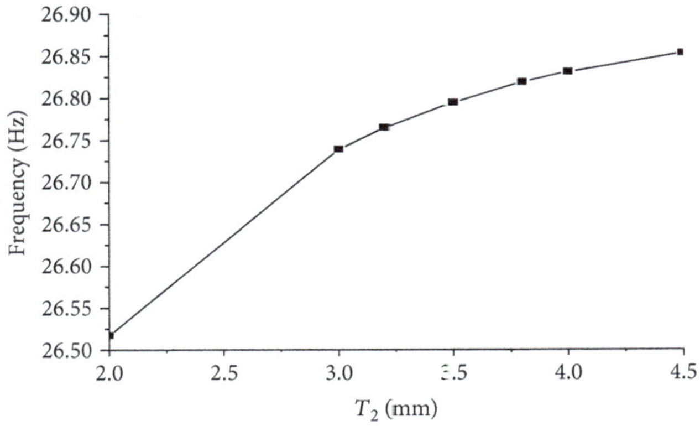


FIGURE 16: Influence of the inertial bracket rib thickness on the first modal frequency of the equipment bay.

frequency decreases with increasing $T_4$. When $T_4$ is 1 mm, the frequency is 26.704 Hz. When $T_4$ is 2.5 mm, the maximum frequency is 26.8 Hz. When $T_4$ is 4 mm, the frequency is 26.778 Hz. Change $T_5$ (with the changing range of 1-2 mm) and keep the other 4 parameters unchanged. The influence of $T_5$ on the first modal frequency of equipment bay is shown in Figure 19. The first modal frequency of the equipment bay increases with increasing $T_5$. When $T_5$ is

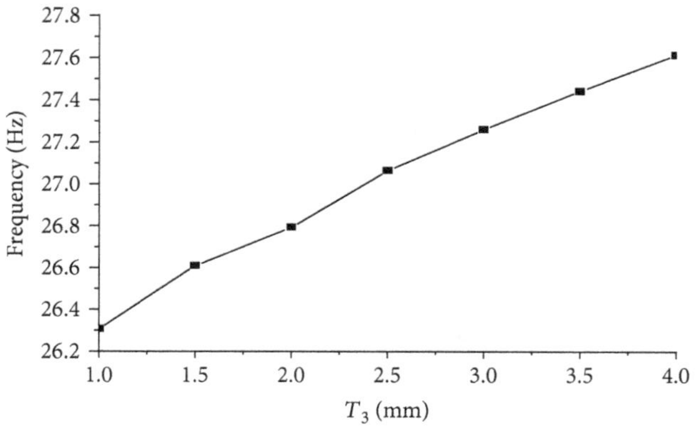


Figure 17: Influence of the inertial bracket side surface thickness on the first modal frequency of the equipment bay.

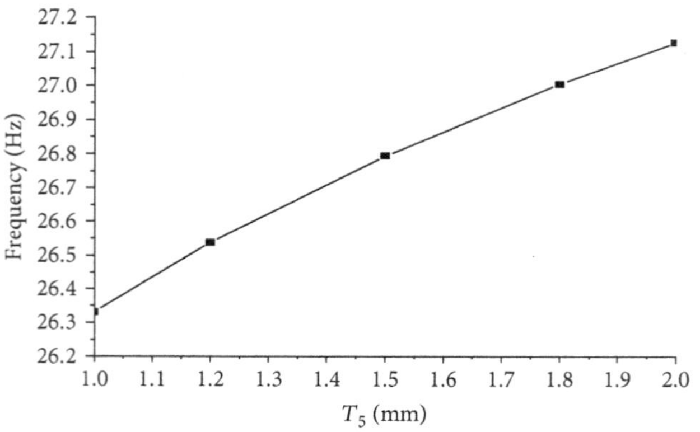


Figure 19: Influence of the stringer thickness on the first modal frequency of the equipment bay.

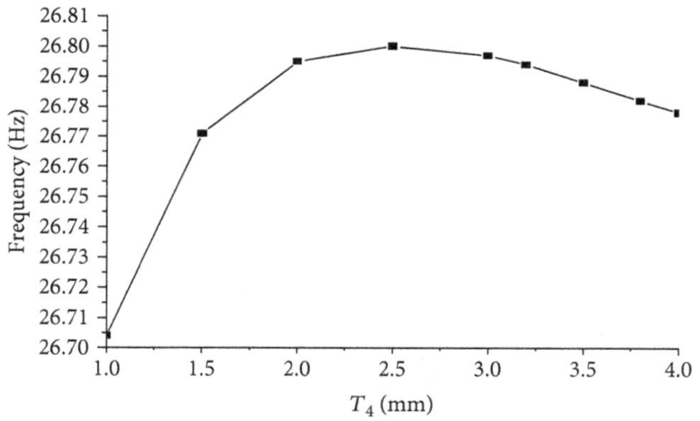


Figure 18: Influence of inertial bracket beam thickness on the first modal frequency of equipment bay.

1 mm, the frequency is 26.331 Hz. When $T_5$ is 2 mm, the frequency is 27.128 Hz.

From the foregoing figures and data, it is known that $T_1$ has the maximum influence on the first modal frequency of the equipment bay, the frequency increases with the increasing $T_1$, $T_3$ and $T_5$ have a relatively high influence on the first modal frequency of the equipment bay, the frequency increases as they increase, $T_2$ has a relatively low influence on the first modal frequency of the equipment bay, and $T_4$ has a minimum influence on the first modal frequency of the equipment bay. Thus, the vibration of the inertial bracket is simplified as the superimposition of two simple circumstances: Rod L is composed of Rod 1 and Rod 2. The shape of Rod L before the application of force is shown in Figure 20(a). When the stiffness of Rod 2 is very high and the stiffness of Rod 1 is relatively low, the shape of Rod L under the effect of Force F is shown in Figure 20(b), with Rod 2 not deformed and Rod 1 bending downward. When the stiffness of Rod 1 is very high and the stiffness of Rod 2 is relatively low, the deformation of Rod L under the effect of Force F is shown in Figure 20(c), with Rod 1 not deformed and Rod 2 bending. However, since the stiffness of Rod 1 and the stiffness of Rod 2 will not differ much, the deformation of Rod L under the effect of Force F is the composition of two circumstances, as is consistent with the vibration of the inertial bracket (Rod 1 is equivalent to the top surface and Rod 2 is equivalent to the side surface). The first modal deformation of the top surface and side surface of the inertial bracket is shown in Figure 21. The comparison of the top surface and side surface before and after deformation proves the result of such composition.

Optimize comprehensively $T_1$, $T_2$, $T_3$, $T_4$, and $T_5$, with the optimizing target: under the condition that the first modal frequency of the equipment bay is more than 26 Hz, the weight of the equipment bay structure is the minimum. First, according to the influence of $T_1$, $T_2$, $T_3$, $T_4$, and $T_5$ on the first modal frequency of the equipment bay, adjust their changing range: the changing range of $T_1$ is 3–4.5 mm, the changing range of $T_2$ is 2–3.5 mm, the changing range of $T_3$ is 2-3 mm, the changing range of $T_4$ is 1–2 mm, and the changing range of $T_5$ is 1.2–2 mm. The result of comprehensive optimization is 3.2 mm for $T_1$, 2 mm for $T_2$, 2.1 mm for $T_3$, 1 mm for $T_4$, and 1.4 mm for $T_5$, the weight of the equipment bay is 162.8 kg, decreasing by 4.9 kg as compared with the initial weight.

## 4. Results and Discussion

Before inertial bracket is added with mass, the vibration of the equipment bay occurs mainly on the mounting panel, and the maximum vibration displacement is 60.2 mm; after the inertial bracket is added with mass, the vibration of the equipment bay occurs mainly on the inertial bracket, and the maximum vibration displacement is 5.32 mm (being 8.8% of the maximum value before mass is added), and with the inertial bracket added with mass, the maximum vibration displacement of the equipment bay is significantly reduced. By adding two supporting plates on both sides of the mounting panel, it can reduce effectively the vibration displacement of the mounting panel. In the further study, it is observed that the influence of the inertial bracket top surface thickness on the first modal frequency of the equipment bay is the highest. The influence of the inertial bracket side surface thickness and stringer thickness on the first modal frequency of the equipment bay is relatively high. The influence of the inertial bracket rib thickness on the first modal frequency of the

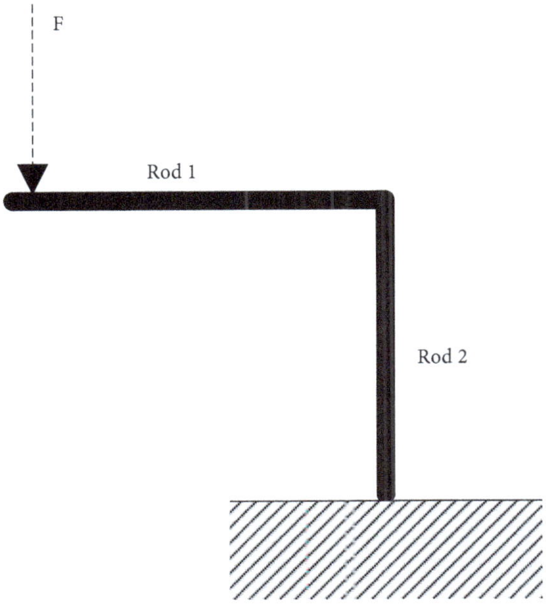


(a) Rod L before application of force

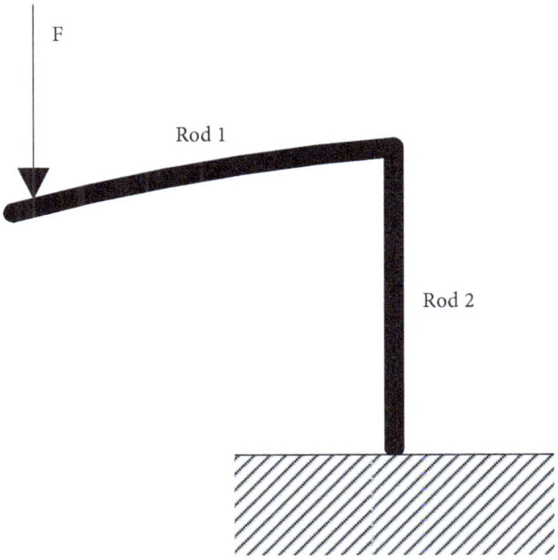


(b) Deformation of Rod L after application of force (the stiffness of Rod 2 is very high)

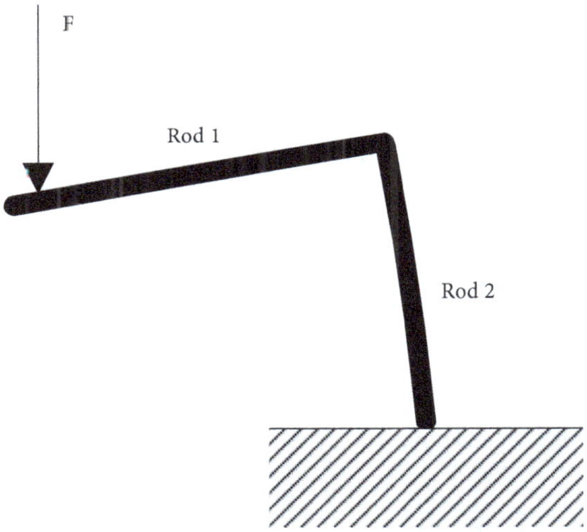


(c) Deformation of Rod L after application of force (the stiffness of Rod 1 is very high)

FIGURE 20: Deformation of Rod L.

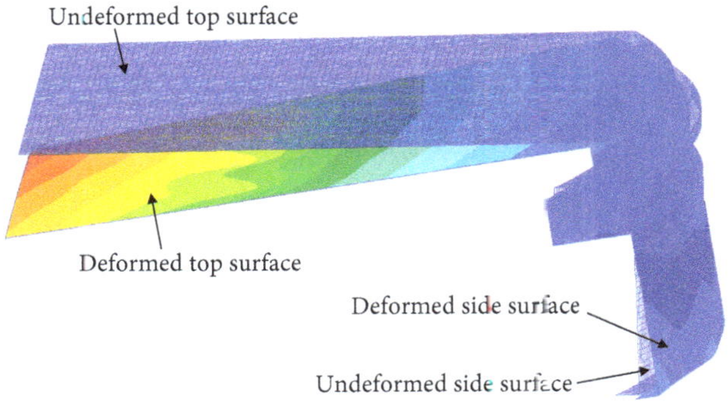


FIGURE 21: First modal deformation of the top surface and side surface of the inertial bracket.

equipment bay is relatively low, and the influence of the inertial bracket beam thickness on the first modal frequency of the equipment bay is the lowest. The vibration of the inertial bracket can be simplified as the deformation of Rod L (when the force is applied to Rod L). In engineering, the stiffness and vibration of the inertial bracket top surface (equivalent to Rod 1 of Rod L) is mostly studied, while the stiffness and vibration of the inertial bracket side surface (equivalent to Rod 2 of Rod L) is less studied. The analysis proves that studies on the stiffness and vibration of the top surface and side surface are all extremely important. The result of the comprehensive optimization of the equipment bay is 3.2 mm for $T_1$, 2 mm for $T_2$, 2.1 mm for $T_3$, 1 mm for $T_4$, and 1.4 mm for $T_5$; the weight of the equipment bay is 162.8 kg, decreasing by 4.9 kg as compared with the initial weight.

## 5. Conclusions

It can greatly reduce the vibration displacement of mounting panels when the inertial bracket is added with mass, and the vibration of the equipment bay with mass mainly occurs on the inertial bracket (without mass, the maximum vibration displacement occurs on the mounting panel). The top surface thickness of the inertial bracket has the maximum influence on the first modal frequency of the equipment bay, the stringer thickness and the side thickness of the inertial bracket have relatively high influence on the first modal frequency, the rib thickness of the inertial bracket has relatively low influence on the first modal frequency, and the beam thickness of the inertial bracket has the minimum influence on the first modal frequency. These results can be used to improve the equipment bay structure and decrease the weight of the equipment bay.

## Conflicts of Interest

The authors declare that there is no conflict of interest regarding the publication of this paper.

## Acknowledgments

This work was supported by the Doctoral Foundation of Nanchang Hangkong University (EA201606185) and Innovation and Entrepreneurship Courses Training Project (KCPY1749 Foundation and Application of Finite Element Method).

## References

[1] Y. G. Choi, K. B. Shin, and W. H. Kim, "A study on size optimization of rocket motor case using the modified 2D axisymmetric finite element model," *International Journal of Precision Engineering and Manufacturing*, vol. 11, no. 6, pp. 901-907, 2010.

[2] T. Engberg and U. A. Korde, "Modeling of the acoustic response of payload bays within launch vehicle fairings," *Journal of Spacecraft and Rockets*, vol. 50, no. 2, pp. 423–432, 2013.

[3] M. N. Mahyari, H. Karimi, H. Naseh, and M. Mirshams, "Numerical and experimental investigation of vortex breaker effectiveness on the improvement in launch vehicle ballistic parameters," *Journal of Mechanical Science and Technology*, vol. 24, no. 10, pp. 1997–2006, 2010.

[4] R. Schwane and Y. Xia, "Time-accurate CFD predictions and data validation for side load generation by flow-structure coupling in over-expanded rocket nozzles," *Journal of Mathematical Modelling and Algorithms*, vol. 4, no. 1, pp. 53-65, 2005.

[5] M. Mense, Y. Pizzo, C. Lallemand, J. C. Loraud, and B. Porterie, "Thermal response of an unprotected structural steel element exposed to a solid rocket propellant fire," *International Journal of Thermal Sciences*, vol. 105, pp. 195–205, 2016.

[6] C. Q. Howard, C. H. Hansen, and A. Zander, "Vibro-acoustic noise control treatments for payload bays of launch vehicles: discrete to fuzzy solutions," *Applied Acoustics*, vol. 66, no. 11, pp. 1235–1261, 2005.

[7] P. R. Salvador and K. G. Xu, "Direct current forcing of an atmospheric multiburner flame for rocket combustor emulation," *Journal of Spacecraft and Rockets*, vol. 55, no. 1, pp. 223–231, 2018.

[8] C. H. Sim, G. S. Kim, D. G. Kim, I. G. Kim, S. H. Park, and J. S. Park, "Experimental and computational modal analyses for launch vehicle models considering liquid propellant and flange joints," *International Journal of Aerospace Engineering*, vol. 2018, no. 2, 12 pages, 2018.

[9] G. Baldesi and M. Toso, "European Space Agency's launcher multibody dynamics simulator used for system and subsystem level analyses," *Ceas Space Journal*, vol. 3, no. 1-2, pp. 27–48, 2012.

[10] V. Hutchinson and J. Olds, "Estimation of launch vehicle propellant tank structural weight using simplified beam approximation," in *40th AIAA/ASME/SAE/ASEE Joint Propulsion Conference and Exhibit*, pp. 1–13, Fort Lauderdale, Florida, July 2004.

[11] R. Marimuthu and B. N. Rao, "An efficient finite element approach to examine the free vibration characteristics of liquid tankages in space launch vehicles," *Meccanica*, vol. 50, no. 5, pp. 1217–1226, 2015.

[12] W. Roh and Y. Kim, "Trajectory optimization for a multi-stage launch vehicle using time finite element and direct collocation methods," *Engineering Optimization*, vol. 34, no. 1, pp. 15–32, 2002.

[13] X. Li and X. Zhang, "The strength analysis of steel sunk screw connections in the rocket," *Acta Astronautica*, vol. 137, pp. 345–352, 2017.

[14] A. Elhefny and G. Liang, "Stress and deformation of rocket gas turbine disc under different loads using finite element modelling," *Propulsion and Power Research*, vol. 2, no. 1, pp. 38–49, 2013.

[15] J. H. Zhang and S. S. Jiang, "Definition of boundary conditions and dynamic analysis of rocket sled and turntable," *Applied Mechanics and Materials*, vol. 52-54, pp. 261–266, 2011.

[16] P. K. Chandana, S. B. Tiwari, and K. N. Vukkadala, "Numerical estimation of sound transmission loss in launch vehicle payload fairing," *Journal of The Institution of Engineers (India): Series C*, vol. 98, no. 4, pp. 471–478, 2017.

# Permissions

All chapters in this book were first published in IJAE, by Hindawi Publishing Corporation; hereby published with permission under the Creative Commons Attribution License or equivalent. Every chapter published in this book has been scrutinized by our experts. Their significance has been extensively debated. The topics covered herein carry significant findings which will fuel the growth of the discipline. They may even be implemented as practical applications or may be referred to as a beginning point for another development.

The contributors of this book come from diverse backgrounds, making this book a truly international effort. This book will bring forth new frontiers with its revolutionizing research information and detailed analysis of the nascent developments around the world.

We would like to thank all the contributing authors for lending their expertise to make the book truly unique. They have played a crucial role in the development of this book. Without their invaluable contributions this book wouldn't have been possible. They have made vital efforts to compile up to date information on the varied aspects of this subject to make this book a valuable addition to the collection of many professionals and students.

This book was conceptualized with the vision of imparting up-to-date information and advanced data in this field. To ensure the same, a matchless editorial board was set up. Every individual on the board went through rigorous rounds of assessment to prove their worth. After which they invested a large part of their time researching and compiling the most relevant data for our readers.

The editorial board has been involved in producing this book since its inception. They have spent rigorous hours researching and exploring the diverse topics which have resulted in the successful publishing of this book. They have passed on their knowledge of decades through this book. To expedite this challenging task, the publisher supported the team at every step. A small team of assistant editors was also appointed to further simplify the editing procedure and attain best results for the readers.

Apart from the editorial board, the designing team has also invested a significant amount of their time in understanding the subject and creating the most relevant covers. They scrutinized every image to scout for the most suitable representation of the subject and create an appropriate cover for the book.

The publishing team has been an ardent support to the editorial, designing and production team. Their endless efforts to recruit the best for this project, has resulted in the accomplishment of this book. They are a veteran in the field of academics and their pool of knowledge is as vast as their experience in printing. Their expertise and guidance has proved useful at every step. Their uncompromising quality standards have made this book an exceptional effort. Their encouragement from time to time has been an inspiration for everyone.

The publisher and the editorial board hope that this book will prove to be a valuable piece of knowledge for researchers, students, practitioners and scholars across the globe.

# List of Contributors

**Zhiyuan Zhang, Runqiang Chi, Baojun Pang and Gongshun Guan**
Hypervelocity Impact Research Center, Harbin Institute of Technology, Harbin 150080, China

**Yun-Hua Wu, Lin-Lin Ge, Bing Hua, Zhi-Ming Chen and Feng Yu**
Micro-Satellite Research Center, Nanjing University of Aeronautics and Astronautics, Nanjing 210016, China

**Feng Wang**
Research Center of Satellite Technology, Harbin Institute of Technology, Harbin 150001, China

**Anwar Ali and Haider Ali**
Electrical Technology Department, University of Technology (UoT), Nowshera, Pakistan

**Shoaib Ahmed Khan and M. Usman Khan**
National University of Computer & Emerging Sciences Peshawar Campus, Pakistan

**M. Rizwan Mughal**
Department of Electrical Engineering, Institute of Space Technology, Islamabad, Pakistan
Department of Electronics and Nanoengineering, School of Electrical Engineering, Aalto University, FI-00076 AALTO, 02150 Espoo, Finland

**Jaan Praks**
Department of Electronics and Nanoengineering, School of Electrical Engineering, Aalto University, FI-00076 AALTO, 02150 Espoo, Finland

**Xin Luo**
School of Computer, China University of Geosciences, Wuhan 430074, China

**Maocai Wang and Guangming Dai**
School of Computer, China University of Geosciences, Wuhan 430074, China
Hubei Key Laboratory of Intelligent Geo-Information Processing, China University of Geosciences, Wuhan 430074, China

**Xiaoyu Chen**
School of Computer, China University of Geosciences, Wuhan 430074, China
Institute of Computer Sciences, Heidelberg University, 69120 Heidelberg, Germany

**Yuyu Zhao, Hui Zhao, Xin Huo and Yu Yao**
Control and Simulation Center, Harbin Institute of Technology, Harbin 150001, China

**Marco Pallone, Mauro Pontani, Paolo Teofilatto and Angelo Minotti**
University of Rome "La Sapienza", Via Salaria 851/831, 00138 Rome, Italy

**Xinyuan Zhang, Ping Shuai, Liangwei Huang and Shaolong Chen**
Qian Xuesen Laboratory of Space Technology, Beijing 100094, China

**Lihong Xu**
China Academy of Space Technology, Beijing 100094, China

**A. Lai, T.-H. Chou, S.-S. Wei, J.-W. Lin and J.-S. Wu**
Department of Mechanical Engineering, National Chiao Tung University, Hsinchu, Taiwan

**Y.-S. Chen**
National Space Organization, National Applied Research Laboratories, Hsinchu, Taiwan

**Kimiya Komurasaki and Kuniyoshi Tabata**
Department of Aeronautics and Astronautics, The University of Tokyo, 7-3-1, Hongo, Bunkyo, Tokyo 113-8656, Japan

**Dario Modenini, Giacomo Curzi and Paolo Tortora**
Department of Industrial Engineering, University of Bologna, Via Fontanelle 40, 47121 Forlì, Italy

**Zhongjiao Shi, Liangyu Zhao and Yeqing Zhu**
Beijing Institute of Technology, Beijing 100081, China
Key Laboratory of Dynamics and Control of Flight Vehicle, Ministry of Education, Beijing 100081, China

**Bindi You, Zhihui Gao, Jianmin Wen, Yiming Sun, Peibo Hao and Dong Liang**
School of Naval Architecture and Ocean Engineering, Harbin Institute of Technology, Weihai 264209, China

**Renato Volpe**
Department of Mechanical and Aerospace Engineering, University of Rome La Sapienza, Via Eudossiana 18, Rome, Italy

**Marco Sabatini**
Department of Astronautics, Electrical and Energetics Engineering, University of Rome La Sapienza, Via Salaria 851, Rome, Italy

**Giovanni B. Palmerini**
Scuola di Ingegneria Aerospaziale, University of Rome La Sapienza, Via Salaria 851, Rome, Italy

**Lei Feng**
Shanghai Institute of Microsystem and Information Technology, Chinese Academy of Sciences, Shanghai 200050, China
Shanghai Engineering Center for Microsatellite, Shanghai 201203, China

**Guotong Li**
Shanghai Engineering Center for Microsatellite, Shanghai 201203, China
Shanghai Tech University, Shanghai 201203, China

**Xun Wang**
College of Aeronautics and Astronautics, National University of Defense Technology, Changsha 410073, China

**Zhaokui Wang**
School of Aerospace Engineering, Tsinghua University, Beijing 100084, China

**Yulin Zhang**
College of Aeronautics and Astronautics, National University of Defense Technology, Changsha 410073, China
School of Aerospace Engineering, Tsinghua University, Beijing 100084, China

**Liu-Lei Shen, Zhi-Bin Shen, Yan Xie and Hai-Yang Li**
College of Aeronautics and Astronautics, National University of Defense Technology, Changsha, Hunan, China

**Chenghu Li, Guanlin Han and Chunxu Duan**
School of Aircraft Engineering, Nanchang Hangkong University, Nanchang 330063, China

# Index

www.ingramcontent.com/pod-product-compliance
Lightning Source LLC
Chambersburg PA
CBHW082024190326
41458CB00010B/3260
* 9 7 8 1 6 4 7 2 6 1 2 5 2 *